An Introduction to Fire Dynamics

Third Edition

Dougal Drysdale
University of Edinburgh, Scotland, UK

**WILEY**

A John Wiley & Sons, Ltd., Publication

Library of Congress Cataloguing-in-Publication Data
Drysdale, Dougal.
  An introduction to fire dynamics / Dougal Drysdale. – 3rd ed.
    p. cm.
  Includes bibliographical references and index.
  ISBN 978-0-470-31903-1 (pbk.)
  1. Fire. 2. Flame. I. Title.
  QD516.D79 2011
  541′.361 – dc22

                          2011015485

A catalogue record for this book is available from the British Library.

Print ISBN: 9780470319031
ePDF ISBN: 9781119975472
oBook ISBN: 9781119975465
ePub ISBN: 9781119976103
Mobi ISBN: 9781119976110

Typeset in 10/12 Times by Laserwords Private Limited, Chennai, India

MIX
Paper from
responsible sources
FSC
www.fsc.org   FSC® C013604

To my family –
Jude
David, Misol and Manow
Andrew, Catriona, Izzy and Alex
and Peter

Contents

About the Author   xi

Preface to the Second Edition   xiii

Preface to the Third Edition   xv

List of Symbols and Abbreviations   xvii

1   Fire Science and Combustion   **1**
1.1   Fuels and the Combustion Process   2
        1.1.1   The Nature of Fuels   2
        1.1.2   Thermal Decomposition and Stability of Polymers   6
1.2   The Physical Chemistry of Combustion in Fires   12
        1.2.1   The Ideal Gas Law   14
        1.2.2   Vapour Pressure of Liquids   18
        1.2.3   Combustion and Energy Release   19
        1.2.4   The Mechanism of Gas Phase Combustion   26
        1.2.5   Temperatures of Flames   30
        Problems   34

2   Heat Transfer   **35**
2.1   Summary of the Heat Transfer Equations   36
2.2   Conduction   38
        2.2.1   Steady State Conduction   38
        2.2.2   Non-steady State Conduction   40
        2.2.3   Numerical Methods of Solving Time-dependent Conduction
                   Problems   48
2.3   Convection   52
2.4   Radiation   59
        2.4.1   Configuration Factors   64
        2.4.2   Radiation from Hot Gases and Non-luminous Flames   72
        2.4.3   Radiation from Luminous Flames and Hot Smoky Gases 76
        Problems   79

3   Limits of Flammability and Premixed Flames   83
3.1   Limits of Flammability   83
       3.1.1   Measurement of Flammability Limits   83
       3.1.2   Characterization of the Lower Flammability Limit   88
       *3.1.3   Dependence of Flammability Limits on Temperature
                 and Pressure*   91
       3.1.4   Flammability Diagrams   94
3.2   The Structure of a Premixed Flame   97
3.3   Heat Losses from Premixed Flames   101
3.4   Measurement of Burning Velocities   106
3.5   Variation of Burning Velocity with Experimental Parameters   109
       3.5.1   Variation of Mixture Composition   110
       3.5.2   Variation of Temperature   111
       3.5.3   Variation of Pressure   112
       3.5.4   Addition of Suppressants   113
3.6   The Effect of Turbulence   116
       Problems   118

4   Diffusion Flames and Fire Plumes   121
4.1   Laminar Jet Flames   123
4.2   Turbulent Jet Flames   128
4.3   Flames from Natural Fires   130
       4.3.1   The Buoyant Plume   132
       4.3.2   The Fire Plume   139
       4.3.3   Interaction of the Fire Plume with Compartment Boundaries   151
       4.3.4   The Effect of Wind on the Fire Plume   163
4.4   Some Practical Applications   165
       4.4.1   Radiation from Flames   166
       4.4.2   The Response of Ceiling-mounted Fire Detectors   169
       4.4.3   Interaction between Sprinkler Sprays and the Fire Plume   171
       4.4.4   The Removal of Smoke   172
       4.4.5   Modelling   174
       Problems   178

5   Steady Burning of Liquids and Solids   181
5.1   Burning of Liquids   182
       5.1.1   Pool Fires   182
       5.1.2   Spill Fires   193
       5.1.3   Burning of Liquid Droplets   194
       5.1.4   Pressurized and Cryogenic Liquids   197
5.2   Burning of Solids   199
       5.2.1   Burning of Synthetic Polymers   199
       5.2.2   Burning of Wood   209
       5.2.3   Burning of Dusts and Powders   221
       Problems   223

6   Ignition: The Initiation of Flaming Combustion   **225**
6.1   Ignition of Flammable Vapour/Air Mixtures   225
6.2   Ignition of Liquids   235
        6.2.1   Ignition of Low Flashpoint Liquids   241
        6.2.2   Ignition of High Flashpoint Liquids   242
        6.2.3   Auto-ignition of Liquid Fuels   245
6.3   Piloted Ignition of Solids   247
        6.3.1   Ignition during a Constant Heat Flux   250
        6.3.2   Ignition Involving a 'Discontinuous' Heat Flux   263
6.4   Spontaneous Ignition of Solids   269
6.5   Surface Ignition by Flame Impingement   271
6.6   Extinction of Flame   272
        6.6.1   Extinction of Premixed Flames   272
        6.6.2   Extinction of Diffusion Flames   273
        Problems   275

7   Spread of Flame   **277**
7.1   Flame Spread Over Liquids   277
7.2   Flame Spread Over Solids   284
        7.2.1   Surface Orientation and Direction of Propagation   284
        7.2.2   Thickness of the Fuel   292
        7.2.3   Density, Thermal Capacity and Thermal Conductivity   294
        7.2.4   Geometry of the Sample   296
        7.2.5   Environmental Effects   297
7.3   Flame Spread Modelling   307
7.4   Spread of Flame through Open Fuel Beds   312
7.5   Applications   313
        7.5.1   Radiation-enhanced Flame Spread   313
        7.5.2   Rate of Vertical Spread   315
        Problems   315

8   Spontaneous Ignition within Solids and Smouldering Combustion   **317**
8.1   Spontaneous Ignition in Bulk Solids   317
        8.1.1   Application of the Frank-Kamenetskii Model   318
        8.1.2   The Thomas Model   324
        8.1.3   Ignition of Dust Layers   325
        8.1.4   Ignition of Oil – Soaked Porous Substrates   329
        8.1.5   Spontaneous Ignition in Haystacks   330
8.2   Smouldering Combustion   331
        8.2.1   Factors Affecting the Propagation of Smouldering   333
        8.2.2   Transition from Smouldering to Flaming Combustion   342
        8.2.3   Initiation of Smouldering Combustion   344
        8.2.4   The Chemical Requirements for Smouldering   346
8.3   Glowing Combustion   347
        Problems   348

9   The Pre-flashover Compartment Fire   **349**
9.1   The Growth Period and the Definition of Flashover   351
9.2   Growth to Flashover   354
        9.2.1   Conditions Necessary for Flashover   354
        9.2.2   Fuel and Ventilation Conditions Necessary for Flashover   364
        9.2.3   Factors Affecting Time to Flashover   378
        9.2.4   Factors Affecting Fire Growth   382
        Problems   385

10   The Post-flashover Compartment Fire   **387**
10.1   Regimes of Burning   387
10.2   Fully Developed Fire Behaviour   396
10.3   Temperatures Achieved in Fully Developed Fires   404
        10.3.1   Experimental Study of Fully Developed Fires in Single
                    Compartments   404
        10.3.2   Mathematical Models for Compartment Fire Temperatures   406
        10.3.3   Fires in Large Compartments   418
10.4   Fire Resistance and Fire Severity   420
10.5   Methods of Calculating Fire Resistance   427
10.6   Projection of Flames from Burning Compartments   435
10.7   Spread of Fire from a Compartment   437
        Problems   439

11   Smoke: Its Formation, Composition and Movement   **441**
11.1   Formation and Measurement of Smoke   443
        11.1.1   Production of Smoke Particles   443
        11.1.2   Measurement of Particulate Smoke   447
        11.1.3   Methods of Test for Smoke Production Potential   450
        11.1.4   The Toxicity of Smoke   455
11.2   Smoke Movement   459
        11.2.1   Forces Responsible for Smoke Movement   459
        11.2.2   Rate of Smoke Production in Fires   465
11.3   Smoke Control Systems   469
        11.3.1   Smoke Control in Large Spaces   470
        11.3.2   Smoke Control in Shopping Centres   471
        11.3.3   Smoke Control on Protected Escape Routes   473

References   **475**

Answers to Selected Problems   **527**

Author Index   **531**

Subject Index   **545**

About the Author

Dougal Drysdale graduated with a degree in Chemistry from the University of Edinburgh in 1962. He gained a PhD in gas phase combustion from Cambridge University (UK) and after two years' postdoctoral work at the University of Toronto, moved to the University of Leeds to work with the gas kinetics group in the Department of Physical Chemistry. He joined the newly formed Department of Fire Engineering at the University of Edinburgh in 1974 and helped develop the first postgraduate degree programme in Fire Engineering under the leadership of Professor David Rasbash. He was invited to teach Fire Dynamics during the spring semester of 1982 at the Centre for Firesafety Studies, Worcester Polytechnic Institute, MA. The notes from this course formed the first draft of the first edition of *An Introduction to Fire Dynamics*, which was published in 1985.

His research interests include various aspects of fire dynamics, including ignition and the fire growth characteristics of combustible materials, compartment fire dynamics and smoke production in fires. He was a member of the Editorial Board for the third and fourth editions of the SFPE *Handbook of Fire Protection Engineering* and was Chairman of the International Association of Fire Safety Science (IAFSS) from 2002–2005. From 1989–2009 he acted as editor of *Fire Safety Journal*, the leading scientific journal in the field. He has been involved in a number of major public inquiries, including the King's Cross Underground Station fire (London, 1987), the Piper Alpha Platform explosion and fire (North Sea, 1988) and the Garley Building fire (Hong Kong, 1996). More recently, he was a member of the Major Incident Investigation Board which was set up following the explosions and fires at the Buncefield Oil Storage and Transfer Depot (Hemel Hempstead, England, 11 December 2005). He is a Fellow of the Royal Society of Edinburgh, the Institution of Fire Engineers and the Society of Fire Protection Engineers. His awards include: 'Man of the Year' (1983), the Arthur B. Guise Medal (1995) and the D. Peter Lund Award (2009) of the Society of Fire Protection Engineers, the Kawagoe Medal of the International Association for Fire Safety Science (2002), the Rasbash Medal of the Institution of Fire Engineers (2004) and the Sjolin Award of FORUM, the Association of International Directors of Fire Research (2005).

He is married to Judy and has three sons and three grandchildren, all living in Edinburgh. His interests are music, hillwalking, curling and coarse golf.

Preface to the Second Edition

The thirteen years that have elapsed between the appearance of the first and second editions of *Introduction to Fire Dynamics* have seen sweeping changes in the subject and, more significantly, in its application. Fire Engineering – now more commonly referred to as Fire Safety Engineering – was identified in the original preface as 'a relatively new discipline', and of course it still is. However, it is beginning to grow in stature as Fire Safety Engineers around the world begin to apply their skills to complex issues that defy solution by the old 'prescriptive' approach to fire safety. This has been reflected by the concurrent development in many countries of new Codes and Regulations, written in such a way as to permit and promote engineered solutions to fire safety problems. The multi-storey atrium and the modern airport terminal building are but two examples where a modern approach to fire safety has been essential.

Preparing a second edition has been somewhat of a nightmare. I have often said that if the first edition had not been completed in late 1984 it might never have been finished. The increased pace of research in the early 1980s was paralleled by the increasing availability of computers and associated peripherals. The first edition was prepared on a typewriter – a device in which the keyboard is directly connected to the printer. Graphs were plotted by hand. In 1984 I was rapidly being overtaken by the wave of new information, so much so that the first edition was out of date by the time it appeared.

In 1984, the International Association for Fire Safety Science – an organisation which has now held five highly successful international symposia – was still to be launched, and the 'Interflam' series of conferences was just beginning to make an impression on the international scene. The vigour of fire research in the decade after 1985 can be judged by examining the contents of the meetings that took place during this period. The scene has been transformed: the resulting exchange of ideas and information has established fire science as the foundation of the new engineering discipline. This has been largely due to the efforts of the luminaries of the fire research community, including in particular Dr. Philip Thomas, the late Prof. Kunio Kawagoe, Prof. T. Akita, Prof. Jim Quintiere and my own mentor, the late Prof. David Rasbash. They perceived the need for organisations such as the IAFSS, and created the circumstances in which they could grow and flourish.

A second edition has been due for over 10 years, but seemed an impossible goal. Fortunately, my friends and colleagues at Worcester Polytechnic Institute came to the rescue. They took the initiative and put me in purdah for four weeks at WPI, with strict instructions to 'get on with it'. Funding for the period was provided by a consortium, consisting of the SFPE Educational Trust, the NFPA, Factory Mutual Research Corporation, Custer Powell Associates, and the Centre for Firesafety Studies at WPI. I am grateful to them all for making it possible, and to Don and Mickey Nelson for making me feel so welcome in their home. Numerous individuals on and off campus helped me to get things together. There was always someone on hand to locate a paper, plot a graph, discuss a problem, or share a coffee. I am grateful to David Lucht, Bob Fitzgerald, Jonathan Barnett, Bob Zalosh and Nick Dembsey for their help. I am indebted to many other individuals who kindly gave their time to respond to questions and comment on sections of the manuscript. In particular, I would like to thank (alphabetically) Paula Beever, Craig Beyler, John Brenton, Geoff Cox, Carlos Fernandez-Pello, George Grant, Bjorn Karlsson, Esko Mikkola, John Rockett and Asif Usmani. Each undertook to review one or more chapters: their feedback was invaluable. Having said this, the responsibility for any errors of fact or omission is mine and mine alone.

It is a sad fact that I managed to carry out over 50% of the revision in four weeks at WPI, but have taken a further two years to complete the task. I would like to thank my colleagues in the Department of Civil and Environmental Engineering for their support and tolerance during this project. This was particularly true of my secretary, Alison Stirling, who displayed amazing *sang-froid* at moments of panic. However, the person to whom I am most indebted is my wife Judy who has displayed boundless patience, tolerance and understanding. Without her support over the years, neither edition would ever have been completed. She finally pulled the pin from the grenade this time, by organising a 'Deadline Party' to which a very large number of friends and colleagues were invited. Missing this deadline was not an option (sorry, John Wiley). It was a great party!

Preface to the Third Edition

'The thirteen years that have elapsed between the second and third editions of *Introduction to Fire Dynamics* have seen sweeping changes in the subject and, more significantly, in its application.' I admit with some embarrassment that this sentence is virtually identical to the one that opens the preface to the second edition. The number 13 bothers me, not simply because of its association with bad luck, but because 13 years is a long time and it could be argued that enough new research had been published for a new edition to have been compiled by 2005. However, a textbook on Fire Dynamics cannot be a literature review – it should be limited to information and data that are deemed to be well-founded by the fire community. The evolution of research results into accepted knowledge takes time, requiring not only the initial peer-review process but also scrutiny by way of further research and application. The practice of Fire Safety Engineering is based on such knowledge but has been in existence as a recognized professional engineering discipline for a remarkably short period of time. Although it was being developed from the mid-1970s onwards by Margaret Law and others, it was not until c. 1990 that it was pulled into the mainstream with the introduction of regulations permitting the use of performance-based fire safety engineering design. At this time, the underpinning 'fire science' was at a relatively early stage in its development and research into many aspects of fire dynamics was still active.

Indeed, Fire Safety Engineering is very close to its research roots. It is significant that the *Handbook of Fire Protection Engineering*, originally published by the Society of Fire Protection Engineers in 1988, is now in its 4th edition (2008). Its chapters cover all aspects of fire safety engineering, but those that deal with the scientific and engineering fundamentals are *de facto* review articles. The practitioner – and indeed the fire safety engineering student – should be alert to the fact that he/she is working in a field that is still developing and that it is necessary to remain aware of current research activities. Consequently, this book should be regarded as a snapshot of where we are at the end of the first decade of the 21st century.

Compared to the first two editions, the third has been prepared under very different circumstances. On the previous occasions, I had the luxury of working on early drafts while at the Centre for Fire Safety Studies at Worcester Polytechnic Institute, away from the usual demands of academic life at Edinburgh University. The third edition has been written at Edinburgh University, but after retirement. I have been very fortunate to have been immersed in a very active fire research group, the BRE Centre for Fire Safety Engineering, led by José Torero. This has been a source of both inspiration and distraction. With so many new colleagues, I have had a unique opportunity to discuss the contents of the book and develop some new areas that were missing from the second edition. However, I have taken care not to change the style of the text, nor to create a tome which might be seen as an attempt to be a literature review. The first edition was close to being such, but the field has developed so rapidly during the last 25 years that such an approach would have been impossible, even if desirable. I am aware that there are some topics that deserve more emphasis and that some recent research has not been included, but I take full responsibility for the decisions regarding the content. I would welcome comments regarding the content as these would be helpful in planning for a fourth edition. Whether or not I will be the author, time (and John Wiley & Sons) will tell!

I owe a huge debt of gratitude to a large number of people for helping me at various stages along the way. In particular, I would like to thank (in alphabetical order) Cecilia Abecassis-Empis, Ron Alpert, Craig Beyler, Luke Bisby, Ricky Carvel, Carlos Fernandez-Pello, Rory Hadden, Martin Gillie, Richard Hull, Tom Lennon, Agustin Majdalani, Jim Quintiere, Guillermo Rein, Pedro Reszka, Martin Shipp, Albert Simeoni, Mike Spearpoint, Anna Stec, Jose Torero and Stephen Welch. I can only apologize if I have missed anyone from the list. In the prefaces to previous editions I acknowledged many others who helped and inspired me at the relevant periods of time. Their contributions are part of this text and although their names are not included here, their roles should not be forgotten. However, I would like to acknowledge two individuals by name: my original mentor, the late David Rasbash who was responsible for establishing the first postgraduate degree programme in Fire Engineering at Edinburgh University, and Philip Thomas who has made so many outstanding contributions to the field and continues to be a source of inspiration. Finally, I wish to thank my wife Judy and our family for their unfailing support over the years and tolerating my highly erratic working practices.

List of Symbols and Abbreviations

a	Absorptivity
A	Arrhenius factor (Chapter 1)
A_f	Fuel bed area (m^2)
A_t	Total internal surface area of a compartment, including ventilation openings (m^2) (Chapter 10)
A_T	Internal surface area of walls and ceiling, excluding ventilation openings (m^2) (Chapter 10)
A_w	Area of ventilation opening (window or door) (m^2)
b	Plume radius (m) (Section 4.3.1)
b	Conserved variable (Equations (5.20) and (5.21))
b	Stick thickness (m) – applies to Figure 5.20 only
b	$\sqrt{k\rho c}$ J/m^2·s$^{1/2}$·K (Table 10.3 only)
B	Spalding's mass transfer number (Equation (5.22))
B	Width of ventilation opening (m) (Chapter 10)
Bi	Biot number $(hL/k)(-)$ (k is the thermal conductivity of the solid)
c_p	Thermal capacity at constant pressure (J/kg·K) or (J/mol·K)
C	Concentration (Chapter 4)
C	Constant (Equations (7.9) and (7.10))
C_d	Discharge coefficient $(-)$ (Chapters 9 and 10)
C_{st}	Stoichiometric concentration (Table 3.1)
d_b	The required depth of clear air above floor level (m) (Chapter 11)
d_q	Quenching distance (mm)
D	Pipe diameter (m) (Sections 2.3 and 3.3)
D	Pool or fire diameter (m) (Chapters 4 and 5)
D	Depth of compartment (Section 10.3.1)
D	Optical density (decibels) (Chapter 11)
D_m	Specific optical density ((b/m)·m^3/m^2) (Equation (11.5))
D_0	Smoke potential ((db/m)·m^3/g) (Equation (11.6))
$\mathcal{D}$	Diffusion coefficient (m^2/s)
E	Total emissive power of a surface (kW/m^2) (Equation (2.4))
E	Constant (Equation (1.14))

E_A	Activation energy (J/mole)
f	Fraction of heat of combustion transferred from flame to surface (Equation (6.16))
f_{ex}	Excess fuel factor (Equation (10.22))
F	Constant (Equation (1.14))
F	Integrated configuration factor ('finite-to-finite area' configuration factor) (Section 2.4.1)
Fo	Fourier number $(\alpha t / L^2)(-)$
g	Gravitational acceleration constant (9.81 m/s^2)
Gr	Grashof number $(gl^3 \Delta\rho/\rho v^2)$ $(-)$
h	Convective heat transfer coefficient (kW/m$^2 \cdot$K)
h_c	Height of roof vent above floor (m) (Chapter 11)
h_0	Height to neutral plane (m) (Figure 10.4)
h_f	Height above neutral plane (m) (Figure 10.4)
h_k	Effective heat transfer coefficient (kW/m$^2 \cdot$K) (Equation (9.6))
H	Height of ventilation opening (m) (Chapter 10)
h	Planck's constant (Equation (2.52))
I	Intensity of radiation (Equations (2.56), and (2.79), and Section (10.2))
I	Intensity of light (Equation (11.2))
k	Thermal conductivity (kW/m$\cdot$K)
k'	Rate coefficient (Equation (1.1))
κ	Boltzmann's constant (Equations (2.52) and (2.54))
K	'Effective emission coefficient' (m^{-1}) (Equations (2.83), (5.12), (10.27))
l	Flame height or length (m) (Chapter 4)
l	Preheat length (m) (Section 7.2.3)
L	Thickness, or half-thickness (m) as defined locally (Section 2.2)
L	Mean beam length (m) (Section 2.4.2)
L	Lower flammability limit (Chapter 3)
L	Pathlength (m) (Section 11.1)
L_v	Latent heat of evaporation or gasification (J/g)
$\dot{m}$	Rate of mass loss (g/s)
M	Mass of air (kg)
M_f	Mass of fuel (kg)
M_w	Molecular weight
n	An integer
n	Number of moles (Chapter 1)
n_A, n_B	Molar concentration (Equation (1.16))
Nu	Nusselt number $(hL/k)(-)$ (k is the thermal conductivity of the fluid)
O	Opening factor, $A_w H^{1/2}/A_t$ (m$^{1/2}$) (Table 10.3 only)
p	Partial pressure (mm Hg, or atm, as defined locally)
p^0	Equilibrium vapour pressure (mm Hg)
P	Pressure (atm)
P_f	Perimeter of fire (m) (Section 11.2)
Pr	Prandtl number $(v/\alpha)(-)$

q_f	Fire load (Equation (10.40)) (MJ)
Q	Rate of heat transfer (W or kW)
$\dot{Q}_c$	Rate of heat release (W or kW)
$\dot{Q}_{conv}$	Rate of convective heat release from a flame
$\dot{Q}_c'''$	Rate of heat production per unit volume (kW/m^3) (Equation (2.13))
$\dot{Q}^*$	Dimensionless heat release rate ($-$) (Equation (4.3))
r	Radial distance (m) (Equations (2.58), (4.49))
r	Stoichiometric ratio (fuel/air) (Chapters 1 and 10)
r_c	Height of roof vent above virtual source of the fire (m) (Section 11.2)
r_0	Characteristic dimension (m) (Chapters 6 and 8)
R	Ideal gas constant (Table 1.9) (Chapters 1, 6 and 8)
R	Radius of burner mouth (m) (Section 4.1)
R	Regression rate, or burning rate (Section 5.1.1)
Re	Reynolds number $(ux/v)(-)$ (Table 4.4)
S	Surface area (Equation (6.2))
S	Rate of transfer of 'sensible heat', defined in Equation (6.18)
t	Time (seconds, unless otherwise specified)
t^*	Γt(hours) (Table 10.3 only)
t_e	Escape time (Chapters 9 and 11)
t_u	Time to achieve untenable conditions (Chapters 9 and 11)
T	Temperature ($^\circ$C or K)
u	Flow velocity (m/s)
u^*	Dimensionless windspeed (Section 4.3.4)
U	Upper flammability limit (Chapter 3)
v	Linear velocity or flowrate (m/s)
V	Volume (m^3) (Chapters 1, 2 and 6)
V	Flame spread rate (Chapter 7)
W	Width of compartment (m)
W	Mass loss by volatilization (g) (Section 11.1)
x	Distance ($\Delta x =$ thickness) (m)
x_A, x_B	Mole fractions
y	Distance (m)
Y	Mass fraction
z	Distance (m) (e.g., height in fire plume)

Greek symbols

α	Thermal diffusivity ($k/\rho c$) (m^2/s)
α'	Entrainment constant (Section 4.3.1)
α_A, α_B	Activity (Equation (1.17))
β	Coefficient of expansion (Equations (1.12) and (2.41))
β	Cooling modulus ($-$) (Equation (6.26))
γ	Energy modulus ($-$) (Equation (6.30))
γ_A, γ_B	Activity coefficients (Equation (1.17))

γ_i, γ_u	Pettersson's heat transfer coefficient (kW/m^{2}K) (Equations (10.33) and (10.35))
Γ	$[O/b]^2/(0.04/1160)^2$ (−) (Table 10.3 only)
δ_{cr}	Critical value of Frank-Kamanetskii's δ (Equation (6.13))
δ_h	Thickness of hydrodynamic boundary layer (m) (Section 2.3)
δ_θ	Thickness of thermal boundary layer (m) (Section 2.3)
ΔH	Change in enthalpy (kJ/mol)
ΔH_c	Heat of combustion (kJ/mol or kJ/g)
ΔH_f	Heat of formation (kJ/mol)
ΔU	Change in internal energy (kJ/mol)
ε	Emissivity
ξ	Dummy variable (Equation (2.23))
θ	Temperature difference (e.g., $T - T_\infty$)
θ	Dimensionless temperature (Chapter 8)
θ	Angle (Equation (2.56), Section 4.3.4 and 7.1)
κ	Absorption coefficient
κ	Constant (Equation (6.9))
λ	Wavelength (μm)
λ_n	Roots of Equation (2.19)
μ	Absolute or dynamic viscosity (Pa·s or N·s/m^2)
η_{O_2}	Mole fraction of oxygen (Equations (1.24) and (5.30))
ν	Kinematic viscosity (μ/ρ) (m^2/s)
ρ	Density (kg/m^3)
σ	Stefan–Boltzmann constant (5.67 × 10^{-8} W/m^2·K^4)
τ	Slab thickness (m) (Equation (2.21) and Section 6.3.1)
τ'	Length of induction period (s)
ϕ	Configuration factor (Section 2.4.1)
ϕ	Equivalence ratio (Equation 1.29, Section 9.2.1)
χ	Factor expressing combustion efficiency (Chapters 1 and 5)
χ	$\dot{m}_{air}/A_w H^{1/2}$ (Chapter 10)
χ_R	Fraction of heat of combustion lost by radiation (−)

Subscripts

a	Ambient
b	Black body (radiation)
c	Cold (Chapter 2)
c	Convective (with f) (Chapter 6)
C	Combustion
cr	Critical
d	Duration (of burning) (Equation (10.39))
e, E	External
f	Fuel

F	Flame
FO	Flashover
g	Gas
h	Hot
ig	Ignition or firepoint
l	Liquid (Chapter 5)
L	Loss by gas replacement (Equation (10.23)))
m	Mean
max	Maximum
n	Normal to a surface (as in Equation 2.56)
o	Initial value or ambient value
o	Centreline value (buoyant plume (Chapter 4))
ox	Oxygen
p	Constant pressure
p	Pyrolysis (Chapter 7)
pl	Plate (Equation (8.1.3))
R	Radiative
s	Surface
u	Unburnt gas
W	Wall (Equation (10.23))
x	In the x-direction
∞	Final value

Superscripts

$\cdot$	Signifies rate of change as in $\dot{m}$
$\cdot$	Indicates that a chemical species is a free radical (e.g., H, the hydrogen atom) (Chapter 1)
$'$	Single prime (signifies 'per unit width') (Chapter 4)
$''$	Double prime (signifies 'per unit area')
$'''$	Triple prime (signifies 'per unit volume')

List of acronyms and abbreviations

ASET	Available Safe Egress Time
ASTM	American Society for Testing and Materials
BRE	Building Research Establishment (Garston, Watford, UK)
BSI	British Standards Institution
CEN	Comité Européen de Normalisation
CFAST	Consolidated model of Fire And Smoke Transport
CFD	Computational Fluid Dynamics
CIB	Conceil Internationale du Bâtiment
CSTB	Centre Scientifique et Technique du Bâtiment (France)
DIN	Deutsches Institut für Normung

ECSC   European Coal and Steel Community
FDS   Fire Dynamics Simulator (developed at NIST)
FMRC   Factory Mutual Research Corporation (Norwood, MA, USA). Now
            FMGlobal
FPA   Fire Protection Association
*FRS   Fire Research Station (now part of BRE, see above)
FTA   Flammability Testing Apparatus (developed at FMGlobal)
*IAFSS   International Association for Fire Safety Science
ISO   International Organization for Standardization
LFL   Lower Flammability Limit
NBS   National Bureau of Standards (now NIST)
NFPA   National Fire Protection Association (1 Batterymarch Park, Quincy, MA,
            USA)
NIST   National Institute for Standards and Technology (Building and Fire Research
            Laboratory, Gaithersburg, MD, USA)
RSET   Required Safe Egress Time
SBI   Single Burning Item
SFPE   Society of Fire Protection Engineers (Bethesda, MD, USA)
UFL   Upper Flammability Limit

*Frequent references are made in this text to the Fire Research Notes (from FRS) and the
proceedings of the triennial IAFSS symposia. These are available on the IAFSS website
http://www.iafss.org. Note that the Proceedings of the Symposia are now referred to as individual
volumes of "Fire Safety Science".

1

Fire Science and Combustion

As a process, fire can take many forms, all of which involve chemical reactions between combustible species and oxygen from the air. Properly harnessed, it provides great benefit as a source of power and heat to meet our industrial and domestic needs, but, unchecked, it can cause untold material damage and human suffering. In the United Kingdom alone, direct losses probably exceed £2 billion (2010 prices), while over 400 people die each year in fires. According to the UK Fire Statistics (Department for Communities and Local Government, 2009), there were 443 fatalities in 2007, continuing a downward trend from over 1000 in 1979. In real terms, the direct fire losses may not have increased significantly over the past two decades, but this holding action has been bought by a substantial increase in other associated costs, namely improving the technical capability of the Fire Service and the adoption of more sophisticated fire protection systems.[1]

Further major advances in combating unwanted fire are unlikely to be achieved simply by continued application of the traditional methods. What is required is a more fundamental approach that can be applied at the design stage rather than tacitly relying on fire incidents to draw attention to inherent fire hazards. Such an approach requires a detailed understanding of fire behaviour from an engineering standpoint. For this reason, it may be said that a study of fire dynamics is as essential to the fire protection engineer as the study of chemistry is to the chemical engineer.

It will be emphasized at various places within this text that although 'fire' is a manifestation of a chemical reaction, the mode of burning may depend more on the physical state and distribution of the fuel, and its environment, than on its chemical nature. Two simple examples may be quoted: a log of wood is difficult to ignite, but thin sticks can be ignited easily and will burn fiercely if piled together; a layer of coal dust will burn relatively slowly, but may cause an explosion if dispersed and ignited as a dust cloud. While these are perhaps extreme examples, they illustrate the complexity of fire behaviour in that their understanding requires knowledge not only of chemistry but also of many subjects normally associated with the engineering disciplines (heat transfer, fluid dynamics, etc.). Indeed, the term 'fire dynamics' has been chosen to describe the subject of fire

---

[1] The total cost associated with fire in England and Wales in 2004 was estimated to be £7.03 billion. This figure includes costs of fire protection, the Fire and Rescue Service (including response), property damage and lost business, as well as the economic costs associated with deaths and injuries and the prosecution of arsonists (Office of the Deputy Prime Minister, 2006).

An Introduction to Fire Dynamics, Third Edition. Dougal Drysdale.
© 2011 John Wiley & Sons, Ltd. Published 2011 by John Wiley & Sons, Ltd.

behaviour as it implies inputs from these disciplines. However, it also incorporates parts of those subjects which are normally associated with the terms 'fire chemistry' and 'fire science'. Some of these are reviewed in the present chapter, although detailed coverage is impossible. It is assumed that the reader has some knowledge of elementary chemistry and physics, including thermodynamics: references to relevant texts and papers are given as appropriate.

1.1 Fuels and the Combustion Process

Most fires involve combustible solids, although in many sectors of industry, liquid and gaseous fuels are also to be found. Fires involving gases, liquids and solids will be discussed in order that a comprehensive picture of the phenomenon can be drawn. The term 'fuel' will be used quite freely to describe that which is burning, whatever the state of matter, or whether it is a 'conventional' fuel such as LPG or an item of furniture within a room. With the exception of hydrogen gas, to which reference is made in Chapter 3, all fuels that are mentioned in this text are carbon-based. Unusual fire problems that may be encountered in the chemical and nuclear industries are not discussed, although the fire dynamics will be similar if not identical. General information on problems of this type may be gleaned from the *National Fire Protection Handbook* (NFPA, 2008) and other sources (e.g., Meidl, 1970; Stull, 1977; Mannan, 2005).

1.1.1 The Nature of Fuels

The range of fuels with which we are concerned is very wide, from the simplest gaseous hydrocarbons (Table 1.1) to solids of high molecular weight and great chemical complexity, some of which occur naturally, such as cellulose, and others that are man-made (e.g., polyethylene and polyurethane) (Table 1.2). All will burn under appropriate conditions, reacting with oxygen from the air, generating combustion products and releasing heat. Thus, a stream or jet of a gaseous hydrocarbon can be ignited in air to give a flame, which is seen as the visible portion of the volume within which the oxidation process is occurring. Flame is a gas phase phenomenon and, clearly, flaming combustion of liquid and solid fuels must involve their conversion to gaseous form. For burning liquids, this process is normally simple evaporative boiling at the surface,[2] but for almost all solids, chemical decomposition or *pyrolysis* is necessary to yield products of sufficiently low molecular weight that can volatilize from the surface and enter the flame. As this requires much more energy than simple evaporation, the surface temperature of a burning solid tends to be high (typically 400°C) (Table 1.2). Exceptions to this rule are those solids which sublime on heating, i.e., pass directly from the solid to the vapour phase without chemical decomposition. There is one relevant example, hexamethylenetetramine (also known as methenamine), which in pill form is used as the ignition source in ASTM D2859-06 (American Society for Testing and Materials, 2006). It sublimes at about 263°C (Budavari, 1996).

The composition of the volatiles released from the surface of a burning solid tends to be extremely complex. This can be understood when the chemical nature of the solid is

---

[2] Liquids with very high boiling points ($\geq 250°C$) may undergo some chemical decomposition (e.g., cooking oil).

Table 1.1   Properties of gaseous and liquid fuels[a]

Common name[b]	Formula	Melting point ($^\circ$C)	Boiling point ($^\circ$C)	Density (liq) (kg/m^3)	Molecular weight
Hydrogen	H_2	-259.3	-252.8	70	2
Carbon monoxide	CO	-199	-191.5	422	28
Methane	CH_4	-182.5	-164	466	16
Ethane	C_2H_6	-183.3	-88.6	572	30
Propane	C_3H_8	-189.7	-42.1	585	44
n-Butane	$n\text{-}C_4H_{10}$	-138.4	-0.5	601	58
n-Pentane	$n\text{-}C_5H_{12}$	-130	36.1	626	72
n-Hexane	$n\text{-}C_6H_{14}$	-95	69.0	660	86
n-Heptane	$n\text{-}C_7H_{16}$	-90.6	98.4	684	100
n-Octane	$n\text{-}C_8H_{18}$	-56.8	125.7	703	114
iso-Octane[c]	iso-C_8H_{18}	-107.4	99.2	692	114
n-Nonane	$n\text{-}C_9H_{20}$	-51	150.8	718	128
n-Decane	$n\text{-}C_{10}H_{22}$	-29.7	174.1	730	142
Ethylene (ethene)	C_2H_4	-169.1	-103.7	(384)	28
Propylene (propene)	C_3H_6	-185.2	-47.4	519	42
Acetylene (ethyne)	C_2H_2	-80.4	-84	621	26
Methanol	CH_3OH	-93.9	65.0	791	32
Ethanol	C_2H_5OH	-117.3	78.5	789	46
Acetone	$(CH_3)_2CO$	-95.3	56.2	790	58
Benzene	C_6H_6	5.5	80.1	874	78

[a]Data from Lide (1993/94).
[b]It should be noted that IUPAC (International Union of Pure and Applied Chemistry) has defined a standard chemical nomenclature which is not used rigorously in this text. 'Common names' are used, although the IUPAC nomenclature will be given where appropriate. See, for example, 'iso-octane' and 'ethylene' in this table.
[c]2,2,4-Trimethyl pentane.

considered. All those of significance are polymeric materials of high molecular weight, whose individual molecules consist of long 'chains' of repeated units which in turn are derived from simple molecules known as monomers (Billmeyer, 1971; Open University, 1973; Hall, 1981; Friedman 1989; Stevens, 1999). Of the two basic types of polymer (addition and condensation), the addition polymer is the simpler in that it is formed by direct addition of monomer units to the end of a growing polymer chain. This may be illustrated by the sequence of reactions:

$$R^\bullet + CH_2 = CH_2 \rightarrow R \cdot CH_2 \cdot CH_2^\bullet \tag{1.R1a}$$

$$R \cdot CH_2 \cdot CH_2^\bullet + CH_2 = CH_2 \rightarrow R \cdot CH_2 \cdot CH_2 \cdot CH_2 \cdot CH_2^\bullet \tag{1.R1b}$$

etc., where $R^\bullet$ is a free radical or atom, and $CH_2{=}CH_2$ is the monomer, ethylene. This process is known as polymerization and in this case will give polyethylene, which has the idealized structure:

$$R\underbrace{(CH_2 \cdot CH_2)}_{\text{monomer unit}}{}_n R'$$

Table 1.2 Properties of some solid fuels[a]

	Density (kg/m³)	Heat capacity (kJ/kg·K)	Thermal conductivity (W/m·K)	Heat of combustion (kJ/g)	Melting point (°C)
Natural polymers					
Cellulose	V[b]	~1.3	V	16.1	chars
Thermoplastic polymers					
Polyethylene					
Low density	940	1.9	0.35	46.5	
High density	970	2.3	0.44	46.5	130–135
Polypropylene					
Isotactic	940	1.9	0.24	46.0	186
Syndiotactic				46.0	138
Polymethylmethacrylate	1190	1.42	0.19	26.2	~160
Polystyrene	1100	1.2	0.11	41.6	240
Polyoxymethylene	1430	1.4	0.29	15.5	181
Polyvinylchloride	1400	1.05	0.16	19.9	–
Polyacrylonitrile	1160–1180	–	–	–	317
Nylon 66	~1200	1.4	0.4	31.9	250–260
Thermosetting polymers					
Polyurethane foams	V	~1.4	V	24.4	–
Phenolic foams	V	–	V	17.9	chars
Polyisocyanurate foams	V	–	V	24.4	chars

[a]From Brandrup and Immergut (1975) and Hall (1981). Heats of combustion refer to CO_2 and H_2O as products.
[b]V = variable.

in which the monomer unit has the same complement and arrangement (although not the same chemical bonding) of atoms as the parent monomer, $CH_2=CH_2$: n is the number of repeated units in the chain and is known as the degree of polymerization, which may be anything from a few hundred to several tens of thousands (Billmeyer, 1971). This type of polymerization relies on the reactivity of the carbon–carbon 'double bond'. In contrast, the process of polymerization which leads to the formation of a 'condensation polymer' involves the loss of a small molecular species (normally H_2O) whenever two monomer units link together. (This is known as a condensation reaction.) Normally, two distinct monomeric species are involved, as in the production of Nylon 66 from hexamethylene diamine and adipic acid.[3] The first stage in the reaction would be:

$$NH_2(CH_2)_6 \, NH_2 + HO \cdot CO \cdot (CH_2)_4 \cdot CO \cdot OH$$
$$\underset{\text{hexamethylene diamine}}{} \qquad \underset{\text{adipic acid}}{} \tag{1.R2}$$

$$\rightarrow NH_2(CH_2)_6 \, NH \cdot CO \cdot (CH_2)_4 \cdot CO \cdot OH + H_2O$$

---

[3] The IUPAC systematic names of these two compounds are: diaminohexane and butane-1,4-dicarboxylic acid, respectively.

The formula of Nylon 66 may be written in the format used above for polyethylene, namely:

$$H \{\!\!-\!\!\{ NH \cdot (CH_2)_6 \cdot NH \cdot CO \cdot (CH_2)_4 \cdot CO \}\!\!-\!\!\}_n OH$$

It should be noted that cellulose, the most widespread of the natural polymers occurring in all higher plants (Section 5.2.2), is a condensation polymer of the monosaccharide D-Glucose ($C_6H_{12}O_6$). The formulae for both monomer and polymer are shown in Figure 5.11.

An essential feature of any monomer is that it must contain two reactive groups, or 'centres', to enable it to combine with adjacent units to form a linear chain (Figure 1.1(a)). The length of the chain (i.e., the value of n in the above formulae) will depend on conditions existing during the polymerization process: these will be selected to produce a polymer of the desired properties. Properties may also be modified by introducing branching into the polymer 'backbone'. This may be achieved by modifying the conditions in a way that will induce branching to occur spontaneously (Figure 1.1(b)) or by introducing a small amount of a monomer which has three reactive groups (unit B in Figure 1.1(c)). This can have the effect of producing a cross-linked structure whose physical (and chemical) properties will be very different from an equivalent unbranched, or only slightly branched, structure (Stevens, 1999). As an example, consider the expanded polyurethanes. In most flexible foams the degree of cross-linking is very low, but by increasing it substantially (e.g., by increasing the proportion of trifunctional monomer, B in Figure 1.1(c)), a polyurethane suitable for rigid foams may be produced.

With respect to flammability, the yield of volatiles from the thermal decomposition of a polymer is much less for highly cross-linked structures since much of the material forms an involatile carbonaceous char, thus effectively reducing the potential supply of gaseous fuel to a flame. An example of this can be found in the phenolic resins, which on heating to a temperature in excess of 500°C may yield up to 60% char (Madorsky, 1964). The structure of a typical phenolic resin is shown in Figure 1.2. A natural polymer that exhibits a high degree of cross-linking is lignin, the 'cement' that binds the cellulose structures together in higher plants, thus imparting greater strength and rigidity to the cell walls.

Synthetic polymers may be classified into two main groups, namely thermoplastics and thermosetting resins (Table 1.2). A third group – the elastomers – may be distinguished on the basis of their rubber-like properties (Billmeyer, 1971; Hall, 1981; Stevens, 1999),

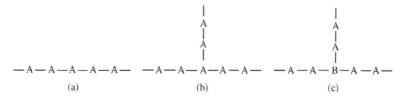

Figure 1.1 Basic structure of polymers: (a) straight chain (e.g., polymethylene, with A=CH₂); (b) branched chain, with random branch points (e.g., polyethylene, with A=CH₂–CH₂, see text); (c) branched chain, involving trifunctional centres (e.g., polyurethane foams in which the straight chains (— A— A—, etc.) correspond to a co-polymer of tolylene di-isocyanate and a polymer diol and B is a trihydric alcohol)

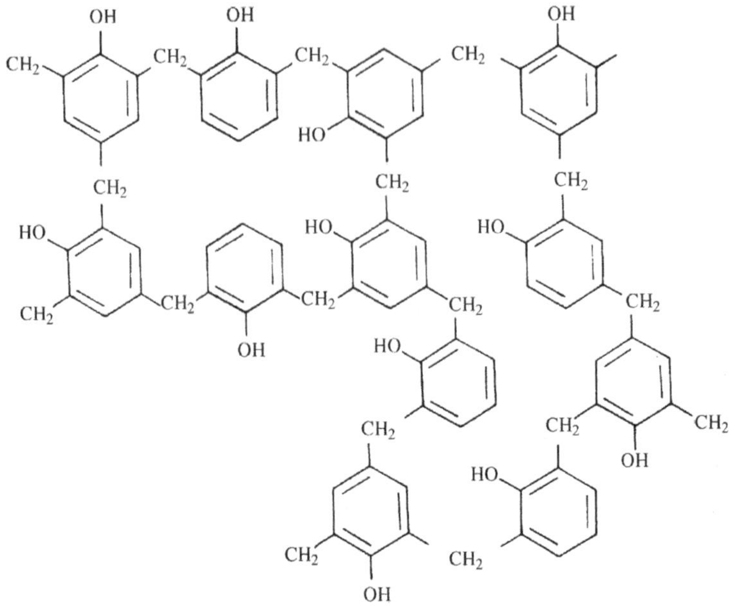

Figure 1.2   Typical cross-linked structure to be found in phenol formaldehyde resins

but will not be considered further here. From the point of view of fire behaviour, the main difference between thermoplastics and thermosetting polymers is that the latter are cross-linked structures that will not melt when heated. Instead, at a sufficiently high temperature, many decompose to give volatiles directly from the solid, leaving behind a carbonaceous residue (cf. the phenolic resins, Figure 1.2), although with polyurethanes, the initial product of decomposition is a liquid. On the other hand, the thermoplastics will soften and melt when heated, which will modify their behaviour under fire conditions. Fire spread may be enhanced by falling droplets or the spread of a burning pool of molten polymer (Section 9.2.4). This is also observed with flexible polyurethane foams, although in this case the liquid melt is a product of the decomposition process.

1.1.2   Thermal Decomposition and Stability of Polymers

The production of gaseous fuel (volatiles) from combustible solids almost invariably involves thermal decomposition, or pyrolysis, of polymer molecules at the elevated temperatures which exist at the surface (Kashiwagi, 1994; Hirschler and Morgan, 2008). Whether or not this is preceded by melting depends on the nature of the material (Figure 1.3 and Table 1.3). In general, the volatiles comprise a complex mixture of pyrolysis products, ranging from simple molecules such as hydrogen and ethylene, to species of relatively high molecular weight which are volatile only at the temperatures existing at the surface where they are formed, when their thermal energy can overcome the cohesive forces at the surface of the condensed fuel. In flaming combustion most of these will be consumed in the flame, but under other conditions (e.g., pyrolysis without combustion following exposure to an external source of heat or, for some materials,

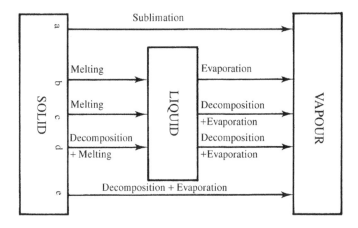

Figure 1.3 Different modes in which fuel vapour is generated from a solid (Table 1.3)

Table 1.3 Formation of volatiles from combustible solids (Figure 1.3). It should be noted that pyrolysis is often enhanced by the presence of oxygen (Cullis and Hirschler, 1981; Kashiwagi and Ohlemiller, 1982)

Designation (Figure 1.3)	Mechanism	Examples
a	Sublimation	Methenamine (see text)
b	Melting and evaporation without chemical change	Low molecular weight paraffin waxes, although the mechanism likely to involve (b) *and* (c)
c	Melting, then decomposition (pyrolysis) followed by evaporation of low molecular weight products	Thermoplastics; high molecular weight waxes, etc.
d	Decomposition to produce molten products,[a] which decompose (pyrolyse) further to yield volatile species	Polyurethanes
e	Decomposition (pyrolysis) to give volatile species directly and (commonly) a solid residue (char)	Cellulose; most thermosetting resins (not standard polyurethanes)

[a]The initial decomposition may also produce species which can volatilize directly.

smouldering combustion (Section 8.2)), the high boiling liquid products and tars will condense to form an aerosol smoke as they mix with cool air.

At high temperatures, a small number of addition polymers (e.g., polymethylmethacrylate, known by the acronym PMMA) will undergo a reverse of the polymerization process (Equations (1.R1a) and (1.R1b)), known as 'unzipping' or 'end-chain scission', to give high yields of monomer in the decomposition products (Table 1.4).[4] This behaviour is a

---

[4] It may be noted at this point that there is no exact equivalent to the 'unzipping' process in the pyrolysis of condensation polymers (compare Equations (1.R1) and (1.R2)).

Table 1.4 Yield of monomer in the pyrolysis of some organic polymers in a vacuum (% total volatiles) (from Madorsky, 1964)

Polymer	Temperature range (°C)	Monomer (%)
Polymethylene[a]	325–450	0.03
Polyethylene	393–444	0.03
Polypropylene	328–410	0.17
Polymethylacrylate	292–399	0.7
Polyethylene oxide	324–363	3.9
Polystyrene	366–375	40.6
Polymethylmethacrylate	246–354	91.4
Polytetrafluoroethylene	504–517	96.6
Poly α-methyl styrene	259–349	100
Polyoxymethylene	Below 200	100

[a] An unbranched polyethylene.

direct result of the chemical structure of the monomer units, which favours the 'unzipping' process: with PMMA this is to the exclusion of any other decomposition mechanism (Madorsky, 1964). It should be contrasted with the pyrolysis of, for example, polyethylene, in which the monomer structure allows the chains to break at random points along their length, causing the average chain length (defined by n, the degree of polymerization) and hence the molecular weight to decrease very rapidly. This leads to the formation of smaller molecules that allow the polymer to soften and melt, producing a mobile liquid at the temperature of decomposition. On the other hand, by 'unzipping', the average molecular weight (determined by n) of the PMMA molecules decreases very slowly and the polymer does not melt and flow (although, given time, it will soften). It is for this reason that PMMA is the polymer most commonly used in experimental work (e.g., see Figure 5.10).

In addition to 'end-chain scission' and 'random-chain scission', which are described above, two other decomposition mechanisms may be identified, namely chain stripping and cross-linking (Wall, 1972; Cullis and Hirschler, 1981; Hirschler and Morgan, 2008). Chain stripping is a process in which the polymer backbone remains intact but molecular species are lost as they break away from the main chain. One relevant example is the thermal decomposition of polyvinyl chloride (PVC), which begins to lose molecular HCl (hydrogen chloride) at about 250°C, leaving behind a char-like residue:

$$R\text{+}(CH_2 - CHCl)_n R' \rightarrow R\text{+}(CH = CH)_n R' + nHCl$$

$$\text{polyvinylchloride} \qquad \text{residue} \tag{1.R3}$$

Although the residue will burn at high temperatures (giving much smoke), hydrogen chloride is a very effective combustion inhibitor and its early release will tend to extinguish a developing flame. For this reason, it is said that PVC has a very low 'flammability', or potential to burn. This is certainly true for rigid PVC, but the flexible grades commonly used for electrical insulation, for example, contain additives (specifically, plasticizers) which make them more flammable. However, even the 'rigid' grades will burn if the ambient conditions are right (Section 5.2.1).

Polymers which undergo cross-linking during pyrolysis tend to char on heating. While this should reduce the amount of fuel available for flaming combustion, the effect on flammability is seldom significant for thermoplastics (cf. polyacrylonitrile, Table 1.2). However, as has already been noted, charring polymers like the phenolic resins do have desirable fire properties. These are highly cross-linked in their normal state (Figure 1.2), and it is likely that further cross-linking occurs during pyrolysis.

In subsequent chapters, it will be shown that some of the fire behaviour of combustible materials can be interpreted in terms of the properties of the volatiles, specifically their composition, reactivity and rate of formation. Thermal stability can be quantified by determining how the rate of decomposition varies with temperature. These results may be expressed in a number of ways, the most common arising from the assumption that the pyrolysis proceeds according to a simple kinetic scheme such that:

$$\dot{m} = \frac{dm}{dt} = -k' \cdot m \qquad (1.1)$$

where m represents the mass (or more correctly, the concentration) of the polymer. While this is a gross simplification, it does permit k', the rate coefficient, to be determined – although this is of little direct value *per se*. However, it allows the temperature dependence of the process to be expressed in a standard form, using the Arrhenius expression for the rate coefficient, i.e.

$$k' = A \exp(-E_A/RT) \qquad (1.2)$$

where E_A is the activation energy (J/mol), R is the universal gas constant (8.314 J/K.mol) and T is the temperature (K). The constant A is known as the pre-exponential factor and in this case will have units of s^{-1}. Much research has been carried out on the thermal decomposition of polymers (Madorsky, 1964; National Bureau of Standards, 1972; Cullis and Hirschler, 1981; Hirschler and Morgan, 2008), but in view of the chemical complexity involved, combined with problems of interpreting data from a variety of sources and experimental techniques, it is not possible to use such information directly in the present context. Some activation energies which were derived from early studies (Madorsky, 1964) are frequently quoted (e.g., Williams, 1974b, 1982) and are included here only for completeness (Table 1.5). However, without a knowledge of A (the pre-exponential factor), these do not permit relative rates of decomposition to be assessed.

Of more immediate value is the summary presented by Madorsky (1964), in which he collates data on relative thermal stabilities of a range of organic polymers, expressed as the temperature at which 50% of a small sample of polymer will decompose in 30 minutes (i.e., the temperature at which the half-life is 1800 s). (This tacitly assumes first-order kinetics, as implied in Equation (1.1).) A selection of these data is presented in Table 1.6. They allowed Madorsky (1964) to make some general comments on polymer stability, which are summarized in Table 1.7. It is possible to make limited comparison of the information contained in these tables with data presented in Chapter 5 (Table 5.11) on the heats of polymer gasification (Tewarson and Pion, 1976). However, it must be borne in mind that the data in Table 1.6 refer to 'pure' polymers while those in Table 5.11 were obtained with commercial samples, many of which contain additives that will modify their behaviour.

Table 1.5   Activation energies for thermal decomposition of some organic polymers in vacuum (from Madorsky, 1964)

Polymer	Molecular weight	Temperature range ($^{\circ}$C)	Activation energy (kJ/mole)
Phenolic resin	–	332–355	18
Polymethylmethacrylate	15 000	225–256	30
Polymethylacrylate	–	271–286	34
Cellulose triacetate	–	283–306	45
Polyethylene oxide	10 000	320–335	46
Cellulose	–	261–291	50
Polystyrene	230 000	318–348	55
Poly α-methyl styrene	350 000	229–275	55
Polypropylene	–	336–366	58
Polyethylene	20 000	360–392	63
Polymethylene	High	345–396	72

With modern analytical equipment, it is possible to obtain much more detailed information about the decomposition of polymeric materials and their additives. Thermogravimetric analysis (TGA) can be used to investigate the rate of mass loss for a small sample as a function of temperature, while differential scanning calorimetry (DSC) provides information on the amount of energy exchanged during the decomposition process, also as a function of temperature (see Cullis and Hirschler, 1981). By coupling a mass spectrometer to TGA equipment it is possible to identify the decomposition products as they are formed. This technique is particularly useful in examining the way in which a flame retardant influences the decomposition mechanism.

While at first sight the composition of the volatiles might seem of secondary importance to their ability simply to burn as a gaseous mixture, such a view does not permit detailed understanding of fire behaviour. The reactivity of the constituents will influence how easily flame may be stabilized at the surface of a combustible solid (Section 6.3.2), while their nature will determine how much soot will be produced in the flame. The latter controls the amount of heat radiated from the flame to the surroundings and the burning surface (Sections 2.4.3, 5.1.1 and 5.2.1), and also influences the quantity of smoke that will be released from the fire (Section 11.1.1). Thus, volatiles containing aromatic species such as benzene (e.g., from the carbonaceous residue which is formed during chain-stripping of PVC, Reaction (1.R3)), or styrene (from polystyrene), give sooty flames of high emissivity (Section 2.4.3), while in contrast polyoxymethylene burns with a non-luminous flame, simply because the volatiles consist entirely of formaldehyde (CH_2O) (Madorsky, 1964), which does not produce soot (Section 11.1.1). It will be shown later how these factors influence the rates of burning of liquids and solids (Sections 5.1 and 5.2). In some cases the toxicity of the combustion products is affected by the nature of the volatiles (cf. hydrochloric acid gas from PVC, hydrogen cyanide from wool and polyurethane, etc.), but the principal toxic species (carbon monoxide) is produced in all fires involving carbon-based fuels, and its yield is strongly dependent on the condition of burning and availability of air (see Section 11.1.4).

Table 1.6 Relative thermal stability of organic polymers based on the temperature at which the half-life $T_h = 30$ min (from Madorsky, 1964)

Polymer	Polymer unit	T_h (°C)
Polyoxymethylene	$-CH_2-O-$	<200
Polymethylmethacrylate A (MW $= 1.5 \times 10^5$)	$-CH_2-C(CH_3)-$ $\quad\quad\quad\mid$ $\quad\quad CO.OCH_3$	283
Poly α-methyl styrene	$-CH_2-C(CH_3)-$ $\quad\quad\quad\mid$ $\quad\quad C_6H_5$	287
Polyisoprene	$-CH_2-C=CH-CH_2-$ $\quad\quad\quad\mid$ $\quad\quad CH_3$	323
Polymethylmethacrylate B (MW $= 5.1 \times 10^6$)	as A	327
Polymethylacrylate	$-CH_2-CH-$ $\quad\quad\quad\mid$ $\quad\quad CO.OCH_3$	328
Polyethylene oxide	$-CH_2-CH_2-O-$	345
Polyisobutylene	$-CH_2-C(CH_3)_2-$	348
Polystyrene (polyvinyl benzene)	$-CH_2-CH-$	364
Polypropylene	$-CH_2-CH-$ $\quad\quad\quad\mid$ $\quad\quad CH_3$	387
Polydivinyl benzene	$-CH_2-CH-$ $-CH_2-HC$	399
Polyethylene[a]	$-CH_2-CH_2-$	406
Polymethylene[a]	$-CH_2-CH_2-$	415
Polybenzyl	$-CH_2-$	430
Polytetrafluoroethylene	$-CF_2-CF_2-$	509

[a]Polyethylene and polymethylene differ only in that polymethylene is a straight chain with no branching at all (as in Figure 1.1(a)). It requires very special processing. Polyethylene normally has a small degree of branching, which occurs randomly during the polymerization process.

Table 1.7 Factors affecting the thermal stability of polymers (from Madorsky, 1964)[a]

Factor	Effect on thermal stability	Examples (with values of T_h,°C)	
Chain branching[b]	Weakens	Polymethylene	(415)
		Polyethylene	(406)
		Polypropylene	(387)
		Polyisobutylene	(348)
Double bonds in polymer backbone	Weakens	Polypropylene	(387)
		Polyisoprene	(323)
Aromatic ring in polymer backbone	Strengthens	Polybenzyl	(430)
		Polystyrene	(364)
High molecular weight[c]	Strengthens	PMMA B	(327)
		PMMA A	(283)
Cross-linking	Strengthens	Polydivinyl benzene	(399)
		Polystyrene	(364)
Oxygen in the polymer backbone	Weakens	Polymethylene	(415)
		Polyethylene oxide	(345)
		Polyoxymethylene	(<200)

[a]While these are general observations, there are exceptions. For example, with some polyamides (nylons), stability decreases with increasing molecular weight (Madorsky, 1964).
[b]'Branching' refers to the replacement of hydrogen atoms linked directly to the polymer backbone by *any* group, e.g., —CH_3 (as in polypropylene) or —C_6H_5 (as in polystyrene). See Figure 1.1.
[c]Molecular weights of PMMA A and PMMA B are 150 000 and 5 100 000, respectively. Kashiwagi and Omori (1988) found significant differences in the time to ignition of two samples of PMMA of different molecular weights (Chapter 6).

1.2 The Physical Chemistry of Combustion in Fires

There are two distinct regimes in which gaseous fuels may burn, namely: (i) in which the fuel is intimately mixed with oxygen (or air) before burning, and (ii) in which the fuel and oxygen (or air) are initially separate but burn in the region where they mix. These give rise to premixed and diffusion flames, respectively: it is the latter that are encountered in the burning of gas jets and of combustible liquids and solids (Chapter 5). Nevertheless, an understanding of premixed burning is necessary for subsequent discussion of flammability limits and explosions (Chapter 3) and ignition phenomena (Chapter 6), and for providing a clearer insight into the elementary processes within the flame (Section 3.2).

In a diffusion flame, the rate of burning is equated with the rate of supply of gaseous fuel which, for gas jet flames (Section 4.1), is independent of the combustion processes. A different situation holds for combustible liquids and solids, for which the rate of supply of volatiles from the fuel surface is directly linked to the rate of heat transfer from the flame to the fuel (Figure 1.4). The rate of burning ($\dot{m}''$) can be expressed quite generally as:

$$\dot{m}'' = \frac{\dot{Q}_F'' - \dot{Q}_L''}{L_v} \text{g/m}^2 \cdot \text{s} \tag{1.3}$$

where $\dot{Q}_F''$ is the heat flux supplied by the flame (kW/m²) and $\dot{Q}_L''$ represents the losses expressed as a heat flux through the fuel surface (kW/m²). L_v is the heat required to

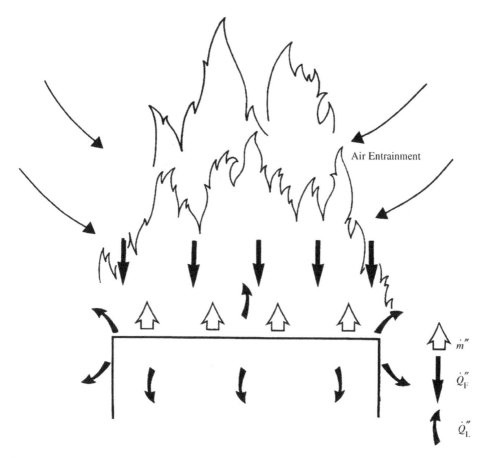

Figure 1.4 Schematic representation of a burning surface, showing the heat and mass transfer processes. $\dot{m}''$, mass flux from the surface; $\dot{Q}''_F$, heat flux from the flame to the surface; $\dot{Q}''_L$, heat losses (expressed as a flux from the surface)

produce the volatiles (kJ/g) which, for a liquid, is simply the latent heat of evaporation (Table 5.9). The heat flux $\dot{Q}''_F$ must in turn be related to the rate of energy release within the flame and the mechanisms of heat transfer involved (see Sections 5.1.1 and 5.2.1).

It will be shown later that the rate at which energy is released in a fire ($\dot{Q}_c$) is the most important single factor that characterizes its behaviour (Babrauskas and Peacock, 1992). It is given by an expression of the form:

$$\dot{Q}_c = \chi \cdot \dot{m}'' \cdot A_f \cdot \Delta H_c \text{ kW} \qquad (1.4)$$

where A_f is the fuel surface area (m^2), ΔH_c (kJ/g) is the heat of combustion of the volatiles and χ is a factor (<1.0) included to account for incomplete combustion (Tewarson, 1982) (Table 5.13). It is now possible to determine the rate of heat release experimentally using the method of oxygen consumption calorimetry (Section 1.2.3), but Equation (1.4) can still be of value when there is limited information available (see Chapter 5).

Closer examination of Equations (1.3) and (1.4) reveals that there are many contributory factors which together determine $\dot{Q}_c$ – including properties relating not only to the material itself (L_v and ΔH_c), but also to the combustion processes within the flame (which in turn determine $\dot{Q}_F''$ and χ). Equation (1.3) emphasizes the importance of the heat transfer terms $\dot{Q}_F''$ and $\dot{Q}_L''$ in determining the rate of supply of fuel vapours to the flame. Indeed, a detailed understanding of heat transfer is a prerequisite to any study of fire phenomena. Consequently, this subject is discussed at some length in Chapter 2, to which frequent reference is made throughout the book. The remainder of this chapter is devoted to a review of those aspects of physical chemistry that are relevant to the understanding of fire behaviour.

1.2.1 The Ideal Gas Law

The release of heat in a fire causes substantial changes in the temperature of the surroundings (Section 10.3) as a result of heat transfer from flames and products of combustion which are formed at high temperatures. Most of the products are gaseous and their behaviour can be interpreted using the ideal gas law:

$$PV = nRT \tag{1.5}$$

where V is the volume occupied by n moles of gas at a pressure P and temperature T (K). (In the SI system, the mole is the amount of a substance which contains as many elementary particles (i.e., atoms or molecules) as $0.012\,kg$ of carbon-12.) In practical terms, the mass of one mole of a substance is the molecular weight expressed in grams. Atomic weights, which may be used to calculate molecular weights, are given in Table 1.8. R is known as the ideal (or universal) gas constant whose value will depend on the units

Table 1.8 Atomic weights of selected elements

	Symbol	Atomic number	Atomic weight
Aluminium	Al	13	27.0
Antimony	Sb	51	121.7
Argon	Ar	18	39.9
Boron	B	5	10.8
Bromine	Br	35	79.9
Carbon	C	6	12.0
Chlorine	Cl	17	35.5
Fluorine	F	9	19.0
Helium	He	2	4.0
Hydrogen	H	1	1.0
Nitrogen	N	7	14.0
Oxygen	O	8	16.0
Phosphorus	P	15	31.0
Sulphur	S	16	32.0

Table 1.9   Values of the ideal gas constant R

Units of pressure	Units of volume	Units of R	Value
N/m^2	m^3	$J/K \cdot mol$	8.31431^a
atm	cm^3	$cm^3 \cdot atm/K \cdot mol$	82.0575
atm	L	$l \cdot atm/K \cdot mol$	0.0820575
atm	m^3	$m^3 \cdot atm/K \cdot mol$	8.20575×10^{-5}†

[a]This is the value applicable to the SI system. However, in view of the variety of ways in which pressure is expressed in the literature, old and new, it is recommended here that the last value (†) is used, with pressure and volume in atmospheres and m^3, respectively.

Table 1.10   Standard atmospheric pressure

Units	Value
Atmospheres	1
Bars	1.01325
Inches of mercury (0°C)	29.9213
Inches of water (4°C)	406.794
kN/m^2 (kPa)	101.325
mm Hg (torr)	760.0

of P and V (Table 1.9). For simplicity, when the ideal gas law is used, pressure should be expressed in atmospheres as data available in the literature (particularly on the vapour pressures of liquids) are presented in a variety of units, including kN/m^2 (or kPa), mm of mercury (mmHg) and bars, all easily converted to atmospheres. Atmospheric pressure expressed in these and other units is given in Table 1.10.

Equation (1.5) incorporates the laws of Boyle ($PV =$ constant at constant temperature) and Gay-Lussac ($V/T =$ constant at constant pressure), and Avogadro's hypothesis, which states that equal volumes of different gases at the same temperature and pressure contain the same number of molecules (or atoms, in the case of an atomic gas such as helium). Setting $P = 1$ atm, $T = 273.17$ K (0°C) and $n = 1$ mole,

$$V = 0.022414 \, m^3 \qquad (1.6)$$

This is the volume that will be occupied by 28 g N_2, 32 g O_2 or 44 g CO_2 at atmospheric pressure and 0°C, assuming that these gases behave ideally. This is not so, but the assumption is good at elevated temperatures. Deviation from ideality increases as the temperature is reduced towards the liquefaction point. Although this clearly applies to a vapour that is in equilibrium with its liquid, Equation (1.5) can be used in a number of ways to interpret and illustrate the fire properties of liquid fuels (Section 6.2).

Table 1.11 Normal composition of the dry atmosphere[a]

Constituent gas[b]	Mole fraction (%)
Nitrogen (N_2)	78.09
Oxygen (O_2)	20.95
Argon (Ar)	0.93
Carbon dioxide (CO_2)	0.03

[a]From Weast (1974/75). It is convenient for many purposes to assume that air consists only of oxygen (21%) and nitrogen (79%). The molar ratio N_2/O_2 is then $79/21 = 3.76$.
[b]Minor constituents include neon ($1.8 \times 10^{-3}\%$), helium ($5.2 \times 10^{-4}\%$), krypton ($1 \times 10^{-4}\%$) and hydrogen ($5 \times 10^{-5}\%$).

The density, or concentration, of a gas may be calculated: for example, taking the composition of normal air as given in Table 1.11, it can be shown that one mole corresponds to $M_w = 0.028\ 95$ kg, so that its density at 0°C (273 K) will be:

$$\rho = \frac{n M_w}{V} = \frac{P M_w}{RT} = 1.292\,\text{kg/m}^3 \tag{1.7}$$

(see Table 11.7). The composition of a mixture of gases may also be expressed in terms of partial pressures (P_i) of the components, i, so that:

$$P = \sum_i P_i \tag{1.8}$$

where P is the total pressure. As the volume fraction of oxygen in normal air is 0.2095, its partial pressure will be 0.2095 atm. This can be converted into a mass concentration as before: thus at 273 K

$$\frac{P M_w}{RT} = \frac{0.2095 \times 0.032}{273 R} \tag{1.9}$$

$$= 0.2993\,\text{kg O}_2/\text{m}^3$$

which gives the mass fraction of oxygen in air (Y_{O_2}) as $0.2993/1.2923 = 0.232$, a quantity that is referred to later (e.g., Equation (5.24)).

The effect of increasing the temperature of a volume of gas can be seen by referring to Equation (1.5): if the volume is kept constant then the pressure will rise in direct proportion to the temperature increase (see Section 1.2.5), while if the pressure is held constant, the gas will expand (V increases) and its density will fall. Density (ρ) varies with temperature (at constant pressure) according to Equation (1.7), i.e.

$$\rho = \frac{P M_w}{R} \cdot \frac{1}{T} \tag{1.10}$$

As PM_w/R is constant, the product ρT will be constant. Consequently, we can write

$$\frac{\rho_0 - \rho_\infty}{\rho_0} = \frac{T_\infty - T_0}{T_\infty} \tag{1.11}$$

where the subscripts 0 and ∞ refer to initial (or ambient) and final conditions, respectively. As $T_\infty = P \cdot M_w/R \cdot \rho_\infty$, this can be rearranged to give

$$\frac{\Delta\rho}{\rho_\infty} = \beta\Delta T \tag{1.12}$$

where $\beta = R\rho_0/P \cdot M_w = 3.66 \times 10^{-3}$ K^{-1}, at the reference state of 1 atmosphere and 0°C. β is the reciprocal of 273 K and is known as the coefficient of thermal expansion. It was first derived for gases by Gay-Lussac in 1802.

If there is any density difference between adjacent masses of air, or indeed any other fluid, relative movement will occur. As the magnitude of this difference determines the buoyant force, the dimensionless group which appears in problems relating to natural convection (the Grashof number, see Section 2.3), can be expressed in terms of either $\Delta\rho/\rho_\infty$ or $\beta\Delta T$ (Table 2.4).

In most fire problems, it may be assumed that atmospheric pressure is constant, but it decreases with height (altitude) according to the relationship:

$$\frac{dp}{dy} = -\rho g \tag{1.12a}$$

where y is height (m), ρ is the density of the fluid (in this case, air) (kg/m^3) and g is the acceleration due to gravity (9.81 m/s^2). Using Equation (1.10), and assuming constant temperature, this may be integrated to give:

$$p = p_0 \exp[-g(\rho_0/p_0)y] \tag{1.12b}$$

Substituting $g = 9.81$ m/s^2, $\rho_0 = 1.2$ kg/m^3 (20°C, see Table 11.6) and $p_0 = 1.01 \times 10^5$ Pa (the value for the 'standard atmosphere'), this becomes:

$$p = p_0 \exp[-1.16 \times 10^{-4}y] \tag{1.12c}$$

Thus it is easy to show that at Denver, Colorado, which is 1 mile (1609 m) above sea level, atmospheric pressure is 83.8 kPa (or 631 mm Hg). The significance of this will be discussed in Section 6.2.

For small values of y – for example, corresponding to the vertical dimension of a building – the difference in pressure between the ground and the upper floors will be very small. If we assume as a first approximation that the density of the air (ρ) within the building is constant, Equation (1.12a) can be integrated to give:

$$p_h = p_0 - \rho g h \tag{1.12d}$$

where h is the height of the building (m). The decrease in pressure with height can be ignored for most purposes (for $h = 50$ m, $p_0 - p_h = 0.6$ kPa, less than 1%). However, if the temperatures inside and outside the building differ by a few degrees, the resulting differences in air density will give rise to pressure differentials across the building envelope. This is the cause of the 'stack effect' that will be discussed in Chapter 11 (Section 11.2.1). The same physics applies to a fully developed compartment fire when large temperature differences exist across the compartment boundaries (see Chapter 10). Strong buoyant flows, driven by differences in density between the hot gases and the ambient atmosphere, are responsible for drawing air into the base of the fire and for the expulsion of flame and hot gases from confined locations (Section 10.2).

1.2.2 Vapour Pressure of Liquids

When exposed to the open atmosphere, any liquid which is stable under normal ambient conditions of temperature and pressure (e.g., water, n-hexane) will evaporate as molecules escape from the surface to form vapour. (Unstable liquids, such as LPG, will be discussed briefly in Chapter 5.) If the system is closed (cf. Figure 6.8(a)), a state of kinetic equilibrium will be achieved when the partial pressure of the vapour above the surface reaches a level at which there is no further net evaporative loss. For a pure liquid, this is the saturated vapour pressure, a property which varies with temperature according to the Clapeyron–Clausius equation:

$$\frac{d(\ln p^\circ)}{dT} = \frac{L_v}{RT^2} \tag{1.13}$$

where p° is the equilibrium vapour pressure and L_v is the latent heat of evaporation (Moore, 1972; Atkins and de Paula, 2006). An integrated form of this is commonly used, for example:

$$\log_{10} p^\circ = (-0.2185E/T) + F \tag{1.14}$$

where E and F are constants, T is in Kelvin and p° is in mm Hg. Values of these for some liquid fuels are given in Table 1.12 (Weast, 1974/5).

The equation may be used to calculate the vapour pressure above the surface of a pure liquid fuel to assess the flammability of the vapour/air mixture (Sections 3.1 and 6.2). The same procedure may be employed for liquid fuel mixtures if the vapour pressures of the components can be calculated. For 'ideal solutions' to which hydrocarbon mixtures approximate, Raoult's law can be used. This states that for a mixture of two liquids, A and B:

$$p_A = x_A \cdot p_A^\circ \text{ and } p_B = x_B \cdot p_B^\circ \tag{1.15}$$

where p_A and p_B are the partial vapour pressures of A and B above the liquid mixture, p_A° and p_B° are the equilibrium vapour pressures of pure A and B (given by Equation (1.14)), and x_A and x_B are the respective 'mole fractions', i.e.

$$x_A = \frac{n_A}{n_A + n_B} \text{ and } x_B = \frac{n_B}{n_A + n_B} \tag{1.16}$$

where n_A and n_B are the molar concentrations of A and B in the mixture. (These are obtained by dividing the mass concentrations (C_A and C_B) by the molecular weights $M_w(A)$ and $M_w(B)$.) In fact, very few liquid mixtures behave ideally and substantial deviations will be found, particularly if the molecules of A or B are partially associated

Table 1.12 Vapour pressures of organic compounds (Weast, 1974/75)

Compound	Formula	E	F	Temperature range (°C)
n-Pentane	$n\text{-}C_5H_{12}$	6595.1	7.4897	−77 to 191
n-Hexane	$n\text{-}C_6H_{14}$	7627.2	7.7171	−54 to 209
Cyclohexane	$c\text{-}C_6H_{12}$	7830.9	7.6621	−45 to 257
n-Octane	$n\text{-}C_8H_{18}$	9221.0	7.8940	−14 to 281
iso-Octane (2,2,4-Trimethyl pentane)		8548.0	7.9349	−36 to 99
n-Decane	$n\text{-}C_{10}H_{22}$	10912.0	8.2481	17 to 173
n-Dodecane	$n\text{-}C_{12}H_{26}$	11857.7	8.1510	48 to 346
Methanol	CH_3OH	8978.8	8.6398	−44 to 224
Ethanol	C_2H_5OH	9673.9	8.8274	−31 to 242
n-Propanol	$n\text{-}C_3H_7OH$	10421.1	8.9373	−15 to 250
Acetone	$(CH_3)_2CO$	7641.5	7.9040	−59 to 214
Methyl ethyl ketone	$CH_3CO.CH_2CH_3$	8149.5	7.9593	−48 to 80
Benzene	C_6H_6	8146.5	7.8337	−37 to 290
Toluene	$C_6H_5CH_3$	8580.5	7.7194	−28 to 31
Styrene	$C_6H_5CH=CH_2$	9634.7	7.9220	−17 to 145

Vapour pressures are calculated using the following equation: $\log_{10} p^\circ = (-0.2185E/T) + F$, where p° is the pressure in mm Hg (torr) (Table 1.10), T is the temperature (Kelvin) and E is the molar heat of vaporization. (Note that the temperature range in the table is given in °C.)

in the pure state (e.g., water, methanol) or if A and B are of different polarity (Moore, 1972; Atkins and de Paula, 2006). Partial pressures must then be calculated using the activities of A and B in the solution, thus:

$$p_A = \alpha_A \cdot p_A^\circ \text{ and } p_B = \alpha_B \cdot p_B^\circ \tag{1.17}$$

where:

$$\alpha_A = \gamma_A \cdot x_A \text{ and } \alpha_B = \gamma_B \cdot x_B$$

α and γ being known as the activity and the activity coefficient, respectively. For an ideal solution, $\gamma = 1$. Values for specific mixtures are available in the literature (e.g., Perry and Green, 2007) and have been used to predict the flashpoints of mixtures of flammable and non-flammable liquids from data on flammability limits (Thorne, 1976) (see Section 6.2).

1.2.3 Combustion and Energy Release

All combustion reactions take place with the release of energy. This may be quantified by defining the heat of combustion (ΔH_c) as the total amount of heat released when unit quantity of a fuel (at 25°C and at atmospheric pressure) is oxidized completely. For a hydrocarbon such as propane (C_3H_8), the products would comprise only carbon dioxide and water, as indicated in the stoichiometric equation:

$$C_3H_8 + 5\ O_2 \rightarrow 3\ CO_2 + 4\ H_2O \tag{1.R4}$$

in which the fuel and the oxygen are in exactly equivalent – or stoichiometric – proportions. The reaction is exothermic (i.e., heat is produced) and the value of ΔH_c will depend on whether the water in the products is in the form of liquid or vapour. The difference will be the latent heat of evaporation of water (44 kJ/mol at 25°C). Thus for propane, the two values are:

$$\Delta H_c(C_3H_8) = -2220 \text{ kJ/mol (the gross heat of combustion)}$$

$$\Delta H_c(C_3H_8) = -2044 \text{ kJ/mol (the net heat of combustion)}$$

where the products are liquid water and water vapour, respectively. In flames and fires, the water remains as vapour and consequently it is more appropriate to use the latter value.

The heat of combustion of propane can be expressed as either -2044 kJ/mol or $-(2044/44) = -46.45$ kJ/g of propane (Table 1.13), where 44 is the gram molecular weight of C_3H_8. By convention, these are expressed as *negative* values, indicating that the reaction is exothermic (i.e., energy is released). If the reaction is allowed to proceed at constant pressure, the energy is the result of a change in *enthalpy* (ΔH) of the system as defined by Reaction (1.R4). However, heats of combustion are normally determined at constant volume in a 'bomb' calorimeter, in which a known mass of fuel is burnt completely in an atmosphere of pure oxygen (Moore, 1972; Atkins and de Paula, 2006; Janssens, 2008). Assuming that there is no heat loss (the system is adiabatic), the quantity of heat released is calculated from the temperature rise of the calorimeter and its contents, whose thermal capacities are accurately known. The use of pure oxygen ensures complete combustion and the result gives the heat released at constant volume, i.e., the change in the internal energy (ΔU) of the system defined by Reaction (1.R4). The difference between the enthalpy change (ΔH) and the internal energy change (ΔU) exists because at constant pressure some of the chemical energy is effectively lost as work done ($P\Delta V$) in the expansion process. Thus, ΔH can be calculated from:

$$\Delta H = \Delta U + P\Delta V \tag{1.18}$$

remembering that for exothermic reactions both ΔH and ΔU are negative. The work done may be estimated using the ideal gas law, i.e.

$$PV = nRT \tag{1.5}$$

where n is the number of moles of gas involved. If there is a change in n, as in Reaction (1.R4), then:

$$P\Delta V = \Delta nRT \tag{1.19}$$

where $\Delta n = 7 - 6 = +1$ and $T = 298$ K. It can be seen that in this case the correction is small (~ 2.5 kJ/mol) and may be neglected in the present context, although it is significant given the accuracy with which heats of combustion can now be measured.

Bomb calorimetry provides the means by which heats of formation of many compounds may be determined. Heat of formation (ΔH_f) is defined as the enthalpy change when a compound is formed in its standard state (1 bar and 298 K) from its constituent elements, also in their standard states. That for carbon dioxide is the heat of the reaction

$$C(\text{graphite}) + O_2(\text{gas}) \rightarrow CO_2(\text{gas}) \tag{1.R5}$$

Table 1.13   Heats of combustion^a of selected fuels at 25°C (298 K)

		$-\Delta H_c$ (kJ/mol)	$-\Delta H_c$ (kJ/g)	$-\Delta H_{c,air}$ (kJ/g(air))	$-\Delta H_{c,ox}$ (kJ/g(O_2))
Carbon monoxide	CO	283	10.10	4.10	17.69
Methane	CH_4	800	50.00	2.91	12.54
Ethane	C_2H_6	1423	47.45	2.96	11.21
Ethene	C_2H_4	1411	50.35	3.42	14.74
Ethyne	C_2H_2	1253	48.20	3.65	15.73
Propane	C_3H_8	2044	46.45	2.97	12.80
n-Butane	n-C_4H_{10}	2650	45.69	2.97	12.80
n-Pentane	n-C_5H_{12}	3259	45.27	2.97	12.80
n-Octane	n-C_8H_{18}	5104	44.77	2.97	12.80
c-Hexane	c-C_6H_{12}	3680	43.81	2.97	12.80
Benzene	C_6H_6	3120	40.00	3.03	13.06
Methanol	CH_3OH	635	19.83	3.07	13.22
Ethanol	C_2H_5OH	1232	26.78	2.99	12.88
Acetone	$(CH_3)_2CO$	1786	30.79	3.25	14.00
D-Glucose	$C_6H_{12}O_6$	2772	15.4	3.08	13.27
Cellulose		–	16.09	3.15	13.59
Polyethylene		–	43.28	2.93	12.65
Polypropylene		–	43.31	2.94	12.66
Polystyrene		–	39.85	3.01	12.97
Polyvinylchloride		–	16.43	2.98	12.84
Polymethylmethacrylate		–	24.89	3.01	12.98
Polyacrylonitrile		–	30.80	3.16	13.61
Polyoxymethylene		–	15.46	3.36	14.50
Polyethyleneterephthalate		–	22.00	3.06	13.21
Polycarbonate		–	29.72	3.04	13.12
Nylon 6,6		–	29.58	2.94	12.67

^{a}The initial states of the fuels correspond to their natural states at normal temperature and pressure (298°C and 1 bar). All products are taken to be in their gaseous state – thus these are the net heats of combustion. All these reactions are exothermic, i.e., the heats of combustion are negative. For clarity, the negative signs appear in the titles of each column.

where ΔH_f^{298} (CO_2) $= -393.5\,$kJ/mol. The negative sign indicates that the product (CO_2) is a more stable chemical configuration than the reactant elements in their standard states, which are assigned heats of formation of zero.

If the heats of formation of the reactants and products of any chemical reaction are known, the total enthalpy change can be calculated. Thus for propane oxidation (Reaction (1.R4)):

$$\Delta H_c(C_3H_8) = 3\Delta H_f\ (CO_2) + 4\Delta H_f(H_2O) - \Delta H_f(C_3H_8) - \Delta H_f\ (O_2) \qquad (1.20)$$

in which $\Delta H_f(O_2) = 0$ (by definition). This incorporates Hess' 'law of constant heat summation', which states that the change in enthalpy depends only on the initial and final states of the system and is independent of the intermediate steps. In fact, $\Delta H_c(C_3H_8)$ is

Table 1.14 Standard heats of formation of
some common gases

Compound	Formula	ΔH_f^{298} (kJ/mol)
Water (vapour)	H_2O	-241.826
Carbon monoxide	CO	-110.523
Carbon dioxide	CO_2	-393.513
Methane	CH_4	-74.75
Propane	C_3H_8	-103.6
Ethene	C_2H_4	$+52.6$
Propene	C_3H_6	$+20.7$
Ethyne	C_2H_2	$+226.9$

easily determined by combustion bomb calorimetry, as are $\Delta H_f(CO_2)$ and $\Delta H_f(H_2O)$, and
Equation (1.20) would be used to calculate $\Delta H_f(C_3H_8)$, which is the heat of the reaction

$$3 \text{ C(graphite)} + 4 \text{ H}_2(\text{gas}) \rightarrow C_3H_8(\text{gas}) \tag{1.R6}$$

Values of the heats of formation of some common gaseous species are given in Table 1.14.
Those species for which the values are positive (e.g., ethene and ethyne) are less stable
than the parent elements and are known as endothermic compounds. Under appropriate
conditions they can be made to decompose with the release of energy. Ethyne (acetylene),
which has a large positive heat of formation, can decompose with explosive violence.

Values of heat of combustion for a range of gases, liquids and solids are given in
Table 1.13: these all refer to normal atmospheric pressure (101.3 kPa) and an ambient
temperature of 298 K (25°C) and to complete combustion. It should be noted that the
values quoted for ΔH_c are the net heats of combustion, i.e., water as a product is in the
vapour state. They differ from the gross heats of combustion by the amount of energy
corresponding to the latent heat of evaporation of the water (2.44 kJ/g (44 kJ/mol) at 25°C).
Furthermore, it is not uncommon in fires for the combustion process to be incomplete,
i.e., χ in Equation (1.4) is less than unity. The actual heat released could be estimated
by using Hess' law of constant heat summation if the composition of the combustion
products was known. The oxidation of propane could be written as a two-stage process,
involving the reactions:

$$C_3H_8 + \tfrac{7}{2}O_2 \rightarrow 3 \text{ CO} + 4 \text{ H}_2O \tag{1.R7}$$

and

$$CO + \tfrac{1}{2}O_2 \rightarrow CO_2 \tag{1.R8}$$

Reaction (1.R4) can be obtained by adding together (1.R7) and three times (1.R8). Then
by Hess' law

$$\Delta H_{R4} = \Delta H_{R7} + 3\Delta H_{R8} \tag{1.21}$$

where ΔH_{R7} is the heat of reaction (1.R7), $\Delta H_{R4} = \Delta H_c(C_3H_8)$ and $\Delta H_{R8} = \Delta H_c(CO)$. As these heats of combustion are both known (Table 1.13), ΔH_{R7} can be shown to be:

$$\Delta H_{R7} = 2044 - 3 \times 283 = 1195 \text{ kJ/mol}$$

Consequently, if it was found that the partial combustion of propane gave only H_2O with CO_2 and CO in the ratio 4:1, the actual heat released per mole of propane burnt would be:

$$\Delta H = \frac{4\Delta H_c(C_3H_8) + \Delta H_{R7}}{5} \tag{1.22}$$

or

$$\Delta H = \Delta H_c(C_3H_8) - \tfrac{3}{5}\Delta H_c(CO) \tag{1.23}$$

which give the same answer (-1874.2 kJ/mol).

The techniques of thermochemistry provide essential information about the amount of heat liberated during a combustion process that has gone to completion. In principle, a correction can be made if the reaction is incomplete, although the large number of products of incomplete combustion that are formed in fires make this approach cumbersome, and effectively unworkable. Yet, information on the rate of heat release in a fire is often required in engineering calculations (Babrauskas and Peacock, 1992), for example in the estimation of flame height (Section 4.3.2), the temperature under a ceiling (Section 4.3.4) or the flashover potential of a room (Section 9.2.2). Until relatively recently, it was common practice to calculate the rate of energy release using Equation (1.4), taking an appropriate value of $\dot{m}''$ and assuming a value for χ to account for incomplete combustion, but an experimental method is now available by which $\dot{Q}_c$ can be determined. This relies on the fact that the heat of combustion of most common fuels is constant if it is expressed in terms of the oxygen, or air consumed. Taking Reaction (1.R4) as an example, it can be said that 2044 kJ are evolved for each mole of propane burnt, or for every five moles of oxygen consumed. The heat of combustion could then be quoted as $\Delta H_{c,ox} = -408.8$ kJ/mol or $(-408.8/32) = -12.77$ kJ/g, where 32 is the molecular weight of oxygen. $\Delta H_{c,ox}$ is given in Table 1.13 for a range of fuels and is seen to lie within fairly narrow limits. Huggett (1980) concluded that typical organic liquids and gases have $\Delta H_{c,ox} = -12.72 \pm 3\%$ kJ/g of oxygen (omitting the reactive gases ethene and ethyne), while polymers have $\Delta H_{c,ox} = -13.02 \pm 4\%$ kJ/g of oxygen (omitting polyoxymethylene). Consequently, if the rate of oxygen consumption can be measured, the rate of heat release can be estimated directly. This method is now widely used both in fire research and in routine testing (see Sections 5.2 and 9.2.2). It is the basis on which the cone calorimeter (Babrauskas, 1992a, 2008b) is founded, and has been adopted in other laboratory-scale apparatuses, such as the flammability test apparatus designed by the Factory Mutual Research Corporation (now FM Global) (Tewarson, 2008; ASTM, 2009). The technique is also used in large-scale equipment (Babrauskas, 1992b) such as the Nordtest/ISO Room Fire Test (Sundström, 1984; Nordtest, 1986), the ASTM room (ASTM 1982) and the furniture calorimeter (Babrauskas et al., 1982; Nordtest, 1991) (see Figure 9.15(a)). These are designed to study the full-scale fire behaviour of wall lining materials, items of furniture and other commodities. The common feature of all these apparatuses is the system for measuring the rate at which oxygen has been consumed in the fire. This involves a hood and duct assembly – the combustion products flowing through a duct of known cross-sectional area and in which careful measurements are made of temperature and velocity, and of the concentrations of oxygen, carbon monoxide and carbon dioxide. If the combustion process is complete (i.e., the only products are water and carbon dioxide), the rate of heat release may be calculated from the expression

$$\dot{Q}_c = (0.21 - \eta_{O_2}) \cdot V \cdot 10^3 \cdot \rho_{O_2} \cdot \Delta H_{c,ox} \qquad (1.24)$$

where V is the volumetric flow of air (m^3/s), ρ_{O_2} is the density of oxygen (kg/m^3) at normal temperature and pressure, and η_{O_2} is the mole fraction of oxygen in the 'scrubbed' gases (i.e., water vapour and acid gases have been removed).

The average value $\Delta H_{c,ox}$ is taken as -13.1 kJ/g(O_2) (see Table 1.13), assuming complete combustion to H_2O and CO_2. Krause and Gann (1980) argue that if combustion is incomplete, i.e., carbon monoxide and soot particles are formed, the effect on the calculated rate of heat release will be small. Their reasoning rests on the fact that if all the carbon was converted to CO, the value used for the heat of combustion ($\Delta H_{c,ox}$) would be no more than 30% too high, while if it all appeared as carbon (smoke particles), it could be no more than 20–25% too low. Given that these factors operate in opposite directions and that in most fires the yield of CO_2 is invariably much higher than that of CO, the resulting error is unlikely to be more than 5%. However, this is not sufficiently accurate for current research and testing procedures, and corrections are necessary, based on the amount of carbon monoxide in the products (see Equations (1.R7), (1.R8), *et seq.*) and the yields of carbon dioxide and water vapour. It is not appropriate to examine this any further in this context as the equations are presented in great detail elsewhere (Janssens, 1991b, 2008; Janssens and Parker, 1992). The question of uncertainty in the calculated heat release values obtained using the cone calorimeter is discussed by Enright and Fleischmann (1999).

Although oxygen consumption calorimetry is predominantly the method of choice, the rate of heat release can also be estimated from measurements of the rates of formation of carbon dioxide and of carbon monoxide. This has been called 'carbon dioxide generation (CDG) calorimetry' (Tewarson and Ogden, 1992). If combustion is complete and the heat of combustion is calculated in terms of the CO_2 produced (i.e., as kJ/g CO_2), values are approximately constant within each generic group of fuels (Tewarson, 2008) – as may be seen in Table 1.15 for a subset of the fuels included in Table 1.13. The rate of heat release can be calculated as:

$$\dot{Q}_c = \frac{d[CO_2]}{dt} \Delta H_{c,CO_2} \qquad (1.25)$$

but a correction for incomplete combustion may be made if the rate of formation of CO is also measured, thus:

$$\dot{Q}_c = \frac{d[CO_2]}{dt} \Delta H_{c,CO_2} + \frac{d[CO]}{dt} \Delta H_{c,CO} \qquad (1.26)$$

$\Delta H_{c,CO}$ also shows a dependence on the fuel type – as can be seen in Table 1.15, in which calculated values of $\Delta H_{c,CO_2}$ and $\Delta H_{c,CO}$ are quoted. Although the technique has an advantage over oxygen consumption calorimetry (OCC) in that the measurement is

Table 1.15 Heats of combustion of selected fuels for carbon dioxide generation (CDG) calorimetry

		$-\Delta H_c$ (kJ/mol fuel)	$-\Delta H_c$ (kJ/gO$_2$)	$-\Delta H_c$ (kJ/gCO$_2$)	$-\Delta H_c$ (kJ/gCO)
Methane	CH_4	800	12.54	18.2	18.5
Ethane	C_2H_6	1423	11.21	16.2	15.3
Ethene	C_2H_4	1411	14.74	16	15.1
Ethyne	C_2H_2	1253	15.73	14.2	12.3
Propane	C_3H_8	2044	12.8	15.5	14.2
n-Butane	n-C_4H_{10}	2650	12.8	15.1	13.6
n-Pentane	n-C_5H_{12}	3259	12.8	14.8	13.2
n-Octane	n-C_8H_{18}	5104	12.8	14.5	12.7
c-Hexane	c-C_6H_{12}	3680	12.8	13.9	11.8
Benzene	C_6H_6	3120	13.06	11.8	8.5
Methanol	CH_3OH	635	13.22	14.4	12.6
Ethanol	C_2H_5OH	1232	12.88	14	11.9
Acetone	$(CH_3)_2CO$	1786	14	13.5	11.2
D-Glucose	$C_6H_{12}O_6$	2772	13.27	10.5	6.4

potentially more accurate,[5] the uncertainty in the values of $\Delta H_{c,CO_2}$ and $\Delta H_{c,CO}$ makes OCC a more reliable method to use. Nevertheless, CDG calorimetry has its value as a method to check the OCC results, particularly if there are insufficient oxygen measurements. For example, CDG calorimetry has been used successfully to derive a record of the rate of heat release in a full-scale tunnel fire for which the oxygen measurements were incomplete (Grant and Drysdale, 1997). In these calculations, $\Delta H_{c,CO_2}$ and $\Delta H_{c,CO}$ were taken as 12.5 kJ/g CO_2 and 7.0 kJ/g CO, respectively for the mixed load of fuels involved (Tewarson, 1996).

The heat of combustion may also be expressed in terms of air 'consumed'. Reaction (1.R4) may be modified to include the nitrogen complement, thus:

$$C_3H_8 + 5\ O_2 + 5 \times 3.76\ N_2 \rightarrow 3\ CO_2 + 4\ H_2O + 18.8\ N_2 \qquad (1.R9)$$

as the ratio of nitrogen to oxygen in air is approximately 3.76 (Table 1.11). Repeating the calculation as before, 2044 kJ are evolved when the oxygen in 23.8 moles of air is consumed. Thus, $\Delta H_{c,air} = 85.88$ kJ/mol or (85.88/28.95) = 2.97 kJ/g, where 28.95 is the 'molecular weight' of air (Section 1.2.1). Values quoted in Table 1.13 cover a wide variety of fuels of all types and give an average of 3.03 ($\pm 2\%$) kJ/g if carbon monoxide and the reactive fuels ethene and ethyne are discounted. A value of 3 kJ/g is a convenient figure to select and is within 12% for the one polymer that appears to behave significantly differently (polyoxymethylene). This may be used to estimate the rate of heat release in a fully developed, ventilation-controlled compartment fire if the rate of air

---

[5] In oxygen consumption calorimetry, the heat release rate is based on relatively small changes in the concentration of oxygen, requiring analytical equipment of the highest standard. Concentrations of CO_2 and CO are much easier to determine as the 'zero value' is vanishingly small.

inflow is known or can be calculated, and it is assumed that all the oxygen is consumed within the compartment boundaries (Section 10.3.2).

While this discussion has focused on the determination of rate of heat release, it should be noted that measurements obtained in the cone calorimeter (and the FTA) can be used to calculate an effective heat of combustion of the fuel (kJ/g) simply as the ratio of the total heat release (kJ) to the total mass loss (g). A refinement is to calculate instantaneous values of the effective heat of combustion from the ratio of the rate of heat release (RHR) to the mass loss rate (MLR) determined at specific times during a test, i.e.

$$\Delta H_c = \frac{RHR}{MLR} \tag{1.27}$$

In general, this is constant for a single, uniform material (e.g., PMMA) but is not the case for char-forming materials such as wood. In these circumstances, the instantaneous values of ΔH_c reveal clearly the difference between the burning of the volatiles and the char (see Section 5.2.2). It also offers a method of gleaning information about the decomposition of materials that have complex pyrolysis mechanisms, such as polyurethane foams (Bustamente *et al.*, 2009).

Stoichiometric equations such as (1.R9) can be used to calculate the air requirements for the complete combustion of any fuel. For example, polymethylmethacrylate has the empirical formula $C_5H_8O_2$ (Table 1.2), identical to the formula of the monomer. The stoichiometric equation for combustion in air may be written:

$$C_5H_8O_2 + 6\ O_2 + 6 \times 3.76\ N_2 \rightarrow 5\ CO_2 + 4\ H_2O + 22.56\ N_2 \tag{1.R10}$$

which shows that 1 mole of PMMA monomer unit requires 28.56 moles of air. Introducing the molecular weights of $C_5H_8O_2$ and air (100 and 28.95, respectively), it is seen that 1 g of PMMA requires 8.27 g of air for stoichiometric burning to CO_2 and water. (In the same way, it can be shown that 15.7 g of air are required to 'burn' 1 g of propane to completion: see Problem 1.11). More generally, we can write:

$$1\ \text{kg fuel} + r\ \text{kg air} \rightarrow (1 + r)\ \text{kg products}$$

where r is the stoichiometric air requirement for the fuel in question. This will be discussed further in Section 3.5.1 and the concept applied in Sections 10.1 and 10.2.

The stoichiometric air requirement can be used to estimate the heat of combustion of any fuel, if this is not known. Taking the example of PMMA, as $\Delta H_{c,air} = 3$ kJ/g, then ΔH_c(PMMA) $= 3 \times 8.27 = 24.8$ kJ/g, in good agreement with the value quoted in Table 1.13.

1.2.4   The Mechanism of Gas Phase Combustion

Chemical equations such as (1.R4) and (1.R7) define the stoichiometry of the complete reaction but hide the complexity of the overall process. Thus, while methane will burn in a flame to yield carbon dioxide and water vapour, according to the reaction

$$CH_4 + 2\ O_2 \rightarrow CO_2 + 2\ H_2O \tag{1.R11}$$

Table 1.16 Mechanism of the gas-phase oxidation of methane (after Bowman, 1975)

	CH_4	+	M	=	$^{\bullet}CH_3$	+	$H^{\bullet}$	+	M	a
	CH_4	+	$^{\bullet}OH$	=	$^{\bullet}CH_3$	+	H_2O			b
	CH_4	+	$H^{\bullet}$	=	$^{\bullet}CH_3$	+	H_2			c
	CH_4	+	$^{\bullet}O^{\bullet}$	=	$^{\bullet}CH_3$	+	$^{\bullet}OH$			d
	O_2	+	$H^{\bullet}$	=	$^{\bullet}O^{\bullet}$	+	$^{\bullet}OH$			e
	$^{\bullet}CH_3$	+	O_2	=	CH_2O	+	$^{\bullet}OH$			f
	CH_2O	+	$^{\bullet}O^{\bullet}$	=	$^{\bullet}CHO$	+	$^{\bullet}OH$			g
	CH_2O	+	$^{\bullet}OH$	=	$^{\bullet}CHO$	+	H_2O			h
	CH_2O	+	$H^{\bullet}$	=	$^{\bullet}CHO$	+	H_2			i
	H_2	+	$^{\bullet}O^{\bullet}$	=	$H^{\bullet}$	+	$^{\bullet}OH$			j
	H_2	+	$^{\bullet}OH$	=	$H^{\bullet}$	+	H_2O			k
	$^{\bullet}CHO$	+	$^{\bullet}O^{\bullet}$	=	CO	+	$^{\bullet}OH$			l
	$^{\bullet}CHO$	+	$^{\bullet}OH$	=	CO	+	H_2O			m
	$^{\bullet}CHO$	+	$H^{\bullet}$	=	CO	+	H_2			n
	CO	+	$^{\bullet}OH$	=	CO_2	+	$H^{\bullet}$			o
$H^{\bullet}$ +	$^{\bullet}OH$	+	M	=	H_2O	+	M			p
$H^{\bullet}$ +	$H^{\bullet}$	+	M	=	H_2	+	M			q
$H^{\bullet}$ +	O_2	+	M	=	$HO_2^{\bullet}$	+	M			r

This reaction scheme is by no means complete. Many radical–radical reactions, including those of the HO_2 radical, have been omitted.
M is any 'third body' participating in radical recombination reactions (p–r) and dissociation reactions such as a.

the mechanism by which this takes place involves a series of elementary steps in which highly reactive molecular fragments (atoms and free radicals), such as $H^{\bullet}$, $^{\bullet}OH$ and $^{\bullet}CH_3$ take part (Table 1.16) (Griffiths and Barnard, 1995; Simmons, 1995; Griffiths, 2008). While these have only transient existences within the flame, they are responsible for rapid consumption of the fuel (reactions b–d in Table 1.16). Their concentration is maintained because they are continuously regenerated in a sequence of chain reactions, e.g.

$$CH_4 + {}^{\bullet}OH \rightarrow H_2O + {}^{\bullet}CH_3 \tag{1.R12}$$

$$^{\bullet}CH_3 + O_2 \rightarrow CH_2O + {}^{\bullet}OH \tag{1.R13}$$

(Table 1.16, b and f), although they are also destroyed in chain termination reactions such as p and q (Table 1.16).

The rate of oxidation of methane may be equated to its rate of removal by reactions b–d. This may be written:[6]

$$-\frac{d[CH_4]}{dt} = k_b[CH_4][{}^{\bullet}OH] + k_c[CH_4][H^{\bullet}] + k_d[CH_4][{}^{\bullet}O^{\bullet}]$$

$$= (k_b[{}^{\bullet}OH] + k_c[H^{\bullet}] + k_d[{}^{\bullet}O^{\bullet}])[CH_4] \tag{1.28}$$

---

[6] An introduction to reaction kinetics may be found in any textbook on physical chemistry, e.g., Moore (1972) or Atkins and de Paula (2006).

where the square brackets indicate concentration, and k_b, k_c and k_d are the appropriate rate coefficients (cf. Equation (1.1)). Clearly, the rate of removal of methane depends directly on the concentrations of free atoms and radicals in the reacting system. This in turn will depend on the rates of initiation (reaction a) and termination (m, n, p and q), but will be enhanced greatly if the branching reaction (e) is significant, i.e.

$$O_2 + H^{\bullet} \rightarrow {}^{\bullet}O^{\bullet} + {}^{\bullet}OH \qquad (1.R14)$$

This has the effect of increasing the number of radicals in the system, replacing one hydrogen atom by three free radicals, as can be seen by examining the fate of the oxygen atoms by reactions d, g and j (Table 1.16). In this respect the hydrogen atom is arguably the most important of the reactive species in the system. If other molecules compete with oxygen for H-atoms (e.g., reactions c, i and n), then the branching process (i.e., the multiplication in number of free radicals) is held in check. No such check exists in the H_2/O_2 reaction in which the oxygen molecule is effectively the only gaseous species with which the hydrogen atoms can react (e.g., Dixon-Lewis and Williams, 1977; Simmons, 1995, 2008):

$$O_2 + H^{\bullet} \rightarrow {}^{\bullet}O^{\bullet} + {}^{\bullet}OH \qquad (1.R14)$$

$$H_2 + {}^{\bullet}O^{\bullet} \rightarrow H^{\bullet} + {}^{\bullet}OH \qquad (1.R15)$$

$$H_2 + {}^{\bullet}OH \rightarrow H^{\bullet} + H_2O \qquad (1.R16)$$

(Reaction r in Table 1.16 is relatively unimportant in flames.) Consequently, under the appropriate conditions (Section 3.1), the rate of oxidation of hydrogen in air is very high, as indicated by its maximum burning velocity, $S_u = 3.2$ m/s, which is more than eight times greater than that for methane (0.37 m/s) (Table 3.1).

Species that react rapidly with hydrogen atoms, effectively replacing them with atoms or radicals that are considerably less reactive, can inhibit gas phase oxidation. Chlorine- and bromine-containing compounds can achieve this effect by giving rise to hydrogen halides (HCl or HBr) in the flame. Reactions such as

$$HBr + H^{\bullet} \rightarrow H_2 + Br^{\bullet} \qquad (1.R17)$$

replace hydrogen atoms with relatively inactive halogen atoms and thereby reduce the overall rate of reaction dramatically. For this reason, many chlorine- and bromine-containing compounds are found to be valuable fire retardants (Lyons, 1970; Simmons, 1995; Lewin and Weil, 2001) and chemical extinguishants (see Section 3.5.4).

Returning to the discussion of methane oxidation, it is seen that the molecule CH_2O (formaldehyde) is formed in Reaction (1.R13) (Table 1.16, f) as an intermediate. Under conditions of complete combustion this would be destroyed in reactions g–i (Table 1.16), but if the reaction sequence was interrupted as a result of chemical or physical quenching, some formaldehyde could survive and appear in the products. Similarly carbon monoxide – the most ubiquitous product of incomplete combustion – will also be released if the concentration of hydroxyl radicals (${}^{\bullet}OH$) is not sufficient to allow the reaction

$$CO + {}^{\bullet}OH \rightarrow CO_2 + H^{\bullet} \qquad (1.R18)$$

to proceed to completion. This equation represents the only significant reaction by which carbon monoxide is oxidized, and indeed is the principal source of carbon dioxide in any combustion system (e.g., Baulch and Drysdale, 1974).

The complexity of the gas phase oxidation process is shown clearly in Table 1.16 for the simplest of the hydrocarbons, methane, although even this reaction scheme is incomplete (for example, Bowman (1975) lists 30 reactions). The complexity increases with the size and structure of the fuel molecule and consequently the number of partially oxidized species that may be produced becomes very large. Most studies of gas phase combustion have been carried out using well-mixed (*premixed*) fuel/air mixtures (Chapter 3), but in natural fires mixing of fuel vapours and air is an integral part of the burning process (Chapter 4): these flames are known as *diffusion flames*. Consequently, the combustion process is much less efficient, burning occurring only in those regions where there is enough fuel and oxidant present and the temperature is sufficiently high. In a free-burning fire in the open, more air is entrained into the flame than is required to burn all the vapours (Steward, 1970; Heskestad, 1986; see Section 4.3.2). Despite this, some products of incomplete combustion survive the flame and are released to the atmosphere. Not all the products are gaseous: some are minute carbonaceous particles formed within the flame under conditions of low oxygen and high temperature (e.g., Rasbash and Drysdale, 1982). These make up the 'particulate' component of smoke which reduces visibility remote from the fire (Section 11.1), and can be formed even under 'well-ventilated' conditions, depending on the nature of the fuel. For example, polystyrene produces a great deal of black smoke because of the presence of phenyl groups (C_6H_5) in the polymer molecule (Table 1.6 *et seq.*). This is a very stable structure, which is not only resistant to oxidation, but ideally suited to act as the building unit for the formation of carbonaceous (or 'soot') particles within the flame (Section 11.1.1).

If the supply of air to a fire is restricted in some way – for example, if it is burning in an enclosed space, or 'compartment', with restricted ventilation (see Figure 9.3) – then the yield of incompletely burned products will increase (e.g., Rasbash, 1967; Woolley and Fardell, 1982; Gottuk and Lattimer, 2008). It is convenient to introduce here the concept of 'equivalence ratio', a term normally associated with premixed fuel/air mixtures. It is given by the expression

$$\phi = \frac{(\text{fuel}/\text{air})_{\text{actual}}}{(\text{fuel}/\text{air})_{\text{stoich}}} \tag{1.29}$$

where $(\text{fuel}/\text{air})_{\text{stoich}}$ is the stoichiometric fuel/air ratio.[7] The terms 'lean' and 'rich' refer to the situations where $\phi < 1$ and $\phi > 1$, respectively. In diffusion flames, assuming that the rate of fuel supply is known, a value can only be assigned to ϕ if the rate of air supply into the flame can be deduced or can be measured. This cannot be done meaningfully for free-burning diffusion flames in the open,[8] but the concept has been used to interpret the results of experimental studies of the composition of the smoke layer under a ceiling, at least up to the condition known as 'flashover' (Beyler, 1984b; Gottuk *et al.*, 1992b; Pitts, 1994). In these, careful attention was paid to the dependence of the yields of

[7] The quantity $(\text{fuel}/\text{air})_{\text{stoich}}$ is equal to the reciprocal of the stoichiometric factor r (i.e., $1/r$) in the unnumbered equation on page 29. This unfortunate discrepancy reflects the fact that fire science and combustion science developed in very different ways.
[8] Estimates have been made for turbulent diffusion flames by Stewart (1970) (see Section 4.3.2).

partially burned products (in particular, carbon monoxide) on the 'equivalence ratio' (ϕ) (Section 9.2.1). In general, high yields of carbon monoxide are to be associated with high equivalence ratios: there is competition for the hydroxyl radicals between CO (see (1.R18)) and other partially burned products (e.g., reaction h in Table 1.16), which causes an effective reduction in the rate of conversion of CO to CO_2 (e.g., Pitts, 1994). This conversion is also suppressed in the presence of soot particles, which are now known to react with hydroxyl radicals (Neoh *et al.*, 1984; Puri and Santoro, 1991) (Section 11.1).

1.2.5 Temperatures of Flames

In a fire, the total amount of heat that can be released is normally of secondary impor-tance to the rate at which it is released (Babrauskas and Peacock, 1992). If the heat of combustion is known, the rate may be calculated from Equation (1.4), provided that the product $\dot{m}'' \cdot A_f \cdot \chi$ is known. This is seldom the case for fires burning in enclosures, but the rate may be estimated if the rate of air inflow ($\dot{m}_{air}$) is known. Then, assuming that all the oxygen is consumed within the enclosure, the rate of heat release is:

$$\dot{Q}_c = \dot{m}_{air} \cdot \Delta H_{c,air} \tag{1.30}$$

The temperatures achieved depend on $\dot{Q}_c$ and the rate of heat loss from the vicinity of the reacting system (see Section 10.3.2). The only situation where it is reasonable to ignore heat loss (at least to a first approximation) is in premixed burning, when the fuel and air are intimately mixed and the reaction rates are high, independent of diffusive or mixing processes. This is the 'adiabatic' model, in which it is assumed that none of the heat generated within the system is lost to the environment, thus producing the maximum theoretical rise in temperature. Taking as an example a flame propagating through a stoichiometric mixture of propane in air (see Figure 3.14), then it is possible to estimate the adiabatic flame temperature, assuming that all the energy released is taken up by the combustion products. From Table 1.13, $\Delta H_c(C_3H_8) = -2044.3$ kJ/mol. The oxidation reaction in air is given by Reaction (1.R9): the combustion energy raises the temperature of the products CO_2, H_2O and N_2, whose final temperatures can be calculated if heat capacities of these species are known. These may be obtained from thermochemical tables (e.g., Chase, 1998) or other sources (e.g., Lewis and von Elbe, 1987) (Table 1.17). It is assumed that nitrogen is not involved in the chemical reaction but acts only as 'thermal ballast', absorbing a major share of the combustion energy. The energy released in the combustion of 1 mole of propane is thus taken up by 3 moles of CO_2, 4 moles of H_2O and 18.8 moles of N_2 (Reaction (1.R9)). The total heat capacity of this mixture is 942.5 J/K (per mole of propane burnt) (see Table 1.18), so that the final flame temperature T_f is:

$$T_f = 25 + \frac{2044300}{942.5} = 2194°C \tag{1.31}$$

assuming an initial temperature of 25°C. The result of this calculation is approximate for the following reasons:

(i) The thermal capacity of each gas is a function of temperature, and for simplicity the values used here refer to an intermediate temperature (1000 K).
(ii) The system is not truly adiabatic as radiation losses from the flame zone and its vicinity will tend to reduce the final temperature and cause the temperature to fall in the post-flame gases (Section 3.3; Figure 3.16).

(iii) At high temperature, the products are partially dissociated into a number of atomic, molecular and free radical species. This can be expressed in terms of the equilibria (Friedman, 2008):

$$H_2O \rightleftharpoons H + OH \tag{1.R19a}$$

$$H_2O \rightleftharpoons H_2 + \frac{1}{2}O_2 \tag{1.R19b}$$

$$CO_2 \leftrightarrow CO + \frac{1}{2}O_2 \tag{1.R19c}$$

As each dissociation is endothermic (absorbing energy rather than releasing it), these will depress the final temperature. The effect of dissociation on the calculated temperature becomes significant above $\sim 1700°C$ (2000 K). Provided these three reactions (1.R19a–c) are sufficient to describe dissociation in the system, the stoichiometric reaction for the oxidation of propane (Reaction (1.R9)) can be rewritten:

$$C_3H_8 + 5\,O_2 + 18.8\,N_2 \rightarrow (3-y)CO_2 + y\,CO + (4-x)H_2O + 18.8\,N_2$$

$$+ (x-z)H_2 + \left(\frac{x}{2} + \frac{y}{2} - \frac{z}{2}\right)O_2 + z\,H^\bullet + z\,{}^\bullet OH \tag{1.R20}$$

Table 1.17 Thermal capacities of common gases at 1000 K

	$C_p^{1000\,K}$ (J/mol·K)
Carbon monoxide (CO)	33.2
Carbon dioxide (CO$_2$)	54.3
Water (vapour) (H$_2$O)	41.2
Nitrogen (N$_2$)	32.7
Oxygen (O$_2$)	34.9
Helium (He)	20.8

Table 1.18 Heat capacity of combustion products (stoichiometric propane/air mixture)

$$C_3H_8 + 5\,O_2 + 18.8\,N_2 \rightarrow 3\,CO_2 + 4\,H_2O + 18.8\,N_2$$

Species	Number of moles in products[a]	Thermal capacity at 1000 K	
		C_p (J/mol·K)	nC_p (J/K)[a]
CO$_2$	3	54.3	162.9
H$_2$O	4	41.2	164.8
N$_2$	18.8	32.7	614.8
	Total thermal capacity/mole propane $=$ 942.5 J/K		

[a]Per mole of propane burnt.

The values of x, y and z are unknown but can be determined by calculating the position of these three equilibria using the appropriate thermodynamic data (e.g., Chase, 1998), provided that the final temperature is known. As this is not the case, a trial value is selected and the corresponding concentrations of $H^{\bullet}$, $^{\bullet}OH$, H_2, CO and O_2 are calculated from the appropriate equilibrium constants (Moore, 1974; Lewis and von Elbe, 1987; Atkins and de Paula, 2006). The heat of Reaction (1.R20) is then calculated from the heats of formation of all species present in the products and the resulting (adiabatic) temperature obtained using the method outlined in Table 1.18. This procedure is repeated, replacing the original trial temperature by the calculated temperature, and then, if necessary, reiterated until two successive iterations give temperatures in satisfactory agreement. The procedure is discussed in detail by Friedman (2008), who shows that the adiabatic flame temperature for a stoichiometric propane/air mixture is 1995°C, significantly lower than the value of 2194°C obtained by ignoring dissociation (see Equation (1.28)). Comparisons of these calculated temperatures are given for methane and ethane in Table 1.19. It must be remembered that the actual temperatures will be lower than the calculated adiabatic temperature (with dissociation) because heat losses are ignored: Table 1.19 compares measured and calculated values for the fuels methane, ethane and propane (Lewis and von Elbe, 1987).

Despite the approximate nature of the calculation given above, it has value in estimating whether or not particular mixtures are flammable. There is good evidence that there is a lower limiting (adiabatic) temperature below which flame cannot propagate (Section 3.3). The consequence of this is that not all mixtures of flammable gas and air will burn if subjected to an ignition source. The flammable region is well defined and is bounded by the lower and upper flammability limits, which can be determined experimentally to within a few tenths of 1% (Sections 3.1.1 and Table 3.1). For propane, the lower limit corresponds to 2.2% propane in air.

Assuming that combustion of this mixture will proceed to completion, the oxidation can be written:

$$0.021 \ C_3H_8 + 0.979(0.21O_2 + 0.79 \ N_2) \rightarrow \text{products}(CO_2, H_2O, O_2 \text{ and } N_2) \quad (1.R21)$$

Table 1.19 Comparison of adiabatic flame temperatures calculated for stoichiometric hydrocarbon/air mixtures and measured flame temperatures for near-stoichiometric mixtures (Lewis and von Elbe, 1987)

Fuel	Diluent	Adiabatic flame temperature (°C) (no dissociation)	Adiabatic flame temperature (°C) (with dissociation)[a]	% fuel	Measured flame temperature (°C)
Methane	air	2116	1950	10.0	1875
Ethane	air	2173	1988	5.8	1895
Propane	air	2194	1995	4.15	1925
n-Butane	air	2199	–	3.2	1895
i-Butane	air	2192	–	3.2	1900

[a]The calculated adiabatic flame temperatures refer to the stoichiometric mixtures, in which the % fuel values are slightly higher than the percentages used in the experimental measurements (9.5%, 5.7%, 4.0%, 3.1% and 3.1% for CH_4, C_2H_6, C_3H_8, n-C_4H_{10} and i-C_4H_{10}, respectively).

Dividing through by 0.021 gives:

$$C_3H_8 + 9.790\ O_2 + 36.829\ N_2 \rightarrow 3\ CO_2 + 4\ H_2O + 4.790\ O_2 + 36.829\ N_2 \quad (1.R22)$$

As the original mixture was 'fuel lean', the excess oxygen will contribute to the total thermal capacity of the mixture of products. Using the method outlined in Table 1.18, the final (adiabatic) flame temperature can be shown to be 1228°C (1501 K), well below that at which the effect of dissociation is significant. If the adiabatic flame temperature is calculated for the limiting mixtures of a number of n-alkanes, they are found to fall within a fairly narrow band (1600 ± 100 K) (see Table 1.20). There is evidence to suggest that the same value also applies to the upper flammability (fuel-rich) limit (Mullins and Penner, 1959; Stull, 1971), but it cannot be derived by the same method as the lower limit because the products will contain a complex mixture of pyrolysis and partially oxidized products from the parent fuel.

It should be noted that the temperature increases reported above will be accompanied by expansion of the gases. Using the ideal gas law (Section 1.2.1), it can be seen that a seven-fold increase in temperature (e.g., 300 K to 2100 K) will be accompanied by a seven-fold increase in volume, neglecting any change in the number of moles in the system:

$$\frac{V_2}{V_1} = \frac{n_2}{n_1} \times \frac{T_2}{T_1} \quad (1.32)$$

where the subscripts 1 and 2 refer to the initial and final states, assuming that $P_1 = P_2$. However, if the volume remains constant there will be a similar, corresponding rise in pressure. Such large increases will be generated very rapidly if a flammable vapour/air mixture is ignited within a confined space (Chapter 3). This will almost certainly cause structural damage to a building unless measures have been incorporated to prevent the build-up of pressure. One such technique is the provision of explosion relief in the form of weakened panels in the building envelope that will fail easily before pressures capable of damaging the rest of the structure have been reached (Bartknecht, 1981; Drysdale and Kemp, 1982; Harris, 1983; Foster, 1998; Zalosh, 2008).

Table 1.20 Calculated adiabatic flame temperatures (T_f) for limiting mixture of n-alkanes

		LEL (%)	ΔH_c (kJ/mol)	T_f (K)
Methane	CH_4	5	800	1446
Ethane	C_2H_6	3	1423	1502
Propane	C_3H_8	2.1	2044	1501
n-Butane	n-C_4H_{10}	1.8	2650	1619
n-Pentane	n-C_5H_{12}	1.4	3259	1585
n-Hexane	n-C_6H_{14}	1.2	3857	1578
n-Heptane	n-C_7H_{16}	1.05	4466	1592
n-Octane	n-C_8H_{18}	0.95	5104	1626
n-Decane	n-$C_{10}H_{22}$	0.75	6282	1595

Problems

1.1 Calculate the vapour densities (kg/m^3) of pure carbon dioxide, propane and butane at 25°C and atmospheric pressure. Assume ideal gas behaviour.

1.2 Assuming ideal gas behaviour, what will the final volume be if 1 m^3 of air is heated from 20°C to 700°C at constant pressure?

1.3 Calculate the vapour pressure of the following pure liquids at 0°C: (a) n-octane; (b) methanol; (c) acetone.

1.4 Calculate the vapour pressures of n-hexane and n-decane above a mixture at 25°C containing 2% n-C_6H_{14} + 98% n-$C_{10}H_{22}$, by volume. Assume that the densities of pure n-hexane and n-decane are 660 and 730 kg/m^3, respectively and that the liquids behave ideally.

1.5 Calculate the vapour pressures of iso-octane and n-dodecane above a mixture at 20°C containing 5% i-C_8H_{18} and 95% n-$C_{12}H_{26}$, by volume. The densities of the pure liquids are 692 and 749 kg/m^3, respectively. Assume ideal behaviour.

1.6 Work out the enthalpy of formation of propane at 25°C (298 K) from Equation (1.20) using data contained in Tables 1.13 and 1.14.

1.7 Given that the stoichiometric reaction for the oxidation of n-pentane is:

$$n\text{-}C_5H_{12} + 8\ O_2 \rightarrow 5\ CO_2 + 6\ H_2O$$

calculate ΔH_f^{298} (C_5H_{12}) from data in Tables 1.13 and 1.14.

1.8 Calculate the enthalpy change (20°C) in the oxidation of n-pentane to carbon monoxide and water, i.e.

$$n\text{-}C_5H_{12} + 5\tfrac{1}{2}\ O_2 \rightarrow 5\ CO + 6\ H_2O$$

1.9 The products of the partial combustion of n-pentane were found to contain CO_2 and CO in the ratio of 4:1. What is the actual heat released per mole of n-pentane burnt if the only other product is H_2O?

1.10 Express the result of Problem 1.9 in terms of heat released (a) per gram of n-pentane burnt; (b) per gram of air consumed.

1.11 Calculate the masses of air required to burn completely 1 g each of propane (C_3H_8), pentane (C_5H_{12}) and decane ($C_{10}H_{22}$).

1.12 Calculate the adiabatic flame temperatures for the following mixtures initially at 25°C assuming that dissociation does not occur:
(a) stoichiometric n-pentane/oxygen mixture;
(b) stoichiometric n-pentane/air mixture;
(c) 1.5% n-pentane in air (lower flammability limit, see Chapter 3).

2

Heat Transfer

An understanding of several branches of physics is required in order to be able to interpret fire phenomena (see Di Nenno *et al*., 2008). These include fluid dynamics, and heat and mass transfer. In view of its importance, the fundamentals of heat transfer will be reviewed in this chapter. Other topics are required in later chapters, but the essentials are introduced in context and key references given (e.g., Batchelor, 1967; Landau and Lifshitz, 1987; Tritton, 1988; Kandola, 2008). In this chapter, certain heat transfer formulae are derived for use later in the text, although their relevance may not be immediately obvious. There are many good textbooks that deal with heat transfer in great depth, several of which have been referred to during the preparation of this chapter (e.g., Rohsenow and Choi, 1961; Holman, 1976; Pitts and Sissom, 1977; Incropera *et al*., 2007; Welty *et al*., 2008). It is recommended that such texts are used to provide the detail which cannot be included here.

There are three basic mechanisms of heat transfer, namely conduction, convection and radiation. While it is probable that all three contribute in every fire, it is often found that one predominates at a given stage, or in a given location. Thus, conduction determines the rate of heat flow in and through solids. It is important in problems relating to ignition and spread of flame over combustible solids (Chapters 6 and 7), and to fire resistance, where knowledge of heat transfer through compartment boundaries and into elements of the structure is required (Chapter 10). Convective heat transfer is associated with the exchange of heat between a gas or liquid and a solid, and involves movement of the fluid medium (e.g., cooling by directing a flow of cold air over the surface of a hot solid). It occurs at all stages in a fire but is particularly important early on when thermal radiation levels are low. In natural fires, the movement of gases associated with this transfer of heat is determined by buoyancy, which also influences the shape and behaviour of diffusion flames (Chapter 4). The buoyant plume will be discussed in Section 4.3.1.

Unlike conduction and convection, radiative heat transfer requires no intervening medium between the heat source and the receiver. It is the transfer of energy by electromagnetic waves, of which visible light is the example with which we are most familiar. Radiation in all parts of the electromagnetic spectrum can be absorbed, transmitted or reflected at a surface, and any opaque object placed in its way will cast a shadow. It becomes the dominant mode of heat transfer in fires as the fuel bed diameter increases beyond about 0.3 m, and determines the growth and spread of fires in compartments. It is the mechanism by which objects at a distance from a fire are heated

An Introduction to Fire Dynamics, Third Edition. Dougal Drysdale.
© 2011 John Wiley & Sons, Ltd. Published 2011 by John Wiley & Sons, Ltd.

to the firepoint condition, and is responsible for the spread of fire through open fuel beds (e.g., forests) and between buildings (Law, 1963). A substantial amount of heat released in flames is transmitted by radiation to the surroundings. Most of this radiation is emitted by minute solid particles of soot which are formed in almost all diffusion flames – this is the source of their characteristic yellow luminosity. The effect of thermal radiation from flames, or indeed from any heated object, on nearby surfaces can only be ascertained by carrying out a detailed heat transfer analysis. Such analyses are required to establish how rapidly combustible materials that are exposed to thermal radiation will reach a state in which they can be ignited and will burn (Chapter 6), or in the case of structural elements how rapidly they will begin to lose their strength, etc. (Chapter 10).

2.1 Summary of the Heat Transfer Equations

At this point, it is necessary to introduce the basic equations of heat transfer, as it is impossible to examine any one of the mechanisms in depth in isolation from the others. Detailed discussion will follow in subsequent sections, and major review articles may be found in the *SFPE Handbook* (Atreya, 2008; Rockett and Milke, 2008; Tien *et al.*, 2008).

(a) *Conduction (Section 2.2)*. Conduction is the mode of heat transfer associated with solids. Although it also occurs in fluids, it is normally masked by convective motion in which heat is dissipated by a mixing process driven by buoyancy. It is common experience that heat will flow from a region of high temperature to one of low temperature; this flow can be expressed as a heat flux, which in one direction is given by:

$$\dot{q}''_x = -\kappa \frac{\Delta T}{\Delta x} \tag{2.1}$$

where ΔT is the temperature difference over a distance Δx.
   In differential form:

$$\dot{q}''_x = -k \frac{dT}{dx} \tag{2.2}$$

where $\dot{q}''_x = (dq_x/dt)/A$, A being the area (perpendicular to the x-direction) through which heat is being transferred. This is known as Fourier's law of heat conduction. The constant k is the thermal conductivity and has units of W/m·K when $\dot{q}''$ is in W/m², T is in °C (or K), and x is in m. Typical values are given in Table 2.1. These refer to specific temperatures (0 or 20°C) as thermal conductivity is dependent on temperature. Data on k as a function of T are available for many pure materials (e.g., Kaye and Laby, 1986), but such information for combustible solids and building materials is fragmentary (Abrams, 1979; Kodur and Harmathy, 2008).
   As a general rule, materials which are good thermal conductors are also good electrical conductors. This is because heat transfer can occur as a result of interactions involving free electrons whose movement constitutes an electric current when a voltage is applied. In insulating materials, the absence of free electrons means that heat can only be transferred as mechanical vibrations through the structure of the molecular lattice, which is a much less efficient process.

Table 2.1 Thermal properties of some common materials[a]

Material	k (W/m·K)	c_p (J/kg·K)	ρ (kg/m^3)	α (m^2/s)	$k\rho c_p$ (W^2·s/m^{4}K^2)
Copper	387	380	8940	1.14×10^{-4}	1.3×10^9
Steel (mild)	45.8	460	7850	1.26×10^{-5}	1.6×10^8
Brick (common)	0.69	840	1600	5.2×10^{-7}	9.3×10^5
Concrete	0.8–1.4	880	1900–2300	5.7×10^{-7}	2×10^6
Glass (plate)	0.76	840	2700	3.3×10^{-7}	1.7×10^6
Gypsum plaster	0.48	840	1440	4.1×10^{-7}	5.8×10^5
PMMA[b]	0.19	1420	1190	1.1×10^{-7}	3.2×10^5
Oak[c]	0.17	2380	800	8.9×10^{-8}	3.2×10^5
Yellow pine[c]	0.14	2850	640	8.3×10^{-8}	2.5×10^5
Asbestos	0.15	1050	577	2.5×10^{-7}	9.1×10^4
Fibre insulating board	0.041	2090	229	8.6×10^{-8}	2.0×10^4
Polyurethane foam[d]	0.034	1400	20	1.2×10^{-6}	9.5×10^2
Air	0.026	1040	1.1	2.2×10^{-5}	–

[a] From Pitts and Sissom (1977) and others. Most values for 0 or 20°C. Figures have been rounded off. Other compendia of data are to be found in most heat transfer texts (e.g., Incropera *et al.*, 2007; Welty *et al.*, 2008).
[b] Polymethylmethacrylate. Values of k, c_p and ρ for other plastics are given in Table 1.2.
[c] Properties measured perpendicular to the grain.
[d] Typical values only.

(b) *Convection (Section 2.3)*. As indicated above, convection is that mode of heat transfer to or from a solid involving movement of a surrounding fluid. The empirical relationship first discussed by Newton is:

$$\dot{q}'' = h\,\Delta T \ \text{W/m}^2 \tag{2.3}$$

where h is known as the convective heat transfer coefficient. This equation defines h which, unlike thermal conductivity, is not a material constant. It depends on the characteristics of the system, the geometry and orientation of the solid and the properties of the fluid, including the flow parameters. In addition, it is also a function of ΔT. The evaluation of h for different situations has been one of the major problems in heat transfer and fluid dynamics. Typical values lie in the range 5–25 W/m^2·K for free convection and 10–500 W/m^2·K for forced convection in air.

(c) *Radiation (Section 2.4)*. According to the Stefan–Boltzmann equation, the total energy emitted by a body is proportional to T^4, where T is the temperature in Kelvin. The total emissive power is expressed as:

$$E = \varepsilon\sigma T^4 \ \text{W/m}^2 \tag{2.4}$$

where σ is the Stefan–Boltzmann constant (5.67×10^{-8} W/m^{2}K^4) and ε is a measure of the efficiency of the surface as a radiator, known as the emissivity. The perfect emitter – the 'black body' – has an emissivity of unity. The intensity of radiant energy ($\dot{q}''$) falling on a surface remote from the emitter can be found by using the appropriate 'configuration factor' ϕ, which takes into account the geometrical relationship between the emitter and the receiver:

$$\dot{q}'' = \phi \varepsilon \sigma T^4 \tag{2.5}$$

These concepts will be developed in detail in Section 2.4.

2.2 Conduction

While many common problems involving heat conduction are essentially steady state (e.g., thermal insulation of buildings), most of those related to fire are transient and require solutions of time-dependent partial differential equations. Nevertheless, a system of this type will move towards an equilibrium that will be achieved provided there is no variation in the heat source or in the integrity of the materials involved. Indeed, as the steady state is the limiting condition, it can be used to solve a number of problems, many of which will be discussed in subsequent chapters. Thus, it is worthwhile to consider steady state before examining transient conduction.

2.2.1 Steady State Conduction

Consider the heat loss through an infinite, plane slab, or wall, of thickness L, whose surfaces are at temperatures T_1 and T_2 ($T_1 > T_2$) (Figure 2.1). In this idealized model the heat flow is unidimensional. Integrating Fourier's equation (Equation (2.2)) leads to:

$$\dot{q}''_x \int_O^L \mathrm{d}x = -k \int_{T_1}^{T_2} \mathrm{d}T \tag{2.6}$$

provided that k is independent of temperature. This gives:

$$\dot{q}''_x = \frac{k}{L}(T_1 - T_2) \tag{2.7}$$

(If k is a function of temperature within the range of interest, then $k\mathrm{d}T$ must be integrated between T_1 and T_2.)

If the wall is composite, consisting of various layers as shown in Figure 2.2, the net heat flux through the wall at the steady state can be calculated by equating the steady state heat fluxes across each layer. Assuming that there is a temperature difference between

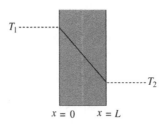

T_1 ----------

---------- T_2

$x = 0$ $x = L$

Figure 2.1 The infinite plane slab. In this example, the surfaces are at temperatures T_1 and T_2, as indicated

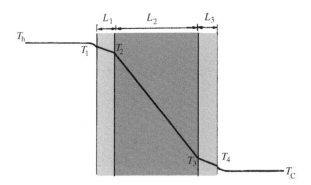

Figure 2.2 The infinite plane composite wall. The temperature of the air in contact with each surface is shown. T_1, T_2 etc. refer to the temperatures of the boundaries in the steady state

each surface and the adjacent air, as shown, then if h_h and h_c are the convective heat transfer coefficients at the inner and outer (hot and cold) surfaces (Equation (2.3)):

$$\dot{q}_x'' = h_h(T_h - T_1) = \frac{k_1}{L_1}(T_1 - T_2) = \frac{k_2}{L_2}(T_2 - T_3)$$

$$= \frac{k_3}{L_3}(T_3 - T_4) = h_c(T_4 - T_c) \tag{2.8}$$

Only the air temperatures T_h and T_c are known, but the heat flux across each layer gives the following series of relationships:

$$T_h - T_1 = \dot{q}_x''/h_h$$

$$T_1 - T_2 = \dot{q}_x''L_1/k_1$$

$$T_2 - T_3 = \dot{q}_x''L_2/k_2$$

$$T_3 - T_4 = \dot{q}_x''L_3/k_3$$

$$T_4 - T_c = \dot{q}_x''/h_c$$

which, if added together and rearranged, give:

$$\dot{q}_x'' = \frac{T_h - T_c}{\frac{1}{h_h} + \frac{L_1}{k_1} + \frac{L_2}{k_2} + \frac{L_3}{k_3} + \frac{1}{h_c}} \tag{2.9}$$

This expression is similar in form to that which describes the relationship between current (I), voltage (V) and resistance (R) in a simple d.c. series circuit:

$$I = \frac{V}{R} \tag{2.10}$$

In Equation (2.9), the flow of heat ($\dot{q}_x''$) is analogous to the current, while potential difference and electrical resistance are replaced by temperature difference and 'thermal resistance', respectively (Figure 2.3). This analogy will be referred to later (e.g., Chapter 10).

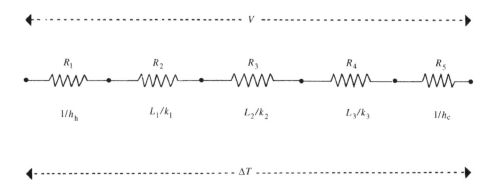

Figure 2.3 The analogy between electrical and thermal resistance

2.2.2 Non-steady State Conduction

Fires are transient phenomena, and the equation of non-steady state heat transfer must be used to interpret not only the details of fire behaviour, such as ignition and flame spread, but also the gross effects such as the response of buildings to developing and fully developed fires. The basic equations for non-steady state conduction can be derived by considering the flow of heat through a small element of volume $dxdydz$ (see Figure 2.4) and the associated heat balance.

Taking flow in the x-direction, the rate of heat transfer through face A is given by:

$$\dot{q}''_x dS = -k\frac{\partial T}{\partial x}dydz \qquad (2.11)$$

where $dS = dydz$, the area of face A. Similarly, the flow of heat *out* through face B is:

$$\dot{q}''_{x+dx} dS = -k\left(\frac{\partial T}{\partial x} + \frac{\partial^2 T}{\partial x^2}dx\right)dydz \qquad (2.12)$$

The difference between Equations (2.11) and (2.12) must be equal to the rate of change in the energy content of the small volume $dxdydz$. This is made up of two terms, namely

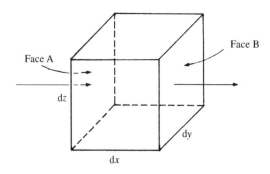

Figure 2.4 Transient heat conduction through an element of volume $dxdydz$

heat storage and heat generation, thus:

$$k\left(\frac{\partial^2 T}{\partial x^2}\right)dxdydz = \rho c\frac{\partial T}{\partial t}dxdydz - \dot{Q}'''dxdydz \tag{2.13}$$

where $\dot{Q}'''$ is the rate of heat release per unit volume and ρ and c are the density and heat capacity, respectively. This simplifies to:

$$\frac{\partial^2 T}{\partial x^2} = \frac{1}{\alpha}\frac{\partial T}{\partial t} - \frac{\dot{Q}'''}{k} \tag{2.14}$$

where $\alpha = k/\rho c$, the 'thermal diffusivity' of the material (Table 2.1), which is assumed to be constant in the above analysis. In most problems, $\dot{Q}''' = 0$, but Equation (2.14) is used in the development of Frank–Kamenetskii's thermal explosion theory, which will be discussed in Section 6.1 in relation to spontaneous ignition. Equation (2.14) would be relevant to any transient heating problem which involves exothermic or endothermic change (e.g., phase changes or chemical decomposition). However, if $\dot{Q}'''$ is zero, Equation (2.14) gives for one dimension:

$$\frac{\partial^2 T}{\partial x^2} = \frac{1}{\alpha}\frac{\partial T}{\partial t} \tag{2.15}$$

For three dimensions, similar energy balances apply in the y- and z-directions and the appropriate equation would be:

$$\frac{\partial^2 T}{\partial x^2} + \frac{\partial^2 T}{\partial y^2} + \frac{\partial^2 T}{\partial z^2} = \nabla^2 T = \frac{1}{\alpha}\frac{\partial T}{\partial t} \tag{2.16}$$

Fortunately, many problems can be reduced to a single dimension. Thus, Equation (2.15) can be applied directly to conduction through materials which may be treated as 'infinite slabs' or 'semi-infinite solids' (see below). Some problems can be reduced to a single dimension by changing to cylindrical or polar coordinates: such applications will be discussed in the analysis of spontaneous ignition (Sections 6.1 and 8.1).

One of the simplest cases to which Equation (2.15) may be applied is the infinite slab,[1] of thickness $2L$ and temperature $T = T_0$, suddenly exposed on both faces to air at a uniform temperature $T = T_\infty$ (Figure 2.5). Writing $\theta = T - T_\infty$, Equation (2.15) becomes:

$$\frac{\partial^2 \theta}{\partial x^2} = \frac{1}{\alpha}\frac{\partial \theta}{\partial t} \tag{2.17}$$

This must be solved with the following boundary conditions:

$$\frac{\partial \theta}{\partial x} = 0 \text{ at } x = 0 \text{ (i.e., at the mid-plane)}$$

$$\theta = \theta_0(= T_0 - T_\infty) \text{ at } t = 0 \text{ (for all } x\text{)}$$

$$\frac{\partial \theta}{\partial x} = -\frac{h}{k}\theta \text{ at } x = \pm L \text{ (at both faces)}$$

---

[1] The infinite slab is a model that allows a problem to be expressed in one dimension. Only the x-direction is relevant.

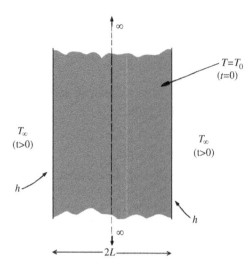

Figure 2.5   Transient heat conduction in a plane infinite slab heated on both faces

where the latter defines the rate of heat transfer through the faces of the slab, h being the convective heat transfer coefficient. The solution is not simple but may be found in any heat transfer text (e.g., Carslaw and Jaeger, 1986):

$$\frac{\theta}{\theta_0} = 2 \sum_{n=1}^{\infty} \frac{\sin \lambda_n L}{\lambda_n L + (\sin \lambda_n L)(\cos \lambda_n L)} \cdot \exp(-\lambda_n^2 \alpha t) \cdot \cos(\lambda_n x) \qquad (2.18)$$

where λ_n are roots of the equation:

$$\cot(\lambda_L) = \frac{\lambda_L L}{\mathrm{Bi}} \qquad (2.19)$$

in which Bi is the Biot number (hL/k).

Examination of Equation (2.18) reveals that the ratio θ/θ_0 is a function of three dimensionless groups: the Biot number, the Fourier number (Fo $= \alpha t/L^2$) and x/L (the distance from the centre line expressed as a fraction of the half-thickness). The Biot number compares the efficiencies with which heat is transferred to the surface by convection from the surrounding air, and from the surface by conduction into the body of the solid, while the Fourier number can be regarded as a dimensionless time variable, which takes into account the thermal properties and characteristic thickness of the body. For convenience, the solutions to Equation (2.18) are normally presented in graphical form, displayed in a series of diagrams (each referring to a given value of x/L) of the ratio θ/θ_0 as a function of Fo for a range of values of Bi. Figure 2.6 shows two such charts, for θ/θ_0 at the surface ($x/L = 1$) and at the centre of the slab ($x/L = 0$).

The form of the temperature profiles within the slab and their variation with time are illustrated schematically in Figure 2.7(a). In a thin slab of a material of high thermal conductivity, the temperature gradients within the slab are much less, and in some circumstances may be neglected (cf. Figure 2.7(b)). This can be seen by comparing the

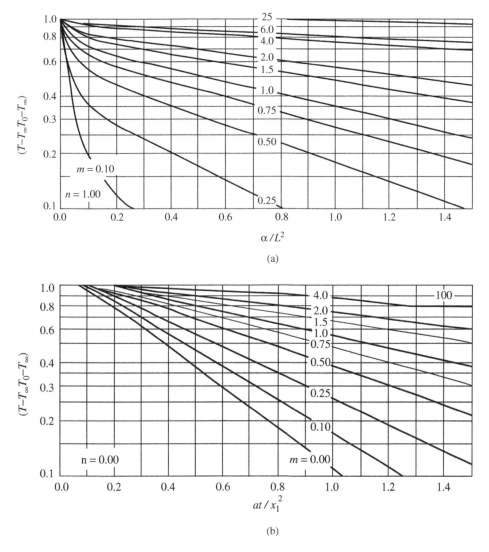

Figure 2.6 Heisler charts for (a) surface temperature and (b) centre temperature of an infinite slab. $n = x/L$ and $m = 1/\text{Bi}$ (Welty *et al.*, 2008). Reproduced by permission of John Wiley & Sons, Inc

values of θ/θ_0 at the surface and at the mid-plane of the infinite slab for different values of Bi (Figure 2.6). Figure 2.8 shows how the ratio $\theta_{x=0}/\theta_{x=L}$ varies with the value of the Biot number: when Bi is vanishingly small, the temperature at the mid-plane of the slab is equal to the surface temperature. If it is less than about 0.1 (i.e., k is large and/or L is small), then temperature gradients within the solid may be ignored (see Figure 2.7(b)). The solid may then be defined as 'thermally thin' and the heat transfer problem can be treated by the 'lumped thermal capacity analysis'. Thus, for a thermally thin slab (or for

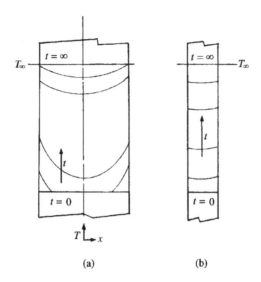

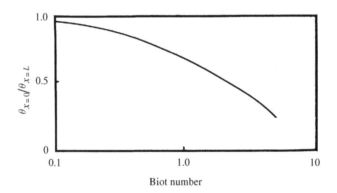

Figure 2.7   (a) Transient temperatures within a thick slab initially at T_0 exposed to an environmental temperature of T_∞: (b) the same for a thin slab. Warren M. Rohsenow, Harry Y. Choi, *Heat, Mass and Momentum Transfer*, © 1961, p. 111. Adapted by permission of Prentice-Hall Inc., Englewood Cliffs, NJ

Figure 2.8   Dependence of the ratio $\theta_{x=0}/\theta_{x=L}$ on Bi for Fo $= 1.0$

that matter any body which complies with this constraint), the energy balance over the time interval dt may be written:

$$Ah(T_\infty - T)\mathrm{d}t = V\rho c\,\mathrm{d}T \tag{2.20}$$

where A is the area through which heat is being transferred and V is the associated volume. This integrates to give:

$$\frac{T_\infty - T}{T_\infty - T_0} = \exp\left(-\frac{2ht}{\tau\rho c}\right) \tag{2.21}$$

where τ, the slab thickness, is equal to $2V/A$ if both faces are heated convectively. A similar model, involving radiant heating on one side and convective cooling at both faces, will be discussed in Section 6.3 in relation to the ignition of 'thin fuels' such as paper and fabrics. This 'lumped thermal capacity' method is a useful approximation, and is relevant to the interpretation of the ignition and flame spread characteristics of thin fuels (Section 7.2). It is also the basis for the response time index (RTI) for sprinklers, which will be discussed in Section 4.4.2.

While the temperature profiles within a 'thick' slab that is being heated symmetrically can be obtained by use of diagrams such as those in Figure 2.6(a) and (b), a problem more relevant to ignition and flame spread is that of a slab heated on one side only, with heat losses potentially at both faces. The limiting case is that of the semi-infinite solid subjected to a uniform heat flux (Figure 2.9). 'Thick' slabs will approximate to this model during the early stages of heating, before heat losses from the rear face have become significant. The relationship between heating time and thickness, which defines the limiting thickness to which this model may be applied, can be derived by considering a semi-infinite slab, initially at a temperature T_0, whose surface is suddenly increased to T_∞. Solving Equation (2.17) (where $\theta = T - T_0$) with the boundary conditions:

$$\theta = 0 \text{ at } t = 0 \text{ for all } x$$

$$\theta = \theta_\infty \text{ at } x = 0 \text{ for } t = 0$$

$$\theta = 0 \text{ as } x \to \infty \text{ for all } t$$

gives (Welty *et al.*, 2008):

$$\frac{\theta}{\theta_\infty} = 1 - \text{erf}\frac{x}{2\sqrt{(\alpha t)}} \tag{2.22}$$

where the error function is defined as:

$$\text{erf } \xi \equiv \frac{2}{\sqrt{\pi}} \int_0^\xi e^{-\eta^2} \cdot d\eta \tag{2.23}$$

While this cannot be evaluated analytically, it is given numerically in handbooks of mathematical functions, as well as in most heat transfer texts (see Table 2.2).

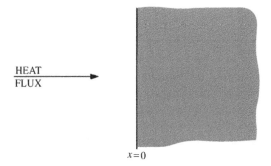

HEAT
FLUX

$x=0$

Figure 2.9   Heat transfer to a semi-infinite solid

Table 2.2 The error function and its compliment

ξ	erf ξ	erfc ξ
0	0	1.0
0.05	0.056372	0.943628
0.1	0.112463	0.887537
0.15	0.167996	0.832004
0.2	0.222703	0.777297
0.25	0.276326	0.723674
0.3	0.328627	0.671373
0.35	0.379382	0.620618
0.4	0.428392	0.571608
0.45	0.475482	0.524518
0.5	0.520500	0.479500
0.55	0.563323	0.436677
0.6	0.603856	0.396144
0.65	0.642029	0.357971
0.7	0.677801	0.322199
0.75	0.711156	0.288844
0.8	0.742101	0.257899
0.85	0.770668	0.229332
0.9	0.796908	0.203092
0.95	0.820891	0.179109
1.0	0.842701	0.157299
1.1	0.880205	0.119795
1.2	0.910314	0.009686
1.3	0.934008	0.065992
1.4	0.952285	0.047715
1.5	0.966105	0.033895
1.6	0.976348	0.023652
1.7	0.983790	0.016210
1.8	0.989091	0.010909
1.9	0.992790	0.007210
2.0	0.995322	0.004678
2.1	0.997021	0.002979
2.2	0.998137	0.001863
2.3	0.998857	0.001143
2.4	0.999311	0.000689
2.5	0.999593	0.000407

Equation (2.22) can be used to define the temperature profiles below the surface of a slab of thickness L, heated instantaneously on one face, until the rear face becomes heated to a temperature significantly above ambient (T_0). If this is set arbitrarily as 0.5% of $T_s - T_0$, i.e., $T = T_0 + 5 \times 10^{-3} (T_s - T_0)$ at $x = L$, where T_s and T are the temperatures of the heated surface and the rear face, respectively. Then substituting in Equation (2.22):

$$1 - \mathrm{erf} \frac{L}{2\sqrt{(\alpha t)}} = 5 \times 10^{-3} \tag{2.24a}$$

which gives:

$$\frac{L}{2\sqrt{(\alpha t)}} \approx 2 \tag{2.24b}$$

This indicates that a wall, or slab, of thickness L can be treated as a semi-infinite solid with little error, provided that $L > 4\sqrt{(\alpha t)}$. In many fire engineering problems involving transient surface heating, it is adequate to assume 'semi-infinite behaviour' if $L > 2\sqrt{(\alpha t)}$ (e.g., Williams, 1977; McCaffrey *et al.*, 1981). The quantity $\sqrt{(\alpha t)}$ is the characteristic thermal conduction length and may be used to estimate the thickness of the heated layer in some situations (Chapters 5 and 6).

If the above model is modified to include convective heat transfer from a stream of fluid at temperature T_∞ to the surface of the semi-infinite solid (initially at temperature T_0) (Figure 2.9), then Equation (2.17) must be solved with the following boundary conditions:

$$\theta = 0 \text{ at } t = 0 \text{ for all } x$$

$$\frac{\partial\theta}{\partial x} = -\frac{h}{k}(\theta_\infty - \theta_s) \text{ at } x = 0 \text{ for all } t$$

The solution, which is given by Carslaw and Jaeger (1986), is:

$$\frac{\theta}{\theta_\infty} = \frac{T - T_0}{T_\infty - T_0} = \operatorname{erfc}\left(\frac{x}{2\sqrt{(\alpha t)}}\right)$$

$$- \exp\left(\frac{xh}{k} + \frac{\alpha t}{(k/h)^2}\right) \cdot \operatorname{erfc}\left(\frac{x}{2\sqrt{(\alpha t)}} + \frac{\sqrt{(\alpha t)}}{k/h}\right) \tag{2.25}$$

(Note that $\operatorname{erfc}(\xi) = 1 - \operatorname{erf}(\xi)$ (Table 2.2).)

The variation of surface temperature (T_s) with time under a given imposed (convective) heat flux can be illustrated by setting $x = 0$ in Equation (2.25), i.e.

$$\frac{\theta_s}{\theta_\infty} = 1 - \exp\left(\frac{\alpha t}{(k/h)^2}\right) \cdot \operatorname{erfc}\left(\frac{\sqrt{(\alpha t)}}{k/h}\right) \tag{2.26}$$

and plotting θ_s/θ_∞ against time. Figure 2.10 shows that the rate of change of surface temperature depends strongly on the value of the ratio $k^2/\alpha = k\rho c$, a quantity known as the 'thermal inertia'. The surface temperature of materials with low thermal inertia (such as fibre insulating board and polyurethane foam) rises quickly when heated. The relevance of this to the ignition and flame spread characteristics of combustible solids will be discussed in Chapters 6 and 7.

Equation (2.26) is cumbersome to use, but it is possible to derive a simplified expression if the surface is exposed to a constant heat flux, $\dot{Q}_R''$. The heat flux (flow) through the semi-infinite solid:

$$\dot{q}'' = -k\frac{\partial\theta}{\partial x} \tag{2.27}$$

obeys a differential equation identical in form to Equation (2.15) (Carslaw and Jaeger, 1986):

$$\frac{\partial^2\dot{q}''}{\partial x^2} = \frac{1}{\alpha}\frac{\partial\dot{q}''}{\partial t} \tag{2.28}$$

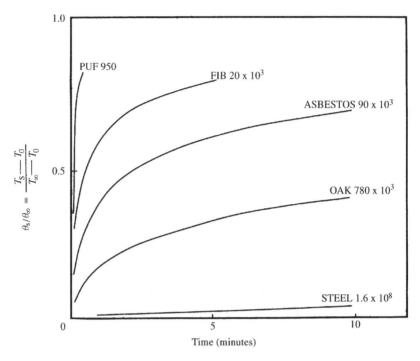

Figure 2.10 Effect of thermal inertia ($k\rho c_p$) on the rate of temperature rise at the surface of a semi-infinite solid. FIB = fibre insulating board; PUF = polyurethane foam. Figures are values of $k\rho c_p$ in W²s/m⁴·K². (From Equation (2.26), with $h = 20$ W/m² · K. T_s = surface temperature)

For $x > 0$ and $t > 0$, the boundary condition $\dot{q}'' = \dot{Q}''_R$ at $x = 0$ and $t > 0$ gives the solution:

$$T - T_0 = \theta = \frac{2\dot{Q}''_R}{k}\left\{\left(\frac{\alpha t}{\pi}\right)^{1/2}\exp\left(-\frac{x^2}{4\alpha t}\right) - \frac{x}{2}\text{erfc}\left(\frac{x}{2\sqrt{\alpha t}}\right)\right\} \tag{2.29}$$

The value of θ at the surface ($x = 0$) then becomes:

$$\theta_s = T_s - T_0 = \frac{2\dot{Q}''_R}{k}\left(\frac{\alpha t}{\pi}\right)^{1/2} \tag{2.30}$$

Although the derivation ignores heat losses from the surface to the surroundings, this equation is nevertheless useful, as will be seen in Chapter 6.

2.2.3 Numerical Methods of Solving Time-dependent Conduction Problems

Where they exist, analytical solutions to transient heat conduction problems are not simple, even if they apply to simple geometries with well-defined boundary conditions (cf.

Equations (2.18) and (2.25)). While basic equations can be written for complex geometries and boundary conditions, they may be intractable and require numerical solution. In general, these require lengthy, iterative calculations, but the availability of personal computers now provides the opportunity to undertake complex heat transfer calculations with ease. Simple problems can be solved using spreadsheets, or a few lines of code; more complex problems – particularly if non-uniform temperature distributions are involved – will require the use of available software packages such as TASEF (Sterner and Wickstrom, 1990), which was developed to calculate temperatures in structures exposed to fire, or versatile mathematical packages such as MATLAB$^®$ (e.g., Pratap, 2006; Moore, 2007).

As an example of a simple numerical solution, consider a steel plate, or bulkhead, which forms the boundary between two compartments, both initially at a temperature T_0 (see Figure 2.11(a)). Suppose the air in one of the compartments is suddenly increased to a temperature T_h: how rapidly will the temperature of the plate increase, and what will its final temperature be? If it is assumed that the bulkhead behaves as an infinite plate (thus reducing the problem to one dimension) of thickness Δx, and that the temperature of the plate remains uniform at all times (i.e., the Biot number is low), the problem becomes quite simple. Indeed, calculating the *final* temperature is trivial, as it is

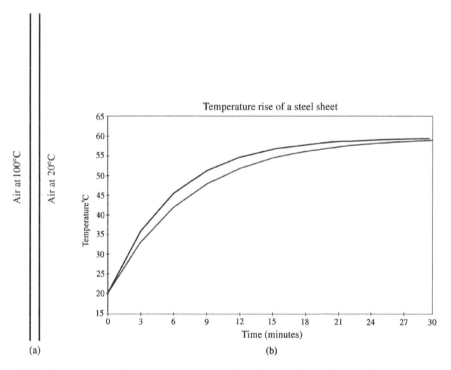

(a)   (b)

Figure 2.11 A comparison between the analytical and numerical solutions for transient, one-dimensional heat transfer (Table 2.3). (a) A 5 mm infinite steel sheet with initial temperature 20°C and convective heat transfer coefficient 0.02 kW/m^{2}K. (b) Lower line: analytical solution (Equation (2.34)); upper line: numerical solution with $\Delta t = 180$ s (Equation (2.38))

a steady state problem. Thus, if the final temperature of the steel is T_s, the steady state is expressed as:

$$\text{Heat in (kW/m}^2) = \text{Heat out (kW/m}^2)$$

$$h(T_h - T_s) = h(T_s - T_0) \tag{2.31}$$

where h is the convective heat transfer coefficient, assumed to be independent of temperature, and radiative heat transfer is neglected. If T_h and T_0 are 100°C and 20°C, respectively, the final steel temperature is 60°C, independent of the value of Δx or h.

To calculate how quickly the plate heats up, an analytical solution for the rate of temperature rise may be obtained from the equation:

$$\frac{dT_s}{dt} = \frac{1}{\rho c_p \Delta x} [h(T_h - T_s) - h(T_s - T_0)] \tag{2.32}$$

(which reduces to Equation (2.31) at the steady state ($dT_s/dt = 0$)). This can be integrated to give an analytical solution, as follows:

$$\int_{T_0}^{T_s} \frac{dT_s}{(T_h - 2T_s + T_0)} = \frac{h}{\rho c_p \Delta x} \int_0^t dt \tag{2.33}$$

$$T_s = \frac{1}{2}(T_h + T_0) - \frac{1}{2}(T_h - T_0)\exp\left[-\frac{2ht}{\rho c_p \Delta x}\right] \tag{2.34}$$

Equation (2.34) is shown in Table 2.3 and Figure 2.11(b) and should be compared with Equation (2.21).

For the numerical approach, Equation (2.32) is cast in finite difference form:

$$\Delta T = \frac{1}{\rho c_p \Delta x} [h(T_h - T_s) - h(T_s - T_0)]\Delta t \tag{2.35}$$

Table 2.3 Comparison of analytical and numerical solutions to the rate of temperature rise of the steel plate shown in Figure 2.11(a) according to Equations (2.34) and (2.38), respectively. $T_h = 100°C$, $T_0 = 20°C$, $h = 0.02\,\text{kW/m}^2\cdot\text{K}$, $\rho = 7850\,\text{kg/m}^3$, $c_p = 0.46\,\text{kJ/kg}\cdot\text{K}$ and $\Delta x = 0.005\,\text{m}$. For the numerical solution, the results for two values of Δt (180 s and 20 s) are shown

Time	Analytical solution (Eq. (2.34))	Numerical solution (Eq. (2.38))	
t (s)	$T_s(t)$(°C)	$T_s(t)$(°C) ($\Delta t = 180$ s)	$T_s(t)$(°C) ($\Delta t = 20$ s)
0	20.00	20.00	20.00
180	33.15	35.95	33.40
360	41.98	45.54	42.31
540	47.91	51.31	48.23
720	51.88	54.77	52.17
900	54.55	56.86	54.80
1080	56.34	58.11	57.24
1260	57.55	58.86	57.70
1440	58.35	59.32	58.47
1620	58.90	59.59	58.98
1800	59.26	59.75	59.32

This can be rearranged to show that for any time interval Δt:

$$\text{Heat in} - \text{Heat out} = \text{Heat stored} \tag{2.36}$$

Writing $\Delta T = T_s(t + \Delta t) - T_s(t)$, Equation (2.35) becomes:

$$h(T_h - T_s(t))\Delta t - h(T_s(t) - T_0)\Delta t = \rho c_p \Delta x (T_s(t + \Delta t) - T_s(t)) \tag{2.37}$$

where $T_s(t)$ and $T_s(t + \Delta t)$ are the temperatures of the steel plate at time t and time $t + \Delta t$, respectively. Equation (2.37) can be rearranged to give $T_s(t + \Delta t)$:

$$T_s(t + \Delta t) = T_s(t) + \frac{h\Delta t}{\rho c_p \Delta x}(T_h - 2T_s(t) + T_0) \tag{2.38}$$

expressing the steel temperature at time $t + \Delta t$ in terms of the temperature at the beginning of the timestep, at time t.

This can be solved iteratively, with initial values $t = 0$ and $T_s(0) = T_0$, calculating T_s at times $t = n\Delta t$, where $n = 0, 1, 2, \ldots, m$, where m defines the total time of exposure. A spreadsheet or a simple program may be used. The analytical and numerical solutions are compared in Figure 2.11(b), in which a timestep of 180 s is used in the latter. The accuracy of the numerical solution can be improved by reducing the timestep, as illustrated in Table 2.3.

The same technique can be used for more complex problems. A (thermally) thick plate, or wall, acting as a barrier between two compartments would not be at uniform temperature; there would be a temperature gradient (cf. Figure 2.2 for the steady state). The numerical solution requires that the wall is represented by a series of thin, parallel elements of equal thickness, as shown in Figure 2.12. Transient heat transfer through the plate is then calculated iteratively, by considering adjacent elements and applying the

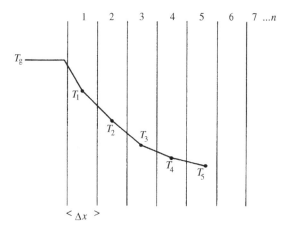

Figure 2.12 Thermal conduction into an infinite, plane slab, divided into equal elements 1, 2, $\ldots, n$ for numerical analysis

logic of Equation (2.36). Thus during the unsteady, heating-up stage the 'internal element' numbered 3 in Figure 2.12 gains heat from element 2, and loses heat to element 4. Thus:

$$\frac{k}{\Delta x}(T_2 - T_3) - \frac{k}{\Delta x}(T_3 - T_4) = \rho c_p \Delta x (T_3(t + \Delta t) - T_3(t)) \tag{2.39}$$

which can be arranged to give:

$$T_3(t + \Delta t) = T_3(t) + \frac{k}{\rho c_p (\Delta x)^2}(T_2 - 2T_3 + T_4) \tag{2.40}$$

This equation is the basis for numerical solutions of complex problems such as the heat losses from a fire compartment during the immediate post-flashover stage, when assumptions can be made about the rate of heat release within the compartment (Section 10.3.2) and the temperature variation has to be calculated (Pettersson *et al.*, 1976). If necessary, the method can be adapted to two and three dimensions, although for convergence the Fourier number (Fo $= \alpha \Delta t/(\Delta x)^2$) must be less than 0.5 for a one-dimensional problem, 0.25 for two dimensions and 0.16 for three dimensions (e.g., Pitts and Sissom, 1977).

2.3 Convection

'Convection' is associated with the transfer of heat by the motion of a fluid. The motion may arise naturally as a consequence of temperature gradients in the fluid which generate buoyancy-driven flows. This is commonly referred to as 'free' or 'natural' convection to distinguish it from 'forced' convection when external forces (such as those provided by a fan or blower) are involved. In fires, we are mostly concerned with free convection, but it is important to distinguish between free convective flows when there is no adjacent surface, such as the plume created above a localized heat source (see Figure 4.10(a)), and free convective flows that are bounded by a surface. The associated boundary layer flow is as illustrated in Figures 2.13–2.15.

If a fire is burning in an open area (see Figure 4.9(a)), most of the heat that is released is carried away from the burning surfaces by a buoyancy-induced, free convective flow. This will be discussed in Sections 4.3 *et seq.*, but here we shall consider boundary layer flows and how they apply to the transfer of heat between a solid surface and the surrounding fluid. The convective flow may be free, or natural (as in Figure 2.15, where the fluid adjacent to the surface becomes heated), or effectively 'forced' if the fluid is flowing as a continuous stream past the target surface. A relevant example of the latter would be the transfer of heat to the fusible element ('link') of a sprinkler head that is exposed to a ceiling jet (Figure 4.24), i.e., the flow of hot combustion products that forms under a ceiling when a fire plume is deflected horizontally (Section 4.4.2). Whether 'free' or 'forced', the rate of heat transfer is given by Equation (2.3) and the convective heat transfer coefficient is defined as $h = \dot{q}''/\Delta T$. The challenge is to determine an appropriate value for h which is known to depend on the fluid properties (thermal conductivity, density and viscosity), the flow parameters (velocity and nature of the flow) and the geometry of the surface (dimensions and angle to the flow). As will be seen below, h can be expressed in terms of certain dimensionless groups that allow the physical properties of the fluid and the flow velocity to be taken into account (see Tables 2.4 and 2.5).

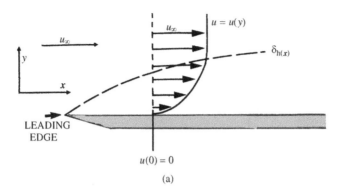

(a)

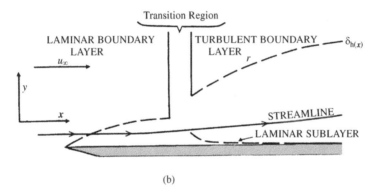

(b)

Figure 2.13 (a) The hydrodynamic boundary layer at the leading edge of a flat plate (isothermal system); (b) as (a) but showing the onset of turbulence and the associated laminar sublayer. Reproduced by permission of Gordon and Breach from Kanury (1975)

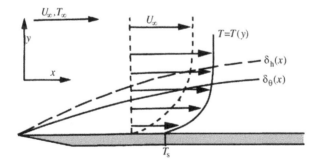

Figure 2.14 The hydrodynamic $(-\,-\,-\,-)$ and thermal (———) boundary layers at the leading edge of a flat plate (non-isothermal system). Reproduced by permission of Gordon and Breach from Kanury (1975)

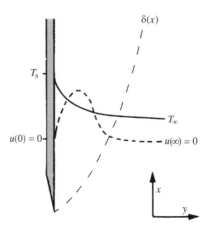

Figure 2.15 Free convective boundary layer at the surface of a vertical flat plate. Reproduced by permission of Gordon and Breach from Kanury (1975)

Table 2.4 Dimensionless groups

	Group	Physical interpretation	References
Biot	$\mathrm{Bi} = \dfrac{hl}{k}$	$\dfrac{\text{internal resistance to heat conduction}}{\text{external resistance to heat conduction}}$	Section 2.2.2
Fourier	$\mathrm{Fo} = \dfrac{\alpha t}{l^2}$	dimensionless time for transient conduction	Section 2.2.2
Froude	$\mathrm{Fr} = \dfrac{u_\infty^2}{lg}$ $= \dfrac{u_\infty^2 \rho}{lg \Delta\rho}$	$\dfrac{\text{inertia forces}}{\text{gravity forces}}$	Section 4.4.4
Grashof	$\mathrm{Gr} = \dfrac{gl^3 \beta \Delta T}{\nu^2}$ $\equiv \dfrac{gl^3 \Delta\rho}{\rho \nu^2}$	$\dfrac{\text{buoyancy forces} \times \text{inertia forces}}{(\text{viscous forces})^2}$ $= \mathrm{Re} \cdot \dfrac{\text{buoyancy forces}}{(\text{viscous forces})^2}$	Equation (2.49)
Lewis	$\mathrm{Le} = \dfrac{D}{\alpha}$	$\dfrac{\text{mass diffusivity}}{\text{thermal diffusivity}}$	Section 5.1.2
Nusselt	$\mathrm{Nu} = \dfrac{hl}{k}$	ratio of temperature gradients (non-dimensionalized heat transfer coefficient)	Section 2.3
Prandtl	$\mathrm{Pr} = \dfrac{\mu c_p}{k}$ $= \dfrac{\nu}{\alpha}$	$\dfrac{\text{momentum diffusivity}}{\text{thermal diffusivity}}$	Section 2.3
Reynolds	$\mathrm{Re} = \dfrac{\rho u_\infty l}{\mu}$	$\dfrac{\text{inertia forces}}{\text{viscous forces}}$	Section 2.3

Table 2.5 Some recommended convective heat transfer correlations[a,b] (Kanury, 1975; Williams, 1982)

Nature of the flow and configuration of the surface	$\overline{\mathrm{Nu}} = \frac{hl}{k}$
Forced convection	
Laminar flow, parallel to a flat plate of length l ($20 < \mathrm{Re} < 3 \times 10^5$)	$0.66\,\mathrm{Re}^{1/2}\mathrm{Pr}^{1/3}$
Turbulent flow, parallel to a flat plate of length l ($\mathrm{Re} > 3 \times 10^5$)	$0.037\,\mathrm{Re}^{4/5}\mathrm{Pr}^{1/3}$
Flow round a sphere of diameter l (general equation)	$2 + 0.6\,\mathrm{Re}^{1/2}\mathrm{Pr}^{1/3}$
Natural convection	
Laminar: natural convection at a vertical flat plate of length l (Figure 2.15) ($10^4 < \mathrm{Gr} \cdot \mathrm{Pr} < 10^9$)	$0.59\,(\mathrm{Gr}\cdot\mathrm{Pr})^{1/4}$
Turbulent: natural convection at a vertical flat plate of length l (Gr.Pr $> 10^9$)	$0.13\,(\mathrm{Gr}\cdot\mathrm{Pr})^{1/3}$
Laminar: natural convection at a hot horizontal plate of length l (face up) ($10^5 < \mathrm{Gr} \cdot \mathrm{Pr} < 2 \times 10^7$)	$0.54\,(\mathrm{Gr}\cdot\mathrm{Pr})^{1/4}$
Turbulent: natural convection at a hot horizontal plate of length l (face up) ($2 \times 10^7 < \mathrm{Gr} \cdot \mathrm{Pr} < 3 \times 10^{10}$)	$0.14\,(\mathrm{Gr}\cdot\mathrm{Pr})^{1/3}$
Vertical parallel plates, separation l:	
Gr $< 2 \times 10^3$	1
$2 \times 10^3 < \mathrm{Gr} < 2.1 \times 10^5$	$0.2\,(\mathrm{Gr}\cdot\mathrm{Pr})^{1/4}$
$2.1 \times 10^5 < \mathrm{Gr} < 1.1 \times 10^7$	$0.071\,(\mathrm{Gr}\cdot\mathrm{Pr})^{1/3}$
Laminar free convection around a heated horizontal cylinder $10^3 < \mathrm{GrPr} < 10^9$	$0.525\,(\mathrm{Gr}\cdot\mathrm{Pr})^{1/4}$

[a]The expressions give average values for the Nusselt number. $\mathrm{Re} = ul/v$, $\mathrm{Pr} = v/\alpha$, $\mathrm{Gr} = gl^3\beta\Delta T/v^2$. Williams (1982) suggests that in most fire problems the Prandtl number can be assumed to be unity ($\mathrm{Pr} = 1$). Note that $\mu (= v\rho)$ and k are temperature-dependent.
[b]Typically, h takes values in the range 5–50 W/m^2·K and 25–250 W/m^2·K for natural convection and forced convection in air, respectively (Welty et al., 2008).

The transfer process occurs close to the surface within a region known as the boundary layer, whose structure determines the magnitude of h. Consider first an isothermal system in which an incompressible fluid is flowing with a free stream velocity u_∞ across a rigid flat plate, parallel to the flow (Figure 2.13). Given that the layer of fluid next to the plate will be stationary ($u(0) = 0$), there will be a velocity gradient perpendicular to the surface described by an equation $u = u(y)$. At a large distance from the plate, $u = u_\infty = u(\infty)$. By definition, the boundary layer is taken to extend from the surface to the point at which $u(y) = 0.99u_\infty$. For small values of x, i.e., close to the leading edge of the plate (see Figure 2.13(a)), the flow within the boundary layer is laminar. This develops into turbulent flow beyond a transition regime, although a laminar sublayer always exists close to the surface, i.e., where y is small (Figure 2.13(b)). As with flow in a pipe, the nature of the flow may be determined by examining the magnitude of the 'local' Reynolds number, $\mathrm{Re}_x = xu_\infty\rho/\mu$, where μ is the absolute viscosity. If $\mathrm{Re}_x < 2 \times 10^5$, then the flow is laminar, while the boundary layer will be turbulent if $\mathrm{Re}_x > 3 \times 10^6$. Between these limits, the layer may be laminar or turbulent. This may be compared with pipe flow where laminar behaviour is observed for Re < 2300 (Re $= Du\rho/\mu$, where D is the pipe diameter[2]).

Figure 2.13 shows the hydrodynamic boundary layer for an isothermal system. Its thickness (δ_h) also depends on the Reynolds number and can be approximated by:

$$\delta_h \approx l\left(\frac{8}{\mathrm{Re}_l}\right)^{1/2} \tag{2.41}$$

for laminar flow, where l is the value of x at which δ_h is measured (Figure 2.13(a)) and Re_l is the local Reynolds number (Kanury, 1975).

If the fluid and the plate are at different temperatures, a thermal boundary layer will exist, as shown in Figure 2.14. The rate at which heat is transferred between the fluid and the surface will then depend on the temperature gradient within the fluid at $y = 0$, i.e.

$$\dot{q}'' = -k\left(\frac{\partial T}{\partial y}\right)_{y=0} \tag{2.42}$$

where k is the thermal conductivity of the fluid. This is Fourier's law of heat conduction (Equation (2.2)) applied to the sublayer of fluid adjacent to the surface. Following Kanury (1975), Equation (2.42) can be approximated by the expression:

$$\dot{q}'' \approx \frac{k}{\delta_\theta}(T_\infty - T_s) \tag{2.43}$$

where δ_θ is the thickness of the thermal boundary layer, and T_∞ and T_s are the temperatures of the main stream fluid and the surface of the plate, respectively. The ratio of the thickness of the two boundary layers (δ_θ/δ_h) is dependent on the Prandtl number, $\mathrm{Pr} = \nu/\alpha$, a dimensionless group (see Table 2.4), which relates the 'momentum diffusivity' and 'thermal diffusivity' of the fluid, where $\nu(= \mu/\rho)$ is the kinematic viscosity. These determine the structures of the hydrodynamic and thermal boundary layers, respectively. Thus, for laminar flow, Kanury (1975) derives the approximate expression:

$$\frac{\delta_\theta}{\delta_h} \approx (\mathrm{Pr})^{-1/3} \tag{2.44}$$

Combining Equations (2.3), (2.41), (2.43) and (2.44) gives:

$$h \approx k/(l(8/\mathrm{Re})^{1/2} \cdot (\mathrm{Pr}^{-1/3})) \tag{2.45}$$

By convention, h is normally expressed as a multiple of k/l, where l is the characteristic dimension of the surface. Thus, Equation (2.45) would be expressed in dimensionless form as:

$$\mathrm{Nu} = \frac{hl}{k} \approx 0.35\,\mathrm{Re}^{1/2}\,\mathrm{Pr}^{1/3} \tag{2.46}$$

---

[2] There are advantages in using dimensionless numbers (such as the Reynolds number) as they allow us to scale many problems in fluid dynamics. Thus, for pipe flow, as long as Re is conserved, the flow characteristics will be unchanged, regardless of the diameter of the pipe or the nature of the fluid (See also Figure 2.16.).

where Nu is the local Nusselt number. This has the same form as the Biot number but differs in that k (the thermal conductivity) refers to the fluid rather than the solid. There is considerable advantage to be gained in expressing the convective heat transfer coefficient in this dimensionless form. It allows heat transfer data from geometrically similar situations to be correlated, thus providing the means by which small-scale experimental results can be used to predict large-scale behaviour. This is a powerful technique to which reference will be made below.

Detailed analysis of the boundary layers can be found in most texts on momentum and heat transfer (e.g., Rohsenow and Choi, 1961; Welty *et al.*, 2008). Thus, the exact solution to the problem outlined above (laminar flow over a flat plate) is given by:

$$\text{Nu} = 0.332 \, \text{Re}^{1/2} \, \text{Pr}^{1/3} \tag{2.47}$$

with which the approximate solution (Equation (2.46)) is in satisfactory agreement. As the Prandtl number does not vary significantly – indeed, it is frequently assumed to be unity in many combustion problems (Kanury, 1975; Williams, 1982) – these equations can be rearranged to show that $h \propto u^{1/2}$ for these conditions. This result is used in Section 4.4.2 in relation to the response of heat detectors to fires.

For turbulent flow, the temperature gradient at $y = 0$ is much steeper than for laminar flow and the Nusselt number is given by:

$$\text{Nu} = 0.037 \, \text{Re}^{4/5} \, \text{Pr}^{1/3} \tag{2.48}$$

Expressions for other geometries under both laminar and turbulent forced flow conditions may be found in the literature (e.g., Kanury, 1975; Williams, 1982; Incropera *et al.*, 2008) (Table 2.5).

In natural or free convection, the hydrodynamic and thermal boundary layers are inseparable as the flow is created by buoyancy induced by the temperature difference between the boundary layer and the ambient fluid. Analysis introduces the Grashof number, which is essentially the ratio of the upward buoyant force to the resisting viscous drag:

$$\text{Gr} = \frac{gl^3(\rho_\infty - \rho)}{\rho v^2} = \frac{gl^3 \beta \Delta T}{v^2} \tag{2.49}$$

where g is the gravitational acceleration constant. The convective heat transfer coefficient is found to be a function of the Prandtl and Grashof numbers. Thus, for a vertical plate (Figure 2.15):

$$\text{Nu} = \frac{hl}{k} = 0.59 \, (\text{Gr} \cdot \text{Pr})^{1/4} \tag{2.50}$$

provided that the flow is laminar ($10^4 < \text{Gr} \cdot \text{Pr} < 10^9$). For turbulent flow ($\text{Gr} \cdot \text{Pr} > 10^9$):

$$\text{Nu} = 0.13 \, (\text{Gr} \cdot \text{Pr})^{1/3} \tag{2.51}$$

Expressions for other configurations are given in Table 2.5.

There are significant advantages in identifying and applying the dimensionless groups relevant to problems in fluid dynamics. To illustrate how powerful this approach can

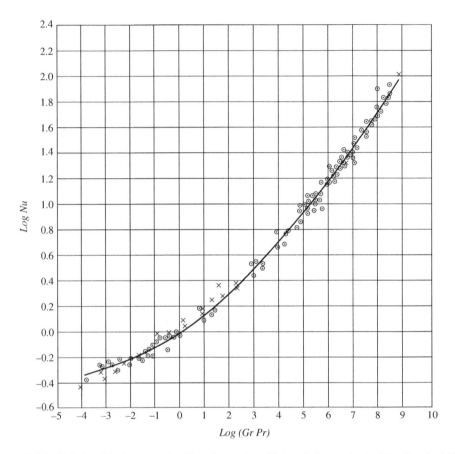

Figure 2.16   Relationship between the Nusselt number (Nu) and the product of the Grashof (Gr) and Prandtl (Pr) numbers for convective heat transfer from horizontal cylinders. Circles are experimental points for gases (air, hydrogen and carbon dioxide) with cylinder diameters from 0.4 cm to 25 cm and temperatures up to c. 1600°C. The crosses are experimental points for liquids (alcohol, aniline, carbon tetrachloride, olive oil and water) with cylinder diameters from 0.6 cm to 5 cm and temperatures up to c. 65°C. Reproduced from Fishenden and Saunders (1950), by permission of the Design Council

be, consider Figure 2.16, which is taken from Fishenden and Saunders (1950). It relates to the problem of convective heat transfer from horizontal cylinders. Experimental values of the heat transfer coefficient for free convective heat loss were calculated from $h = \dot{q}''/\Delta T$ (see Equation (2.3)) and expressed as Nusselt numbers (Nu). The data refer to horizontal cylinders of a range of diameters, for a range of temperatures and for a variety of fluids, both gaseous and liquid. As may be seen from Figure 2.16 a plot of log(Nu) vs. log(Gr·Pr) reveals an excellent correlation, right up to the limit of the onset of turbulence (Gr · Pr > 10^8) (see Table 2.5). The non-dimensional groups relevant to fire problems will be discussed in Chapter 4.

2.4 Radiation

As indicated earlier, thermal radiation involves transfer of heat by electromagnetic waves confined to a relatively narrow 'window' in the electromagnetic spectrum (Figure 2.17). It incorporates visible light and extends towards the far infra-red, corresponding to wavelengths between $\lambda = 0.4$ and $100\,\mu m$.[3] As a body is heated and its temperature rises, it will lose heat partly by convection (if in a fluid such as air) and partly by radiation. Depending on the emissivity and the value of h (the convective heat transfer coefficient), convection predominates at low temperatures ($<$ c. 150–200°C), but above c. 400°C, radiation becomes increasingly dominant. At a temperature of around 550°C, the body emits sufficient radiation within the optical region of the spectrum for a dull red glow to be visible. As the temperature is increased further, colour changes are observed which can be used to give a rough guide to the temperature (Table 2.6). These

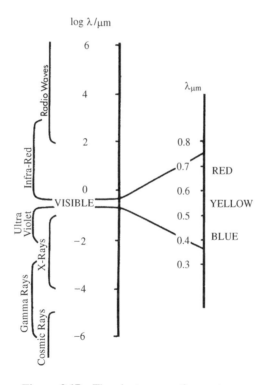

Figure 2.17 The electromagnetic spectrum

---

[3] Flames emit a small amount of radiation in the ultra-violet, at wavelengths less than $0.4\,\mu m$. This is insignificant in heat transfer terms, but highly sensitive UV detectors may be used in specialized applications to detect fire at a very early stage (e.g., in large aircraft hangars).

Table 2.6   Visual colour of hot objects

Temperature (°C)	Appearance
550	First visible red glow
700	Dull red
900	Cherry red
1100	Orange
1400	White

changes are due to the changing spectral distribution with temperature, illustrated in Figure 2.18(a) for an ideal emitter, i.e., a black body. These curves are described by Planck's distribution law, which embodies the fundamental concept of the quantum theory – i.e., that electromagnetic radiation is discontinuous, being emitted in discrete amounts known as 'quanta':

$$E_{b,\lambda} = \frac{2\pi c^2 h \lambda^{-5}}{\exp(ch/\lambda \kappa T) - 1} \tag{2.52}$$

where $E_{b,\lambda}$ is the total amount of energy emitted per unit area by a black body within a narrow band of wavelengths (between λ and $\lambda + d\lambda$), c is the velocity of light, h is Planck's constant, k is Boltzmann's constant and T is the absolute temperature. The maximum moves to shorter wavelengths as the temperature increases according to Wien's law:

$$\lambda_{max} T = 2.9 \times 10^3 \mu m \cdot K \tag{2.53}$$

Thus, at 1000 K, the maximum is at $2.9 \, \mu m$, as shown in Figure 2.18(a).

Integrating Equation (2.52) between $\lambda = 0$ and $\lambda = \infty$ gives the total emissive power of a black body as:

$$E_b = \int_0^\infty E_{b,\lambda} \cdot d\lambda = \frac{2\pi^5 \kappa^4 T^4}{15 c^2 h^3} \tag{2.54}$$

Comparing this with the relationship derived semi-empirically by Stefan and Boltzmann (Equation (2.4) (with $\varepsilon = 1$)) shows that the constant σ is a function of three fundamental physical constants, c, h and k. Consequently, its value is known very accurately ($\sigma = 5.67 \times 10^{-8}$ W/m^2·K^4).

The emissivity of a real surface is less than unity ($\varepsilon < 1$) and may depend on wavelength. Thus, it should be defined as:

$$\varepsilon_\lambda = \frac{E_\lambda}{E_{b,\lambda}} \tag{2.55}$$

where E_λ is the emissive power of the real surface between λ and $\lambda + d\lambda$. The variation of monochromatic emissive power with λ for a fictitious 'real body' is shown in Figure 2.18(b). However, it is found convenient to introduce the concept of a 'grey body' (or an 'ideal, non-black' body) for which ε is independent of wavelength. While this is an approximation, it permits simple use of the Stefan–Boltzmann equation (Equation (2.4)). Typical values of ε for solids are given in Table 2.7. Kirchhoff's law states that these are

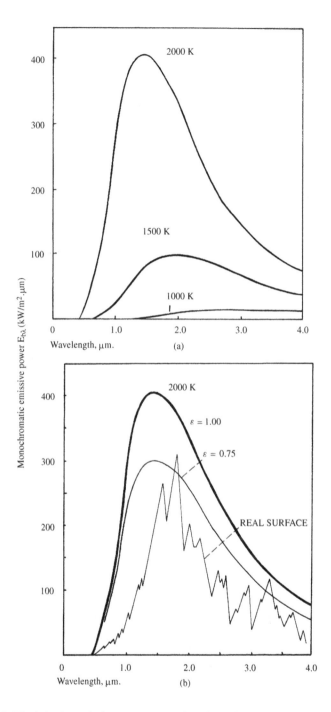

Figure 2.18 (a) Black body emissive power as a function of wavelength and temperature: (b) comparison of emissive power of ideal black bodies and grey bodies with that of a 'real' surface. Adapted from Gray and Muller (1974) by permission

Table 2.7 Emissivities $(\varepsilon)^a$

Surface	Temperature (°C)	Emissivity
Stainless steel, polished	100	0.074
Steel, polished	425–1025	0.14–0.38
Cast iron, polished	200	0.21
Rough steel plate	38–370	0.94–0.97
Asbestos board	24	0.96
Brick, rough red	20	0.93
Brick, glazed	1100	0.75
Fire brick	1000	0.75
Concrete tiles[b]	1000	0.63
Plaster	10–90	0.91
Oak, planed	20	0.9

[a]From table 23.4 in Welty *et al.* (2008). Source: table of normal total emissivities compiled by H.C. Hottel in McAdams (1954).
[b]The value for concrete tiles is taken from Incropera *et al.* (2007).

equal to their absorptivities, as dictated by the first law of thermodynamics; thus a black body is a perfect absorber, with $a = 1$.

$$E = \varepsilon \sigma T^4 \tag{2.4}$$

Equation (2.4) gives the total radiation emitted by unit area of a grey surface into the hemisphere above it. It can be used without modification to calculate radiative heat loss from a surface but as the radiation is diffuse, calculation of the rate of heat transfer to nearby objects requires a method of calculating the amount of energy being radiated in any direction. To enable this to be done, the intensity of normal radiation (I_n) is defined as 'the energy radiated per second per unit surface area per unit solid angle from an element of surface within a small cone of solid angle with its axis normal to the surface'. Lambert's cosine law can then be used to calculate the emission intensity in a direction θ to the normal (Figure 2.19), i.e.

$$I = I_n \cos \theta \tag{2.56}$$

which applies only to diffuse emitters. The relationship between I_n and E may be found by considering the thermal radiation emitted from a small element of surface area dA_1 through the solid angle $d\omega$ obtained by rotating the vectors defined by the angles θ and $\theta + d\theta$ through an angle of 360° with the normal to the surface as the axis (Figure 2.20). From the definition of I_n and Lambert's cosine law:

$$dE = I_n \cos \theta dA_1 \cdot d\omega \tag{2.57}$$

where the differential solid angle $d\omega$ is, by definition:

$$d\omega = dA_2/r^2 \tag{2.58}$$

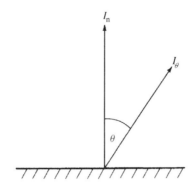

Figure 2.19   The intensity of normal radiation (I_n)

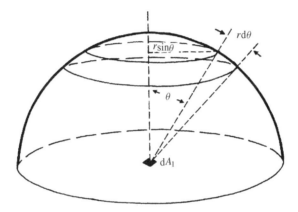

Figure 2.20   Derivation of the relationship between I_n and E (Equations (2.56)–(2.62))

and

$$dA_2 = 2\pi r \sin\theta \cdot r \, d\theta \tag{2.59}$$

Substituting Equations (2.58) and (2.59) into Equation (2.57) gives

$$dE = 2\pi I_n \sin\theta \cos\theta d\theta dA_1 \tag{2.60}$$

Expressing this as a heat flux from dA_1 and integrating from $\theta = 0$ to $\theta = \pi/2$, gives:

$$E = 2\pi I_n \int_0^{\pi/2} \sin\theta \cos\theta d\theta \tag{2.61}$$

$$= \pi I_n \tag{2.62}$$

This equation relates the emissive power to the intensity of normal radiation (see also Tien *et al.*, 2008).

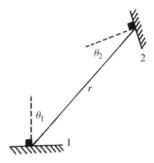

Figure 2.21 Derivation of the configuration factor ϕ for a small element of surface at 2 exposed to a radiating surface at 1 (Equations (2.63)–(2.66))

2.4.1 Configuration Factors[4]

Equation (2.4) gives the total heat flux emitted by a surface. In order to calculate the radiant intensity at a point distant from the radiator, a geometrical – or 'configuration' – factor must be used. Consider two surfaces, 1 and 2, of which the first is radiating with an emissive power E_1 (Figure 2.21). The radiant intensity falling on a small element of surface dA_2 on surface 2 is obtained by calculating the amount of energy from a small element of surface dA_1 that is transmitted through the solid angle subtended by dA_2 at dA_1:

$$d\dot{q} = I_n dA_1 \cos\theta_1 \cdot \frac{dA_2 \cos\theta_2}{r^2} \tag{2.63}$$

The incident radiant flux at dA_2 is then

$$d\dot{q}'' = \frac{d\dot{q}}{dA_2} = I_n dA_1 \cos\theta_1 \frac{\cos\theta_2}{r^2} \tag{2.64}$$

But $(dA_1 \cos\theta_1)/r^2$ is the solid angle subtended by dA_1 at dA_2. Integrating over A_1, and setting $I_n = E/\pi$,

$$\dot{q}'' = E \cdot \int_0^{A_1} \frac{\cos\theta_1 \cos\theta_2}{\pi r^2} \cdot dA_1 \tag{2.65}$$

$$= \phi E \tag{2.66}$$

where ϕ is known as the configuration factor. Values may be derived for various shapes and geometries from tables and charts in the literature (McGuire, 1953; Rohsenow and Choi, 1961; Hottel and Sarofim, 1967; Tien *et al.*, 2008).

Figure 2.22 is a nomogram from which the configuration factor ϕ may be derived for the geometry shown in Figure 2.23(a), i.e., a receiving element dA lying on the perpendicular to one corner of a radiant rectangle. This method of presentation allows advantage to be taken of the fact that configuration factors are additive. Thus the element dA in Figure 2.23(b) views four rectangles, A, B, C and D. The configuration factors for

---

[4] Also known as "view factors"

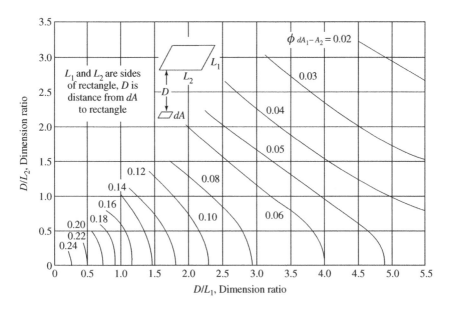

Figure 2.22 Configuration factor ϕ for direct radiation from a rectangle to a parallel small element of surface dA lying on a perpendicular to a corner of the radiator (Figure 2.23(a)) (Hottel, 1930). Reproduced by permission of John Wiley & Sons, Inc

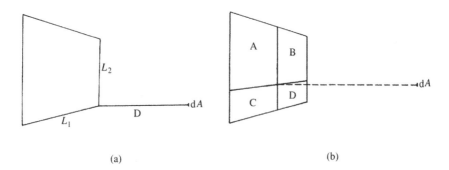

Figure 2.23 (a) Receiver element dA lying on the perpendicular from a corner of a parallel rectangle (see Figure 2.22). (b) Receiver element dA lying on the perpendicular from a point on the radiant rectangle, to illustrate that the configuration factors of rectangles A–D are additive (Equation (2.67))

each can be read from Figure 2.22 (or from Table 2.8 (McGuire, 1953)) and the total configuration factor obtained as their sum:

$$\phi_{\text{total}} = \phi_{\text{A}} + \phi_{\text{B}} + \phi_{\text{C}} + \phi_{\text{D}} \tag{2.67}$$

This may be used to estimate heat fluxes on surfaces exposed to radiation from a fire. In the UK, permissible building separation distances are calculated on the basis that the

Table 2.8 Values of $\phi(\alpha, S)$ for various values of α and S^a

α	$S = 1$	$S = 0.9$	$S = 0.8$	$S = 0.7$	$S = 0.6$	$S = 0.5$	$S = 0.4$	$S = 0.3$	$S = 0.2$	$S = 0.1$
2.0	0.178	0.178	0.177	0.175	0.172	0.167	0.161	0.149	0.132	0.102
1.0	0.139	0.138	0.137	0.136	0.133	0.129	0.123	0.113	0.099	0.075
0.9	0.132	0.132	0.131	0.130	0.127	0.123	0.117	0.108	0.094	0.071
0.8	0.125	0.125	0.124	0.122	0.120	0.116	0.111	0.102	0.089	0.067
0.7	0.117	0.116	0.116	0.115	0.112	0.109	0.104	0.096	0.083	0.063
0.6	0.107	0.107	0.106	0.105	0.103	0.100	0.096	0.088	0.077	0.058
0.5	0.097	0.096	0.096	0.095	0.093	0.090	0.086	0.080	0.070	0.053
0.4	0.084	0.083	0.083	0.082	0.081	0.079	0.075	0.070	0.062	0.048
0.3	0.069	0.068	0.068	0.068	0.067	0.065	0.063	0.059	0.052	0.040
0.2	0.051	0.051	0.050	0.050	0.049	0.048	0.047	0.045	0.040	0.032
0.1	0.028	0.028	0.028	0.028	0.028	0.028	0.027	0.026	0.024	0.021
0.09	0.026	0.026	0.026	0.026	0.025	0.025	0.025	0.024	0.022	0.019
0.08	0.023	0.023	0.023	0.023	0.023	0.023	0.022	0.022	0.020	0.017
0.07	0.021	0.021	0.021	0.021	0.020	0.020	0.020	0.019	0.018	0.016
0.06	0.018	0.018	0.018	0.018	0.018	0.017	0.017	0.017	0.016	0.014
0.05	0.015	0.015	0.015	0.015	0.015	0.015	0.015	0.014	0.014	0.013
0.04	0.012	0.012	0.012	0.012	0.012	0.012	0.012	0.012	0.011	0.010
0.03	0.009	0.009	0.009	0.009	0.009	0.009	0.009	0.009	0.009	0.008
0.02	0.006	0.006	0.006	0.006	0.006	0.006	0.006	0.006	0.006	0.006
0.01	0.003	0.003	0.003	0.003	0.003	0.003	0.003	0.003	0.003	0.003

$^a S = L_1/L_2$ and $\alpha = (L_1 \times L_2)/D^2$ (see Figure 2.22). From McGuire (1953). Reproduced by permission of The Controller, HMSO. © Crown copyright.

exterior of one building must not be exposed to a heat flux of more than $1.2\,\text{W/cm}^2$ ($12\,\text{kW/m}^2$) if an adjacent building is involved in fire (Law, 1963). This level of radiant flux is commonly assumed to be the minimum necessary for the pilot ignition of wood (Section 6.3). The radiating surfaces of a building are taken to be windows and exterior woodwork,[5] and the separation distance is worked out on the basis of the *maximum* heat flux to which an adjacent building may be exposed. Figure 2.24 illustrates the locus of a given value of ϕ for a building on fire. The individual radiators (windows, etc.) may achieve temperatures of up to $1100°\text{C}$ (1373 K), corresponding to a maximum emissive power of $20\,\text{W/cm}^2$ if $\varepsilon = 1$, unless the fire load is small (in which case it does not burn for a sufficient length of time for these temperatures to be achieved) or the fire is fuel-controlled (Chapter 10). Law (1963) assumes that the radiating areas will have an emissive power of $17\,\text{W/cm}^2$ unless the latter conditions hold: then $8.5\,\text{W/cm}^2$ is used. It should be noted that in the absence of external combustible cladding, only openings are considered as 'radiators'. If large flames are projecting from the openings, as will occur particularly with underventilated compartment fires (Section 10.6), much higher heat fluxes may be expected on target surfaces (Lougheed and Yung, 1993).

To illustrate how configuration factors may be used, consider the side of a building, $5.0\,\text{m}$ long by $3.0\,\text{m}$ high, with two windows, each 1 m by 1 m, located symmetrically,

---

[5] This approach has not been extended to other forms of combustible cladding.

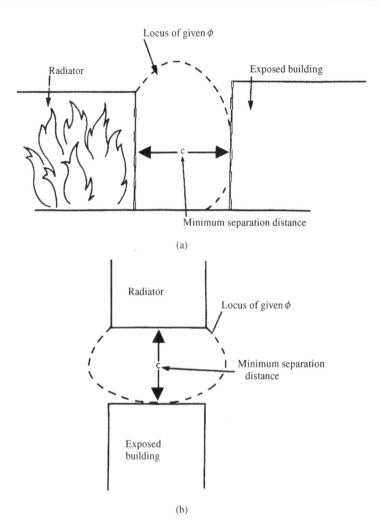

Figure 2.24 Locus of a given configuration factor for a particular radiator shown in (a) elevation and (b) plan view (Law, 1963). Reproduced by permission of The Controller, HMSO © Crown copyright

as shown in Figure 2.25. To calculate the maximum incident heat flux at a distance of 5 m from the wall if the building compartment is on fire (i.e., no external cladding is involved), only the rectangle ABCD enclosing the windows need be considered. At 5 m distance on the axis of symmetry where the heat flux will be greatest (see Figure 2.24):

$$\phi_{ABCD} = 4\phi_{AKHG} = 4(\phi_{AEFG} - \phi_{KEFH}) \tag{2.68}$$

Using Table 2.8, it is found that:

$$\phi_{AEFG} = 0.009 \text{ and } \phi_{KEFH} = 0.003$$

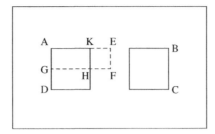

Figure 2.25 Calculation of the configuration factor for the face of a building with two windows, symmetrically located (see Equation (2.68)), where the location of maximum incident heat flux lies on the perpendicular to the point marked 'F'

Therefore:

$$\phi_{ABCD} = 4 \times 0.006 = 0.024$$

Assuming that the emissive power of each window is $17\,W/cm^2$ (after Law, 1963):

$$\dot{q}''_{max}(5\text{ m}) = 0.024 \times 17\ W/cm^2$$

$$= 0.41\ W/cm^2$$

A simpler solution is possible for symmetrical geometries similar to Figure 2.25. Since only 67% of the area ABCD is radiating, the 'average emissive power' for area ABCD is $0.67 \times 17\ W/cm^2 = 11.4\ W/cm^2$. Then, as the configuration factor for AEFG is 0.009:

$$\dot{q}''_{max}(5\text{ m}) = 4 \times 0.009 \times 11.4 = 0.41\ W/cm^2$$

It must be emphasized that the configuration factor defined by Equations (2.65) and (2.66), i.e.

$$\phi = \int_0^{A_1} \frac{\cos\theta_1 \cos\theta_2}{\pi r^2} \cdot dA_1 \tag{2.69}$$

allows the radiant heat *flux* at a point to be calculated at a distance r from a radiator. In Schaum's terminology (Pitts and Sissom, 1977), this would be a 'finite-to-infinitesimal area' configuration factor, and is useful in certain problems relating to ignition (Chapter 6) and for evaluating situations in which people might be exposed to levels of radiant heat (Table 2.9). However, if it is required to calculate the energy exchange between two surfaces, then Equation (2.63) must be integrated twice, over the areas of both surfaces, A_1 and A_2 (see Figure 2.21). The rate of radiant heat transfer to surface 2 from surface 1 is then given by:

$$\dot{Q}_{1,2} = F_{1,2}A_1\varepsilon_1\sigma T_1^4 \tag{2.70}$$

where:

$$F_{1,2} = \frac{1}{A_1} \int_{A_1} \int_{A_2} \frac{\cos\theta_1 \cos\theta_2}{\pi r^2} dA_1 dA_2 \tag{2.71}$$

Table 2.9 Effects of thermal radiation

Radiant heat flux (kW/m^2)	Observed effect
0.67	Summer sunshine in UK[a]
1	Maximum for indefinite skin exposure
6.4	Pain after 8 s skin exposure[b]
10.4	Pain after 3 s exposure[a]
12.5	Volatiles from wood may be ignited by pilot after prolonged exposure (see Section 6.3)
16	Blistering of skin after 5 s[b]
29	Wood ignites spontaneously after prolonged exposure[a] (see Section 6.4)
52	Fibreboard ignites spontaneously in 5 s[a]

[a]D.I. Lawson (1954).
[b]S.H. Tan (1967).
The data quoted for human exposure are essentially in agreement with information given by Purser (2008) and Beyler (2008).

which is called the 'integrated configuration factor', or 'finite-to-finite area configuration factor'. It is commonly given the symbol F and the product $F_{1,2}A_1$ is known as the 'exchange area'. By symmetry:

$$F_{1,2}A_1 = F_{2,1}A_2 \tag{2.72}$$

Values of the integrated configuration factor are available in the literature in the form of charts and tables. From Figure 2.26, values of $F_{1,2}$ can be deduced for radiation exchange between two parallel rectangular plates. Figure 2.27 can be applied to plates at right angles to each other. As with the configuration factor ϕ, integrated configuration factors (F) are additive and can be manipulated to obtain an integrated configuration factor for more complex situations such as that shown in Figure 2.28. They must be used when a full heat transfer analysis of an enclosed space is required. Their application to fire problems is discussed by Steward (1974a) and Tien et al. (2008):

It is important to remember that radiative heat transfer is a two-way process. Not only will the receiver radiate but also the emitting surface will receive radiation from its surroundings, including an increasing contribution from the receiver as its temperature rises. This can best be illustrated by an example: consider a vertical steel plate, 1 m square, which is heated internally by means of electrical heating elements at a rate corresponding to 50 kW (Figure 2.29(a)). The final temperature of the plate (T_p) can be calculated from the steady state heat balance, Equation (2.73):

$$50\,000 = 2\varepsilon\sigma(T_p^4 - T_0^4) + 2h(T_p - T_0) \tag{2.73}$$

where T_0 is the ambient temperature, 25°C (298 K). The factor of 2 appears because the plate is losing heat from both surfaces. It is assumed that the plate is sufficiently thin for

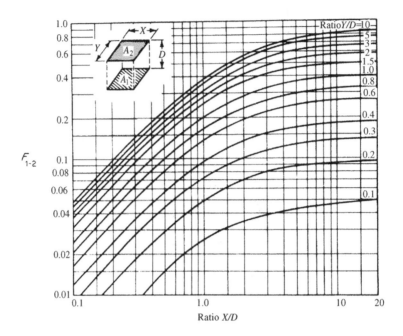

Figure 2.26 View factor for total radiation exchange between two identical, parallel, directly opposed flat plates (Hamilton and Morgan, 1952)

heat losses from the edges to be ignored. Equation (2.73) can be reduced to:

$$2\varepsilon\sigma T_p^4 + 2hT_p - (50\,000 + 596h) = 0 \tag{2.74}$$

as $2\varepsilon\sigma T_0^4 \ll 50\,000$. Equation (2.74) may be solved for T_p with $\varepsilon = 0.85$ and $h = 12$ W/m$^2\cdot$K, using the Newton–Raphson method (Bajpai *et al.*, 1990) to give $T_p = 793$ K (520°C). If a second steel plate, 1 m square but with no internal heater, is suspended vertically 0.15 m from the first (Figure 2.29(b)) then, ignoring reflected radiation, the following two steady state equations can be written.
For plate 1:

$$50\,000 + A_2 F_{2,1}\varepsilon^2\sigma T_2^4 + (1 - A_2 F_{2,1})\varepsilon\sigma T_0^4 = 2A_1 h(T_1 - T_0) + 2A_1\varepsilon\sigma T_1^4 \tag{2.75}$$

and for plate 2:

$$A_1 F_{1,2}\varepsilon^2\sigma T_1^4 + (1 - A_1 F_{1,2})\varepsilon\sigma T_0^4 = 2A_2 h(T_2 - T_0) + 2A_2\varepsilon\sigma T_2^4 \tag{2.76}$$

Of the two terms expressing radiative heat gain on the left-hand sides of Equations (2.75) and (2.76), the first contains ε^2, which is equivalent to the product: (emissivity of emitter) × (absorptivity of receiver). The second refers to radiation from the surroundings at ambient temperature and can be ignored. From Figure 2.26, $A_1 F_{1,2} = A_2 F_{2,1} \approx 0.75$. The above equations then become:

$$9.639T_1^4 + 2.4 \times 10^9 T_1 - 3.072T_2^4 - 5.715 \times 10^{12} = 0 \tag{2.77}$$

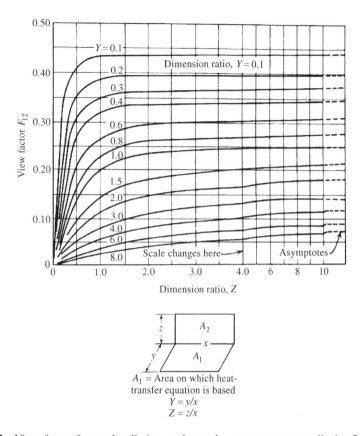

Figure 2.27 View factor for total radiation exchange between two perpendicular flat plates with a common edge (Hamilton and Morgan, 1952)

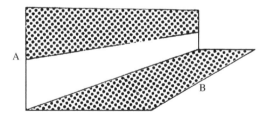

Figure 2.28 View factors for surfaces A and B can be calculated from Figure 2.27 (see text)

and

$$3.072T_1^4 - 9.639T_2^4 - 2.4 \times 10^9 T_2 + 7.152 \times 10^{11} = 0 \qquad (2.78)$$

These give $T_1 = 804$ K (531°C) and $T_2 = 526$ K (253°C), thus illustrating the results of cross-radiation in confined situations. This general effect is even more significant at temperatures associated with burning, and is extremely important in fire growth and

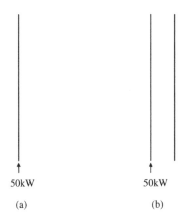

50kW   50kW

(a)   (b)

Figure 2.29 (a) Heat losses from a vertical, internally heated flat plate (Equations (2.73) and (2.74)); (b) heat losses and radiation exchange between two vertical, flat plates, one of which is internally heated (Equations (2.75)–(2.78))

spread, particularly in spaces such as ducts, ceiling voids and even gaps between items of furniture (Section 9.2.4).

2.4.2 Radiation from Hot Gases and Non-luminous Flames

Only gases whose molecules have a dipole moment can interact with electromagnetic radiation in the 'thermal' region of the spectrum (0.4–100 μm). Thus, homonuclear diatomic molecules such as N_2, O_2 and H_2 are completely transparent in this range, while heteronuclear molecules such as CO, CO_2, H_2O and HCl absorb (and emit) in certain discrete wavelength bands (Figure 2.30). Such species do not exhibit the continuous absorption that is characteristic of 'black' and 'grey' bodies (Figure 2.18), and absorption (and emission) occurs throughout the volume of the gas: consequently, the radiative properties depend on its depth or 'path length'.

Consider a monochromatic beam of radiation of wavelength λ passing through a layer of gas (Figure 2.31). The reduction in intensity as the beam passes through a thin layer dx is proportional to the intensity $I_{\lambda x}$, the thickness of the layer (dx) and the concentration of absorbing species within that layer (C), i.e.

$$dI_\lambda = \kappa_\lambda C I_{\lambda x} \, dx \tag{2.79}$$

where κ_λ, the constant of proportionality, is known as the monochromatic absorption coefficient. Integrating from $x = 0$ to $x = L$ gives:

$$I_{\lambda L} = I_{\lambda 0} \exp(-\kappa_\lambda C L) \tag{2.80}$$

where $I_{\lambda 0}$ is the incident intensity at $x = 0$. This is known as the Lambert–Beer law.[6]

---

[6] Also known as Bouguer's law.

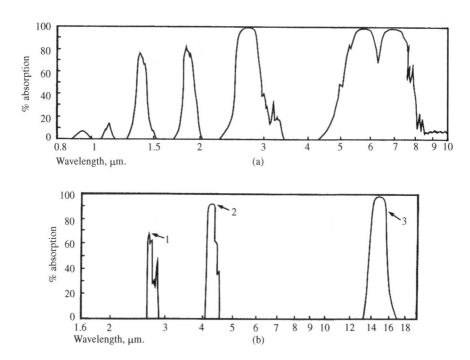

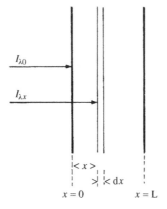

Figure 2.30 Absorption spectra of (a) water vapour, 0.8–10 μm, at atmospheric pressure and 127°C (thickness of layer 104 cm); (b) carbon dioxide, 1.6–20 μm, at atmospheric pressure: curve 1, thickness of layer 5 cm; curves 2 and 3, thickness of layer 6.3 cm. Adapted from Kreith (1976)

Figure 2.31 Absorption of monochromatic radiation in a layer of absorbing medium

The monochromatic absorptivity is then:

$$a_\lambda = \frac{I_{\lambda 0} - I_{\lambda L}}{I_{\lambda 0}} = 1 - \exp(-\kappa_\lambda C L) \qquad (2.81)$$

which, by Kirchhoff's law, is equal to the monochromatic emissivity, ε_λ, at the same wavelength λ. Equation (2.81) shows that as $L \to \infty$, a_λ and ε_λ approach a value of unity.

A volume of gas containing carbon dioxide and water vapour does not behave as a 'grey' body as the emissivity is strongly dependent on wavelength (Figure 2.30): radiation is emitted in discrete bands (Figure 2.30). Hottel and Egbert (1942) developed an empirical method by which an 'equivalent grey body' emissivity of a volume of hot gas containing these species could be worked out. (Other gases were included in the original work, but only CO_2 and H_2O are relevant here.) The procedure is based on a series of careful measurements of the radiant heat output from hot carbon dioxide and water vapour (separately and together) at various uniform temperatures and partial pressures with different geometries of radiating gas. As emissivity at a single wavelength is known to depend on both the concentration of the emitting species and the 'path length' through the radiating gas as viewed by the receiver (cf. Figure 2.31), Hottel's first step was to determine the effective total emissivity of CO_2 and water vapour as a function of temperature for a range of values of the product pL, where p is the (partial) pressure of the emitter and L is the mean equivalent beam length, which depends on the geometry of the volume of gas (see Table 2.10). The results are shown in Figure 2.32, in which the values of the product pL have been converted from Hottel's original units of atmosphere-feet to atmosphere-metres (Edwards, 1985; Tien *et al.*, 2008). If the partial pressure of the emitting species and the mean beam length are known, then the effective 'grey body'

Table 2.10 Mean equivalent beam length (L) for a gaseous medium emitting to a surface (Gray and Muller, 1974)[a]

Shape	L
Right circular cylinders	
1. Height $=$ diameter (D), radiating to:	
(a) centre of base	$0.7D$
(b) whole surface	$0.6D$
2. Height $= 0.5D$, radiating to:	
(a) end	$0.43D$
(b) side	$0.46D$
(c) whole surface	$0.45D$
3. Height $= 2D$, radiating to:	
(a) end	$0.60D$
(b) side	$0.76D$
(c) whole surface	$0.73D$
Sphere, diameter D, radiating to:	
entire surface	$0.64D$

[a]These values correspond to the optically thick limit (see Tien *et al.*, 2008). They will be approximately 10% higher for an optically thin gas.

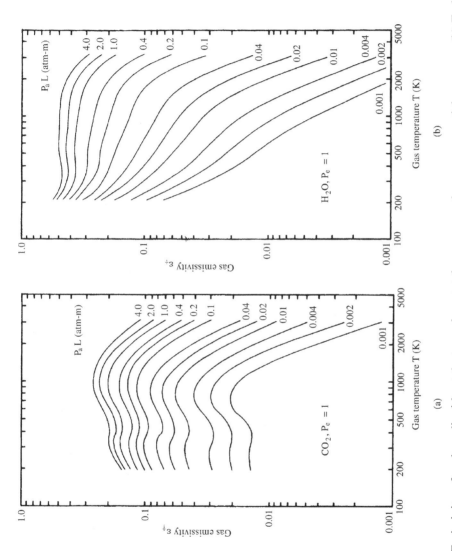

Figure 2.32 (a) Emissivity of carbon dioxide at 1 atmosphere total pressure and near zero partial pressure. (b) Emissivity of water vapour at 1 atmosphere total pressure and near zero partial pressure. From Edwards (1985). Reproduced by permission of the Society of Fire Protection Engineers

emissivity can be obtained at any temperature up to c. 3000 K. While these diagrams apply to gas mixtures at a total pressure of 1 atmosphere, an effect known as pressure broadening, which depends on the partial pressures of emitting species, influences the emission and must be taken into account in any accurate analyses. In addition, a correction for the overlap of the 4.4 µm band of CO_2 and the 4.8 µm band of H_2O is necessary. These modifications are described in detail in most heat transfer texts (e.g., Gray and Muller, 1974; Edwards, 1985; Welty $et\ al.$, 2008), but will not be considered here. The example that follows is included in order to introduce the concepts and illustrate the fact that non-luminous flames have very low emissivities. Errors arising from neglect of the corrections are not significant in this context.

Consider a small fire involving a 0.3 m diameter pool of methanol. The flame, which will be non-luminous, can be approximated by a cylinder of height 0.3 m. The mean beam length for radiation falling at the centre of the surface of the pool then becomes 0.7×0.3 m $= 0.21$ m (Table 2.10). Furthermore, if the composition of the flame is taken to be that produced by burning a limiting methanol/air mixture (6.7% methanol in air), then the partial pressures of carbon dioxide and water vapour will be $p_c = 0.065$ atm and $p_w = 0.13$ atm, respectively, although these are likely to be overestimates. Multiplying these by the mean beam length (giving 0.014 and 0.027 atm·m, respectively) and assuming a uniform temperature of 1200°C (Rasbash $et\ al.$, 1956), ε_c and ε_w can be evaluated from Figure 2.31(a) and 2.31(b), respectively – thus, approximately $\varepsilon_c = 0.04$ and $\varepsilon_w = 0.04$. The resultant emissivity will be:

$$\varepsilon_g = 0.04 + 0.04 = 0.08 \tag{2.82}$$

although this must be regarded as highly approximate as the assumptions are somewhat arbitrary. However, it is not too dissimilar to the emissivities observed by Rasbash $et\ al.$ (1956). The low emissivity of the flame has a significant effect on the burning behaviour of this particular fuel. This will be discussed in Section 5.1.1.

Substantial progress has been made in developing methods of calculating the emissivity of non-luminous combustion gases (de Ris, 1979). While a discussion of these techniques is beyond the scope of the present text, it is worth mentioning that they reveal that the empirical method outlined above gives acceptable values of emissivity up to about 1000 K. Above this temperature and particularly at long path lengths, Hottel's method apparently underestimates the emissivity, probably as a result of overestimating the overlap correction. However, these methods apply to gases of uniform temperature and composition. If, as in a flame, the outer regions are cooler than the inner, then there will be some re-absorption of radiation and a consequent reduction of the emissive power. The effect is much more pronounced with luminous flames and layers of hot smoke (Orloff $et\ al.$, 1979; Grosshandler and Modak, 1981). It should also be noted that water vapour in the atmosphere will attenuate the intensity of radiation at a distance from a large fire (e.g., a pool fire, Section 4.4.1), an effect that will be enhanced on days of high humidity.

2.4.3 Radiation from Luminous Flames and Hot Smoky Gases

With few exceptions (e.g., methanol and paraformaldehyde), liquids and solids burn with luminous diffusion flames. The characteristic yellow luminosity is the net effect of emission from minute carbonaceous particles, known as 'soot', with diameters of the order of 10–100 nm which are formed within the flame, mainly on the fuel side of the reaction zone (Section 11.1). These may be consumed as they pass through the oxidative region of the flame, but otherwise they will escape from the flame tip to form smoke (Section 11.1). The propensity of different fuels – gases, liquids and solids – to produce soot can be assessed by measuring the laminar flame 'smoke point', i.e., the minimum height of a laminar diffusion flame (Section 4.1) at which smoke is released (Section 11.1 and Table 11.1). The smaller the value of the smoke point, the greater the tendency for soot to form in the flame. While within the flame, individual soot particles attain high temperatures and each will act as a minute black (or 'grey') body. The resulting emission spectrum from the flame will be continuous, and the net emissive power will be a function of the particle concentration and the flame thickness (or mean beam length, L). The smoke point has been found to relate inversely to the proportion of the heat of combustion that is lost by radiation from the flame. This holds for a range of fuels (gases, liquids and solids) (Markstein, 1984; de Ris and Cheng, 1994) and for both laminar and turbulent flames (Markstein, 1984). Generally speaking, as the presence of soot particles in the flame provides the mechanism for radiative heat loss, the 'sootier' the flame, the lower its average temperature – see Table 2.11 (de Ris, 1979) and Table 5.4 (Rasbash et al., 1956).

By Kirchhoff's law, the emissivity can be expressed in terms of a relationship identical in form to that for monochromatic absorptivity (Equation (2.81)), thus:

$$\varepsilon = 1 - \exp(-KL) \tag{2.83}$$

where K is an effective emission coefficient, proportional inter alia to the soot particle concentration, which can be determined as the 'particulate volume fraction', f_v (Section 11.1.1). A few empirical values of K are available in the literature (Table 2.11) and permit approximate values of emissive power to be calculated, provided the flame temperature is known or can be measured. However, this simple method for calculating emissivity cannot be used for large fires as it contains an implicit assumption that the flame is uniform with respect to both temperature and soot concentrations. There is now clear evidence that this is not the case. This has come from more detailed studies of the structure of the flames, mapping not only the temperature fields but also the local concentrations of the soot particles (e.g., Sivanathu and Gore, 1992).

Theory indicates that provided the soot particle diameter is less than the radiation wavelength (mostly $\lambda > 1\,\mu m$ (10^3 nm)), the emission coefficient will be proportional to the

Table 2.11 Emissivities (ε) and emission coefficients (K) for four thermoplastics (de Ris, 1979)

Fuel[a]	Flame temperature (K)	Emissivity (ε)	Emission coefficient K (m^{-1})	Carbon appearing as soot (%)
Polyoxymethylene	1400	0.05	–	0
PMMA	1400	0.26	1.3	0.30
Polypropylene	1350	0.59	1.8	5.5
Polystyrene	1190	0.81	5.3	18

[a]The fuel beds were 0.305 m square, except for the PMMA experiments (0.305 m × 0.311 m).

'soot volume fraction' (f_v), which is the proportion of the flame volume occupied by particulate matter. This can be determined using sophisticated optical techniques. Markstein (1979) and Pagni and Bard (1979) obtained data on f_v as a function of height above horizontal slabs of burning PMMA, which showed clearly that f_v decreases with height from a maximum close to the fuel surface (Brosmer and Tien, 1987). Furthermore, due to the non-uniformity of temperature there is some attenuation of radiation from large flames since the outside perimeter is at a lower temperature, and the cooler soot, while still radiating, will absorb radiation from the hotter regions within. The consequence of this is that for hydrocarbon pool fires, the radiative fraction (normally 0.3–0.5, depending on the fuel) is found to decrease as the pool diameter is increased above 2 m (see Figure 4.35, Koseki, 1989). This makes accurate modelling of thermal radiation from large pool fires very difficult, particularly as the 'mean beam length' cannot be assumed equal to the pool – or tank – diameter. Simple examples of calculation of heat flux at a distance are given in Chapter 4, but these will inevitably overestimate the heat flux as they do not take into account the non-uniformity of temperature.

For smaller fires ($D < 1$ m, perhaps), emission from the soot particles is superimposed on emission from the molecular species H_2O and CO_2. This is shown clearly for the wood crib flames of Figure 2.33. Hydrocarbons (gases, liquids and solids) are much sootier, and the black-body background will tend to dominate the radiation, particularly if the hydrocarbons have aromatic character (e.g., polystyrene). Good progress has been made towards modelling the radiant output from flames of this nature in which both the soot emission and the molecular emissions are taken into account. However, further discussion of this is beyond the scope of this text, and the reader is referred elsewhere (de Ris, 1979; Mudan, 1984; Moss, 1995; Tien *et al.*, 2008; Yeoh and Yuen, 2009).

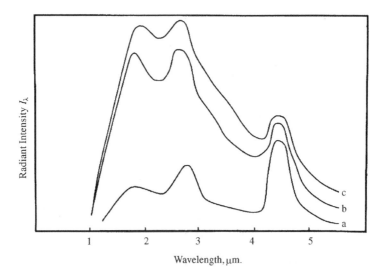

Figure 2.33 Spectra of flames at different thicknesses above wooden cribs (Hägglund and Persson, 1976a). The emission 'peaks' are due to H_2O, H_2O and CO_2, respectively, reading from left to right. Compare with Figure 2.30. Reproduced by permission of the Swedish Fire Protection Ass./FoU-Brand

For convenience and simplicity, it is sometimes assumed that a luminous flame behaves as a 'grey body', i.e., the emissivity is independent of the wavelength. However, the dominance of the CO_2 and H_2O emissions in the early stages of a fire provides an opportunity to design infra-red detectors that can distinguish a flame unambiguously from a hot surface. This can be done by using sensors which can compare the intensity of emission at 4.4 μm with that at c. 3.8 μm, outside the CO_2 band; a significantly stronger signal at 4.4 μm will be recognized as the presence of flame, at least during the early stages when the flame is still relatively 'thin'.

If the flame is thick ($L > 1$ m) and luminous (e.g., hydrocarbon flames), it is common to assume black-body behaviour, i.e., $\varepsilon = 1$. This was not found to be the case for 2 m thick flames from free-burning wood crib fires (Figure 2.33).

Whether the emissivity of a flame is assumed or calculated on the basis of an empirical or theoretical equation, Equation (2.4) is still applicable. If the flame temperature is known, then the emissive power can be calculated using the mean equivalent beam length (Table 2.10), but before the radiant heat flux at a distance can be estimated, a configuration factor must be derived. This is generally obtained by assuming that the flame can be approximated by a simple geometrical shape, such as a rectangle of height between 1.5 and 2 times the fuel bed diameter[7] (Section 4.3.2) and working out the appropriate configuration factor using Figure 2.22 or Table 2.8. This type of model has been used to predict levels of radiant heat at various locations in a petrochemical plant during an emergency, such as a tank fire or emergency flaring (Robertson, 1976; Mannan, 2005).

Radiation from hot smoke is now known to be an important contributory factor to the development of fire within enclosed spaces. During the growth period of a compartment fire, hot smoky gases accumulate under the ceiling, radiating to the lower levels and thereby enhancing the onset of fully developed burning (Chapter 9). The smoke layer is non-homogeneous and re-absorption of radiation in the lower layers is significant. This has been modelled successfully by Orloff et al. (1979) and others, and will be discussed further in Chapter 9.

Problems

2.1 Consider a steel barrier, 5 mm thick, separating two compartments which are at temperatures of 100°C and 20°C, respectively. Calculate the rate of heat transfer through the barrier under steady state conditions if the thermal conductivity of the steel is 46 W/m·K and the convective heat transfer coefficient is 8 W/m²·K.

2.2 Using the example given in Problem 2.1, calculate the temperatures of the two exposed surfaces (the 'hot' and 'cold' sides of the barrier). Check the Biot number to demonstrate that this is an example of a 'thermally thin' material. What would the surface temperatures be if the barrier was 50 mm thick? Calculate the Biot number for this situation. What are your conclusions?

2.3 Calculate the total steady state heat loss through a 200 mm external brick wall which measures 8 m ×4 m high and contains one single-glazed window, 3 m by 1.5 m, located centrally. The glass is 3 mm thick. Ignore the effects of the window

---

[7] Simplified geometric shapes for flames such as those emerging from openings can be assumed provided other relevant parameters are taken into account (see, for example, Law and O'Brien, 1981).

frame. The inside and outside temperatures are 25°C and 0°C, respectively. Assume a convective heat transfer coefficient of 8 W/m²·K on both sides of the wall.

2.4 What would the steady state heat loss through the wall described in Problem 2.3 be if the following modifications are made:

(a) The window is replaced by a double-glazed unit comprising two 2 mm thick sheets of glass separated by a 2 mm air gap. (Assume that heat transfer across the gap is by conduction through the air.)

(b) The brick is lined on the inside by 10 mm fibre insulating board and on the outside by 12 mm pine boards.

(c) The wall was constructed as a cavity wall, filled with polyurethane foam. Assume that the cavity and both courses of brick are 100 mm across.

2.5 A fire in a room rapidly raises the temperature of the surface of the walls and maintains them at 1000°C for a prolonged period. On the other side of one wall is a large warehouse, whose ambient temperature is normally 20°C. If the wall is solid brick, 200 mm thick, and retains its integrity, what would be the steady state temperature of the surface of the wall on the warehouse side? Assume that the thermal conductivity of the brick is independent of temperature and that the convective heat transfer coefficient at the wall is 12 W/m²·K.

2.6 What would the steady state temperature at the inner surface of the warehouse wall (described in the previous problem) be if several sheets of fibre insulation board (total thickness 0.3 m) were stacked vertically against the wall? (Assume perfect contact between the boards and the wall.)

2.7 Take the model described in Problem 2.1, with both sides of the barrier at 20°C. Assume that the temperature of the air on one side is suddenly increased to 150°C. Calculate numerically the temperature of the steel after 10 minutes using a timestep of 60 s. Compare your answer with the value of T_s taken from the analytical solution derived in Section 2.2.3 (Equation (2.34)).

2.8 A vertical sheet of cotton fabric, 0.6 mm thick, is immersed in a stream of hot air at 150°C. Calculate how long it will take to reach 100°C if the heat transfer coefficient is $h = 20$ W/m².K and ρ and c are 300 kg/m³ and 1400 J/kg, respectively. Assume the initial temperature to be 20°C.

2.9 What temperature will a heavy cotton fabric, 1.0 mm thick, reach in 5 seconds if it is exposed to a hot air stream at 500°C? Use the data given in Problem 2.8.

2.10 Vertical slabs of Perspex (PMMA) are exposed on both sides to a stream of hot air at 200°C. Using the Heisler charts (Figure 2.6), derive the surface temperature after 30 seconds for thicknesses of: (a) 3 mm; (b) 5 mm; (c) 10 mm. Assume that the initial temperature is 20°C.

2.11 Estimate the mid-plane temperatures for the slabs of PMMA described in Problem 2.10, for the same conditions.

2.12 Which of the following could be treated as semi-infinite solids if exposed to a convective heat flux for 30 seconds: (a) 6 mm PMMA; (b) 40 mm PUF; (c) 10 mm fibre insulating board?

2.13 Using Equation (2.26), calculate the surface temperature of a thick slab of (a) yellow pine; (b) fibre insulating board, after 15 s exposure to a steady stream of air at 300°C. Take the heat transfer coefficient to be 15 W/m^2·K. Assume the slabs to behave as semi-infinite solids, initially at a temperature of 20°C.

2.14 A horizontal steel plate measuring 25 cm × 25 cm lies horizontally on an insulating pad and is maintained at a temperature of 150°C. What would the rate of heat loss be if the surface was cooled by:
(a) natural convection: (Nu = 0.54 (GrPr)$^{1/4}$)?
(b) a laminar flow of air ($u = 5$ m/s): (Nu = 0.66 Re$^{1/2}$Pr$^{1/3}$)?
(c) a turbulent flow of air ($u = 40$ m/s): (Nu = 0.037 Re$^{4/5}$Pr$^{1/4}$)?
(Ignore heat losses from the edge of the plate and to the insulating pad.) The air temperature is 20°C and the kinematic viscosity of air is $v = 18 \times 10^{-6}$ m^2/s. Would radiative heat losses be significant?

2.15 Calculate the configuration factor for an element of surface at S, parallel to the radiator ABCD, as shown in the figure below. Dimensions are as follows: $a = 2$ m, $b = 1$ m, $c = 2.5$ m, $d = 0.5$ m and $e = 6$ m.

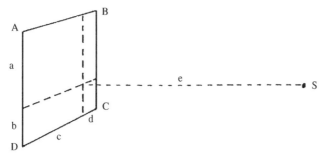

2.16 A building is totally involved in fire, but there is no external flaming. One wall, measuring 20 m long by 10 m high, has four symmetrically placed windows, each 6 m long by 2 m high (see figure below). Assuming that the windows act as blackbody radiators at 1000°C, calculate the radiant heat flux 10 m from the building, where the radiant heat flux is a maximum. Assume that the receiver is parallel to the face of the building.

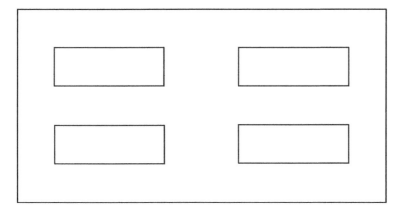

2.17 Taking the building fire described in Problem 2.16, what will the radiant heat flux be at a point 10 m from the building at ground level on a line perpendicular to the mid-point of the base of the wall? Assume that the receiver is parallel to the face of the building.

2.18 A steel plate, measuring 1 m ×1 m, is held vertically. It contains electrical heating elements which are capable of raising its temperature as high as 800°C. Given that the convective heat transfer coefficient can be calculated from:

$$Nu = 0.59 \; (GrPr)^{1/4}(10^4 < GrPr < 10^9)$$

$$Nu = 0.13 \; (GrPr)^{1/3}(GrPr > 10^9)$$

calculate the total rate of heat loss from the plate when its temperature is maintained at 200°C, 400°C, 600°C and 800°C, in each case comparing the radiative and convective components. Assume that the emissivity of the steel is 0.85 and that the atmospheric temperature is 20°C. The kinematic viscosity of air is approximately 18×10^{-6} m²/s at 20°C: use this value for the question, although it is known to increase with temperature (e.g., see Welty *et al.*, 2008).

2.19 The plate described in Problem 2.18 is heated electrically at a rate corresponding to 25 kW. Using the data provided, calculate the steady state temperature of the plate. (Hint: use Newton's method to solve the equation.)

3

Limits of Flammability and Premixed Flames

In premixed burning, gaseous fuel and oxidizer are intimately mixed prior to ignition. Ignition requires that sufficient energy is supplied in a suitable form, such as an electric spark, to initiate the combustion process which will then propagate through the mixture as a flame, or 'deflagration' (Chapter 1). The rate of combustion is typically high, determined by the chemical kinetics of oxidation rather than by the relatively slow mixing of fuel and oxidizer which determines the structure and behaviour of diffusion flames (Chapter 4). However, before premixed flames are discussed further, it is appropriate to examine flammability limits in some detail and identify the conditions under which mixtures of gaseous fuel and air, or any other oxidizing atmosphere, will burn.

3.1 Limits of Flammability

3.1.1 Measurement of Flammability Limits

Although it is common practice to refer to gases and vapours such as methane, propane and acetone as 'flammable', their mixtures with air will only burn if the fuel concentration lies within well-defined limits, known as the lower and upper flammability limits (abbreviated to LFL and UFL). For methane, these are 5% and 15% by volume (i.e., molar proportions), respectively. The most extensive review of the flammability of gases and vapours is that of Zabetakis (1965) which, despite its age, remains the standard reference. It is based largely on a collection of data obtained with an apparatus developed at the US Bureau of Mines (Coward and Jones, 1952) (Figure 3.1). Although there are certain disadvantages in this method, these data are considered to be the most reliable that are available. Alternative methods do exist (e.g., Sorenson *et al.*, 1975; Hirst *et al.*, 1981/82), but none has been used extensively enough to provide a challenge to the Bureau of Mines apparatus.

In this method the experimental criterion used to determine whether or not a given mixture is flammable, is its ability to propagate flame. The apparatus, which is shown schematically in Figure 3.1, consists of a vertical tube 1.5 m long and 0.05 m internal diameter, into which premixed gas/air mixtures of known composition can be introduced. An ignition source, which may be a spark or a small flame, is introduced to the lower end

An Introduction to Fire Dynamics, Third Edition. Dougal Drysdale.
© 2011 John Wiley & Sons, Ltd. Published 2011 by John Wiley & Sons, Ltd.

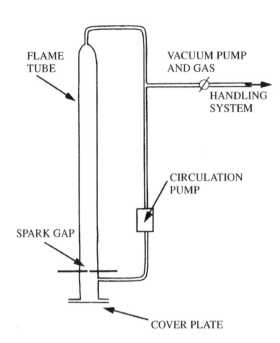

Figure 3.1 The essential features of the US Bureau of Mines apparatus for determining limits of flammability of gases and vapours (Coward and Jones, 1952). (Not to scale.) The circulation pump is necessary to ensure complete mixing of the gases within the flame tube prior to ignition

of the tube, which is first opened by the removal of a cover plate. The mixture is deemed flammable if flame propagates upwards by at least 75 cm. The limits are established experimentally by a process of 'bracketting' and defined as:

$$L = \frac{1}{2}(L_{nf} + L_f) \tag{3.1a}$$

where L_{nf} is the highest concentration of fuel *in air* that is non-flammable and L_f is the lowest concentration of fuel *in air* that is flammable. L_{nf} and L_f are obtained by carrying out a series of measurements in which the fuel concentration is varied around the LFL until an acceptably small difference between the values is obtained. The same procedure is adopted for the UFL, which is calculated as:

$$U = \frac{1}{2}(U_{nf} + U_f) \tag{3.1b}$$

where U_{nf} is the lowest concentration of fuel *in air* that is non-flammable and U_f is the highest concentration of fuel in air that is flammable (Zabetakis, 1965). The limits are normally expressed in terms of volume percentage at 25°C, although they are functions of temperature and pressure (Section 3.1.3). Flammability limit data are given in a number of publications (Fire Protection Association, 1972; Lewis and von Elbe, 1987; Yaws, 1999, NFPA, 2008), but most incorporate data from Zabetakis' review (1965) (Table 3.1). It should be noted that Yaws' collection of data is the most extensive, incorporating information on a much wider range of compounds.

Table 3.1 Flammability data for gases and vapours

	Lower flammability limit $(L)^a$			$\dfrac{L}{C_{st}}$	Upper flammability limit $(U)^a$		$\dfrac{U}{C_{st}}$	S_u^b	Minimum ignition energy^b	Minimum quenching distance^b
	% Vol	g/m^3	kJ/m^3		% Vol	g/m^3		(m/s)	(mJ)	(mm)
Hydrogen	4.0^c	3.6	435	0.13	75	67	2.5	3.2	0.01	0.5
Carbon monoxide	12.5	157	1591	0.42	74	932	2.5	0.43	–	–
Methane	5.0	36	1906	0.53	15	126	1.6	0.37	0.26	2.0
Ethane	3.0	41	1952	0.53	12.4	190	2.2	0.44	0.24	1.8
Propane	2.1	42	1951	0.52	9.5	210	2.4	0.42	0.25	1.8
n-Butane	1.8	48	2200	0.58	8.4	240	2.7	0.42	0.26	1.8
n-Pentane	1.4	46	2090	0.55	7.8	270	3.1	0.42	0.22	1.8
n-Hexane	1.2	47	2124	0.56	7.4	310	3.4	0.42	0.23	1.8
n-Heptane	1.05	47	2116	0.56	6.7	320	3.6	0.42	0.24	1.8
n-Octane	0.95	49	2199	0.58	–	–	–	–	–	–
n-Nonane	0.85	49	2194	0.58	–	–	–	–	–	–
n-Decane	0.75	48	2145	0.56	5.6	380	4.2	0.40	–	–
Ethene	2.7	35	1654	0.41	36	700	5.5	>0.69	0.12	1.2
Propene	2.4	46	2110	0.54	11	210	2.5	0.48	0.28	–
Butene-1	1.7	44	1998	0.50	9.7	270	2.9	0.48	–	–
Acetylene	2.5	29	1410	0.32	(100)	–	–	1.7	0.02	
Methanol	6.7	103	2141	0.55	36	810	2.9	0.52	0.14	1.5
Ethanol	3.3	70	1948	0.50	19	480	2.9	–	–	–
n-Propanol	2.2	60	1874	0.49	14	420	3.2	0.38	–	–
Acetone	2.6	70	2035	0.52	13	390	2.6	0.50	1.1	–
Methyl ethyl ketone	1.9	62	1974	0.52	10	350	2.7	–	–	–
Diethyl ketone	1.6	63	2121	0.55	–	–	–	–	–	–
Diethyl ether	1.9	64		0.56	36	1088	11			–
Benzene	1.3	47	1910	0.48	7.9	300	2.9	0.45	0.22	1.8

^{a}Data from Zabetakis (1965). Mass concentration values are approximate and refer to 0°C $(L(g/m^3) \approx 0.45\, M_w L\ (vol\%))$. C_{st} is the stoichiometric concentration.
^{b}Data from various sources including Kanury (1975) and Mannan (2005). There is uncertainty with some of these data (Harris, 1983; Mannan, 2005). S_u is the fundamental burning velocity (see Section 3.4).
^{c}See Section 3.5.4.

It is worth emphasizing that if the concentration of a flammable gas in air is below the lower flammability limit, the mixture cannot burn, whereas if the concentration is above the upper flammability limit, mixing with air can bring it within the limits and hence create a hazardous situation. For this reason, the lower limit is a better measure of the risk associated with a flammable gas: this issue is addressed again in Section 3.2.4.

Earlier studies showed that the tube diameter has an effect on the result, although it is small for the lower limit if the diameter is 5 cm or more (Figure 3.2). The closing of

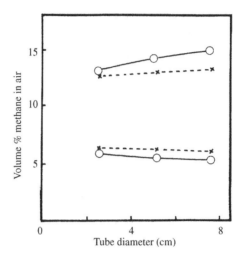

Figure 3.2 Variation of observed flammability limits for methane/air mixtures. o, upward propagation; ×, downward propagation (Linnett and Simpson, 1957)

the limits in narrower tubes can be explained in terms of heat loss to the wall: indeed, if the diameter is reduced to the quenching diameter, flame will be unable to propagate even through the most reactive mixture (Section 3.3).[1] The limits quoted in the literature (Zabetakis, 1965) refer to upward propagation of flame. These are slightly wider than the limits for downward propagation (Figure 3.2), which may be determined using a modified apparatus. The difference arises because the buoyant movement of the burnt gases acts in opposition to the downward propagating flame, creating instability. However, the same behaviour can be observed in unconfined mixtures. Following central ignition, flame will propagate spherically while at the same time the growing volume of burnt gas will rise, causing distortion. Even in this situation it is possible to observe upward (and horizontal) propagation without flame travelling downwards through a near-limit mixture (e.g., Sapko *et al.*, 1976; Roberts *et al.*, 1980; Hertzberg, 1982). This may be explained in terms of the buoyancy-induced upward movement of the hot, burnt gases, a process that may have a contributory role to play in defining the lower limit (see Section 3.3).

Very small amounts of energy are sufficient to ignite flammable vapour/air mixtures. Figure 3.3 shows how the minimum spark energy capable of igniting a mixture varies with composition. The minimum on this curve – known as the minimum ignition energy (MIE) – corresponds to the most reactive mixture, which normally lies just on the fuel-rich side of stoichiometry. Typical values of MIE are quoted in Table 3.1. As no mixture can be ignited by a spark of energy less than the minimum ignition energy, it is possible to design certain items of low power electrical equipment which are intrinsically safe and may be used in locations where there is a risk of a flammable atmosphere being formed (British Standards Institution, 2002; National Fire Protection Association, 2008). This can be achieved by designing the equipment and circuits in such a way that even the worst fault condition cannot ignite a stoichiometric mixture of the specified gas in air.

---

[1] The flammability limits of vapours of low reactivity will be much more sensitive to the diameter and must be determined in significantly wider tubes or large vessels. This is illustrated in Section 6.2.

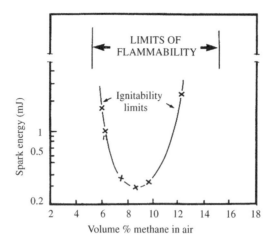

Figure 3.3 Ignitability curve and limits of flammability for methane/air mixtures at atmospheric pressure and 26°C (Zabetakis, 1965)

Limits of ignitability, which vary with the strength of the ignition source, can be distinguished from limits of flammability (Figure 3.3). The latter must be determined using an ignition source that is sufficiently large to ignite near-limit mixtures. However, as the limits vary significantly with initial temperature (Figure 3.4), flame may propagate a short distance in a mixture that is technically 'non-flammable' under ambient conditions if the ignition source is large enough to cause a local rise in temperature. Thus the criterion for flammability in the US Bureau of Mines apparatus is propagation of flame at least half-way up the flame tube (Figure 3.1). At this point it is assumed that the flame will be propagating into a mixture which has not been affected by the ignition source.

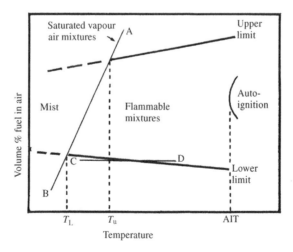

Figure 3.4 Effect of initial temperature on the limits of flammability of a flammable vapour/air mixture at a constant initial pressure (Zabetakis, 1965)

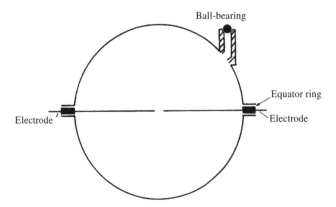

Figure 3.5 The essential features of the spherical pressure vessel, volume 6 litres, used to determine limits of flammability of gases and vapours (Hirst *et al.*, 1981/82). The equator ring also carries the gas inlet port and the outlet port to a vacuum line. Pressure rise can be detected by means of a pressure transducer or a low pressure relief valve (as shown)

Various criticisms can be made of the US Bureau of Mines apparatus, relating mainly to procedure. In its original form, it is unsatisfactory for examining the effects that some additives (such as the halons, which act as chemical inhibitors – see Sections 1.2.4 and 3.5.4) have on the limits, as the method used to prepare the mixture is cumbersome. Furthermore, heavier-than-air mixtures tend to 'slump' from the tube when the cover plate is removed. This will affect the local concentration of vapour at the lower end of the tube where the ignition source is located.

The need to make accurate measurements of the flammability limits of mixtures containing chemical extinguishants has led to the examination of a new method which relies not on flame propagation as the criterion of flammability, but on pressure rise inside a 6 litre spherical steel vessel (Figure 3.5). This is a very sensitive indicator as outside the limit the pressure rise is effectively zero, provided that the energy dissipated by the ignition source is not excessive. In general, data obtained with this apparatus agree quite well with flame tube results for the lower limit, although the upper limit values are less satisfactory (see Hirst *et al.*, 1981/82). However, the lower flammability limit obtained for hydrogen in air using the pressure criterion (~8%) is much higher than that based on observations of flame propagation (~4%). This behaviour is likely to be unique, arising from the high molecular diffusivity of hydrogen combined with buoyancy effects to produce finger-like flamelets capable of propagating vertically, but consuming little fuel (Hertzberg, 1982; Lewis and von Elbe, 1987). Further study is required to establish whether or not the new method is a more relevant and reliable means of measuring the limits.

3.1.2 *Characterization of the Lower Flammability Limit*

When expressed as volume percentage, the flammability limits of the members of a homologous series[2] decrease monotonically with increasing molecular weight, or carbon

---

[2] The term 'homologous series' applies to a series of organic compounds that have a similar general formula (e.g., C_nH_{2n+2} for the alkanes methane, ethane, etc.) and consequently have similar chemical properties.

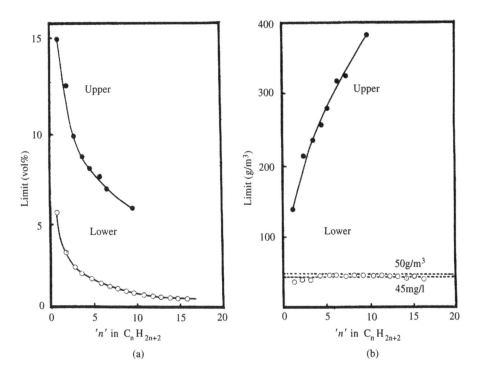

Figure 3.6 Variation of the lower (○) and upper (●) flammability limits with carbon number for the n-alkanes. Limits expressed as (a) percentage by volume; (b) mass concentration (g/m^3). (Data from Zabetakis, 1965)

number (Figure 3.6(a)). However, a different picture is found if the limits are converted to mass concentration. Figure 3.6(b) shows that the lower flammability limits of the C_4–C_{10} straight chain alkanes correspond to ~48 g/m^3 (see also Table 3.1). There is no corresponding rule for the upper flammability limit, as is clear from Figure 3.6(b). As the heat of combustion of the alkanes per unit mass is approximately constant (~45 ± 1 kJ/g, see Table 1.13), the lower limit corresponds to an 'energy density' of ~2160 kJ/m^3, or 48.4 kJ per mole of the lower limit mixture. Examination of the lower limits of a range of hydrocarbons and their oxygenated derivatives suggests that the concept of a 'critical energy density' may be more widely applicable. Most lie in the range 2050 ± 150 kJ/m^3 (see Table 3.1), the exceptions being the alkenes and alkynes, although the higher homologues in the alkene series (e.g., butene-1) tend to fall within the same spread. It should be borne in mind that the uncertainty in the measured values of the lower limits from which these figures are derived is likely to be at least ±5%.

These various criteria offer methods of assessing whether or not the concentration of a mixture of gases in air will be above or below the LFL. For a mixture of alkanes in air at ambient temperature, the lower flammability limit corresponds to:

$$\sum_i^n C_i = 48 \text{ g/m}^3 \tag{3.2a}$$

where C_i is the mass concentration of component i at 25°C. This can also be expressed as:

$$\sum_i \frac{C_i}{48} = \sum_i \frac{l_i}{L_i} = 1 \tag{3.2b}$$

where l_i is the percentage composition (molar proportion) of component i in the vapour/air mixture, and L_i is the corresponding value for its lower flammability limit. If the lower flammability limit of the mixture of hydrocarbons in air is L_m, then writing $l_i = L_m . f_i$, where f_i is the proportion of hydrocarbon i in the original hydrocarbon mixture (by volume):

$$L_m \sum_i \frac{f_i}{L_i} = 1 \tag{3.2c}$$

or in the form normally quoted:

$$L_m = \frac{100}{\sum_i \frac{P_i}{L_i}} \tag{3.2d}$$

in which P_i is the percentage composition of component i in the mixture such that $\Sigma_i P_i = 100$. This is Le Chatelier's law (Le Chatelier and Boudouard, 1898; Coward and Jones, 1952). As an example, it can be used to calculate the lower flammability limit of a mixture of hydrocarbon gases containing 50% propane, 40% n-butane and 10% ethane: thus

$$L_m = \frac{100}{\frac{50}{2.1} + \frac{40}{1.8} + \frac{10}{3.0}} = 2.0\%$$

If the adiabatic flame temperatures for the limiting mixtures are calculated using the method outlined in Section 1.2.5, the results suggest that – at least for the alkanes – there is a limiting flame temperature of 1500–1600 K, below which flame cannot exist. This is discussed in Section 1.2.5 (see Table 1.19). Although the unsaturated hydrocarbons give lower values, the concept of a limiting temperature may be used to check whether or not lean gas mixtures are flammable. For example, consider the mixture comprising 2.5% butane, 20% carbon dioxide and 77.5% air. The components are in the ratio 1:8:31; thus the overall reaction may be expressed as:

$$C_4H_{10} + 8\,CO_2 + 31(0.21\,O_2 + 0.79\,N_2) = 12\,CO_2 + 5\,H_2O + 0.02\,O_2 + 24.49\,N_2$$

The thermal capacity of the final products is 1659 J/K per mole of butane. Combining this with the heat of combustion of butane (2650 kJ/mol) gives a temperature rise of 1597 degrees, i.e., an adiabatic flame temperature of 1890 K, well above the limiting value. While little reliance can be placed on the absolute value of this figure, it is sufficiently large to indicate that the mixture should be considered flammable. Of course, this type of calculation can be turned round to estimate how much carbon dioxide would be required to 'inert' a stoichiometric butane/air mixture (i.e., render it non-flammable under ambient conditions), using a conservative value for the limiting flame temperature (e.g., 1500 K).

It is interesting to note that for many of the gases and vapours listed in Table 3.1, the ratio of L to the percentage concentration in the stoichiometric mixture (C_{st}) is approximately constant, $L \approx 0.55\,C_{st}$ at 25°C. This would suggest that about 45% of the heat of combustion released during stoichiometric burning would have to be removed to quench (extinguish) the flame. This is relevant to the understanding of how flame arresters operate (see Sections 3.3 and 6.6.1).

3.1.3 Dependence of Flammability Limits on Temperature and Pressure

As the initial temperature is increased, the limits widen, as illustrated schematically in Figure 3.4. The line AB in this diagram identifies the saturated vapour pressure so that the region to its left corresponds to an aerosol mist or droplet suspension. The limits are continuous across this boundary: thus the lower limit for tetralin (1, 2, 3, 4-tetrahydronaphthalene, $C_{10}H_{12}$) mist at 20°C corresponds to a concentration of 45–50 g/m^3, in close agreement with the lower limit concentrations observed for hydrocarbon gases and vapours (Table 3.1). However, if the droplet diameter is increased above ~10 μ, the lower limit appears to decrease: coarse droplets will tend to fall into an upward-propagating flame, effectively increasing the local concentration (Figure 3.7) (Burgoyne and Cohen, 1954).

A vapour/air mixture that is non-flammable under ambient conditions may become flammable if its temperature is increased: compare points C and D in Figure 3.4, which refer to the same mixture at different temperatures. The lower limit decreases with rising temperature simply because less combustion energy needs to be released to achieve the

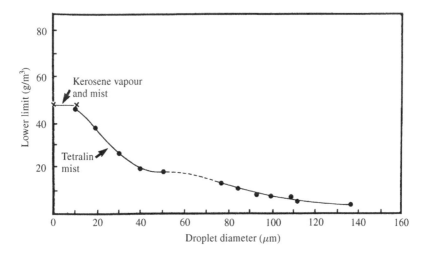

Figure 3.7 Variation of the lower flammability limit of tetralin mist as a function of droplet diameter (Burgoyne and Cohen, 1954, by permission)

limiting flame temperature (T_{lim}): consequently, a lower concentration of fuel in air will be sufficient for flame to propagate through the mixture. In terms of the changes in enthalpy:

$$\frac{L_{25}}{100} \cdot \Delta H_c = c_p(T_{lim} - 25) \tag{3.3a}$$

$$\frac{L_T}{100} \cdot \Delta H_c = c_p(T_{lim} - T) \tag{3.3b}$$

where L_{25} and L_T are the lower limits (% by volume) at 25°C and T°C, respectively. ΔH_c is the heat of combustion (J/mol) and c_p is the thermal capacity of the products (J/K). Dividing Equation (3.3b) by (3.3a) gives:

$$\frac{L_T}{L_{25}} = \frac{T_{lim} - T}{T_{lim} - 25}$$

$$= 1 - \frac{T - 25}{T_{lim} - 25} \tag{3.3c}$$

Taking $T_{lim} = 1300°C$ (Zabetakis, 1965), Equation (3.3c) can be rearranged to give the lower limit at any temperature T:

$$L_T = L_{25}(1 - 7.8 \times 10^{-4}(T - 25)) \tag{3.3d}$$

While this is only an approximation, since the temperature dependence of ΔH_c and c_p is neglected, it agrees satisfactorily with the empirical relationship quoted by Zabetakis *et al.* (1959):

$$L_T = L_{25}(1 - 7.21 \times 10^{-4}(T - 25)) \tag{3.3e}$$

which is based on work by Burgess and Wheeler. The upper limit is found to obey a similar relationship, namely:

$$U_T = U_{25}(1 + 7.21 \times 10^{-4}(T - 25)) \tag{3.4}$$

provided that the mixture does not exhibit cool flame[3] formation at temperature T. Equation (3.4) suggests that the upper limit is also determined by a limiting temperature criterion, a view expressed by Mullins and Penner (1959). It is not possible to calculate the adiabatic flame temperature at the upper limit by the simple method outlined above as the products will contain a complex mixture of pyrolysed and partially oxidized species as well as H_2O and CO_2. However, Stull (1971) has shown theoretically that the flame temperature at the upper limit is approximately the same as the lower limit (Section 1.2.3).

Pressure has little effect on the limits in the sub-atmospheric range, provided that it is not less than 75–100 mm Hg (approximately 0.1 atm) (Figure 3.8(a)). Thus it is possible to determine flammability limits at reduced pressures and apply the results to ambient conditions. In this way the flammability limits of vapours which would be supersaturated

---

[3] The temperatures of 'cool flames' are significantly lower than those of deflagrations (certainly $\ll 1000°C$). They are transient events associated with rich fuel/air mixtures and are observed to occur at c. 300°C. The phenomenon has complex kinetic origins, as discussed by Griffiths and Barnard (1995) (see also Griffiths, 2004).

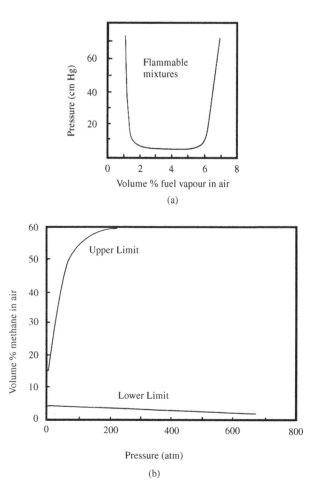

Figure 3.8 Variation of flammability limits with pressure: (a) gasoline vapour in air at reduced pressures (reprinted with permission from Mullins and Penner, copyright 1959 Pergamon Press); (b) methane in air at super-atmospheric pressures (Zabetakis, 1965)

at atmospheric pressure can be derived. This is entirely consistent with the thermal nature of the limit. Thus, n-decane is quoted as having $L_{25} = 0.75\%$, corresponding to a vapour pressure of 5.7 mm Hg under a standard atmosphere (760 mm Hg). At 25°C the saturated vapour pressure of n-decane is 1.77 mm Hg (Table 1.10) or 0.24% by volume at a total pressure of 760 mm Hg. Thus, under normal conditions the vapour/air mixture at the surface of a pool of n-decane at 25°C will be non-flammable (Section 6.2.2). However, if the atmospheric pressure is reduced to 236 mm Hg and the liquid temperature remains constant, the mixture becomes flammable as the decane vapour now constitutes 0.75% of the total pressure. Such dramatic changes in pressure occur in fuel tanks of aircraft following take-off. With kerosene as the fuel, the free space above the liquid surface will contain a vapour/air mixture which is non-flammable at sea level but will become flammable when the aircraft climbs above a certain height. Of course, at high altitude

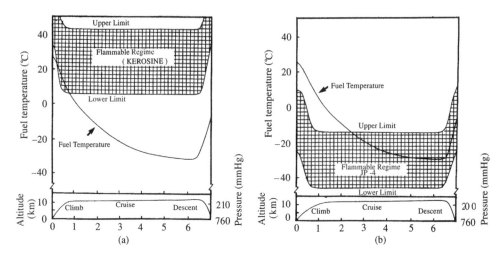

Figure 3.9 Changing regimes of flammability during aircraft flight (in hours): (a) kerosene; (b) JP-4 fuel. Reproduced with permission from Ministry of Aviation (1962)

the temperature of the fuel will gradually fall as the ambient temperature will be low: eventually, the vapour pressure of the liquid will decrease until the mixture is no longer flammable (Figure 3.9(a)) (see Equation (1.14)). However, while the mixture lies within the limits of flammability, there is a potential explosion hazard. Highly flammable liquids such as JP-4, whose vapour pressure lies above the upper flammability limit under ambient conditions, will not present a hazard during the early stages of flight. However, during a flight at high altitude, its temperature will decrease and the hazard may develop as the vapour pressure falls and persists as the plane descends and ambient pressure increases (Figure 3.9(b)).[4]

Substantial increases of pressure above atmospheric produce significant changes in the upper limit (Figure 3.8(b)), while the lower limit is scarcely affected. Thus, at an initial pressure of 200 atm, the upper and lower limits for methane/air mixtures are approximately 60% and 4%, respectively (cf. 15% and 5% at 1 atm).

3.1.4 Flammability Diagrams

Information is readily available in the literature on flammability limits of vapour/air mixtures (e.g. Table 3.1), but in some circumstances it is necessary to know the regimes of flammability associated with more complex combinations of gases, such as hydrocarbon, oxygen and nitrogen. Similarly, it may be necessary to record the effects of adding flame inhibitors (see Section 1.2.4) to flammable vapour/air mixtures, presenting the results in a way that would be useful to a fire protection engineer or a plant operator. As an example, consider the three-component mixture of methane, oxygen and nitrogen. The flammable regime – which must be established experimentally – can be presented on a triangular

---

[4] This behaviour is also discussed by Boucher (2008).

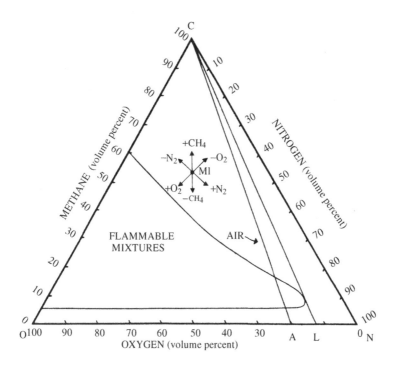

Figure 3.10 Flammability diagram for the three-component system methane/oxygen/nitrogen at atmospheric pressure and 26°C (after Zabetakis, 1965). Points on the line CA correspond to methane/air mixtures

diagram as shown in Figure 3.10, or, as the third component (e.g., oxygen) is a dependent variable, displayed using rectangular coordinates as shown in Figure 3.11.

Thus in Figure 3.10, each axis represents one of the three component gases and the region of flammability is defined by the locus of points corresponding to the limits. Thus, the mixture marked M1 is non-flammable. 'Air' corresponds to the line running from the top apex, C (fuel = 100%) to the lower axis at the point A where the fuel concentration is zero: at every point on this line, the ratio of O_2 to N_2 is 21:79, i.e., the proportions corresponding to normal air. (By drawing a line from 21% on the oxygen axis, parallel to the methane axis (OC), the concentration of nitrogen can be read directly.) The 'air line' CA intersects the envelope of the flammability region at two points corresponding to 5% and 15% methane, i.e., the lower and upper limits of methane in air. The flammability limits of methane in pure oxygen can be obtained from the diagram by examining the intersections of the flammability envelope with the axis OC on which $N_2 = 0\%$. These are 5% and 60%, respectively. It is significant that the lower flammability limit of methane in oxygen is the same as that in air. This is because the heat capacities of nitrogen and oxygen are very similar (Table 1.16) and at the fuel-lean limit the excess oxygen acts only as thermal 'ballast' (Section 1.2.3).

Another important observation that can be made from these diagrams is that there is a minimum oxygen concentration below which methane will not burn. The corresponding O_2/N_2 mixture is given by the line CL which forms a tangent to the tip of the flammability

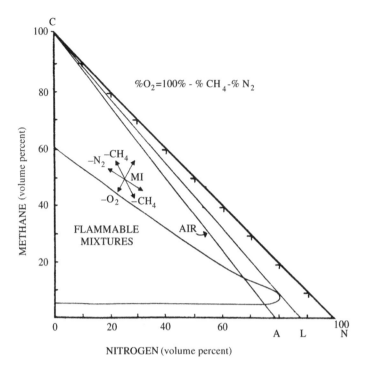

Figure 3.11 Flammability diagram for $CH_4/O_2/N_2$ at atmospheric pressure and 26°C. The information contained in this diagram is identical to that contained in Figure 3.10 (after Zabetakis, 1965)

region. Oxygen concentrations falling to the right of this line, where the ratio of O_2 to N_2 is less than 13:87, will not support methane combustion at ambient temperatures.

Figure 3.11 displays exactly the same information as shown in Figure 3.10. This method of presentation has the advantage that it requires conventional graph paper. Such diagrams may be used to decide how spaces containing mixtures of flammable gas, oxygen and nitrogen can be inerted safely. The point corresponding to M1 in Figures 3.10 and 3.11 corresponds to a non-flammable mixture consisting of 50% CH_4, 25% O_2 and 25% N_2. If this represents a mixture that is normally flowing through an item of chemical plant, then part of the shutdown procedure would be to replace the mixture by air. However, as a general principle, one should avoid flammable mixtures inside the equipment. If the flowing mixture was gradually replaced by air, this could be represented on these diagrams by a straight line joining M1 and A. It is clear that there would be a period of time when a flammable mixture would exist in the system until CH_4 fell below 5%. A similar effect would occur if the flow of fuel was simply turned off (consider the straight line joining M1 and 50% O_2 on the oxygen axis. The correct procedure would be to reduce the oxygen flow (and/or increase the nitrogen flow) until the O_2/N_2 ratio lies to the right of the line CL in Figure 3.10. Then the methane flow can be stopped safely, and the item purged with air as soon as the concentration of CH_4 falls below 5%.

This type of plot may also be used to compare the effect of adding different gases to vapour/air mixtures. Figure 3.12 shows the reduction in the limits of methane in air

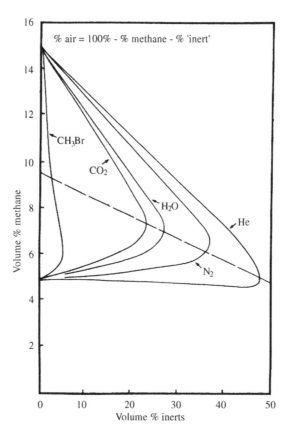

Figure 3.12 Flammability limits of various methane/air/inert gas mixtures at atmospheric pressure and 26°C (after Zabetakis, 1965). The dashed line corresponds to stoichiometric methane/air mixtures

caused by the addition of a number of gases, including helium, nitrogen, water vapour and carbon dioxide. Of these, carbon dioxide is clearly the most effective in rendering the mixture non-flammable, which can be understood in terms of the different heat capacities of the four gases (Table 1.16). On the other hand, methyl bromide (CH_3Br) is much more efficient than CO_2. Only c. 3% is sufficient to render a stoichiometric methane/air mixture non-flammable, while about 23% of carbon dioxide is required to do the same job. This is because methyl bromide acts as a chemical inhibitor, producing HBr in the flame which suppresses the oxidation reactions (Section 1.2.4) by removing hydrogen atoms (Reaction 1.R17). The unusual shape of this particular curve indicates that methyl bromide is itself combustible (Wolfhard and Simmons, 1955).

3.2   The Structure of a Premixed Flame

A premixed flame can be studied experimentally by stabilizing it on a gas burner. The simplest is the Bunsen burner operating with full aeration (Figure 3.13(a)): the characteristic

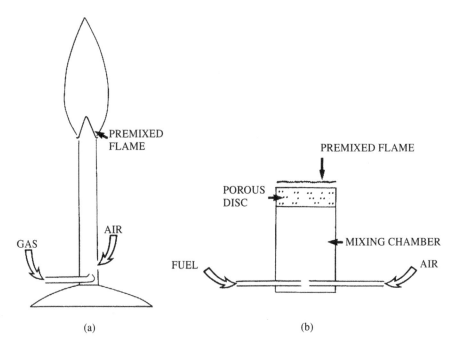

Figure 3.13 (a) Premixed flame on a Bunsen burner with full aeration. (b) Flat premixed flame stabilized on a porous disc (e.g., Botha and Spalding, 1954)

blue cone is a premixed flame which, although fixed in space, is propagating against the gas flow. However, the porous disc burner developed by Botha and Spalding (1954) is more suitable for experimental work as it produces a stationary flat flame, ideal for measurement (Figure 3.13(b)). By use of suitable probes, temperature and concentration profiles through the flame can be obtained (e.g., Fristrom and Westenberg, 1965; Gaydon and Wolfhard, 1979; Vagelopoulos and Egolfopoulos, 1998). These are similar to those illustrated in Figure 3.14, which shows principally the variation of temperature through the flame. The leading edge is located at $x = 0$. Three distinct zones may be identified, as follows:

(i) A pre-heat zone in which the temperature of the unburnt gases rises to some arbitrary value, T_{ig} (see below).
(ii) The reaction zone, in which most of the combustion takes place.
(iii) The post-flame region, characterized by high temperature and radical recombination, leading to local equilibrium. Cooling will subsequently occur.

Of these, zone (ii) is the visible part of the 'flame' and is about 1 mm thick for common hydrocarbon fuels at ambient pressure, but less for highly reactive species such as hydrogen and ethylene.

The thickness of the pre-heat zone ((i) above) can be estimated from an analysis of the temperature profile (Figure 3.15). If it is assumed that no oxidation occurs at temperatures below T_{ig} (a convenient but ill-defined 'ignition temperature'), the following quasi-steady

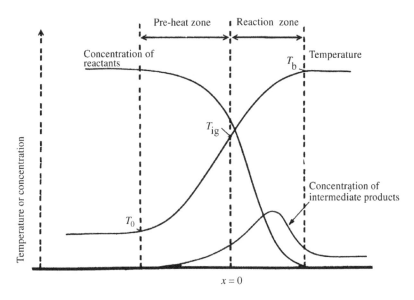

Figure 3.14 Temperature and concentration profiles through a plane combustion wave. Reproduced by permission of Academic Press from Lewis and von Elbe (1987)

state equation may be written to describe conduction of heat ahead of the leading edge of the flame (which lies at $x = 0$):

$$k \left(\frac{d^2 T}{dx^2} \right) - \rho_u S_u c_p \left(\frac{dT}{dx} \right) = 0 \tag{3.5}$$

where ρ_u is the density of the unburnt gas at the initial temperature T_0 and S_u is the rate at which unburnt gas flows into the flame (the burning velocity, see Section 3.4). This equation should be compared with Equation (2.15), with $dt = dx/S_u$. Integration between the limits $x = -\infty$ to $x = 0$ (i.e., from T_0 to T_{ig}) gives:

$$k \left(\frac{dT}{dx} \right) = \rho_u S_u c_p (T_{ig} - T_0) \tag{3.6}$$

Setting dT/dx equal to $(T_{ig} - T_0)/\eta_0$ as shown in Figure 3.15, where η_0 is taken to be a first-order approximation to the thickness of the pre-heat zone, gives:

$$\eta_0 = \frac{k}{\rho_u S_u c_p} \tag{3.7a}$$

(Lewis and von Elbe, 1987). The actual value of the zone thickness depends on how the leading edge of the pre-heat zone is defined. Gaydon and Wolfhard (1979) identify it as the point at which $(T - T_0) = 0.01 \times (T_{ig} - T_0)$. Equation (3.5) must be integrated twice to give:

$$\eta_0' = \frac{4.6 k}{\rho_u S_u c_p} \tag{3.7b}$$

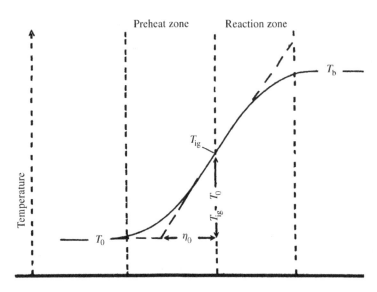

Figure 3.15 Analysis of the temperature profile in the pre-heat zone of the combustion wave. Reproduced by permission of Academic Press from Lewis and von Elbe (1987)

While this is still very approximate – with ambient values of k, ρ_u, S_u and c_p, $\eta_0' \approx$ 0.3 mm, compared to experimental values of 1 mm – the derivation emphasizes the existence and importance of the pre-heat zone. If a premixed flame comes sufficiently close to a solid surface to disturb the pre-heat zone, then heat will be transferred to the surface, and the flame will be cooled (Section 3.3), and ultimately quenched.

However, in free propagation through a quiescent mixture, the velocity with which a flame will travel depends on the efficiency with which heat is transferred ahead of zone (ii). (This general concept can be applied to almost all types of flame spread, and will be discussed at length in Chapter 7.) While a comprehensive analysis of premixed flame propagation must incorporate the four conservation equations of mass, momentum, energy and chemical species (e.g., Williams, 1988; Kuo, 2005), an approximate solution may be gained for an infinite, plane adiabatic wave from the energy equation on its own (cf. Equation (2.14)), i.e.

$$k\left(\frac{d^2T}{dx^2}\right) - \rho u c_p \left(\frac{dT}{dx}\right) - \dot{Q}''' = 0 \qquad (3.8)$$

where $\dot{Q}'''$ is the rate of heat release per unit volume and u is the linear flowrate of gas into the combustion zone which matches the opposed burning velocity, cf. Equation (3.5). If the reaction rate is strongly temperature-dependent (i.e., $\dot{Q}''' \propto \exp(E_A/RT)$ where the activation energy, E_A, is large), it is possible to neglect combustion in the 'pre-heat' zone (zone (i)), where the temperature is less than the fictitious ignition temperature (T_{ig} in Figure 3.15). This done, zones (i) and (ii) can be treated separately to derive expressions for the temperature gradient at $x = 0$ (where $T = T_{ig}$) which are then equated to each other. Using this procedure, which is attributed to Zeldovich and Frank-Kamenetskii (e.g., see Kanury (1975)), an expression for S_u, the 'fundamental burning velocity', can be derived:

$$S_u = \left(\frac{2k}{\rho_0^2 c_p^2 (T_F - T_0)} \cdot \dot{Q}_{ave}''' \right)^{0.5} \tag{3.9}$$

In this equation the subscript zero refers to initial conditions, T_F is the flame temperature and $\dot{Q}_{ave}'''$ is the average rate of heat release in the reaction zone (ii). The quantity S_u is the velocity with which the flame propagates into the unburnt mixture, provided that there is no turbulence in the system (Section 3.6).

A more fundamental analysis of flame propagation confirms that for any given mixture there is one and only one burning velocity – an eigenvalue – and that mass diffusion must also be taken into account (e.g., Dixon-Lewis, 1967). Thus, it is found that burning velocity is a maximum for a slightly fuel-rich mixture, consistent with experimental observation. This occurs because highly mobile hydrogen atoms diffuse ahead of the reaction zone and thus contribute significantly to the mechanism of propagation. Their concentration is higher in the fuel-rich flame.

The relationship between the fundamental burning velocity S_u and the parameters k, c_p, T_F and T_0, indicated in Equation (3.9), is consistent with the observations that will be discussed in Section 3.4. It also predicts that the burning velocity is directly proportional to the square root of the reaction rate, which is incorporated in $\dot{Q}'''$: consequently, $S_u \propto (\exp(-E_A/RT))^{0.5}$, or $S_u \propto \exp(-E_A/2RT)$. This result will be used in Section 3.3.

3.3 Heat Losses from Premixed Flames

The existence of flammability limits is not predicted by the theories outlined in the previous section. Spalding (1957) and Mayer (1957) pointed out that this discrepancy could be resolved by incorporating heat losses into the model. To examine Mayer's argument, consider the temperature profile through the adiabatic flame front shown in Figure 3.16: if there is any heat loss, there will be a decrease in temperature and a consequent fall in the rate of heat release.

Mayer begins his analysis by comparing the temperature profiles normal to the flame front of adiabatic and non-adiabatic flames (Figure 3.16). Without heat losses, the flame

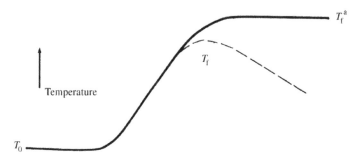

Figure 3.16 Temperature profiles through adiabatic (———) and non-adiabatic (– – – –) premixed flames (after Mayer, 1957)

temperature reaches the adiabatic value, T_F^a, but in reality it is likely to deviate from the adiabatic curve in zone (ii), thereafter passing through a maximum and decaying slowly in the 'post-flame gases' (Figure 3.16). Because the reaction rate is strongly dependent on temperature (proportional to $\exp(-E_A/RT)$), any reduction in T_F will be accompanied by a substantial decrease in the rate of heat generation. A flame will be unable to propagate if the rate of heat loss exceeds the rate of heat production.

Applying a crude 'lumped thermal capacity' model to the propagating flame, the adiabatic case can be described by:

$$\rho_0 S_u^a c_p (T_F^a - T_0) = \rho_0 S_u^a \Delta H_c \tag{3.10}$$

which rearranges to give:

$$\rho_0 S_u^a (c_p T_0 + \Delta H_c) - \rho_0 S_u^a c_p T_F^a = 0 \tag{3.11}$$

where T_F^a is the adiabatic flame temperature and ΔH_c is the heat of combustion.

For the non-adiabatic flame a heat loss term $\mathcal{L}$ (λ, T_F) must be included, thus:

$$\rho_0 S_u (c_p T_0 + \Delta H_c) - \rho_0 S_u c_p T_F - \mathcal{L}(\lambda, T_F) = 0 \tag{3.12}$$

where λ is a 'heat loss parameter'. The following relationships are substituted into Equation (3.12):

$$S_u = B \exp(-E/2RT) \tag{3.13}$$

where B is assumed to be a constant (see Section 3.2.1), and

$$\Delta H_c = c_p (T_F^a - T_0) \tag{3.14}$$

(from Equation (3.10)), to give:

$$\mathcal{L}(\lambda, T_F) = \rho_0 c_p B (T_F^a - T_F) \exp(-E/2RT_F) \tag{3.15}$$

The form of this relationship is illustrated in Figure 3.17. It gives the rate of heat loss which is being experienced by a flame at a given temperature. Thus, when $\mathcal{L} = 0$, the temperature is equal to the adiabatic flame temperature, T_F^a. The relationship also indicates that there is a maximum heat loss which the flame is able to sustain: this can be used to interpret flame quenching and limits of flammability.

(a) *Convective heat losses and quenching diameters.* The ability of a flame to propagate along a narrow duct or tube will depend on the extent of heat losses to the walls. Consider a flame propagating through a flammable mixture contained within a narrow, circular pipe of internal diameter D (Figure 3.18). If the flame thickness is approximated by $\delta \approx k/\rho_0 S_u c_p$ (see Equation (3.7a)), the heat transferred by convection from the flame to the walls per unit surface area of flame $(\pi D^2/4)$ is given by:

$$\dot{q}_{conv} = h(T_F - T_0) \cdot \frac{\pi D \delta}{\pi D^2/4} \tag{3.16}$$

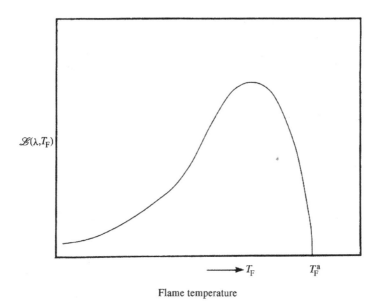

$\mathscr{L}(\lambda,T_\mathrm{F})$

T_F T_F^a

Flame temperature

Figure 3.17   Relationship between heat loss and premixed flame temperature according to Equation (3.15) (after Mayer, 1957)

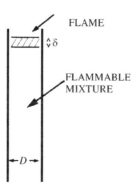

FLAME

$\updownarrow\,\delta$

FLAMMABLE MIXTURE

$\leftarrow D \rightarrow$

Figure 3.18   Propagation of a premixed flame in a tube or duct (Mayer, 1957)

where $\pi D\delta$ is the area of contact between the flame and the pipe, and the heat transfer coefficient is:

$$h = \mathrm{Nu} \cdot k/D \tag{3.17}$$

(Section 2.3). For this configuration, Nu = 3.65 (Mayer, 1957). Writing $\alpha = k/\rho_0 c_\mathrm{p}$, as before (Section 2.2.2):

$$\dot{q}_\mathrm{conv} = 14.6\frac{k\alpha}{D^2 S_\mathrm{u}}(T_\mathrm{F} - T_0) \tag{3.18}$$

This indicates that $\dot{q}_{conv}$ will increase with decreasing pipe diameter, but decrease with increasing T_F, as S_u is strongly temperature-dependent (Equation (3.13)).

If $\dot{q}_{conv}$ for a pipe, diameter D, is plotted on the same diagram as $\mathcal{L}$ (Figure 3.19), any intersection of the curves represents a quasi-equilibrium situation, which defines the temperature of the flame that will propagate through the mixture confined by the pipe. Although there are normally two intersections, only the right-hand one corresponds to stable propagation. Consider the intersection at b_1. A slight decrease in temperature gives $\dot{q}_{conv} > \mathcal{L}$, which results in cooling of the system, while a slight increase causes a continuing rise in temperature as $\mathcal{L} > \dot{q}_{conv}$. If the same arguments are applied to the intersection a_1, it can be seen that any perturbation is self-correcting and the system will always return to the starting point (a_1).

The three convective heat loss curves shown in Figure 3.19 correspond to three pipe diameters which decrease in the sequence D_1, D_2 and D_Q, and which identify the limiting pipe diameter (D_Q) below which $\dot{q}_{conv} > \mathcal{L}$ for all values of T_F. If $D < D_Q$, flame cannot propagate as heat losses to the walls of the tube are too great. However, while the heat loss mechanism can account qualitatively for this phenomenon, it is likely that the surface will also be responsible for the loss of free radicals from the reaction zone by destroying those which migrate to the surface. At present it is not possible to quantify the relative importance of these two mechanisms in physical quenching. Conceptually, D_Q is related to the quenching distance that is of relevance to the design of flame arresters and explosion-proof equipment.

(b) *Radiative heat losses.* In the previous paragraphs, it was tacitly assumed that radiative heat losses were negligible compared with convective loss to the tube walls. If the flammable vapour/air mixture is unconfined so that a propagating flame will not come

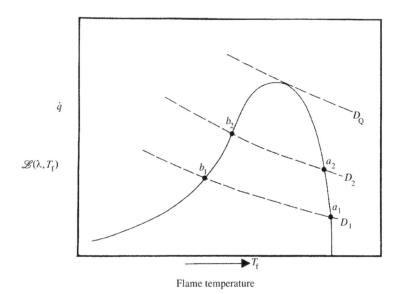

Figure 3.19 Quenching of a flame in a narrow pipe (Mayer, 1957). Solid line, $\mathcal{L}$, heat losses from the flame; dashed lines, convective heat losses for three different values of tube diameter. $D_1 > D_2 > D_Q$ (schematic)

into contact with any surfaces, then radiation will be the *only* mechanism by which heat can be lost. Radiative loss from the reaction zone occurs in two ways, namely:

(i) indirectly, by conduction into the post-flame gases which cool by radiation ($\dot{q}_{\text{rad(i)}}$); and
(ii) directly, by radiation from the reaction zone ($\dot{q}_{\text{rad(ii)}}$).

Of these, the latter is relatively unimportant as the reaction zone is very thin and the average concentrations of the emitting species (principally CO_2 and H_2O, see Section 2.4.2) are low. The following expression, derived from Equation (3.8) with $\dot{Q}''' = 0$, may be written for the post-flame gases:

$$\rho_0 S_u c_p \left(\frac{dT}{dx}\right) - k \left(\frac{d^2 T}{dx^2}\right) = -R_l(T, C_n) \tag{3.19}$$

where R_l, the volumetric heat loss, is a function of temperature and concentration ($n = CO_2$ and H_2O). Close to the downstream edge of the flame, where the temperature is a maximum, the flow or convective term on the left-hand side of Equation (3.19) dominates the conduction term, which is ignored, to allow the temperature gradient at this 'hot boundary' to be approximated by

$$\frac{dT}{dx} = -\frac{R_l(T, C_n)}{\rho_0 S_u c_p} \tag{3.20}$$

The heat loss from the reaction zone to the post-flame gases then becomes

$$\dot{q}_{\text{rad(i)}} = -k_f \left(\frac{dT}{dx}\right) = k_f \frac{R_l(T, C_n)}{\rho_0 S_u c_p} \tag{3.21}$$

Using Hottel's charts for the emissivity of water vapour and carbon dioxide (Section 2.4.2), Mayer showed that R_l could be expressed as:

$$R_l = 1.7 \times 10^{-6}(p_{CO_2} + 0.18 p_{H_2O})T^2 \tag{3.22}$$

where the partial pressures are in atmospheres.

Following the argument developed above for convective heat losses, steady propagation of flame through an unconfined flammable mixture can be represented by the stable intersection of the curves $\dot{q}'_{\text{rad}}$ and $\mathcal{L}$ (Figure 3.20), where $\dot{q}_{\text{rad}} = \dot{q}_{\text{rad(i)}}$. As in this case it is composition that is changed, changes in both the $\mathcal{L}$ versus T_F and $\dot{q}_{\text{rad}}$ versus T_F curves must be considered, as illustrated schematically in Figure 3.20. The intersection between $\mathcal{L}$ and $\dot{q}_{\text{rad}}$ defines the temperature of the steady state flame propagating through a particular flammable vapour/air mixture. If the concentration of the vapour is reduced, then clearly $\mathcal{L}$ will change but so will $\dot{q}_{\text{rad}}$, as the partial pressures of the products CO_2 and H_2O will also be reduced. The limit mixture corresponds to that in which there is a tangency condition between the corresponding curves – the point marked 'P' in Figure 3.20.

While this model shows how the existence of the limits may be explained in terms of the rates of heat production and loss, other factors are also likely to be significant. Thus, it has been suggested that buoyancy is capable of creating sufficient instability at

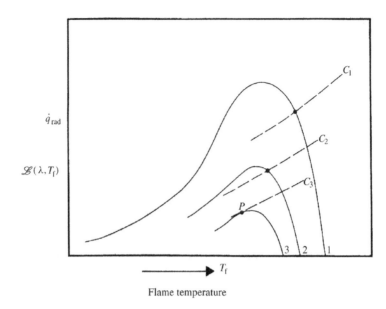

Flame temperature

Figure 3.20 Flammability limit of an unconfined premixed flame with radiation losses (Mayer, 1957). The fuel concentrations decrease in the sequence $C_1 > C_2 > C_3$ (schematic). Solid lines, $\mathcal{L}$, heat losses from the flame; dashed lines, radiative heat losses for three different values of fuel concentration

the leading edge of an upward-propagating flame to cause extinction in a limiting mixture (Lovachev *et al.*, 1973; Hertzberg, 1982). Further refinement will be necessary before there is a detailed understanding of the importance of these contributory mechanisms (Williams, 1988).

3.4 Measurement of Burning Velocities

The fundamental burning velocity (S_u) is defined as the rate at which a plane (i.e., flat) combustion wave will propagate into a stationary, quiescent flammable mixture of infinite extent. Normally the maximum value is quoted for a given flammable gas as this is a measure of its reactivity and to some extent determines the violence of any confined deflagration in which it might be involved (Bartknecht, 1981; Harris, 1983; Lewis and von Elbe, 1987).

Burning velocity must be distinguished from 'flame speed', which is a measure of the rate of movement of flame with respect to a fixed observer. To illustrate the difference, consider a flammable mixture confined to a tube or duct of length l, one end of which is closed (Figure 3.21). Following ignition at the closed end, flame will propagate along the duct, reaching the open end in time t. The average flame speed is then l/t. However, this is substantially greater than the burning velocity as the flammable mixture ahead of the flame front is set into motion by the expansion of the burnt gas behind the flame front.[5]

---

[5] In the US Bureau of Mines apparatus (Figure 3.1), the mixture is ignited at the open end of the tube and the hot combustion products are vented directly. The flame propagates into a static mixture.

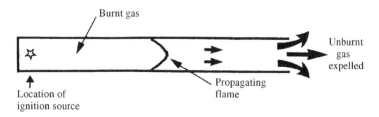

Figure 3.21 Propagation of premixed flame through a flammable mixture in a duct following ignition (⚝) at the closed end

Flame speed cannot be converted into a burning velocity simply by taking the rate of movement of the unburnt mixture into account; the flame front is not planar, there will be heat losses to the walls and (most significantly) the unburnt gas will become increasingly turbulent as it flows along the duct ahead of the flame (see Section 3.6).

It is difficult to devise an experiment for measuring S_u in which interaction between the flame and the apparatus does not influence the result. Several methods are available (e.g., Gaydon and Wolfhard, 1979; Kuo, 2005), although it is necessary to correct the final result to take such interactions into account. Perhaps the simplest method of estimating burning velocity is that using a device similar to the Bunsen burner but which has the capability of producing laminar flows (Re < 2300) of a gas/air mixture whose composition can be varied. Within certain limits of mixture composition and flowrate, a flame can be established which takes the form of a cone sitting at the open end of the vertical burner tube (Figure 3.13(a)). The flame front is in a state of quasi-equilibrium with the flowing mixture, adopting the configuration in which the local burning velocity is equal to the local flowrate vector perpendicular to the flame front. If S_u is the burning velocity, U is the average (laminar) flowrate parallel to the tube axis, and θ is the half-angle of the 'cone' (Figure 3.22), then:

$$S_u = U \sin \theta \tag{3.23}$$

However, this method underestimates S_u by at least 25% for the following reasons:

(a) No account is taken of the velocity distribution across the diameter of the tube.
(b) Heat transfer from the flame zone to the unburnt gas will cause the flow lines to diverge from being parallel to the tube axis close to the flame front (i.e., the measured value of θ is too small).
(c) Heat losses from the edge of the cone to the burner rim, while stabilizing the flame, also affect the flame shape.
(d) The flame front is not planar.

Corrections can be made for (a) and (b) (Andrews and Bradley, 1972; Gaydon and Wolfhard, 1979): it is more difficult to compensate for the effects of (c) and (d), although their effects may be reduced by using a wider tube.

The burner developed by Botha and Spalding (1954) (Figure 3.13(b)) is capable of producing flat, laminar flames and allows the amount of heat transferred from the flame to the burner to be measured. The burner comprised a water-cooled, sintered metal disc

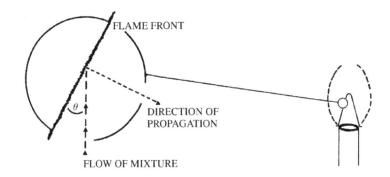

Figure 3.22   An approximate determination of burning velocity: the cone angle method

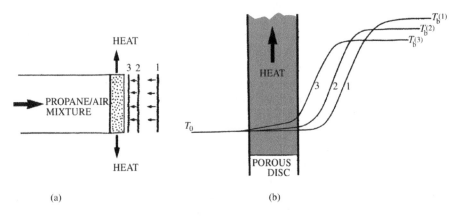

Figure 3.23   Determination of burning velocity – Botha and Spalding's porous burner: (a) showing different positions of the flame front corresponding to different flowrates of fuel/air mixture through the porous disc; (b) showing schematically the temperature distributions for different positions of the flame front in (a) (after Botha and Spalding, 1954, by permission)

through which a fuel/air mixture of known composition could be made to flow at selected linear flowrates. Flat flames could be stabilized, the heat transferred to the burner decreasing as the flowrate was increased, the flame establishing itself at greater distances from the burner surface. This is shown schematically in Figure 3.23.

The rate of heat transfer to the burner was obtained by measuring the increase in temperature of the flow of cooling water. This was determined as a function of the flowrate for each premixture studied and a value for the burning velocity obtained by extrapolating the heat loss to zero (Figure 3.24). An extrapolation is necessary because the flame becomes unstable and lifts off when the flowrate exceeds a critical value. Botha and Spalding obtained 0.42 m/s as the maximum burning velocity for propane/air mixtures by this method.[6] There is some uncertainty in this value, arising mainly from the fact that the measurements of heat transfer to the burner were not of high accuracy. Other

---

[6] This should be compared with $S_u = 0.3$ m/s obtained using the 'cone angle method' discussed above.

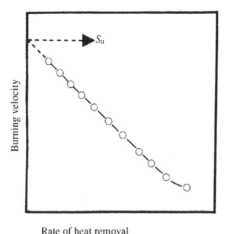

Rate of heat removal

Figure 3.24 Determination of burning velocity (Botha and Spalding, 1954). Extrapolation of burning rate to zero heat loss (reproduced by permission)

techniques are available for measuring burning velocity (Gaydon and Wolfhard, 1979; Kuo, 2005), the most widely accepted of which is the 'spherical bomb' method of Lewis and co-workers (Manton *et al.*, 1953; Lewis and von Elbe, 1987). The mixture is contained within a 15 litre sphere capable of withstanding explosion pressures and ignited centrally by means of a spark. Analysis of the rate of pressure rise and the rate of propagation of the spherical flame front allows S_u to be calculated. Manton *et al.* (1953) obtained a value of 0.404 m/s, in good agreement with that reported by Botha and Spalding (1954).

However, more precise values are required to enable advanced chemical kinetics models – such as CHEMKIN – to be thoroughly validated. For this reason, more sophisticated experimental techniques for determining S_u are being sought. For example, Vagelopoulos and Egolfopoulos (1998) have developed a technique in which a flat flame is formed by a jet impinging on a flat plate. The velocity of the mixture entering the flame front is measured directly by means of laser Doppler velocimetry.[7] Further discussion of this is beyond the scope of this text, but the results appear to be slightly lower than the values quoted in Table 3.1. For example, the maximum burning velocities of methane and propane were found to be 0.37 and 0.41 m/s, respectively. Clearly, the available data on burning velocities of flammable gases and vapours still need careful evaluation as considerable variation is still to be found in the published literature (see, for example, Andrews and Bradley (1972), Tseng *et al.* (1993) and Kuo (2005)).

3.5 Variation of Burning Velocity with Experimental Parameters

Many studies have been made of the variation of burning velocity with experimental parameters such as flammable gas concentration and temperature (Kanury, 1975; Gaydon and Wolfhard, 1979; Lewis and von Elbe, 1987). The main conclusions are outlined in the following sections.

---

[7] By definition, this is equal to the burning velocity. See also Dong *et al.* (2002).

3.5.1   Variation of Mixture Composition

The burning velocity (S_u) of fuel/air mixtures is a maximum for mixtures slightly on the fuel-rich side of stoichiometric, i.e., ϕ (Equation (1.26)) is slightly greater than 1.0 (see also Section 3.2). Mixtures close to the flammability limits have finite burning velocities and there is no evidence that S_u tends to zero at the limit. This is consistent with the concept that the limit represents a criticality at which the rate of heat generation within the flame cannot sustain the heat losses (Mayer, 1957; Spalding, 1957) (Section 3.3) and with observations that a limiting flame temperature exists (White, 1925). Lovachev *et al.* (1973) have pointed out that buoyancy-induced instabilities at the leading edge of a near-limit flame may be sufficient to cause flame extinction (Hertzberg, 1982), but the importance of this mechanism has not been quantified.

Figure 3.25 shows typical results for the variation of burning velocity with mixture composition for methane/air and propane/air mixtures. (Similar, more extensive data sets are reported by Vagelopoulos and Egolfopoulos (1998).) Values quoted in the literature (see Table 3.1) refer to the maximum values measured, i.e., to slightly fuel-rich mixtures. It might be anticipated that the maximum would be observed for the stoichiometric mixture for which the flame temperature is a maximum, but the fact that it does not suggests that flame propagation cannot be explained entirely in terms of heat transfer, as discussed earlier (Section 3.2).

Burning velocity is increased if the proportion of oxygen in the atmosphere is increased. Thus the maximum burning rate for methane/air mixtures increases from about 0.37 m/s to over 3.25 m/s as nitrogen in the air is replaced by oxygen (Figure 3.26). This is consistent with our observations of the temperatures achieved in the combustion zone (Chapter 1), if Equation (3.13) holds (Sections 3.2 and 3.3).

The mechanism by which a flammable vapour/air mixture may be rendered non-flammable by adding a gas such as N_2 or CO_2 that is 'inert' from the combustion

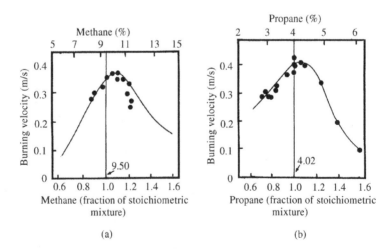

Figure 3.25 Variation of burning velocity with composition: (a) methane/air mixtures; (b) propane/air mixtures. Reproduced by permission of Academic Press from Lewis and von Elbe (1987)

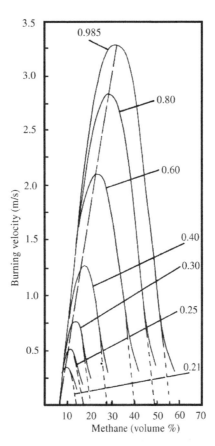

Figure 3.26 Variation of burning velocity with composition: methane in oxygen/nitrogen mixtures (Lewis and von Elbe, 1987). Numbers refer to the ratio $O_2/(O_2+ N_2)$

standpoint can be interpreted in terms of a limiting flame temperature. However, if the nitrogen in the air is replaced completely by another gas (e.g., CO_2 or He) then the burning velocity will be changed. For example, as the heat capacity of CO_2 is over 60% higher than that of N_2 at 1000 K, replacing N_2 by CO_2 will result in a reduction in the burning velocity because the flame temperature will be significantly less. On the other hand, the burning velocity of a stoichiometric mixture of a flammable gas in a 21/79 mixture of O_2/He is much higher than in air because helium has a low thermal capacity and a much higher thermal conductivity (see Equation (3.9), Section 3.2). Oxygen/helium mixtures are used in certain diving applications: in such an environment, a deflagration following the ignition of a leak of flammable gas or vapour would be more violent than if the atmosphere were normal air.

3.5.2 *Variation of Temperature*

The burning velocities quoted in Table 3.1 refer to unburnt mixtures at ambient temperature (20–25°C). However, S_u is increased at higher initial temperatures, as illustrated

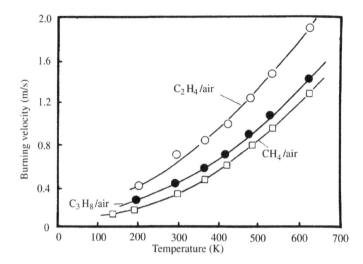

Figure 3.27 Variation of burning velocity with temperature according to Dugger *et al.* (1955). Reproduced by permission of Gordon and Breach from Kanury (1975)

for methane, propane and ethylene in Figure 3.27. Zabetakis (1965) quotes the following expression for methane, propane, *n*-heptane and iso-octane in the range 200–600 K:

$$S_u = 0.1 + 3 \times 10^{-6} T^2 \quad m/s \tag{3.24}$$

At 300 K, this gives 0.37 m/s, close to the currently accepted value of S_u for methane, but somewhat less than that for propane (0.42 m/s). At temperatures above 800 K, mixtures of gaseous fuel and air will undergo thermal degradation and slow oxidation ('preflame reactions'), thus changing the chemical composition of the mixture. Under these conditions the measured burning velocity will be less than that based on an extrapolation from lower temperatures (e.g., from Equation (3.24) or Figure 3.27). If the temperature is sufficiently high, the flammable mixture may ignite spontaneously (see Figure 3.4). Some auto-ignition temperatures – which refer to near-stoichiometric mixtures for which the AIT is a minimum – are shown in Table 6.3; the phenomenon is discussed in Section 6.1.

3.5.3 Variation of Pressure

There is no simple relationship between burning velocity and pressure. Lewis (1954) assumed that a proportionality of the form $S_u \propto p^n$ would hold, where p is the pressure, and determined the value of n for a range of gases and oxygen concentrations using the spherical bomb method described above. He found that n depended strongly on the value of S_u, being zero for burning velocities in the range 0.45–1.0 m/s (Figure 3.28). For $S_u < 0.45$ m/s, the dependence (i.e., n) is negative, while for $S_u > 1.0$ m/s, the dependence is positive. Thus, S_u for methane/air mixtures will decrease with pressure while that for methane/O_2 will increase. Note, however, that the effect is small. Doubling the pressure of a stoichiometric methane/oxygen mixture increases the burning velocity by a factor of only 1.07.

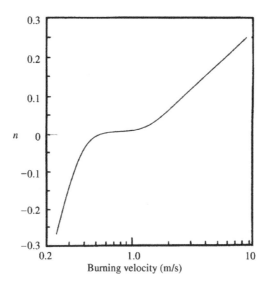

Figure 3.28   Influence of pressure on flame speed (Lewis, 1954; Kanury, 1975). n refers to the exponent in the proportionality $S_u \propto p^n$, where p is the initial pressure

3.5.4   Addition of Suppressants

A flammable mixture may be rendered non-flammable by the addition of a sufficient amount of a suitable suppressant. Additives such as nitrogen and carbon dioxide act as inert diluents, increasing the thermal capacity of the mixture (per unit mass of fuel) and thereby reducing the flame temperature, ultimately to below the limiting value when flame propagation will not be possible (Section 3.1.4). This is illustrated in Figure 3.29(a), which shows the variation of flame temperature, determined by infra-red radiance measurements, as nitrogen is added to stoichiometric methane/air mixtures (Hertzberg *et al.*, 1981). The limiting flame temperature (1500–1600 K) corresponds with 35–38% N_2, in agreement with values calculated on the basis that the lower flammability limit is determined by a critical limiting temperature of this magnitude. Consequently, it is anticipated that the burning velocity at the limit will be similar to that of a limiting methane/air mixture.

However, if chemical inhibitors are present in the unburnt vapour/air mixture, there will be significant reduction in burning velocity without a corresponding reduction in flame temperature. Halogen-containing species are particularly effective in this respect. For example, Simmons and Wolfhard (1955) found that the addition of 2% methyl bromide to a stoichiometric mixture of ethylene and air reduced the burning velocity from 0.66 m/s to 0.25 m/s. These species exert their influence by inhibiting the oxidation chain reactions, reacting with the chain carriers (in particular hydrogen atoms) and replacing them by relatively inert atoms or radicals. As the branching reaction:

$$O_2 + H^\bullet = {}^\bullet O^\bullet + OH^\bullet$$

is largely responsible for maintaining the high reaction rate (Section 1.2.2), any reduction in hydrogen atom concentration will have a very significant effect on the overall

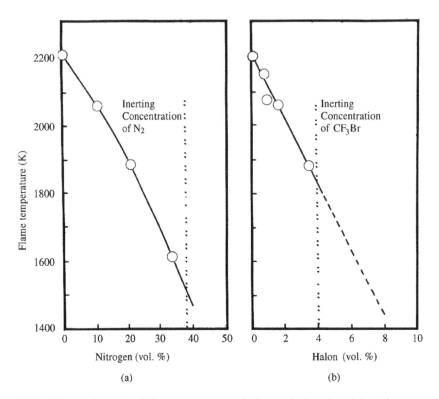

Figure 3.29 Measured premixed flame temperature during explosions in a 3.7 m diameter sphere, stoichiometric methane/air mixtures with addition of (a) nitrogen and (b) CF$_3$Br (Hertzberg, 1982). Reproduced by permission of University of Waterloo Press

reaction rate. Thus, because relatively small amounts of these agents are required, the associated change in heat capacity is small (in relative terms), even at the peak concentration. Consequently, the flame temperature at the limit is greater than 1500–1600 K. This is shown in Figure 3.29(b), in which the measured flame temperature is plotted against concentration of Halon 1301 (bromotrifluoromethane) for a range of fuel/air concentrations. A stoichiometric CH$_4$/air mixture is rendered non-flammable by the addition of 4% 1301, although the flame temperature at the limit is >1800 K. However, there is some dispute over the interpretation of these data. Hertzberg's experiments involved spark ignition of the gaseous mixtures inside a 3.66 m diameter sphere. There is evidence to suggest that the halon is very efficient at suppressing ignition by a small spark: somewhat higher concentrations are required to suppress flame propagation when a larger source of ignition – such as a flame – is used. Hertzberg (1982) suggests that 8% of Halon 1301 is necessary to inert a stoichiometric methane/air mixture under these conditions. If this is correct, it would suggest that chemical inhibition may be of less importance than is currently assumed (at least for CF$_3$Br).

Sawyer and Fristrom (1971) have used the effect on S_u as a means of assessing the relative efficiencies of a range of inhibitors. However, normal practice is to determine the effect of an inhibitor on the flammability limits of suitable gases or vapours. The 'peak concentration' is determined from a flammability diagram such as those shown in Figures 3.12 and 3.30, and refers to the minimum concentration of the agent which is capable of rendering the most reactive vapour/air mixture non-flammable. Some typical values are shown in Table 3.2. Unfortunately, many of these inhibitors (particularly Halon 1211 and Halon 1301) have now been banned as they have been shown to survive for long enough in the atmosphere to be harmful to the ozone layer (Montreal Protocol, 1987): they are now used only where there is an unacceptably high risk, and there is no alternative. In view of this problem, halon replacements have been sought: a review of the current situation appears in the latest edition of the *NFPA Handbook* (Di Nenno and Taylor, 2008).

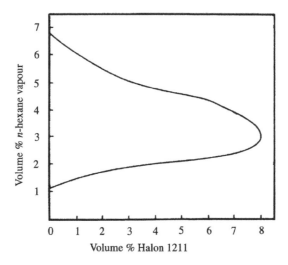

Figure 3.30 Flammability envelope for the addition of CF_2BrCl to a stoichiometric *n*-hexane/air mixture (Hirst *et al.*, 1981/82)

Table 3.2 Peak concentrations[a] of various halons in *n*-hexane/air mixtures (Hirst *et al.*, 1981/82)

Halon number[b]	Common name	Formula	Boiling point (°C)	Peak concentration (%)
1211	BCF	CF_2ClBr	−4.0	8.1
1301	BTM	CF_3Br	−57.6	8.0
1202	DDM	CF_2Br_2	24.4	5.4
2402	DTE	$C_2F_4Br_2$	47.5	5.2

[a]The term 'peak concentration' is the minimum concentration required to inhibit combustion of any fuel/air mixture. It is illustrated in Figure 3.30.
[b]The halon number consists of four digits, referring to the numbers of carbon, fluorine, chlorine and bromine atoms in the molecule, respectively. Thus CF_3Br is Halon 1301.

3.6 The Effect of Turbulence

The previous sections in this chapter have dealt with deflagration in quiescent fuel/air mixtures, involving the propagation of flame into a stationary mixture of infinite extent. The rate at which a flame propagates into the quiescent mixture is referred to as the laminar burning velocity, S_u. If the unburned mixture is turbulent, then the rate of propagation of the flame into the unburned mixture will be greater than S_u and does not have a unique value. Thus, strictly speaking, 'turbulent burning velocity' is not analogous to 'laminar burning velocity'. Bradley (1993) has discussed the value of the turbulent burning velocity as a meaningful parameter.

The effect of turbulence is of considerable importance regarding the behaviour of gas explosions, but is difficult to quantify. In a series of measurements of flame speed using a Bunsen burner technique, Damkohler (1940) found that the rate of propagation was independent of the Reynolds number of the unburnt mixture for Re < 2300, but increased as $\sqrt{Re}$ for 2300 < Re < 6000, then as Re for Re > 6000 (Re is defined in Table 2.4). Similar results were obtained by Rasbash and Rogowski (1960) in studies of flame propagation in ducts. The mechanism is understood to involve an increase in the efficiency of the transport processes (transfer of heat and reactive species) as a result of eddy mixing at the flame front. As these control the rate of propagation (Section 3.2), the rate of burning in turbulent mixtures is high.

The rate of pressure rise following the ignition of a flammable vapour/air mixture in an enclosed space (see Section 1.2.6) is increased substantially if there is turbulence in the flammable mixture ahead of the flame. This is observed if there are obstacles in the path of a propagating flame, but in addition substantial turbulence will be generated if the enclosure is subdivided into compartments (rooms) linked by open doors. As well as increasing the pressure in the adjacent space, there will be increased turbulence as unburnt mixture is pushed through the openings ahead of the propagating flame, as illustrated schematically in Figure 3.31. This is sometimes referred to as 'pressure piling': very rapid and unpredictable rates of pressure rise can be generated in this way (e.g., Harris, 1983). While it is a recognised problem in industry (Phylaktou and Andrews, 1993), it also contributes to the severity of gas explosions in buildings. The first recognized example in the UK was the Ronan Point gas explosion that led to the partial collapse of a multi-storey apartment building in London in 1968 (Rasbash, 1969; Rasbash et al., 1970)[8] Indeed, a great deal can be learned about the dynamics of gas explosions by careful investigation of incidents in which buildings have suffered damaged (Foster, 1998).

A rather similar situation is encountered in chemical plant and in typical modules found on offshore oil production platforms. Because these structures are 'open' to the atmosphere, it was tacitly assumed that ignition of a release of flammable gas or vapour would not produce significant overpressures. However, there are numerous obstacles, formed by pipework and items of equipment, that will create turbulence in advance of a propagating flame, thereby increasing the burning rate and violence of the explosion (e.g., Rasbash, 1986). The initial explosion which led to the loss of the Piper Alpha platform in the North Sea in July 1988 (Cullen, 1989) produced high overpressures on account of flame acceleration due to the turbulence induced in this manner. The effect has been

---

[8] Ronan Point collapse. http://news.bbc.co.uk/onthisday/hi/dates/stories/may/16/newsid_2514000/2514277.stm

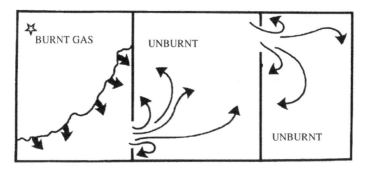

Figure 3.31 Development of an explosion in a multi-chambered compartment, showing development of turbulence ahead of the flame front. Ignition at ✵

demonstrated experimentally (Harrison and Eyre, 1987) and modelled with a considerable degree of success using computational fluid dynamics (Hjertager, 1993).

If a mixture is contained in a pipe or a duct of sufficient length (see Figure 3.21), ignition at the closed end will cause the unburnt gas to be expelled towards the open end, generating turbulent pipe flow in the process. This will lead to flame acceleration, which may be sufficient to produce a shock wave. Compression by the shock is adiabatic, generating temperatures that may be high enough to initiate combustion immediately behind the shock front, forming a self-sustaining detonation. The detonation wave will propagate through the mixture at a very high velocity, typically in excess of 1800 m/s, and generating very high overpressures. This is a very simplistic explanation of a complex process known as 'Deflagration to Detonation Transition' (DDT), but it will only occur for mixtures that lie within the limits of detonability. These are analogous to and lie within the limits of flammability for flammable gases and vapours, Table 3.3 (Lewis and von Elbe, 1987). For detonation to develop in pipes or ducts, a minimum 'run-up' length can be identified (Health and Safety Executive, 1980). This may be as much as 60 pipe diameters for alkanes, but is substantially less for more reactive gases such as ethylene and hydrogen. Bends or obstacles will reduce the run-up length, regardless of the nature of the gas involved as they induce further turbulence and, with it, further flame acceleration. This has been studied in some detail (Rasbash and Rogowski, 1962; Rasbash, 1986). The onset of detonation may be prevented by providing suitably spaced vents to relieve the

Table 3.3 Comparison of limits of flammability and detonability[a]

Mixture	LFL (%)	UFL (%)	Lower limit of detonability (%)	Upper limit of detonability (%)
Hydrogen/air	4.0	75	18.3	59
Acetylene/air	2.5	(100)	4.2	50
Diethylether ether/air	1.9	36	2.8	4.5

[a]The limits of detonability are taken from Lewis and von Elbe (1987). There are few data available for fuel/air mixtures (mainly for fuel/oxygen mixtures).

pressure, but once a detonation wave has become established, conventional vents provide no protection as it travels at speeds greater than that of sound. Instead, the containing pipe or duct will be shattered close to the point at which the detonation starts, where pressures in excess of 1 MPa (10 bar) would be anticipated.

There have been a small number of reported incidents in which a major explosion has occurred following the ignition of a large cloud of flammable gas or vapour in the open. These events are commonly referred to as 'unconfined vapour cloud explosions' (Strehlow, 1973; Gugan, 1979; Zalosh, 2008), although it is clear that in each case there was some degree of confinement and/or obstruction. The pressures generated in such explosions could only have been produced by flame acceleration, achieving velocities of several hundred metres per second. This requires turbulence generation in the unburnt gas ahead of the flame front, which requires obstacles such as pipework in chemical plant (e.g., the Flixborough explosion of 1974) or vegetation, either in the form of trees (e.g., the Ufa explosion of 1993) or hedgerows (e.g., Buncefield, 2005). It has been suggested that such 'clouds' can detonate under certain conditions (e.g., Burgess and Zabetakis, 1973), but only recently has sufficient data become available from a single incident (the Buncefield explosion of 12 December 2005) to put this to the test (Steel Construction Institute, 2009[9]).

These events should be distinguished from BLEVEs (Boiling Liquid, Expanding Vapour Explosions), in which the flammable material is released suddenly following the violent rupture of a pressurized storage vessel which has been exposed to fire for a prolonged period (e.g., Feyzin in 1966, Crescent City, Illinois in 1970 (Strehlow, 1973)). BLEVEs are discussed briefly in Section 5.1.4.

Problems

3.1 Calculate the lower flammability limit of a mixture containing 84% methane, 10% ethane and 6% propane.

3.2 Given that the lower flammability limit of n-butane (n-C_4H_{10}) in air is 1.8% by volume, calculate the adiabatic flame temperature at the limit. (Assume the initial temperature to be 20°C.)

3.3 Calculate the lower flammability limit of propane in a mixture of (a) 21% oxygen + 79% helium and (b) 21% oxygen + 79% carbon dioxide, assuming a limiting adiabatic flame temperature of 1600 K. (Initial temperature 20°C.)

3.4 Using the result of Problem 3.2, calculate by how much a stoichiometric n-butane/air mixture would have to be diluted by (a) nitrogen (N_2), (b) carbon dioxide (CO_2), to render the mixture non-flammable.

3.5 Calculate by how much a stoichiometric propane/air mixture would have to be diluted by (a) carbon dioxide (CO_2), (b) bromotrifluoromethane (CF_3Br, Halon 1301), to render the mixture non-flammable. Assume that the limiting adiabatic flame temperature is 1600 K and that the heat capacity of CF_3Br is 101 J/mol K at 1000 K. It is found experimentally that only 5% CF_3Br is required to inert a stoichiometric propane/air mixture. Explain why this differs from your answer.

---

[9] See http://news.hse.gov.uk/2009/06/25/rr718-buncefield-explosion-mechanism-phase-1-volumes-1-and-2/

3.6 Calculate the lower and upper flammability limits of propane at 200°C and 400°C.

3.7 Calculate the range of temperatures within which the vapour/air mixture above the liquid surface in a can of n-hexane at atmospheric pressure will be flammable.

3.8 Calculate the range of ambient pressures within which the vapour/air mixture above the liquid surface in a can of n-decane (n-$C_{10}H_{22}$) will be flammable at 25°C.

3.9 Given that the lower and upper flammability limits of butane in air and in oxygen are 1.8% and 8.4%, and 1.8% and 49%, respectively, and that the 'limiting oxygen index' for butane is 13%, sketch the flammability limits for the $C_4H_{10}/O_2/N_2$ system, using rectangular coordinates.

3.10 By inspection of Equation (3.9) *et seq.*, how will S_u vary with thermal conductivity (k), thermal capacity (c) and temperature? Compare your conclusions with the empirical results discussed in Section 3.5.

4

Diffusion Flames and Fire Plumes

The principal characteristic of the diffusion flame is that the fuel and oxidizer (air) are initially separate and combustion occurs in the zone where the gases mix. The classical diffusion flame can be demonstrated using a simple Bunsen burner (Figure 3.13(a)) with the air inlet port closed. The stream of fuel issuing from the burner chimney mixes with air by entrainment and diffusion and, if ignited, will burn wherever the concentrations of fuel and oxygen are within the appropriate (high temperature) flammability limits (Section 3.1.3). The appearance of the flame will depend on the nature of the fuel and the velocity of the fuel jet with respect to the surrounding air. Thus, hydrogen burns with a flame that is almost invisible, while all hydrocarbon gases yield flames which have the characteristic yellow luminosity arising from incandescent carbonaceous particles formed within the flame (Section 2.4.3). Laminar flames are obtained at low flowrates. Careful inspection reveals that just above the burner rim, the flame is blue, similar in appearance to a premixed flame. This zone exists because some premixing can occur close to the rim where flame is quenched (Section 3.3a). At high flowrates, the flame will become turbulent (Section 4.2), eventually 'lifting off' when flame stability near the burner rim is lost due to excess air entrainment at the base of the flame. The momentum of the fuel vapour largely determines the behaviour of these types of flame, which are often referred to as 'momentum jet flames'. A typical example to be found in the chemical industry is the flare stack that is used in emergencies to release excess gaseous products from an item of chemical plant and so prevent dangerous pressure excursions.

In contrast, flames associated with the burning of condensed fuels (i.e., solids and liquids) are dominated by buoyancy, the momentum of the volatiles rising from the surface being relatively unimportant. If the fuel bed is less than 0.05 m in diameter, the flame will be laminar, the degree of turbulence increasing as the diameter of the fuel bed is increased, until for diameters greater than 0.3 m buoyant diffusion flames with fully developed turbulence are observed (Section 5.1.1).

This chapter deals principally with flames from burning liquids and solids, although much of our knowledge comes from studies of flames produced on flat, porous bed gas burners such as those used by McCaffrey (1979), Cox and Chitty (1980) and Zukoski (1981a): these are designed to give a low momentum source of fuel vapour. The relative importance of momentum (or inertia) and buoyancy in the flame will determine the type of fire, and the Froude number (Fr) may be used as a means of classification. It is a measure

An Introduction to Fire Dynamics, Third Edition. Dougal Drysdale.
© 2011 John Wiley & Sons, Ltd. Published 2011 by John Wiley & Sons, Ltd.

of the relative importance of inertia and buoyancy in the system, and is conveniently expressed as:

$$Fr = U^2/gD \qquad (4.1)$$

where U is the velocity of the gases, D is a characteristic dimension (normally taken as the diameter of the burner) and g is the acceleration due to gravity. Turbulent jet flames have high Froude numbers, based on the exit velocity of the fuel from a pipe or orifice. With natural fires, the initial velocity of the vapours in general cannot be measured, but can be derived from the rate of heat release[1] ($\dot{Q}_c$). Assuming a circular fuel bed of diameter D (area $\pi D^2/4$), fuel density ρ and heat of combustion of the fuel vapour of ΔH_c, the initial velocity of the fuel vapours can be expressed as:

$$U = \frac{\dot{Q}_c}{\Delta H_c \rho (\pi D^2/4)} \qquad (4.2)$$

Comparing the above two expressions, it can be seen that the Froude number is proportional to $\dot{Q}_c^2/D^5$, a scaling criterion that will be encountered below. (It was first identified by Thomas *et al.* (1961) using dimensional analysis.) Indeed, a 'dimensionless heat release rate' ($\dot{Q}_c^*$), introduced in the 1970s by Zukoski (1975) and others, is the square root of a Froude number expressed in terms of the heat release rate of a fire. It is used to classify

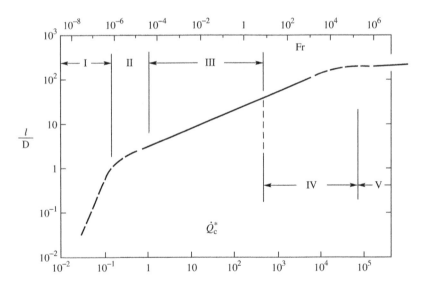

Figure 4.1 Schematic diagram showing flame length (l) as a function of the fuel flowrate parameters, expressed as $\dot{Q}_c^*$. The extreme right-hand region (V) corresponds to the fully turbulent jet fire (cf. Figure 4.7), dominated by the momentum of the fuel. Regions I and II correspond to buoyancy-driven turbulent diffusion flames (cf. Figure 4.8). Adapted from Zukoski (1986). Reprinted by permission

---

[1] The term 'rate of energy release' is more satisfactory in this context, but 'rate of heat release' has come to be the accepted terminology.

fire types and correlate aspects of fire behaviour (McCaffrey, 1995), such as flame height (Figures 4.1 and 4.17). It is given by:

$$\dot{Q}_c^* = \frac{\dot{Q}_c}{\rho_\infty c_p T_\infty \sqrt{gD} \cdot D^2}$$

(4.3)

Heskestad (1981) recommends the use of an alternative form of modified Froude number which takes into account the stoichiometry of the reaction (see Equation (4.39)).

4.1 Laminar Jet Flames

When a jet of gas issues into a still atmosphere, air is entrained as a result of shear forces between the jet and the surrounding air (cf. Section 4.3.1). The resulting flame will be laminar provided that the Reynolds number at the origin is less than ~2000. However, the shear forces cause instability in the gas flow which gives rise to flame flicker (Gaydon and Wolfhard, 1979). For hydrocarbon diffusion flames on a Bunsen burner, the flickering has a frequency of 10–15 Hz. This can be virtually eliminated if the surrounding air is made to move concurrently with and at the same linear velocity as the gas jet. Burke and Schumann (1928) chose to work with this arrangement in their classic study of laminar diffusion flames. They enclosed the burner tube inside a concentric cylinder carrying the flow of air: by varying the relative diameters of the tubes they were able to establish 'over-ventilated' and 'under-ventilated' flames as shown in Figure 4.2. These studies established that combustion occurred in the fuel/air mixing zone and suggested that the flame structure could be analysed on the assumption that the burning rate was controlled by the rate of mixing rather than by the chemical kinetics.

The rate of diffusion of one gas into another can be described by Fick's law (Incropera *et al.*, 2007; Welty *et al.*, 2008), which for one dimension is:

$$\dot{m}_i'' = -\mathcal{D}_i \frac{dC_i}{dx}$$

(4.4)

where $\dot{m}_i''$ and C_i are the mass flux and concentration of species i, respectively and $\mathcal{D}_i$ is the diffusion coefficient for species i in the particular gas mixture. It is analogous to Fourier's law of conductive heat transfer in which the heat flux is proportional to the temperature gradient, $\dot{q}_c'' = -k(dT/dx)$; here, mass flux is proportional to the concentration gradient. Transient mass transfer in three dimensions requires solution of the equation

$$\nabla^2 C_i = \frac{1}{\mathcal{D}_i}\left(\frac{\partial C_i}{\partial t}\right)$$

(4.5)

which can be compared with Equation (2.16). As with heat transfer problems, solution of this basic partial differential equation is made easier if the problem is reduced to a single space dimension. This is possible for the diffusion flames illustrated in Figure 4.2 since the model can be described in cylindrical coordinates, i.e.

$$\frac{\partial C(r,y)}{\partial t} = \mathcal{D}\left[\frac{\partial^2 C(r,y)}{\partial r^2} + \frac{1}{r} \cdot \frac{\partial C(r,y)}{\partial r}\right]$$

(4.6)

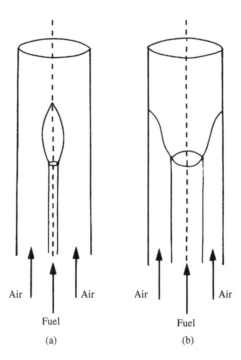

Air   Air   Air   Air

Fuel   Fuel

(a)   (b)

Figure 4.2   Burke and Schumann's study of the structure of diffusion flames: (a) over-ventilated and (b) under-ventilated flames

where the concentration $C(r, y)$ is a function of radial distance from the axis of symmetry (r) and height above the burner rim (y) (the subscript 'i' has been dropped for convenience). Normally this would be applied to the 'infinite cylinder' in which there would be no diffusion parallel to the axis, but here we have a flowing system in which time can be expressed as a distance travelled vertically (y) at a known velocity (u). Thus, as $t = y/u$, Equation (4.6) can be rewritten:

$$u\frac{\partial C(r, y)}{\partial y} = \mathcal{D}\left[\frac{\partial^2 C(r, y)}{\partial r^2} + \frac{1}{r} \cdot \frac{\partial C(r, y)}{\partial r}\right] \qquad (4.7)$$

The solution to this equation will give concentration (e.g., of fuel in air) as a function of height and radial distance from the burner axis (see Figure 4.3). (Axial diffusion will occur but is neglected in this approximate model.) Burke and Schumann (1928) suggested that the flame shape would be defined by the envelope corresponding to $C(r, y) = C_{\text{stoich}}$, where C_{stoich} is the stoichiometric concentration of fuel in air, but to obtain a solution to the above equation the following additional assumptions were necessary:

  (i) the reaction zone (i.e., where $C(r, y) = C_{\text{stoich}}$) is infinitesimally thin;
 (ii) rate of diffusion determines the rate of burning; and
(iii) the diffusion coefficient $\mathcal{D}$ is constant.

Assumptions (i) and (ii) are effectively equivalent and, combined with the basic assumption regarding flame shape, imply that reaction is virtually instantaneous wherever the

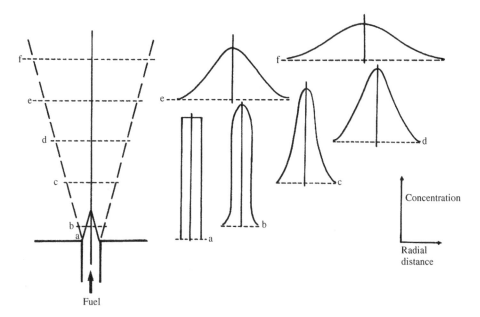

Figure 4.3 Concentration profiles in a jet emerging into an infinite quiescent atmosphere (after Kanury, 1975)

concentration is stoichiometric. This is a gross oversimplification as reaction will occur wherever the mixture is within the limits, which will be wide at the high temperatures encountered in flames (see Figure 3.4). Moreover, diffusion coefficients vary considerably with both temperature and composition of the gas mixture. Nevertheless, the resulting analytical solution to Equation (4.7) accounts very satisfactorily for the shapes of both over-ventilated and under-ventilated flames, as illustrated in Figure 4.4, thus establishing the validity of the proposed basic structure.

A much simpler model was developed by Jost (1939) in which the tip of a diffusion flame was defined as the point on the flame axis ($r = 0$) at which air is first found ($y = l$ in Figure 4.5). He used Einstein's diffusion equation, $x^2 = 2\mathcal{D}t$ (where x is the average distance travelled by a molecule in time t) to establish the (average) time it would take a molecule from the air to diffuse from the rim of the burner to its axis, i.e., $t = R^2/2\mathcal{D}$, where R is the radius of the burner mouth. Considering the concentric burner system of Burke and Schumann in which air and fuel are moving concurrently with a velocity u (and air is in excess, Figure 4.2(a)), in time t the gases will flow through a distance ut. Thus the height of the flame (l), according to the above definition, will be

$$l = \frac{uR^2}{2\mathcal{D}} \tag{4.8}$$

which, if expressed in terms of volumetric flowrate, $\dot{V} = \pi R^2 u$, gives:

$$l = \frac{\dot{V}}{2\pi\mathcal{D}} \tag{4.9}$$

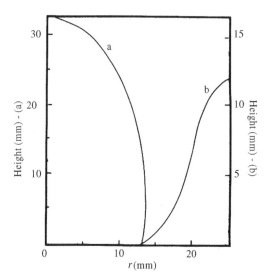

Figure 4.4 Shapes of (a) over-ventilated and (b) under-ventilated diffusion flames according to Equation (4.7), for $C(r, y) = C_{\text{stoich}}$. Reprinted from Burke and Schumann, *Ind. Eng. Chem.*, **20**, 998. Published 1928 American Chemical Society

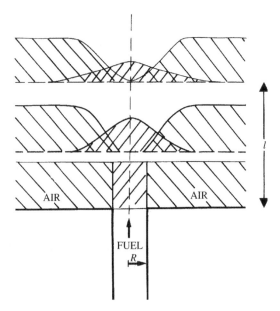

Figure 4.5 Jost's model of the diffusion flame (Jost, 1939)

This equation predicts that the flame height will be proportional to the volumetric flowrate and independent of the burner radius, of which the latter is essentially correct. However, buoyancy influences the height of the laminar flame and the dependence is closer to $\dot{V}^{0.5}$. The predicted inverse dependence on the diffusion coefficient is not observed strictly, but this is not unexpected as $\mathcal{D}$ varies considerably with temperature and with mixture composition. Furthermore, a change in the stoichiometry would be expected to alter the flame height, but this is not incorporated into Jost's model.

The limited success of these simple diffusion models indicates that the underlying assumptions are essentially correct. They can be expressed in a different format by identifying the tip of the flame with the height at which combustion is complete, implying that sufficient air is entrained through the jet boundary in the time interval $t = l/u$ to burn all the fuel issuing from the mouth of the burner during the same period. While this is a useful concept, it is an oversimplification, as will be seen below (Section 4.3.2).

Laminar jet flames are unsuitable for studying the detailed structure of the diffusion flame. Wolfhard and Parker developed a method of producing flat diffusion flames using a burner consisting of two contiguous slots, one carrying the fuel gas and the other carrying the oxidant (Figure 4.6(a)) (Gaydon and Wolfhard, 1979). Provided that there is

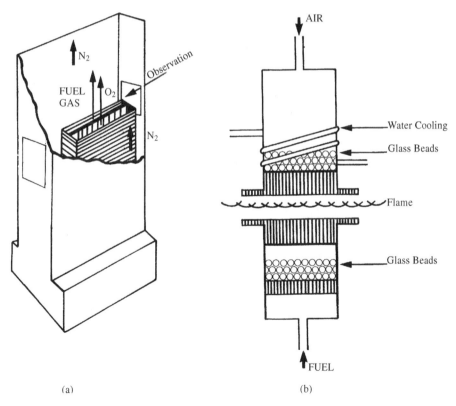

(a)   (b)

Figure 4.6 (a) The Wolfhard–Parker burner for producing flat diffusion flames. (b) The counterflow diffusion flame apparatus. Reproduced with permission from Gaydon and Wolfhard (1979)

a concurrent flow of nitrogen surrounding the burner, this arrangement yields a stable, vertical flame sheet on which various types of measurement can be made on both sides of the combustion zone. Using this device, valuable information has been obtained on the spatial concentrations of combustion intermediates (including free radicals) which has led to a better understanding of the chemical processes within the flame. This type of work has given an insight into the mechanism of smoke (or 'soot') formation in diffusion flames (Kent *et al.*, 1981). However, like the jet flames, this type of flat flame is affected by the presence of the burner rim. This can be avoided by using a counterflow diffusion flame burner in which a flat flame is stabilized in the stagnant layer where diametrically opposed flows of fuel and oxidant meet (Figure 4.6(b)) (see Gaydon and Wolfhard, 1979). This system has been widely used to examine the stability and extinction of diffusion flames of both gaseous and solid fuels (e.g., Williams, 1981, 2000).

4.2   Turbulent Jet Flames

In the previous section it was pointed out that the height of a jet flame will increase approximately as the square root of the volumetric flowrate of the fuel, but this is true only in the laminar regime. Above a certain jet velocity, turbulence begins, initially at the flame tip, and the flame height decreases with flowrate to a roughly constant value for the fully turbulent flame (Figure 4.7). This corresponds to high values of $\dot{Q}_c^*$ (Equation (4.3)). The transition from a laminar to a turbulent flame is observed to occur at a nozzle Reynolds number significantly greater than 2000 (Hottel and Hawthorne, 1949) as it is the local Reynolds number (Re = ux/v) within the flame which determines the onset of turbulence. Re decreases significantly with rise in temperature as a result of the variation in kinematic viscosity (v). Turbulence first appears at the tip of the flame, extending further down towards the burner nozzle as the jet velocity is increased, although never reaching it

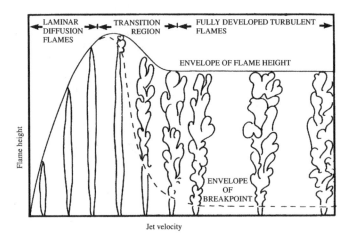

Figure 4.7   Height of momentum jet flames as a function of nozzle velocity, showing transition to turbulence (Hottel and Hawthorne, 1949). © 1949 Williams and Wilkins Co., Baltimore

(Figure 4.7). The decrease in flame height from the maximum inside the laminar region to a constant value in the fully turbulent regime can be understood qualitatively in terms of increased entrainment of air by eddy mixing, which results in more efficient combustion.

Hawthorne *et al.* (1949) derived the following expression theoretically, relating the turbulent flame height l_T to the diameter of the burner jet, d_i, the flame temperature T_F (K) (Table 4.1), the initial temperature T_i (K), and the average molecular weights of air (M_{air}) and the fuel issuing from the jet (M_f):

$$\frac{l_T}{d_i} = \frac{5.3}{C_f}\left[\frac{T_F}{mT_i}\left(C_f + (1 - C_f)\frac{M_{air}}{M_f}\right)\right]^{1/2} \tag{4.10}$$

where m is the molar ratio of reactants to products (both inclusive of nitrogen) for the stoichiometric mixture, and $C_f = (1 + r_i)/(1 + r)$ in which r is the stoichiometric molar air/fuel ratio and r_i is the initial air/fuel ratio, taking into account situations in which there is air in the initial fuel mixture.

This refers to the fully turbulent momentum jet flame in which buoyancy effects are neglected (high Froude number). It is in good agreement with measurements made on the turbulent flames for a range of gases (Lewis and von Elbe, 1987; Kanury, 1975) and shows that the flame height is linearly dependent on nozzle diameter, but independent of the volumetric flowrate. Because combustion is more efficient in these than in laminar diffusion flames, their emissivity tends to be less as a result of the lower yield of carbonaceous particles. The magnitude of the effect depends on the nature of the fuel: for methane and propane, c. 30% of the heat of combustion may be lost by radiation from a laminar diffusion flame, while this may be reduced to only 20% for a turbulent flame (Markstein, 1975, 1976; Delichatsios and Orloff, 1988). The effect is even greater for flames from fuels which have a greater tendency to produce soot, such as ethylene (ethene) and acetylene (ethyne) (Delichatsios and Orloff, 1988).

Table 4.1   Flame temperatures relevant to fully turbulent jet flames (Equation (4.10) (Lewis and von Elbe, 1987)

Fuel	Fuel concentration in air (%)	Flame temperature[a] (°C)
Hydrogen	31.6	2045
Methane	10.0	1875
Ethane	5.8	1895
Propane	4.2	1925
Butane	3.2	1895
Ethylene	7.0	1975
Propylene	4.5	1935
Acetylene	9.0	2325

[a]Determined by the sodium D-line reversal method (see Gaydon and Wolfhard, 1979). Valid for $T_0 = 20°C$.

4.3   Flames from Natural Fires[2]

Turbulent jet flames are associated with high Froude numbers, which correspond to values of $\dot{Q}_c^*$ of the order of 10^6 (Heskestad, 2008), indicating that the momentum of the fuel stream is dominating the behaviour. In natural fires, buoyancy is the predominant driving force, consistent with values of $\dot{Q}_c^*$ that are around six orders of magnitude lower (Figure 4.1). These flames have a much less ordered structure and are more susceptible to external influences (such as air movement) than jet flames. Corlett (1974) drew attention to the existence of a layer of pure fuel vapour above the centre of the surface of a burning liquid when the pool diameter was between 0.03 and 0.3 m (see also Bouhafid *et al.*, 1988). The flames from this size of fire are essentially laminar, but become increasingly turbulent as the diameter is increased. The turbulence aids mixing at low level, but the layer close to the surface will still be fuel rich. Flame shapes are illustrated in Figure 4.8, which shows a progression of increasing fire sizes, corresponding to *decreasing* values of $\dot{Q}_c^*$. Continuous flame cover over the fuel bed (as in Figure 4.8(d)) does not occur for large diameter fires, corresponding to very low values of $\dot{Q}_c^*$. Instead, discrete flames of reduced height (relative to the diameter of the fire) are observed (Figure 4.8(e)). This type of behaviour is typical of large area wildland fires (Heskestad, 1991) and can provide the conditions under which 'fire whirls' may form naturally. The photograph in Figure 4.9 shows a persistent fire whirl which formed during an experimental fire of low $\dot{Q}_c^*$ on

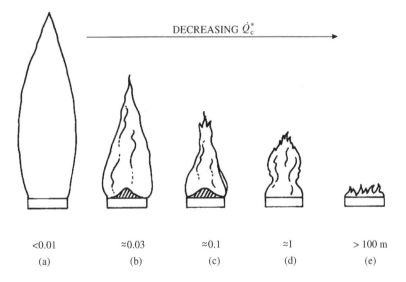

Figure 4.8 Classification of natural diffusion flames as 'structured' (b and c) and 'unstructured' (a, d and e) according to Corlett (1974). (e) flame would be classified as a 'mass fire', but can be modelled on a laboratory scale if $\dot{Q}_c$ is kept small (e.g., Heskestad, 1991). The shaded areas indicate fuel-rich cores

---

[2] In this section, the burning surface is horizontal. The situation in which the burning surface is vertical is discussed in Chapter 7, in the context of flame spread.

Figure 4.9 A fire whirl above a 0.3 m ×0.3 m fire of low $\dot{Q}^*$. This photograph was taken during one of the tests carried out at the Fire Research Station, Borehamwood, as part of a study on the scaling of wildland fires (Thomas *et al.*, 1968)

a 10 ft (0.3 m) square continuous fuel bed (Thomas *et al.*, 1968). Fire whirls were first investigated experimentally by Emmons and Ying (1966), who identified them with the concentration of vorticity in rising, hot gases, similar to the formation of tornadoes. They are associated with very intense combustion and can contribute to the dispersion of fire-brands when they form in wildfires. There have been several studies over the years (e.g., Hassan *et al.*, 2005), but in the opinion of Chuah *et al.* (2009), our understanding of the mechanism is 'still incomplete'.

A related but somewhat different phenomenon has been observed for very large mass fires, with areas of the order of a square kilometre or more – the so-called 'firestorm'. Firestorms developed after some of the major bombing raids during the Second World War (Pitts, 1991), and there was clear evidence that one occurred during the fire that followed the ignition of a large release of propane into a heavily forested area near Ufa, Russia, in June 1989 (Makhviladze and Yakush, 2002). The power of a firestorm is so great that it will create winds of hurricane strength, giving rise to complete destruction within the confines of the burning area. Evidence suggests that a firestorm has a rotary motion, but will not form if there is a significant wind capable of disrupting the natural airflow into the fire (Pitts, 1991).

Porous bed gas burners have been used by several authors to study diffusion flames of the type illustrated by Figure 4.8(b) (Corlett, 1968, 1970; Chitty and Cox, 1979; McCaffrey, 1979; Zukoski *et al.*, 1981a,b; Cetegen *et al.*, 1984; Hasemi and Tokunaga, 1984; Cox and Chitty, 1985; and others). The system has the advantage over fires involving combustible solids and liquids in that the fuel flowrate is an independent variable and the flame can be maintained indefinitely for experimental purposes (e.g., Smith and Cox,

1992). McCaffrey (1979) showed that the 'fire plume' above a 30 cm square burner consisted of three distinct regimes (see Figure 4.10), namely:

(i) The near field, above the burner surface, where there is persistent flame and an accelerating flow of burning gases (the flame zone).
(ii) A region in which there is intermittent flaming and a near-constant flow velocity (the intermittent zone).
(iii) The buoyant plume, which is characterized by decreasing velocity and temperature with height.

While these are inseparable in the fire plume, it is appropriate to consider the buoyant plume on its own since its properties are relevant to other aspects of fire engineering, including fire detection (Section 4.4.2) and smoke movement and control (Sections 11.2 and 11.3). In the next two sections, we shall be discussing the unbounded plume and interactions with ceilings and walls will be considered in Section 4.3.4. The fire plume associated with burning on a vertical surface (i.e., a wall fire) will be considered in Section 4.3.3.1.

4.3.1 The Buoyant Plume

The concept of buoyancy was introduced in Section 2.3 in relation to natural convection. If a density difference exists between adjacent masses of fluid as a result of a temperature gradient, then the force of buoyancy will cause the less dense fluid to rise with respect to its surroundings. The buoyancy force (per unit volume), which is given by $g(\rho_\infty - \rho)$, is resisted by viscous drag within the fluid, the relative magnitude of these opposing forces being expressed as the Grashof number (Equation (2.49)). The term 'buoyant plume' is used to describe the convective column rising above a source of heat. Its structure is determined by its interaction with the surrounding fluid. Intuitively, one would expect the temperature within the plume to depend on the source strength (i.e., the rate of heat release) and the height above the source: this may be confirmed by theoretical analysis.

The mathematical model of the simple buoyant plume is based on a point source as shown in Figure 4.11(a) (Yih, 1952; Morton et al., 1956; Thomas et al., 1963; Heskestad, 1972; Williams, 1982; Heskestad, 2008). The ideal plume in an infinite, quiescent atmosphere would be axisymmetric and extend vertically to a height where the buoyancy force has become too weak to overcome the viscous drag. Under certain atmospheric conditions, a temperature inversion can form that will effectively trap a rising smoke plume, halting its vertical movement and causing it to spread laterally at that level. The same effect can be observed in confined spaces: the commonest example is the stratification of cigarette smoke at relatively low levels in a warm room, under quiescent conditions. In high spaces, such as atria, the temperature at roof level may be sufficient to prevent smoke from a fire at ground level reaching smoke detectors at the ceiling. Heskestad (1989) has reviewed the work of Morton et al. (1956) and shown how their original expressions may be adapted to give the minimum rate of (convected) heat release $\dot{Q}_{conv,min}$ (kW) necessary to ensure that the plume reaches the ceiling (height H (m)):

$$\dot{Q}_{conv,min} = 1.06 \times 10^{-3} H^{5/2} \Delta T_a^{3/2} \tag{4.11}$$

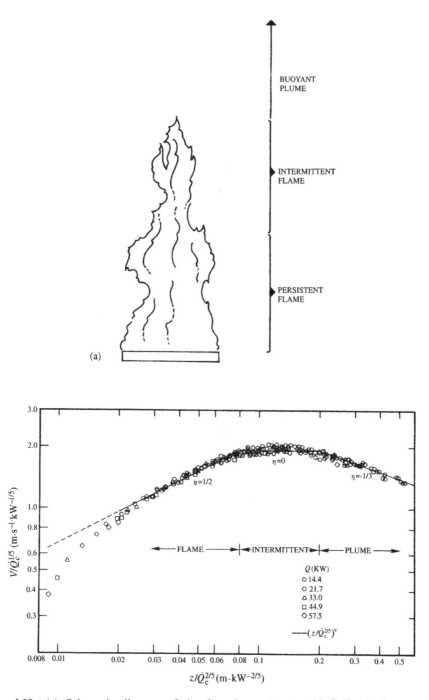

Figure 4.10 (a) Schematic diagram of the fire plume showing McCaffrey's three regimes. (b) Variation of upward velocity (V) with height (z) above the burner surface, plotted as $V/\dot{Q}_c^{1/5}$ versus $z/\dot{Q}_c^{2/5}$ (Table 4.2), where $\dot{Q}_c$ is the nominal rate of heat release (kW) (McCaffrey, 1979)

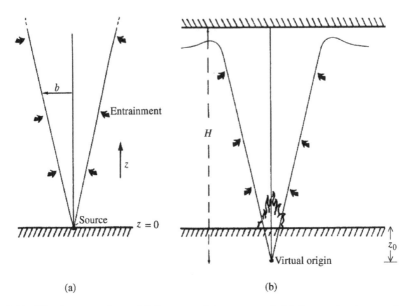

Figure 4.11 The buoyant plume (a) from a point source and (b) from a 'real source', showing interaction with a ceiling

where T_a (the ambient temperature in K) is assumed to increase linearly with height, ΔT_a being the increase in ambient temperature between the level of the fire source and the ceiling. As an example, Heskestad uses this equation to illustrate that in a 50 m high atrium with $\Delta T_a = 5$ K, a minimum convective heat output of 210 kW will be required before smoke will reach a detector mounted on the ceiling.

Cooling of the plume occurs as a result of dilution with ambient air which is entrained through the plume boundary. The decrease in temperature with height is accompanied by broadening of the plume and a reduction in the upward flow velocity. The structure of the plume may be derived theoretically through the conservation equations for mass, momentum and energy, but a complete analytical solution is not possible: simplifying assumptions have to be made. It is likely that detailed solutions would develop Gaussian-like radial distributions of excess temperature (ΔT), density deficit ($\Delta \rho$) and upward velocity (u) through horizontal sections of the plume as a function of height. Morton et al. (1956) and others (see Zukoski, 1995) assumed Gaussian distributions and self-similarity of the radial profiles, but $\Delta \rho$ and ΔT cannot be self-similar unless the plume is 'weak', i.e., $T_0/T \approx 1$. This is clear from the following relationship derived from the ideal gas law:

$$\frac{\Delta \rho}{\Delta \rho_0} = \left(\frac{\Delta T}{\Delta T_0}\right)\left(\frac{T_0}{T}\right) \tag{4.12}$$

This assumption ($T_0/T \approx 1$) is known as the Boussinesq approximation. It allows density differences to be ignored, except in the buoyancy term (see, for example, Quintiere, 2006). Self-similarity cannot hold for 'strong' plumes for these variables. Consequently, it is normal to assume self-similarity either between u and $\Delta\rho$ (Yih, 1952; Morton *et al.*, 1956; Thomas *et al.*, 1963) or between u and ΔT (Zukoski *et al.*, 1981a; Cetegen *et al.*, 1984).

For the present argument,[3] a simpler approach is appropriate in which 'top hat profiles' are assumed, i.e., T, ρ and u are assumed constant across the plume (radius b) at any specified height (z) (Morton *et al.*, 1956; Heskestad, 1972 and 2008).[4] Starting with relationships derived from the conservation equations, a simple dimensional analysis may be applied to obtain the functional relationships between temperature and upward flow velocity on the one hand and source strength and height on the other. For conservation of momentum, the following proportionality may be written for an axisymmetric plume (of radius b at height z above a point source) in an infinite atmosphere (density ρ_∞) if viscous forces are neglected and temperature differences are small:

$$\frac{d}{dz}(\rho_0 u_0^2 b^2) \propto g(\rho_0 - \rho_\infty)b^2 \tag{4.13}$$

where u_0 and ρ_0 are the vertical flow velocity and density on the plume axis at height z above the point source (Figure 4.11(a)). Similarly, for the conservation of mass:

$$\frac{d}{dz}(\rho_0 u_0 b^2) \propto \rho v b \propto \rho u_0 b \tag{4.14}$$

in which the increase in mass flow with height is due to entrainment of air through the plume boundary. The entrainment velocity (v) is assumed to be directly proportional to u_0, i.e., $v = \alpha' u_0$, where α' is the entrainment constant which Morton *et al.* (1956) estimated to be about 0.09 for still air conditions. Any wind, or other air movement, will deflect the plume and effectively increase the entrainment constant (see also Section 4.3.5).

Finally, the conservation of energy may be represented by the following:

$$c_p \rho_0 u_0 b^2 \Delta T_0 \propto \dot{Q}_{conv} \tag{4.15}$$

where ΔT_0 is the temperature excess over ambient on the axis at height z and $\dot{Q}_{conv}$ is the (convective) heat output from the source, i.e., the source strength. Radiative heat losses from the plume rising from a pure heat source are assumed to be negligible.

Heskestad (1972, 1975) assumed that the variables b, u_0 and ΔT_0 are directly proportional to simple powers of z, the height, i.e.

$$b \propto z^s; \quad u_0 \propto z^m \text{ and } \Delta T_0 \propto z^n \tag{4.16}$$

By substituting these three relationships into Equations (4.13), (4.14) and (4.15) and solving for s, m and n, assuming consistency of units, it can be shown that:

$$b \propto z \tag{4.17}$$

$$u_0 \propto A^{1/3} \dot{Q}_{conv}^{1/3} \, z^{-1/3} \tag{4.18}$$

$$\Delta T_0 \propto (A^{2/3} T_\infty / g) \dot{Q}_{conv}^{2/3} \, z^{-5/3} \tag{4.19}$$

---

[3] A more rigorous treatment is given by Zukoski (1995) and Quintiere (2006).

[4] This assumption applies to all the equations that follow, although terminology is used to emphasize that in reality T_0 and u_0 will be at their maximum values on the centreline of the plume.

where $A = g/c_p T_\infty \rho_\infty$ and T_∞ is the ambient air temperature (see also Heskestad, 2008). Applying the buoyant plume model of Morton *et al.* (1956), Zukoski *et al.* (1981a) and Cetegen *et al.* (1984) developed relationships which are in close agreement with Equations (4.17)–(4.19), using self-similarity between velocity (u) and temperature excess (ΔT) profiles. Using a dimensionless heat release rate $\dot{Q}_Z^*$ of the form

$$\dot{Q}_Z^* = \frac{\dot{Q}_c}{\rho_\infty C_p T_\infty Z^2 \sqrt{gZ}} \tag{4.20}$$

(where Z is a characteristic length[5]) they obtained the following expressions:

$$b = C_l z \tag{4.21}$$

$$u_0 = C_v (gz)^{1/2} \dot{Q}_Z^{*1/3} \tag{4.22}$$

$$\Delta T_0 = C_T \dot{Q}_Z^{*2/3} T_\infty \tag{4.23}$$

in which the constants C_l, C_v and C_T were derived from data of Yokoi (1960). These relationships have been reviewed (Beyler, 1986b) and compared with experimental data which have become available on the rate of entrainment into fire plumes (e.g., Cetegen *et al.*, 1984). Beyler recommends the following expression for the centreline temperature rise at height z, assuming that $T_\infty = 293$ K:

$$\Delta T_0 = 26 \frac{\dot{Q}_{conv}^{2/3}}{z^{5/3}} \tag{4.24}$$

where $\dot{Q}_{conv}$ is the rate of heat release corrected for radiative loss (i.e., $\dot{Q}_c(1 - \chi_R)$ and χ_R is the fraction of the total heat released that is lost by radiation. This is normally taken as 0.3 for flames (see Table 5.13), although Quintiere and Grove (1998) have shown that it is a significant parameter and the correct value should be used for each scenario. If it is not possible to correct for radiative losses, then Equation (4.25) may be used:

$$\Delta T_0 = 22 \frac{\dot{Q}_c^{2/3}}{z^{5/3}} \tag{4.25}$$

The above correlations hold remarkably well in the 'far field' of a fire plume (i.e., in the buoyant plume above the flames), particularly where the plume is 'weak' in the sense that the ratio $\Delta T/T_\infty$ is small. However, for a fire it is necessary to introduce a correction for the finite area of the source by identifying the location of a 'virtual origin' (Figure 4.11(b)), defined as the equivalent point source which produces a (buoyant) plume of identical entrainment characteristics to the real plume.

In early work, it was assumed that for a real fire, the virtual origin would lie approximately $z_0 = 1.5 A_f^{1/2}$ m below a fuel bed of area A_f. This was based on the

---

[5] The value chosen for Z depends on the problem. For pool, or pool-like fires, it is taken as the effective diameter. When dealing with a point source, it can be taken as the natural length scale $Z = \left(\frac{\dot{Q}_c}{\rho_\infty c_p T_\infty \sqrt{g}}\right)$ (see Quintiere, (2006) for the derivation).

assumption that the plume spreads with an angle of c. 15° to the vertical (Figure 4.11(b)) (Morton *et al.*, 1956; Thomas *et al.*, 1963). However, studies by Heskestad (1983b), Cetegen *et al.*, (1984), Cox and Chitty (1985) and others (Zukoski, 1995) have shown that the location of the virtual source is dependent on the rate of heat release, as well as the diameter (or area) of the fire. Heskestad (1983b, 2008) recommends the use of the formula

$$\frac{z_0}{D} = -1.02 + 0.083 \frac{\dot{Q}_{\text{conv}}^{2/5}}{D} \tag{4.26}$$

which is based on data from various sources on pool fires of diameters in the range 0.16–2.4 m. It gives a good mean of the other correlations (Figure 4.12), although others have been proposed (e.g., Zukoski (1995) presents a correlation based on flame height rather than heat release rate). In Equations (4.24) and (4.25) (for example), 'the height above the point source' should be replaced by $(z - z_0)$, where z remains the height above the fuel surface.

By implication, the above discussion refers to flat fuel beds (e.g., pool fires, gas burners, etc., on which the correlations are based), and Equation (4.26) will not apply if the 'fire' is three-dimensional, in that the fuel bed has a significant vertical extension (e.g., with wood cribs and fires in multi-tier storage arrays). This is discussed by You and Kung (1985), who quote correlations for two-, three- and four-tier storage (see also Heskestad, 2008).

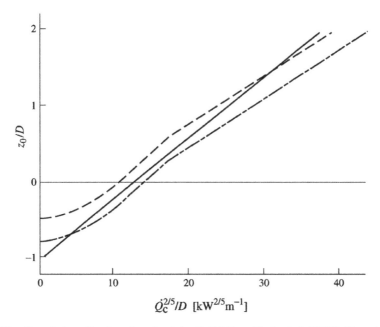

Figure 4.12 Correlations for the virtual origin. Solid line: Heskestad (1986) (Equation (4.26)); dashed line and dotted line: Cetegen *et al.* (1984) with and without a flush floor, respectively. After Heskestad (2008) by permission of the Society of Fire Protection Engineers

The proportionalities indicated in Equations (4.17)–(4.19) provide the basis for the scaling laws which can be used to correlate data and compare behaviour in situations according to the principles of similarity (see Section 4.4.4). As a simple example, consider the temperature at a height H_1 directly above a source of convective heat output $\dot{Q}_{conv1}$. The same temperature will exist at a height H_2 on the centreline of the buoyant plume from a similar source of heat output $\dot{Q}_{conv2}$, provided that:

$$\dot{Q}_{conv2} = \dot{Q}_{conv1} \left(\frac{H_2'}{H_1'}\right)^{5/2} \tag{4.27}$$

where H_1' and H_2' refer to the heights above the respective virtual origins (Figure 4.11(b)). In this way, the product $\dot{Q}_{conv}^{2/3} z^{-5/3}$ in Equation (4.19) ($z = H'$) is constant. Specific examples of the application of this type of analysis are given in Sections 4.3.4 and 4.4.3. In principle, a similar relationship should hold for the concentration of smoke particles. Heskestad (1972) quotes

$$C_0 \propto A^{-1/3} \dot{m} \dot{Q}_{conv}^{-1/3} z^{-5/3} \tag{4.28}$$

where C_0 is the centreline concentration of combustion products and $\dot{m}$ is the rate of burning, expressed as a mass flow. However, as $\dot{Q}_{conv} \propto \dot{m}$ for a given fuel, Equation (4.28) can be rewritten:

$$C_0 \propto A^{-1/3} \dot{Q}_{conv}^{2/3} z^{-5/3} \tag{4.29}$$

showing that the concentration of smoke follows ΔT (compare Equations (4.19) and (4.29)), i.e., if the term $\dot{Q}_{conv}^{2/3} H^{-5/3}$ is maintained constant, the concentration of smoke particles will be the same (for a given fuel bed). This is relevant to the operation of smoke detectors in geometrically similar locations of different heights.

This section has concentrated on axisymmetric plumes from square or circular sources, but other geometries are encountered. Yokoi (1960) and Hasemi (1988) have considered the plumes arising from sources which have length/breadth ratios significantly greater than one (see Section 4.3.2). In general, in the far field, rectangular sources can be approximated by a virtual point source, but the extreme situation – the 'line source' – has relevance to certain problems. For these, Zukoski (1995) has shown how the heat release rate can be specified per unit length and has derived a set of equations similar to (4.21)–(4.23) using a modified version of the heat release group $\dot{Q}_l^*$:

$$\dot{Q}_l^* = \frac{\dot{Q}_c/L}{\rho_\infty c_p T_\infty (gZ)^{1/2} Z} \tag{4.30}$$

where $\dot{Q}_c/L$ is the rate of heat release per unit length of source and Z could be taken as the width of the source (see Equation (4.43)). There are few data available to enable correlations to be tested, yet the behaviour of the line plume is important in defining the entrainment characteristics of a 'spill' plume emerging from a shop at ground level under a balcony and into an atrium or multi-storey shopping centre. The volume of smoke that has to be removed by the smoke control system is determined by the rate of entrainment into the plume: this has to be calculated for designing a smoke control system (see Chapter 11) (Thomas et al., 1998; CIBSE, 2003).

4.3.2 The Fire Plume

The subdivision of the fire plume into three regions was discussed briefly in the introduction to Section 4.3. Flame exists in the near field and the intermittent zone, although it is persistent only in the former. This is illustrated by results of Chitty and Cox (1979), who mapped out regimes of 'equal combustion intensity' throughout a methane diffusion flame above a 0.3 m square porous burner. Using an electrostatic probe, they determined the fraction of time that flame was present at different locations within the fire plume and found that the most intense combustion (defined as flame being present for more than 50% of the time) occurs in the lower region, particularly near the edge of the burner (Figure 4.13). Bouhafid *et al*. (1988) report contours of temperature and concentrations

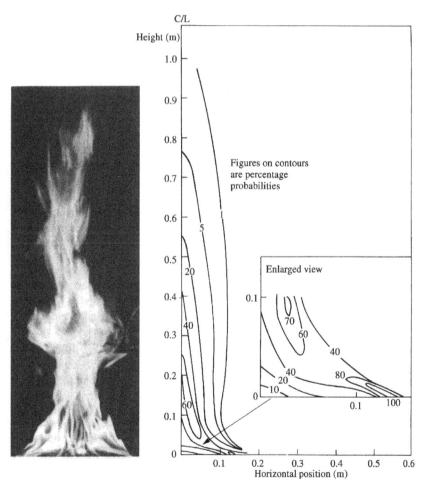

Figure 4.13 Intensity of combustion within a buoyant diffusion flame, shown as probability contours and compared with a typical instantaneous photograph of the flame. 0.3 m square porous burner, $\dot{Q}_c = 47$ kW. Visual flame height 1.0–1.2 m (Chitty and Cox, 1979). Reproduced by permission of The Controller, HMSO. © Crown copyright

of CO, CO_2 and O_2 near the base of the flames above a kerosene pool fire, 0.15 m in diameter, which show similarities to Chitty and Cox's map of combustion intensity (Figure 4.13). The low probability recorded immediately above the central area of the burner, or pool, is consistent with the presence of a cool fuel-rich zone above the fuel surface on which Corlett (1974) commented (see Figure 4.8).

Visual estimates of average flame height are 10–15% greater than the vertical distance on the flame axis to the point where flame intermittency is 50%, as determined photo-graphically (Zukoski *et al.*, 1981a,b). The motion of the intermittent (oscillating) flames occupies a considerable proportion of the fire plume (Figures 4.13 and 4.14) and is quite regular, exhibiting a frequency (f) which is a function of $D^{-1/2}$, where D is the fire diameter (Figure 4.15). Zukoski (1995) suggests

$$f = (0.50 \pm 0.04)(g/D)^{1/2} \text{ Hz} \tag{4.31}$$

which is essentially in agreement with observations made by Pagni (1990) and Hamins *et al.* (1992), as well as the correlation derived by Malalakesera *et al.* (1996) in their review of the pulsation of buoyant diffusion flames.

The phenomenon is illustrated in Figure 4.16, which shows 1.3 s of a cine film of the flame on the 0.3 m square gas burner used by McCaffrey (1979) and Chitty and Cox (1979). The oscillation frequency is 3 Hz, similar to that observed by Rasbash *et al.*

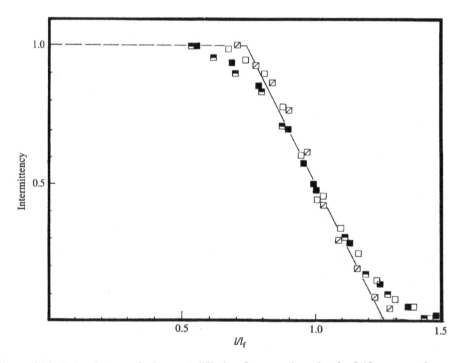

Figure 4.14 Intermittency of a buoyant diffusion flame on the axis of a 0.19 m porous burner. ■, $\dot{Q}_c = 21.1$ kW, $l_f = 0.65$ m; ▣, $\dot{Q}_c = 42.2$ kW; $l_f = 0.90$ m; □, $\dot{Q}_c = 63.3$ kW, $l_f = 1.05$ m; ◪, $Q_c = 84.4$ kW, $l_f = 1.16$ m (Zukoski *et al.*, 1981b)

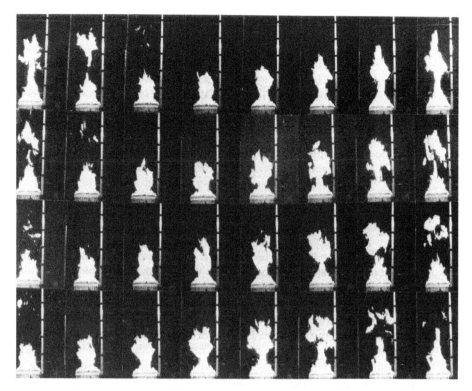

Figure 4.15 Intermittency of a buoyant diffusion flame burning on a 0.3 m porous burner. The sequence represents 1.3 s of cine film, showing 3 Hz oscillation (McCaffrey, 1979)

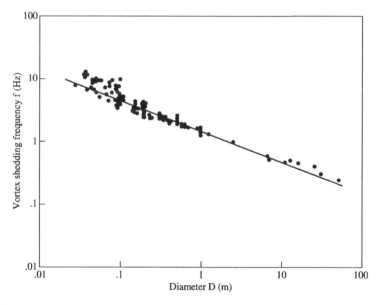

Figure 4.16 Variation of flame oscillation frequency with fire diameter for pool and gas burner fires (Pagni, 1990). From Cox (1995), with permission

Figure 4.17 Schematic diagram of the axisymmetric vortex-like structures in the buoyant diffusion flame. After Zukoski *et al*. (1981a), by permission

(1956) for a 0.3 m diameter petrol fire (Section 5.1.1). The oscillations are generated by instabilities at the boundary layer between the fire plume and the surrounding air although they have their origins low in the flame, close to the surface of the fuel (Weckman and Sobiesiak, 1988). These give rise to disturbances, the largest taking the form of axisymmetric vortex-like structures, or eddies (Figure 4.17). Zukoski *et al*. (1981a,b) have suggested that these play a significant part in determining the rate of air entrainment into the flame. The observed oscillations are a result of these structures rising upwards through the fire plume and burning out, thus exposing the upper boundary of the next vortex structure which becomes the new flame tip. This is referred to as 'eddy shedding' and, for small fires, is responsible for the characteristic 'flicker' that may be used to distinguish infra-red emission from a flame and that from a steady background source (Bryan, 1974; Middleton, 1983).

4.3.2.1 Flame Heights

It is sometimes necessary to know the size of a flame above a burning fuel bed, as this will determine how the flame will interact with its surroundings, in particular whether it will reach the ceiling of a compartment or provide sufficient radiant heat to ignite nearby combustible items. The basic parameters which determine height were first derived by Thomas *et al*. (1961), who applied dimensional analysis to the problem of the free-burning fire, i.e., one in which 'the pyrolysis rate and energy release rate are affected only by the burning of the fuel itself and not by the room environment' (Walton and Thomas, 2008). They assumed that buoyancy was the driving force and that air for combustion of the fuel volatiles was entrained through the flame envelope. The tip of the flame was defined as the height at which sufficient air had entered the flame to burn the volatiles, and the following functional relationship was derived:

$$\frac{l}{D} = f\left(\frac{\dot{m}^2}{\rho^2 g D^5 \beta \Delta T}\right) \tag{4.32}$$

in which l is the flame height above the fuel surface, D is the diameter of the fuel bed, $\dot{m}$ and ρ are the mass flowrate and density of the fuel vapour, ΔT is the average excess

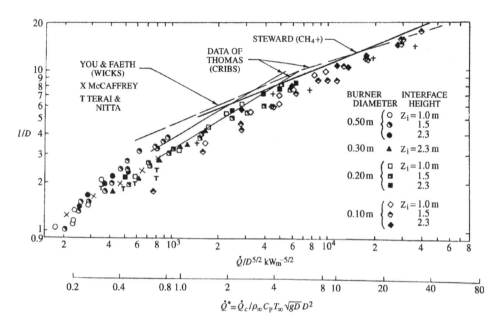

Figure 4.18 Dependence of flame height on heat release parameters (Zukoski *et al.*, 1981a). 'Interface height' refers to the vertical distance from the fire source to the lower boundary of the ceiling layer. Additional data of Thomas *et al.* (1961), Steward (1970), Terai and Nitta (1975), McCaffrey (1979) and You and Faeth (1979). By permission

temperature of the flame, and g and β are the acceleration due to gravity and the expansion coefficient of air, respectively. The group $g\beta\Delta T$ is indicative of the importance of buoyancy, which is introduced into the analysis in terms of the Grashof number (Equation (2.49)). The dimensionless group in Equation (4.32) is a Froude number and may be compared with $\dot{Q}^*$ (Equation (4.3)). It contains the elements of Froude modelling (Section 4.4.5), in which the rate of heat release must scale with $D^{5/2}$.

Data on flame heights can be correlated by using either $(\dot{Q}_c/D^{5/2})$, which has dimensions kW/m$^{5/2}$, or the dimensionless group $\dot{Q}^*$, defined above (Equation (4.3)). An example is shown in Figure 4.18, in which the data obtained by Zukoski *et al.* (1981a) – with three gas burners of different diameters – are compared with data (Terai and Nitta, 1975; McCaffrey, 1979) and correlations of others (Thomas *et al.*, 1961; Steward, 1970; You and Faeth, 1979). The flame heights (l) are normalized against the diameter of the fuel bed, or burner, and $\log(l/D)$ plotted against $\log(\dot{Q}_c/D^{5/2})$ [6]. Figure 4.18 also shows $\log(l/D)$ as a function of $\log(\dot{Q}^*)$. For large values of l/D (>6), the slope of the line is 2/5, indicating that the flame height is virtually independent of the diameter of the burner, or fuel bed, i.e.

$$\frac{l}{D} \propto \left(\frac{\dot{Q}_c}{D^{5/2}}\right)^{2/5} \propto \frac{\dot{Q}_c^{2/5}}{D} \tag{4.33}$$

---

[6] Logically, $\dot{Q}_{conv}$ should be used in these correlations, but it is not clear if it has been used consistently.

giving (from the data on visible flame heights of Zukoski et al. (1981a)):

$$l = 0.23\dot{Q}_c^{2/5} \text{ m} \tag{4.34}$$

if $\dot{Q}_c$ is in kW. This corresponds to values of $\dot{Q}^*$ greater than c. 5.

However, Thomas et al. (1961) found their data for wood crib fires to give values of l/D between 3 and 10, which correlated as follows:

$$\frac{l}{D} \propto \left(\frac{\dot{Q}_c}{D^{5/2}}\right)^{0.61} \tag{4.35}$$

or

$$l \propto \frac{\dot{Q}_c^{0.61}}{D^{0.5}} \tag{4.36}$$

corresponding approximately to a two-thirds power law in the range $0.5 < \dot{Q}^* < 7$. However, for $l/D < 2$, the relationship between l/D and $(Q_c/D^{5/2})$ appeared to be almost linear ($\dot{Q}^* < 0.5$). There is now strong evidence that the slope is changing rapidly in this range of values of $\dot{Q}^*$, and a square law is more appropriate when $\dot{Q}^*$ falls below 0.2 (Figure 4.19) (Zukoski, 1985):

$$\frac{l}{D} \propto \left(\frac{\dot{Q}_c}{D^{5/2}}\right)^2 \propto \dot{Q}^{*2} \tag{4.37}$$

Zukoski (1985) draws attention to this fact, and identifies several regimes in the relationship of l/D to $\dot{Q}^*$ in which different power laws apply. These are summarized in Figure 4.1. Most of the fires of interest in the context of buildings are identified with $\dot{Q}^* < 5$, but in the chemical and process industries, a much wider range is encountered, from large-scale pool fires ($\dot{Q}^* < 1$) to fully turbulent jet fires with $\dot{Q}^* > 10^3$.

In general, the above is in essential agreement with results of McCaffrey (1979) and Thomas et al. (1961). Steward (1970) obtained a substantial amount of data on turbulent diffusion flames and carried out a fundamental analysis of the flame structure based on the conservation equations. One interesting conclusion that he derived from the study was that within its height the momentum jet diffusion flame entrains a much greater quantity of air (400% excess) than is required to burn the fuel gases. For the buoyant diffusion flame, Heskestad (1983a) has correlated data from a wide variety of sources, including pool fires (Section 5.1), using the equation

$$\frac{l}{D} = 15.6N^{1/5} - 1.02 \tag{4.38}$$

in which the non-dimensional number N is derived from a modified Froude number (Heskestad, 1981) and is given by:

$$N = \left(\frac{c_p T_\infty}{g\rho_\infty^2(\Delta H_c/r)^3}\right) \frac{\dot{Q}_c^2}{D^5} \tag{4.39}$$

where c_p is the specific heat of air, ρ_∞ and T_∞ are the ambient air density and temperature, respectively, ΔH_c is the heat of combustion and r is the stoichiometric ratio of air to

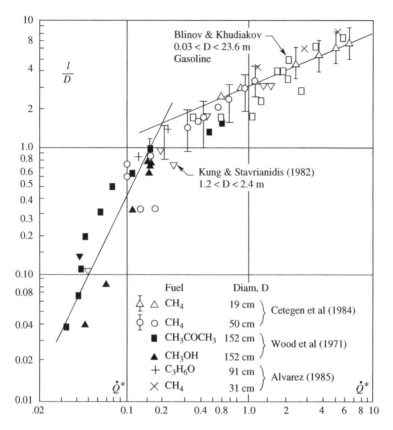

Figure 4.19 Correlation of l/D with $\dot{Q}^*$ for small values of $\dot{Q}^*$. A range of fuels are represented here, and include methane (Cetegen *et al.*, 1984), methanol and acetone (Wood *et al.*, 1971) and gasoline (Blinov and Khudiakov, 1957). Kung and Stavrianidis (1982) used four different fuels: methanol, hydrocarbon and silicone transformer fluids and heptane. Adapted from Zukoski (1986), by permission

volatiles. Given that most of the terms in Equation (4.39) are known ($\Delta H_c/r \approx 3000$ kJ/kg, see Section 1.2.3), Equation (4.38) can be rewritten:

$$l = 0.23\dot{Q}_c^{2/5} - 1.02D \qquad (4.40)$$

$\dot{Q}_c$ in kW, and l and D in m. The correlation is very satisfactory (Figure 4.20), although it has not been tested outside the range $7 < \dot{Q}_c^{2/5}/D < 700$ kW$^{2/5}$/m. It can be expressed in terms of $\dot{Q}^*$ as follows ($0.12 < \dot{Q}^* < 1.2 \times 10^4$) (McCaffrey, 1995):

$$l/D = 3.7\dot{Q}^{*2/5} - 1.02 \qquad (4.41)$$

It captures the change of slope which occurs around $\dot{Q}^* = 1$ (see Figures 4.1, 4.18 and 4.19), but breaks down for $\dot{Q}^* < 0.2$, corresponding to flames with $l/D < 1$ (Figure 4.19). In very low Froude number fires $\dot{Q}^* \leq 0.01$, such as large mass fires (e.g.,

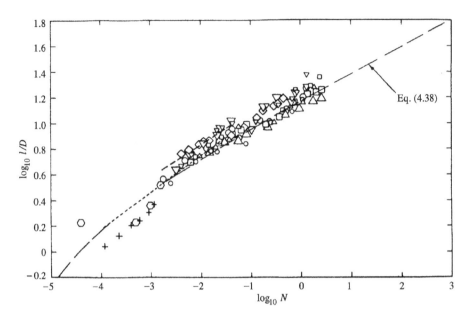

Figure 4.20 Correlation of flame height data from measurements by Vienneau (1964). (○, methane; o, methane + nitrogen; ▽, ethylene; ▿, ethylene + nitrogen; □, propane; ▫, propane + nitrogen; △, butane; ▵, butane + nitrogen; ◇ hydrogen): D'Sousa and McGuire (1977) (_ _ _ _ _, natural gas); Blinov and Khudiakov (1957) (○, gasoline); Hägglund and Persson (1976b) (+, JP-4 fuel); and Block (1970) (_ _ _ _, Equation 4.38)). From Heskestad (1983a), by permission

Figure 4.8(e), $D = O(>100\,\text{m})$ (Corlett, 1974)); the flame envelope breaks up and a number of separate, distinct 'flamelets' are formed (see Heskestad, 1991; also Figure 4.9)). The heights of these flames are much less than the fuel bed diameter (Zukoski, 1995).

The above discussion refers to axisymmetric fire plumes, where the burner/burning surface is square, or circular. There are very limited data on the behaviour of flames from surfaces of other shapes, for example rectangular sources with one side significantly longer than the other. This has been studied by Hasemi and Nishihata (1989), who found that the flame height data could be correlated with a modified $\dot{Q}^*$, given by:

$$\dot{Q}^*_{\text{mod}} = \frac{\dot{Q}_c}{\rho_0 c_p T_0 g^{1/2} A^{3/2} B} \tag{4.42}$$

where A and B are the lengths of the shorter and the longer sides of the rectangular fuel bed, respectively. When $A = B$, this becomes identical to the original definition of $\dot{Q}^*$ (Equation (4.3)), while for the line fire ($B \rightarrow \infty$), expressing the rate of heat release in terms of unit length of burner (or fire) gives:

$$\dot{Q}^*_l = \frac{\dot{Q}_l}{\rho_0 c_p T_0 g^{1/2} A^{3/2}} \tag{4.43}$$

where $\dot{Q}_l$ is the rate of heat per unit length (kW/m). (The value of B in Equation (4.42) becomes 1 m.) Hasemi and Nishihata's flame height correlation is shown in Figure 4.21.

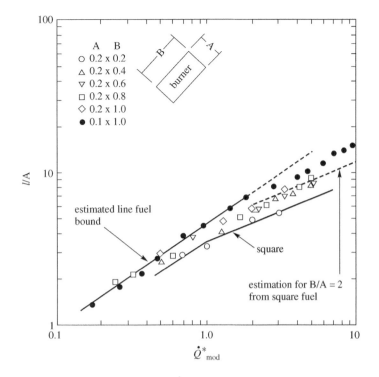

Figure 4.21 Relationship between l/A and $\dot{Q}^*_{mod}$ (Equation (4.42)), from Hasemi and Nishihata (1989). Reprinted by permission

This reveals that flame height is a function of the aspect ratio (A/B), but for a given value of $\dot{Q}^*_{mod}$ it is a maximum for the line fire. Yuan and Cox (1996) derived the following flame height correlation for a line burner:

$$l_f/A = 3.46\dot{Q}^{*2/3}_{mod} \tag{4.44}$$

It holds for values of $\dot{Q}_l > 30$ kW/m and is in good agreement with the data of Hasemi and Nishihata (1989). If $\dot{Q}_l < 30$ kW/m², the flames are laminar and a different correlation applies (the exponent is significantly greater (4/3)).

4.3.2.2 Flame Volume

Orloff and de Ris (1982) carried out a very detailed study of the radiation characteristics of flames above porous burners 0.1–0.7 m in diameter. They used the radiation measurements to define the outline of the flame, which allowed them to calculate the flame volume (V_f). This proved to be directly proportional to the rate of heat release in the range studied (25–250 kW) for two gaseous fuels (methane and propene) and polymethylmethacrylate, yielding the relationship:

$$\dot{Q}''' = \frac{\dot{Q}_c}{V_f} = 1200 \text{ kW/m}^3 \text{ or } 1.2 \text{ MW/m}^3$$

where $\dot{Q}'''$ is the 'power density' of the flame. These authors draw attention to the fact that the principles of Froude modelling predict that $\dot{Q}'''$ should be proportional to $\dot{Q}_c^{-1/5}$, but this weak dependence does not reveal itself over their range of data. However, Cox (1995) quotes $\dot{Q}''' = 0.5\,\text{MW/m}^3$, which may be more consistent with fires two orders of magnitude greater than those studied by Orloff and de Ris (1982).

4.3.2.3 Flame Temperatures and Velocities

Average temperatures and gas velocities on the centreline of axisymmetric buoyant diffusion flames have been measured by McCaffrey (1979) for methane burning on a 0.3 m square porous burner and by Kung and Stavrianides (1982) for hydrocarbon pool fires with diameters of 1.22, 1.74 and 2.42 m. McCaffrey's results clearly delineate the three regions of the fire plume, for each of which there were identifiable correlations between temperature (expressed as $2g\Delta T/T_\infty$), gas velocity (normalized as $u_0/\dot{Q}_c^{1/5}$) and height above the burner surface (z) (normalized as $z/\dot{Q}_c^{2/5}$). These are summarized in Table 4.2 and Figures 4.10 and 4.22. It can be seen that the average temperature is approximately constant in the upper part of the near field (persistent flaming) ($\Delta T = 800°C$ in these flames), but falls in the region of intermittent flaming to $\sim$320°C at the boundary of the buoyant plume. Thus, one would expect the temperature at the average flame height as defined by Zukoski et al. (1981a,b) to lie in the region of 500–600°C. In fact, a temperature of 550°C is sometimes used to define maximum vertical reach – e.g., of flames emerging from the window of a compartment that has undergone flashover (Bullen and Thomas, 1979) (see Section 10.2).

The average centreline velocity within the near field is independent of fire size ($\dot{Q}_c$) but increases as $z^{1/2}$ to a maximum velocity which is independent of z in the intermittent region (Table 4.2). McCaffrey (1979) found that this maximum was directly proportional to $\dot{Q}_c^{1/5}$, an observation which is significant in understanding the interaction between sprinklers and fire plumes. If the fire is too large ('strong source'), the downward

Table 4.2 Summary of centreline data for a buoyant methane diffusion flame on a 0.3 m square porous burner (McCaffrey, 1979) (Figure 4.10 and 4.22). These refer to values of $\dot{Q}^*$ in the range 0.25–1.0

Centreline velocity: $\dfrac{u_0}{Q^{1/5}} = k\left(\dfrac{z}{Q^{2/5}}\right)^{\eta}$

Centreline temperature: $\dfrac{2g\Delta T_0}{T_0} = \left(\dfrac{k}{C}\right)^2\left(\dfrac{z}{\dot{Q}^{2/5}}\right)^{2\eta-1}$

Region[a]	k	η	$z/\dot{Q}^{2/5}$ (m/kW$^{2/5}$)	C
Flame	6.8 m$^{1/2}$/s	1/2	<0.08	0.9
Intermittent	1.9 m/kW$^{1/5}$·s	0	0.08–0.2	0.9
Plume	1.1 m$^{4/3}$/kW$^{1/3}$·s	−1/3	>0.2	0.9

[a] See Figure 4.10(a).

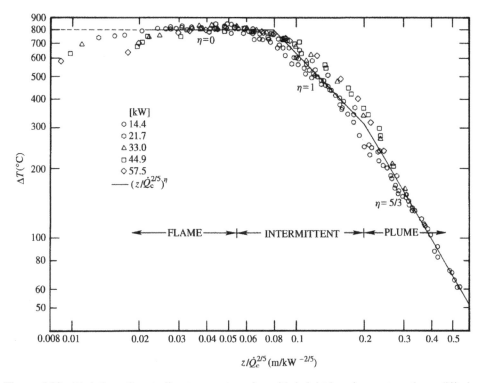

Figure 4.22 Variation of centreline temperature rise with height in a buoyant methane diffusion flame. Scales as $z/\dot{Q}_c^{2/5}$ (Table 4.2) (McCaffrey (1979), by permission). A similar correlation has been demonstrated for a range of hydrocarbon pool fires by Kung and Stavrianides (1982)

momentum of the spray or the terminal velocity of the droplets may be insufficient to overcome the updraft and water will not penetrate to the fuel bed. This is discussed further in Section 4.4.3.

4.3.2.4 Entrainment

The vertical movement of the buoyant gases in the fire plume causes air to be entrained from the surrounding atmosphere (see Equation (4.13) *et seq*.). Not only does this provide air for combustion of the fuel vapours, but it dilutes and cools the fire products as they rise above the flame into the far field, causing a progressive increase in the volume of 'smoke' generated by the fire. In the open (and in the early stages of a fire in a compartment), this will be clear air at normal temperatures: the amount entrained will quickly dominate the upward flow, even below the maximum height of the flame. Ma and Quintiere (2003) have reviewed the available data and found values for the ratio of air entrained below the flame tip to the stoichiometric requirement ranging from 5 (turbulent diffusion flames (Steward, 1970)) to 15 or 20. They suggest a value of 10 ± 5, which should be compared with Heskestad's (1986) conclusion that more than 10 times the stoichiometric air requirement is entrained into the flame below the flame tip.

Regarding the buoyant plume, it is necessary to estimate the upward flow to be able to calculate the rate of accumulation of smoke under a ceiling, or the extraction rate that will be required to maintain the smoke layer at or above a certain critical level (see Chapter 11). Following Heskestad (2008), the upward mass flow at any level in a weak plume (i.e., $\Delta T_0 / T_\infty \ll 1$) may be written:

$$\dot{m}_{\text{ent}} = E' \rho_\infty u_0 b^2 \tag{4.45}$$

where u_0 is the centreline velocity, b is the radius of the plume, defined by $u = 0.5 u_0$, and E' is a proportionality constant. Using Equations (4.21) and (4.22) for b_u and u_0, this becomes for height z:

$$\dot{m}_{\text{ent}} = E \rho_\infty z^2 \sqrt{gz} \dot{Q}_z^{*1/3} \tag{4.46}$$

or

$$\dot{m}_{\text{ent}} = E \left(\frac{g \rho_\infty^2}{c_p T_\infty} \right)^{1/3} \dot{Q}_c^{1/3} z^{5/3} \tag{4.47}$$

where $E = E' C_v C_l^2$, and z is the height above the (virtual) source. Yih (1952) deduced a value of $E = 0.153$ from measurements of the flow above a point source, although subsequently, Cetegen et al. (1984) found $E = 0.21$ to give good agreement with a range of experimental data on the weak plume. They also concluded that it gave a reasonable approximation to the flow in the strongly buoyant region above the flame tip, provided that z was the height above the virtual origin. This was based on results of experiments in which natural gas was burned under a hood from which the fire products were extracted at a rate sufficient to maintain the smoke layer at a constant level. The mass flow into the layer could then be equated to the extract rate. By varying the distance between the burner and the hood, and the rate of burning of fuel, data were gathered on the mass flow as a function of $\dot{Q}^*$ and z. Heskestad (1986) converted Equation (4.47) (with $E = 0.21$) to

$$\dot{m} = 0.076 \dot{Q}_c^{1/3} z^{5/3} \text{ kg/s} \tag{4.48}$$

for standard conditions (293 K, 101.3 kPa),[7] but noted that their analysis was based on similarity between excess temperature and upward velocity. He showed that improved agreement was obtained if the analysis was based on similarity between upward velocity and density deficit, which gave:

$$\dot{m} = 0.071 \dot{Q}_c^{1/3} z^{5/3} [1 + 0.026 \dot{Q}_c^{2/3} z^{-5/3}] \text{ kg/s} \tag{4.49}$$

Figure 4.23 compares these two interpretations of the plume mass flow results of Cetegen et al. (1984). Although there is a significant scatter, Equation (4.49) appears to give better agreement with the experimental data.

The results against which the above equations have been tested were obtained in a carefully controlled, draught-free environment. The amount entrained is influenced significantly by any air movement, created artificially by, for example, an air conditioning

---

[7] In a comprehensive review, Beyler (1986b) noted that the constant derived from Zukoski's work (0.076) was not found by all investigators. Values range from 0.066 (Ricou and Spalding, 1961) to 0.138 (Hasemi and Tokunaga, 1984).

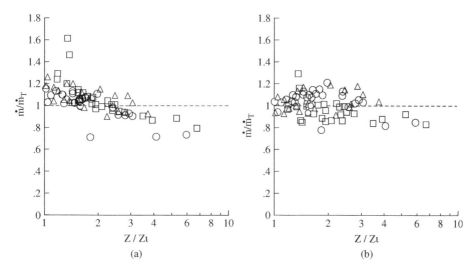

Figure 4.23 Plume mass flows above flames measured by Cetegen *et al.* (1984): (a) according to Equation (4.48), assuming similarity between ΔT and upward velocity; (b) according to Equation (4.49), assuming similarity between $\Delta \rho$ and upward velocity. In (a), Cetegen's formula for the virtual origin is used; in (b), Heskestad's (Equation (4.26)). From Heskestad (1986), by permission of the Combustion Institute

system (Zukoski *et al.*, 1981a), or naturally, if the fire is burning in a confined space and induces a directional flow from an open door, etc. (Quintiere *et al.*, 1981). If a fire is burning against a wall, or in a corner, the entrainment is also affected (see below).

4.3.3 Interaction of the Fire Plume with Compartment Boundaries

With the unconfined axisymmetric plume, there are no physical barriers to limit vertical movement or restrict air entrainment across the plume boundary, but in a confined space the fire plume can be influenced by surrounding surfaces. Thus, if an item is burning against a wall, the area through which air may be entrained is reduced (Figure 4.24); similarly, if the fire plume impinges on a ceiling, it will be deflected horizontally to form a ceiling jet, again with restricted entrainment (Figure 4.25). The consequences regarding flame height (or length) and plume temperatures need to be examined. However, additional effects must be considered, the most important relating to heat transfer to the surfaces involved and how quickly these surfaces (if combustible) will ignite and contribute to the fire growth process if given such exposure to flame (see Section 7.3). This is a topic that is directly relevant to our understanding of fire development in a room (Chapter 9).

4.3.3.1 Walls

If the fire is close to a wall, or in a corner formed by the intersection of two walls, the resulting restriction on free air entrainment will have a significant effect (Figure 4.26). The same three regimes identified in Figure 4.10(a) are observed, but in the buoyant plume

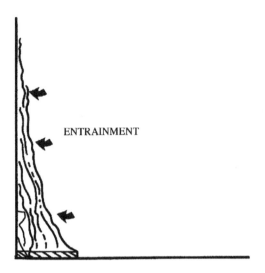

Figure 4.24 Interaction of a flame with a vertical surface

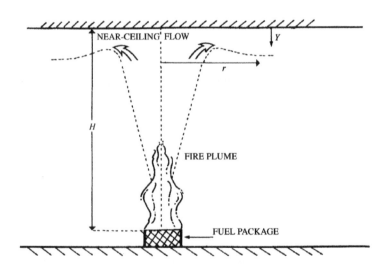

Figure 4.25 The fire plume and its interaction with a ceiling (after Alpert, 1972)

the temperature decreases less rapidly with height as the rate of mixing with ambient air will be less than for the unbounded case (Hasemi and Tokunaga, 1984). Relatively few measurements had been made of the effect on flame height, and it was assumed that the flame would be taller than for an equivalent fire plume burning in the open. A simple model was used to estimate flame height which involved an imaginary 'mirror image' fire source as shown in Figure 4.27. The flame height was assumed to be equal to that produced by the combined 'actual' and 'imaginary' fire sources burning in the open. However, Hasemi and Tokunaga (1984) found evidence that this was not the case

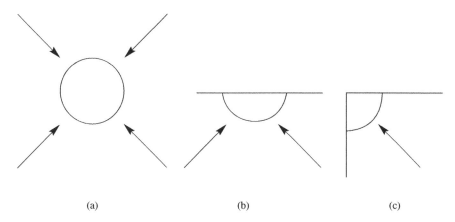

Figure 4.26 Plan view of a fire: (a) free burning; (b) burning against a wall; and (c) burning in a corner. The arrows signify the direction of air entrainment into the flame

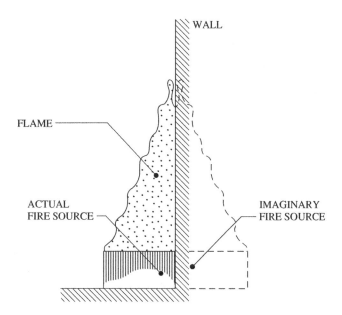

Figure 4.27 Concept of the 'imaginary fire source'. After Hasemi and Tokunaga (1984), reproduced by permission of the International Association for Fire Safety Science

for experimental gas fires ($0.4 < \dot{Q}^* < 2.0$), and that the flame height was similar to that predicted for an open fire, despite a significant reduction (c. 40%) in the amount of air entrained (Zukoski *et al.*, 1981a). This has been confirmed by results of Back *et al.* (1994), who showed that the relationship between flame height (expressed as l/D) and $\dot{Q}^{2/5}/D$ followed Heskestad's correlation satisfactorily (see Equation (4.41) and Figure 4.28).

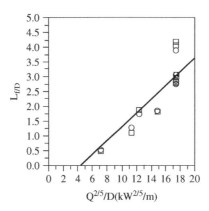

Figure 4.28 Comparison of experimentally determined flame heights for fires against a vertical surface (Figure 4.24) with the Heskestad correlation (Equation (4.40)). Circular symbols are based on videotape analysis and square symbols are based on the height at which the average temperature on the centerline is 500°C. Back *et al*. (1994), reproduced by permission of the Society of Fire Protection Engineers

This suggests that the turbulent structure of the fire plume is altered when it adheres to a vertical surface, enhancing the burning rate of the fuel vapours despite the reduction in the amount of air entrained. On the other hand, if the fire is in a corner, the flame height is increased significantly: Takahashi *et al*. (1997) reported that the flame height is almost doubled ($0.6 < \dot{Q}^* < 4.0$) provided that the (square) burner has been placed exactly in the corner, with no gap, allowing the flame to attach to the vertical surfaces. If there is a gap between the sides of the burner and the walls that is greater than twice the characteristic dimension of the burner, then the height of the flame is not affected by the presence of the walls. They also showed that when the flame was attached to the corner, the upward mass flow rate ($\dot{m}$) at the height corresponding to the flame tip was significantly less than the mass flow rate at the equivalent height for the free burning fire. As less air has been entrained, the temperature will be higher.

Indeed, for a fire against a wall and in a corner (Figure 4.26(b) and (c)), the maximum temperatures as a function of height are predicted more closely by the 'imaginary fire source model' than by assuming free burning of the 'real fire' in the open. This is a direct consequence of the reduction in entrainment which occurs even when the fire plume is simply deflected towards the restricting wall as a result of the directional momentum of the inflowing air (Figure 4.24). As noted above, flame attachment as shown in Figure 4.24 requires that the burning surface is right against the wall as in Figure 4.26(b). If the circular burner illustrated in Figure 4.26(a) was just touching the wall (i.e., the wall is tangential to the edge of the burner), flame attachment would not occur (Zukoski *et al*., 1981; Zukoski, 1995). This point is discussed by Williamson *et al*. (1991) in the context of experimental procedures for testing the flame spread properties of wall lining materials and has been examined further by Lattimer and Sorathia (2003).

If the surface of the wall is combustible, it may ignite and begin to burn, thus allowing flames to spread upwards (Section 7.2.1). Whether or not ignition will occur will depend

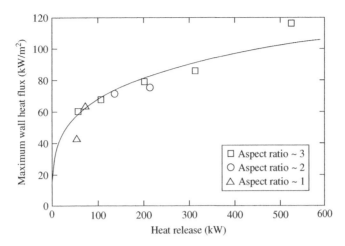

Figure 4.29 Peak wall heat fluxes for square propane burner fires against a flat wall. The aspect ratio refers to the ratio of flame height to burner width (Back *et al.*, 1994). Diagram taken from Lattimer (2008). Reproduced by permission of the Society of Fire Protection Engineers

on the properties of the surface and the rate of heat transfer from the flame (Section 6.3).[8] The latter will depend on the physical characteristics of the flame, which in turn will depend on the dimensions of the fuel bed from which the flames are generated. This subject is reviewed extensively by Lattimer (2008). Particularly important is the thickness of the flame with respect to the wall – greater thickness leads to a higher emissivity (Section 2.4.3, Equation (2.83)) and an increased rate of radiative heat transfer to the surface. This can exceed 100 kW/m^2, depending on the size of the fire and other circumstances (Back *et al.*, 1994), as demonstrated in Figures 4.28 and 4.29. Figure 4.30 shows clearly that the peak heat flux to the wall occurs below $z/L_f = 0.5$, i.e., where flaming is continuous (compare this with Figure 4.14). This also applies to fires in a corner configuration (Hasemi *et al.*, 1996), although the maximum heat fluxes are found at a short distance from the corner itself (10–20 cm in the work of Kokkala (1993) and Lattimer and Sorathia (2003)). This is likely to be a view factor/configuration factor effect if radiation is the dominant mode of heat transfer. As with heat transfer to a plane wall, the peak heat flux occurred where the continuous flame was attached to the walls and appeared to increase as the size of the square burner was increased. In contrast, thin flames – such as those produced from a line burner at the foot of the wall – have much lower emissivities. Nevertheless, high rates of heat transfer can be achieved from a line burner under certain confined conditions, as described by Foley and Drysdale (1995).

If a vertical surface is ignited by a small ignition source and begins to burn, effectively without a supporting fire at the base (as in Figure 4.24), flame will spread upwards as a consequence of heat transfer to the contiguous material above the burning area. Flame spread will be discussed in detail in Chapter 7, but one of the most important parameters

---

[8] Wall lining materials are commonly tested in the 'corner-wall' configuration, with the fire source (usually a sand-bed burner) located at the junction of the two walls (e.g., ISO, 2010; CEN, 2002; NFPA, 2006c).

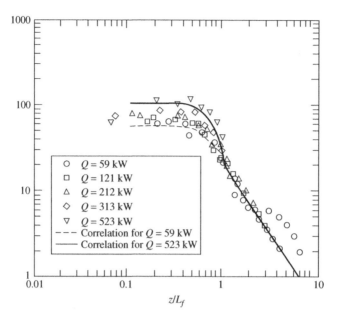

Figure 4.30   Vertical heat flux distribution along the centreline of the flames from square propane burner fires adjacent to a flat wall (Figure 4.24) (Back *et al*., 1994). The height is normalized with the flame length. Diagram taken from Lattimer (2008), with permission

that determines the rate of spread is the flame height (see Section 7.2.1 and Figure 7.9). This correlates with the rate of heat release per unit width, similar to the correlation observed for the line burner (see Figure 4.21 and Equations (4.43) and (4.44)) (Hasemi and Nishihata, 1989; Yuan and Cox, 1996). Delichatsios (1984) deduced that the height of the flame associated with a wall fire would be proportional to the two-thirds power of $\dot{Q}'_l$, i.e.

$$L_f = K\,\dot{Q}_l^{\prime n} \tag{4.50}$$

where L_f is the height of the flame, K is a constant and the exponent $n = 2/3$. This is consistent with a number of studies (e.g., Hasemi, 1984, 1985; Saito *et al*., 1985), but higher values of the exponent have been reported (e.g., Quintiere *et al*., 1986). In these studies the correlations took no account of the height of the burning area, which might be expected to be significant and possibly important in the modelling of upwards flame spread (see Section 7.3). Tsai and Drysdale (2002) varied the area and the aspect ratio (height to width) of the burning surface[9] and found that Equation (4.50) was still valid, with values of n within the range 0.62–0.66 when $\dot{Q}'_l > 30\,\text{kW/m}$. However, the exponent was closer to 1.0 when the flame was laminar, $\dot{Q}'_l < 30\,\text{kW/m}$. The latter observation was also made for line fires by Yuan and Cox (1996).

---

[9] Three widths were used (150 mm, 300 mm and 570 mm) and the aspect ratio (height to width) varied from 0.26 to 2.0.

4.3.3.2 Ceilings

If the vertical extent of a fire plume is limited by the presence of a ceiling, the hot gases will be deflected as a horizontal ceiling jet, defined by Alpert (2008) as 'the relatively rapid gas flow in a shallow layer beneath the ceiling surface that is driven by the buoyancy of the hot combustion products from the plume'. It spreads radially from the point of impingement and provides the mechanism by which combustion products are carried to ceiling-mounted fire detectors. To enable the response of heat and smoke detectors to be analysed, the rate of development and the properties of the ceiling jet must be known. Although the time lag which is associated with the response of any detector can be considered in terms of a transport time lag and a delay to detector operation (Newman, 1988; Mowrer, 1990; see also Custer *et al.*, 2008), it is convenient to consider first the steady state fire and the resulting temperature distribution under the ceiling. Alpert (1972, 2008) has provided correlations based on a series of large-scale tests, carried out at the Factory Mutual Test Centre,[10] in which a number of substantial fires were burned below flat ceilings of various heights, H (Table 4.3): in all cases the flame height was much less than H. The resulting ceiling jet is thus described as 'a weak plume-driven flow field' (Alpert, 2008) and its characteristics are relevant to the early stages of fire growth in an enclosure. Temperatures were measured at different locations under the ceiling (Figure 4.25): it was found that at any radial distance (r) from the plume axis, the vertical temperature distribution exhibited a maximum (T_{max}) close to the ceiling, at $Y \not> 0.01H$ (see Figure 4.25). Below this, the temperature fell rapidly to ambient (T_∞) for $Y \not> 0.125H$. These figures are valid only if horizontal travel is unconfined and a static layer of hot gases does not accumulate beneath the ceiling. This will be achieved to a first approximation if the fire is at least $3H$ distant from the nearest vertical obstruction; however, if confinement occurs by virtue of the fire being close to a wall, or in a corner, the horizontal extent of the free ceiling from the point of impingement would presumably have to be much greater for this condition to hold.

Table 4.3   Summary of the fire tests from which Equations (4.53) and (4.54) were derived (Alpert, 1972)

Fuel	Fuel array size (m)	Fire intensity (MW)	Ceiling height (m)
Heptane spray	3.7 m diameter	7.0–22.8	4.6–7.9
Heptane pan[a]	0.6 × 0.6	1.0	7.6
Ethanol pan	1.0 × 1.0	0.67	8.5
Wood pallets[a]	1.2 × 1.2 × 1.5	4.9	6.1
Cardboard boxes	2.4 × 2.4 × 4.6	3.9	13.7
Polystyrene in cardboard boxes	2.4 × 2.4 × 4.6	98	13.7
PVC in cardboard boxes	2.4 × 2.4 × 4.6	35	13.4
Polyethylene pallets	1.2 × 1.2 × 2.7	4.2–11.4	15.5

[a]Located in a 90° corner.

[10] Now the FM Global Test Center.

Alpert (1972) showed that the maximum gas temperature (T_{max}) near the ceiling at a given radial position r (provided that $r > 0.18H$) could be related to the rate of heat release ($\dot{Q}_c$, kW) by the steady state equation:

$$T_{max} - T_\infty = 5.38\frac{\dot{Q}_c^{2/3}/H^{5/3}}{(r/H)^{2/3}} \tag{4.51}$$

in which H is used as the length scale. This may be rearranged to give the more commonly quoted expression:

$$T_{max} - T_\infty = \frac{5.38(\dot{Q}_c/r)^{2/3}}{H} \tag{4.52}$$

If $r \leq 0.18H$ (i.e., within the area where the plume impinges on the ceiling[11]):

$$T_{max} - T_\infty = \frac{16.9\dot{Q}_c^{2/3}}{H^{5/3}} \tag{4.53}$$

Inspection of Equation (4.52) reveals a lower rate of entrainment into the ceiling jet than into the vertical fire plume: in the latter, the temperature varies as $H^{-5/3}$, while in the horizontal ceiling jet it varies as $r^{-2/3}$. This is consistent with the knowledge that mixing between a hot layer moving on top of a cooler fluid is relatively inefficient. The process is controlled by the Richardson number (Zukoski, 1995), which is the ratio of the buoyancy force acting on the layer to the dynamic pressure of the flow, i.e.

$$Ri = \frac{g(\rho_0 - \rho_{layer})h}{\rho_{layer}V^2}$$

where ρ_{layer} and h are the density and the depth of the layer, respectively, and V is its velocity with respect to the ambient layer below. Mixing is suppressed at high values of Ri. (A detailed analysis of the turbulent ceiling jet has been developed by Alpert (1975b).)

If the fire is by a wall, or in a corner, the temperatures will be greater, due not only to the lower rate of entrainment into the vertical plume, but also to the restriction under the ceiling where the flow is no longer radial and symmetric. This can be accounted for in Equations (4.52) and (4.53) by multiplying $\dot{Q}_c$ by a factor of two or four, respectively.

The dependence of T_{max} on r and H for a 20 MW fire according to these equations is shown in Figure 4.31. Such information may be used to assess the response of heat detectors to steady burning or slowly developing fires (Section 4.4.2), although the results calculated for a 20 MW fire under a 5 m ceiling are not reliable as it is clear that there is flame impingement on the ceiling ($T > 550°C$). Heskestad and Hamada (1993) found that for strong fire plumes (defined in this context as fires for which the ratio of the 'free flame height' (Equation (4.40)) to ceiling height (H in Figure 4.25) is greater than 0.3), a more convincing correlation is obtained if the radial distance is scaled against b (the radius of plume where it impinges on the ceiling) rather than H. However, there is a limit to this correlation which begins to break down for values of $l/H > 2$ when there is significant flaming under the ceiling.

---

[11] Otherwise known as the 'turning region'.

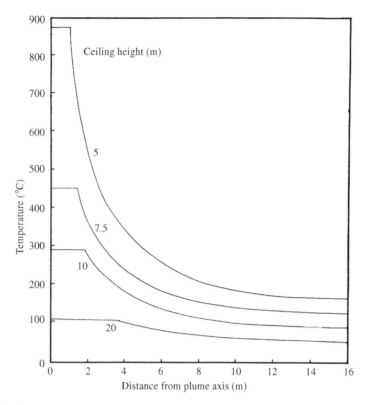

Figure 4.31 Gas temperatures near the ceiling according to Equations (4.52) and (4.53), for a large-scale fire ($\dot{Q}_c$ = 20 MW) for different ceiling heights (see Figure 4.25). Note that the formulae are unlikely to apply for the 5 m ceiling because of flame impingement (after Alpert, 1972)

When $l/H > 1$, the part of the flame that is deflected horizontally will become part of the ceiling jet. Babrauskas (1980a) reviewed contemporary information on flames under non-combustible ceilings and suggested that the length of the horizontal part of the flame (h_r) could be related to the 'cut-off height', h_c (Figure 4.32). On the assumption that the amount of air entrained into the horizontal flame was equivalent to the amount that would have been entrained into the 'cut-off height' (had it remained vertical), he argued that the ratio h_r/h_c would strongly depend on the configuration involved (Table 4.4), the greatest extension occurring when a fire is confined to a corridor and the flames are channelled in one direction, consistent with the results of Hinkley *et al.* (1968) (see below, Figure 4.33). However, subsequent work has shown that this method overestimates horizontal extension, at least for the intermittent flame.

Gross (1989) reported that for the deflection of the intermittent flame (as defined in Figure 4.10(a)), the total flame length (vertical height plus horizontal distance to the flame tip) was significantly less than the total height of the vertical flame in the absence of the ceiling. A similar observation was made by Kokkala and Rinkenen (1987). (The intermittent flame will be 'fuel lean' in that the flow of burning gas contains excess air (Section 4.3.2) and the reduced rate of air entrainment into the horizontal flow is of little

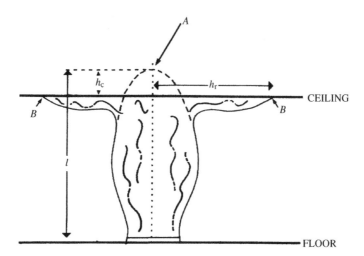

Figure 4.32 Deflection of a flame beneath a ceiling, illustrating Babrauskas' 'cut-off height' h_c. A = location of the flame tip in the absence of a ceiling; B = limit of the flame deflected under the ceiling. After Babrauskas (1980a) by permission

Table 4.4 Flame extension under ceilings: $\dot{Q}_c = 0.5$ MW and $H = 2$ m (Babrauskas, 1980a,b) (Figure 4.32)

Configuration	h_r/h_c
Unrestricted plume, unbounded ceiling	1.5
Full plume, quarter ceiling[a]	3.0
Quarter plume, quarter ceiling	12
Corridor	Dependent on width

[a]Fire located in a corner but not close enough for flame attachment as in Figure 4.27.

consequence.) Heskestad and Hamada (1993) report a minor reduction in the total flame length, expressed as:

$$h_r/h_c \approx 0.95$$

where h_r is the mean radius of the horizontal flame and h_c is the 'cut-off height' as defined in Figure 4.31. The range of values of h_r/h_c was 0.88 to 1.05, suggesting that the effect of horizontal deflection on the total flame length is actually very small.[12] (The behaviour of these flames has also been studied by Hasemi *et al.* (1995) in experiments designed to investigate the heat flux to the ceiling.)

This argument applies to the fuel-lean situation when only the intermittent part of the flame is deflected. If the burning gases are fuel-rich, as will occur if the fire is large in relation to the height of the ceiling (e.g., see Figure 4.33), considerable flame extension

---

[12] Note that the position of the maximum extent of the intermittent flame is difficult to determine with any accuracy.

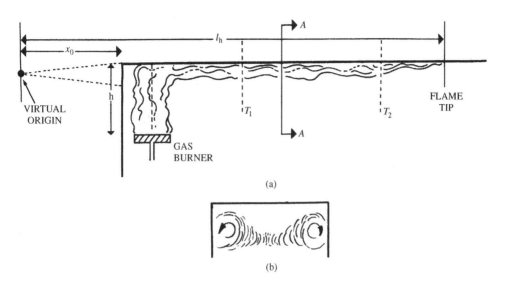

(a)

(b)

Figure 4.33 (a) Deflection of a flame beneath a model of a corridor ceiling (longitudinal section) showing the location of the 'virtual origin'. T_1 and T_2 identify the locations of the vertical temperature distributions shown in Figure 4.34. (b) Transverse section A–A. Not to scale. After Hinkley *et al.* (1968). Reproduced by permission of The Controller, HMSO. © Crown copyright

can occur, depending on the configuration. This was first investigated systematically by Hinkley *et al.* (1968), who studied the deflection of fire plumes produced on a porous bed gas burner, by an inverted channel with its closed end located adjacent to the burner (Figure 4.33). The lining of the channel was non-combustible. This situation models the behaviour of flames under a corridor ceiling and is easier to examine than the unbounded ceiling. The appearance and behaviour of the flames were found to depend strongly on the height of the ceiling above the burner (*h* in Figure 4.33) and on the gas flowrate. For low flowrates (corresponding to fuel-lean flames under the ceiling), the horizontal flame was of limited extent and burned close to the ceiling (cf. Gross, 1989). Alternatively, with high fuel flowrates or a low ceiling (small *h*), a burning, fuel-rich layer was found to extend towards the end of the channel with flaming occurring at the lower boundary. The transition between fuel-lean and fuel-rich burning was related to a critical value of $(\dot{m}'/\rho_0 g^{1/2})d^{3/2} \approx 0.025$, where $\dot{m}'$ was taken as the rate of burning per unit width of channel (g/m.s) and *d* is the depth of the layer of hot gas below the ceiling (m). The difference between these two regimes of burning is illustrated clearly in Figure 4.34, which shows vertical temperature distributions below the ceiling at 2.0 m and 5.2 m from the closed end (Figure 4.33). This flame extension will be even greater if the lining material is combustible, as extra fuel vapour will be evolved from the linings and contribute to the flaming process (Hinkley and Wraight, 1969).

It should be noted that some of the processes involved here are common with the development of a layer of fuel-rich gases under the ceiling of a compartment (room) as a fire approaches flashover (Section 9.2.1). A smoke layer forms and descends as the 'reservoir' formed by the walls and the ceiling fills with hot smoke and combustion

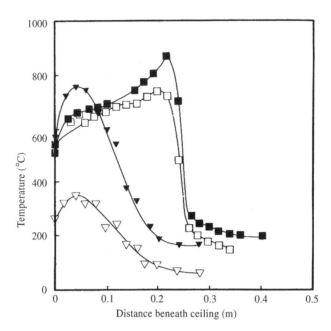

Figure 4.34 Vertical temperature distributions below a corridor ceiling for fuel-lean (▼, ▽) and fuel-rich (■, □) horizontal flame extensions. Closed and open symbols refer to points 2 m and 5 m from the axis of the vertical fire plume, respectively (T_1 and T_2 in Figure 4.33). After Hinkley *et al.* (1968). Reproduced by permission of The Controller, HMSO. © Crown copyright

products. This limits the vertical height through which air can be entrained into the flames, and eventually the flames reaching the ceiling will be very 'fuel rich', with relatively low levels of oxygen. This fuel-rich layer will eventually burn – a process which is associated with the flashover transition.

There have been a number of studies of heat transfer from fire plumes to ceilings (You and Faeth, 1979; You, 1985; Kokkala, 1991; Hasemi *et al.*, 1995; Lattimer, 2008). Hasemi *et al.* (1995) carried out experiments in which porous burners (0.3 m, 0.5 m and 1.0 m square) were located below an unconfined flat ceiling. The rate of heat release and the ceiling height were varied. They found that the heat flux was a maximum within the area of plume impingement, increasing rapidly as the flame height (L_f) exceeded the ceiling height (H) and peaking at approximately 90 kW/m² (the authors quote 80–100 kW/m²). At this point, $L_f/H \sim 2.5$, i.e., the continuous (persistent) region was impinging on the ceiling (see also Lattimer (2008)). In similar experiments, but with a much smaller burner (0.06 m diameter), Kokkala (1991) found the maximum heat flux at the ceiling to be only 60 kW/m² at the stagnation point. In both studies, the heat flux was found to decrease with radial distance from the plume axis.

Hasemi *et al.* (2001) also studied the behaviour of radially symmetric ceiling flames produced from a porous gas burner, flush-mounted at the centre of a square, non-combustible ceiling (1.82 m ×1.82 m). This is not a realistic scenario as ceilings tend not to burn unless there is a fire lower in the room, but the results are informative.

The combustion efficiency of these flames is low and it was found that the flame area increased linearly with the actual (measured) rate of heat release. The maximum rate of heat transfer to the ceiling was less than 30 kW/m^2 close to the burner, falling to about 5 kW/m^2 at the edge of the flame. This is consistent with heat transfer from a thin flame of low emissivity and suggests that flame spread on the underside of a horizontal ceiling would be slow (Section 7.2.1) if only the ceiling had been ignited.

4.3.4   The Effect of Wind on the Fire Plume

If a flame is burning in the open, it will be deflected by any air movement, the extent of which will depend on the wind velocity and the rate of heat release of the fire. Many studies of flame deflection have been carried out in the context of examining the effect of wind on the spread of fire through open fuel beds (e.g., Thomas, 1965) and more recently in tunnel fires (see, for example, Ingason (2005)). In the petrochemical industries, the interest is in how flames deflected by the wind may create hazardous conditions regarding neighbouring items of equipment (Lois and Swithenbank, 1979; Mannan, 2005). This should be taken into account in the layout of plant when the consequences of fire incidents are being considered. A rule of thumb that is commonly used is that a 2 m/s wind will bend the flame 45° from the vertical and for fires near the ground (e.g., bund fires) the flame will tend to hug the ground downwind of the fuel bed, to a distance of ∼0.5D, where D is the fire diameter (Robertson, 1976; Mannan, 2005). This can significantly increase the fire exposure of items downwind, either by causing direct flame impingement, or by increasing the levels of radiant heat flux (Pipkin and Sliepcevich, 1964; Beyler, 2008).

Several correlations have been developed during studies of pool fires burning in the open in which 'flame tilt' (the angle θ between the vertical and the centreline of the flame, Figure 4.35(a)) has been measured. These can be expressed in the following format:

$$\cos(\theta) = d'(u^*)^{e'} \qquad \text{for } u^* \geq 1 \tag{4.54}$$

and

$$\cos(\theta) = 1 \qquad \text{for } u^* < 1 \tag{4.55}$$

where

$$u^* = u_w/u_c \text{ if } u_w \geq u_c \tag{4.56}$$

and

$$u^* = 1 \text{ if } u_w < u_c \tag{4.57}$$

u^* is a dimensionless wind speed, being the ratio of the wind velocity (u_w, m/s) and a characteristic velocity (u_c) which is given by:

$$u_c = (g\dot{m}''D/\rho_a)^{1/3} \tag{4.58}$$

where $\dot{m}''$ is the mass burning rate (g/m^2·s). Table 4.5 gives values of the parameters in this equation for four correlations, but only one is shown in Figure 4.35(b). These are

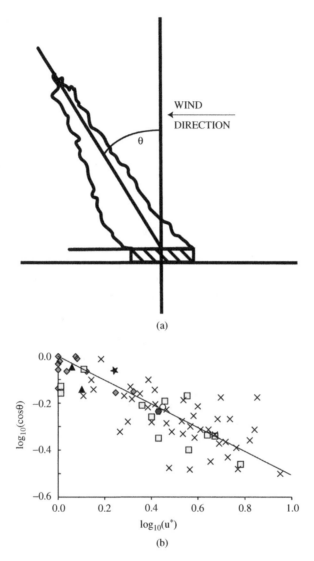

(a)

(b)

Figure 4.35 (a) Deflection of a flame by wind. (b) Relationship between the flame tilt angle (θ) and the non-dimensional wind velocity (Equation (4.56)). Data are as follows: $\times$ LNG pool fires on land (American Gas Association, 1974); $\square$ LNG pool fires on land (Moorhouse, 1982); ($\bullet$), LNG pool fires on land (Minzer and Eyre, 1982); $\circ$ LNG pool fires on land (Minzer and Eyre, 1982); ($\star$), LNG pool fires on water (Raj *et al.*, 1979); $\blacktriangle$ kerosene pool fires on land (Japan Institute for Safety Engineering, 1982); $\diamond$ Munoz *et al.* (2004). Adapted from Mudan (1984). The correlation proposed by the American Gas Association (AGA) is shown

Table 4.5 Parameters for the flame tilt correlations (Equations (4.53))

	d'	e'	Notes
American Gas Association (1974)[a]	1	−0.50	LNG pool fires
Thomas (1965)	0.7	−0.49	Wood crib fires
Moorehouse (1982)	0.86	−0.25	LNG pool fires
Munoz et al. (2004)	0.96	−0.26	Hydrocarbon pool fires

[a] See also Mudan (1984).

compared with pool fire data from various sources, including Munoz et al. (2004). There is a considerable scatter in the data, particularly at $u^* = 1$, when the velocity of the wind is relatively low when compared to the updraught of the fire. This is shown clearly by the data of Munoz et al. (2004) in Figure 4.35, when $u^* = 1$. The correlations suggest that there is no deflection if $u^* \leq 1$ (Equation (4.55)), but it is clear that there can be considerable disturbance of the flame at relatively low wind speeds.

Beyler (2008) has expressed the view that the AGA correlation is probably the most accurate, but at least one other correlation does exist. Welker and Sliepcevich (1966) and Fang (1969) developed correlations that relate the angle of deflection to a Froude number. These are not discussed here, but the reader may wish to consult the review article 'Wind effects on fires' by Pitts (1991).

Air movement tends to enhance the rate of entrainment of air into a fire plume. This is likely to promote combustion within the flame and thus reduce its length, although this remains to be quantified properly. However, an investigation has been carried out on entrainment into flames within compartments during the early stages of fire development to determine the influence of the directional flow of air from the ventilation opening. Quintiere et al. (1981) have shown that the rate may be increased by a factor of two or three, which could have a significant effect on the rate of fire growth.

4.4 Some Practical Applications

Research in fire dynamics has provided concepts and techniques which may be used by the practising fire protection engineer to predict and quantify the likely effects of fire. The results of such research have been drawn together as a series of state-of-the-art reviews in the *SFPE Handbook* (Di Nenno et al., 2008), which currently represents the best available source of background information. A Code of Practice 'Application of fire safety engineering principles to the design of buildings' has been published by the BSI (British Standards Institution, 2001), but in order to be able to translate 'research into practice', a sound understanding of the underlying science is essential. In this section, some of the information which has been presented above is drawn together to illustrate how the knowledge may be applied. However, it should be remembered that this is a developing subject and the reader is encouraged to be continuously on the lookout for new applications of the available knowledge, either to improve existing techniques or to develop new ones. This chapter finishes with a brief review of modelling techniques that are in current use.

4.4.1 Radiation from Flames

It was shown in Sections 2.4.2 and 2.4.3 that the radiant heat flux received from a flame depends on a number of factors, including flame temperature and thickness, concentration of emitting species and the geometric relationship between the flame and the 'receiver'. While considerable progress is being made towards developing a reliable method for calculating flame radiation (Tien *et al.*, 2008), a high degree of accuracy is seldom required in 'real-world' fire engineering problems, such as estimating what level of radiant flux an item of plant might receive from a nearby fire in order that a water spray system can be designed to keep the item cool (e.g., storage tanks in a petrochemical plant).

Beyler (2008) discusses the methods available for calculating the radiation levels from large pool fires in detail. Here, two approximate methods described by Lees (Mannan, 2005) are compared, identifying some of the associated problems. Both of these require a knowledge of the flame height (*l*), which may be obtained from Equation (4.40):

$$l = 0.23\dot{Q}_c^{2/5} - 1.02D \tag{4.40}$$

The total rate of heat release ($\dot{Q}_c$) may be calculated from:

$$\dot{Q}_c = \dot{m}'' \Delta H_c A_f \tag{4.59}$$

where A_f is the surface area of the fuel (m^2). As only a fraction (χ) of the heat of combustion is radiated, it is necessary to incorporate this into the equation, thus:

$$\dot{Q}_r = \chi \dot{m}'' \Delta H_c A_f \tag{4.60}$$

This fraction is sometimes assumed to be 0.3, but studies have shown that it varies not only with the fuel involved, but also with the tank diameter, as shown in Figure 4.36 (Koseki, 1989; see also Beyler, 2008). As noted in Section 2.4.3, the radiative fraction is found to decrease as the tank diameter increases.

In the first method, it is assumed that $\dot{Q}_r$ originates from a point source on the flame axis at a height $0.5l$ above the fuel surface. The heat flux ($\dot{q}_r''$) at a distance R from the point source (P) is then:

$$\dot{q}_r'' = \chi \dot{m}'' \Delta H_c A_f / 4\pi R^2 \tag{4.61}$$

as illustrated in Figure 4.37, where $R^2 = (l/2)^2 + d^2$, d being the distance from the plume axis to the receiver, as shown. However, if the surface of the receiver is at an angle θ to the line-of-sight (PT), the flux will be reduced by a factor $\cos \theta$:

$$\dot{q}_r'' = (\chi \dot{m}'' \Delta H_c A_f \cos \theta) / 4\pi R^2 \tag{4.62}$$

Consider a 10 m diameter gasoline pool fire. Given that this will burn with a regression rate of 5 mm/min (Section 5.1.1, Figure 5.1) corresponding to a mass flowrate of $\dot{m}'' = 0.058$ kg/m$^2 \cdot$ s, then as $\Delta H_c = 45$ kJ/g (Table 1.13), the rate of heat release according to Equation (4.59) will be 206 MW. Using Equation (4.62) with $\chi = 0.3$, the radiant heat flux at a distance d is shown in Figure 4.38. This will be an overestimate as χ will be less than 0.3; also note that Equation 4.62 does not apply at short distances.

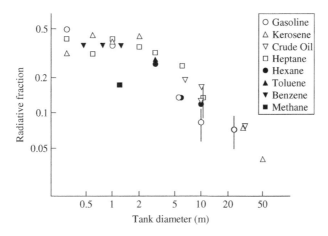

Figure 4.36 Radiative fractions (χ) measured for pool fires of diameters from 0.3 to 50 m. o, gasoline; △, kerosene; ▽, crude oil; □, heptane; ●, hexane; ▲, toluene; ▼, benzene; ■, methanol (Koseki, 1989). Reprinted with permission from NFPA *Fire Technology* (Vol. 25, No. 3) copyright © 1989, National Fire Protection Association, Quincy, MA

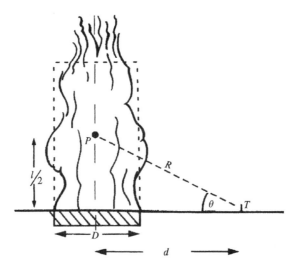

Figure 4.37 Estimating the radiant heat flux received at point T from a pool fire, diameter D. Equivalent point source at P

In the second method, the flame is approximated by a vertical rectangle, $l \times D$, straddling the tank in a plane at right angles to the line of sight. Taking $\chi = 0.3$, the net emissive power of one face of this rectangle would then be:

$$E = \frac{1}{2}(0.3\dot{m}''\Delta H_c A_f \cos\theta/lD)$$

$$= 151 \text{ kW/m}^2 \tag{4.63}$$

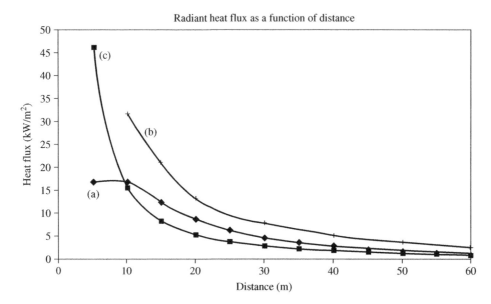

Figure 4.38 Variation of incident radiant heat flux ($\dot{q}''_{r,T}$) with distance from a 10 m diameter pool of gasoline (see Figure 4.37): (a) assuming point source ($\blacklozenge$); (b) assuming that the flame behaves as a vertical rectangle, $l \times D$ (+); and (c) calculated from the correlation by Shokri and Beyler (1989) ($\blacksquare$)

The radiant flux at a distant point can then be obtained from:

$$\dot{q}''_{r,T} = \phi E \tag{4.64}$$

Values of $\dot{q}''_{r,T}$ calculated by this method for the above problems are also shown in Figure 4.38. Higher figures are obtained because the emitter is treated as an extended source: provided that $d > 2D$, values of $\dot{q}''_{r,T}$ are about double those obtained from Equation (4.62). By using Equations (4.63) and (4.64), a very conservative figure is obtained which would result in unnecessary expense as the protection system would then be overdesigned. There are several simplifying assumptions in the above calculations, at least two of which will lead to an overestimate of the heat flux, i.e., that combustion is 100% efficient, and that the flame acts as if it were at a uniform temperature. The former is certainly incorrect, while the latter ignores the fact that the effective radiative fraction will be less than 0.3 for a tank of this size (Figure 4.36). Shokri and Beyler (1989) have reviewed data on radiant fluxes $\dot{q}''_{r,T}$ measured in the vicinity of experimental hydrocarbon pool fires. They estimated the 'effective' emissive power (E_{eff}) of each by calculating the appropriate configuration factor ϕ (assuming the flame to be cylindrical) and using Equation (4.64). Values of E_{eff} in the range 16–90 kW/m^2 were obtained, with a tendency to decrease with increasing pool diameter, as expected due to the increasing amounts of black smoke which envelop the flame at larger diameters (see Figure 4.36).

Shokri and Beyler (1989) found that the following empirical expression (in which d is the distance from the pool centreline to the 'target'):

$$\dot{q}_{r,T}'' = 15.4 \left(\frac{d}{D}\right)^{-1.59} \text{ kW/m}^2 \tag{4.65}$$

was an excellent fit to the data from over 80 of the pool fire experiments examined in their review, which refers to the configuration shown in Figure 4.37 (i.e., vertical, ground-level targets). It can be seen in Figure 4.38 that this predicts lower heat fluxes than the point source method in the far field, but continues to increase monotonically as the distance from the tank decreases. These authors recommend that applying a safety factor of two to this formula will give heat flux values above any of the measurements that were used in its derivation; these values still lie below the heat fluxes obtained using the vertical rectangle approximation.

The above calculations assume that the flames are vertical and are not influenced by wind. If the presence of wind has to be taken into account, then the appropriate flame configuration can be deduced from information presented in Section 4.3.5.

4.4.2 The Response of Ceiling-mounted Fire Detectors

In Section 4.3.4, it was shown that the temperature under a ceiling could be related to the size of the fire ($\dot{Q}_c$), the height of the ceiling (H) and the distance from the axis of the fire plume (r) (Figure 4.31) (Alpert, 1972). Equations (4.52) and (4.53) may be used to estimate the response time of ceiling-mounted heat detectors, provided that the heat transfer to the sensing elements can be calculated. Of course, it is easy to identify the minimum size of fire ($\dot{Q}_{min}$, kW) that will activate fixed-temperature heat detectors, as $T_{max} \geq T_L$, where T_L is the temperature rating. Thus, from Equations (4.52) and (4.53), for $r > 0.18H$:

$$\dot{Q}_{min} = r(H(T_L - T_\infty)/5.38)^{3/2} \tag{4.66}$$

and for $r \leq 0.18H$:

$$\dot{Q}_{min} = ((T_L - T_\infty)/16.9)^{3/2} H^{5/2} \tag{4.67}$$

If the detectors are to be spaced at 6 m centres on a flat ceiling in an industrial building, then the maximum distance from the plume axis to any detector head is $(0.5 \times 6^2)^{1/2}$, or $r = 4.24$ m. Thus, for the worst case:

$$\dot{Q}_{min} = 4.24(H(T_L - T_\infty)/5.38)^{3/2}$$
$$= 0.34(H(T_L - T_\infty))^{3/2} \tag{4.68}$$

showing that for a given sensor, the minimum size of fire that may be detected is proportional to $H^{3/2}$. Substituting $H = 10$ m and $T_\infty = 20°C$, and assuming $T_L = 60°C$, then the minimum size of fire that can be detected in a 10 m high enclosure is 2.7 MW.

However, rapid activation of the detector will require a high rate of heat transfer ($\dot{q}$) to the sensing element of area A, which consequently must be exposed to a temperature significantly in excess of T_L. The rate will be given by (Equation (2.3)):

$$\dot{q} = hA\Delta T$$

where h, the heat transfer coefficient for forced convection, will be a function of the Reynolds and Prandtl numbers (Section 2.3).

The response time (t) of the sensing element can be derived from Equations (2.20) and (2.21), setting T_{max} as the steady fire-induced temperature at the head and T (the temperature of the element at time of response) as T_L: thus,

$$t = \frac{Mc}{A}\frac{1}{h}\ln\left(\frac{T_{max} - T_\infty}{T_{max} - T_L}\right) \tag{4.69}$$

i.e.

$$t = -\frac{Mc}{Ah}\ln\left(1 - \frac{\Delta T_L}{\Delta T_{max}}\right) \tag{4.70}$$

where Mc is the thermal capacity of the element and A is its surface area (through which heat will be transferred) and ΔT_L and ΔT_{max} are $T_L - T_\infty$ and $T_{max} - T_\infty$, respectively. The quantity

$$\tau = \frac{Mc}{Ah} \tag{4.71}$$

is the time constant of the detector, but while Mc/A is readily calculated, it is very difficult to estimate h from first principles. However, it refers to conditions of forced convection so that if the flow is laminar, then according to Equation (2.39) $h \propto Re^{1/2}$, hence $h \propto u^{1/2}$ and $\tau \propto u^{-1/2}$. This leads to the concept of the Response Time Index (RTI), which was originally introduced by Heskestad and Smith (1976) to characterize the thermal response of sprinkler heads. It is determined experimentally in the 'plunge test': this involves suddenly immersing the sprinkler head in a flow of hot air, the temperature (T_0) and flowrate (u_0) of which are known. The RTI is then defined as:

$$RTI = \tau_0 u_0^{1/2} \tag{4.72}$$

where τ_0 is the value of the time constant determined under the standard conditions from Equation (4.71), and t becomes the response time t_0. It is assumed[13] that the product $\tau u^{1/2}$ will be equal to the RTI at any other temperature and flowrate (assumed laminar). Thus, if the value of u under fire conditions can be predicted, the time to sprinkler actuation (t) can be calculated from:

$$\frac{t}{t_0} = \left(\frac{u_0}{u}\right)^{1/2}\left(\frac{\ln(1 - \Delta T_L/\Delta T_{max})}{\ln(1 - \Delta T_L/\Delta T_{max,0})}\right) \tag{4.73}$$

This can be used if there is a fire of constant heat output, giving a temperature (T_{max}) and flowrate (u) at the detector head under a large unobstructed ceiling: the temperatures

---

[13] This assumption has been shown to be invalid if conduction losses from the sprinkler head into the associated pipework are significant. This will be the case for conditions in which the sensing element is heated slowly, e.g., as a result of a slowly developing fire. This is discussed by Heskestad and Bill (1988) and Beever (1990).

can be calculated using Equations (4.52) and (4.53), while the gas velocities can be calculated from:

$$u_{max} = \frac{0.197 \dot{Q}^{1/2} H^{1/2}}{r^{5/6}} \text{ m/s} \quad (4.74)$$

which applies to the ceiling jet ($r > 0.18H$ and $Y \approx 0.01H$, see Figure 4.25), while within the buoyant plume ($r \geq 0.18H$):

$$u_{max} 0.946 \left(\frac{\dot{Q}}{H}\right)^{1/3} \text{ m/s} \quad (4.75)$$

where $\dot{Q}$ is in kW (Alpert, 1972). This work forms the basis for Appendix B of NFPA 72 (Evans and Stroup, 1986; National Fire Protection Association, 2007). However, it is necessary to take into account the growth period of a fire when the temperatures and flowrates under the ceiling are increasing. Beyler (1984a) showed how this could be incorporated by assuming that $\dot{Q} \propto t^2$ (Section 9.2.4) and using ceiling jet correlations developed by Heskestad and Delichatsios (1978): this is included in NFPA 72, but is not discussed further here (see Custer *et al.*, 2008).

It should be remembered that Alpert's equations were derived from steady state fires burning under horizontal ceilings of effectively unlimited extent. They will not apply to ceilings of significantly different geometries. Obstructions on the ceiling should represent no more than 1% of the height of the compartment. Moreover, higher temperatures and greater velocities would be anticipated if the fire were close to a wall or in a corner (Section 4.3.4).

4.4.3 Interaction between Sprinkler Sprays and the Fire Plume

The maximum upward velocity in a fire plume ($u(max)$) is achieved in the intermittent flame, corresponding to $z/\dot{Q}_c^{2/5} = 0.08$ to 0.2 in Table 4.2 (McCaffrey, 1979): thus, McCaffrey's data give

$$u_0(max) = 1.9 \dot{Q}_c^{1/5} \text{ m/s} \quad (4.76)$$

where $\dot{Q}_c$ is in kW.

For a sprinkler to function successfully and extinguish a fire, the droplets must be capable of penetrating the plume to reach the burning fuel surface. Rasbash (1962, 1985) and Yao (1976, 1997) identified two regimes, one in which the total downward momentum of the spray was sufficient to overcome the upward momentum of the plume, while in the other the droplets were falling under gravity. In the gravity regime, the terminal velocity of the water drops will determine whether successful penetration can occur. In Figure 4.39, the terminal velocities for water drops in air at three different temperatures are shown as a function of drop size. For comparison, values of $\dot{Q}$ (MW) (for which $u_0(max)$ correspond to the terminal velocities shown on the left-hand ordinate) are given on the right-hand ordinate, referring to McCaffrey's methane flames. Thus, in the 'gravity regime', drops less than 2 mm in diameter would be unable to penetrate vertically into the fire plume above a 4 MW fire. This can be overcome by generating sufficient momentum at the point of discharge but this will be at the expense of droplet size. Penetration may

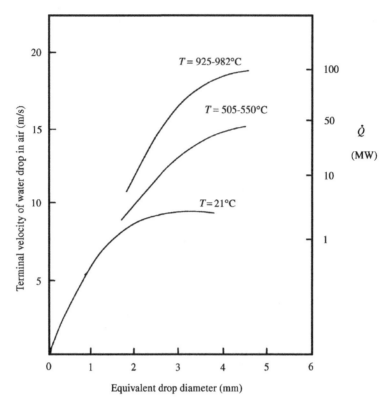

Figure 4.39 Terminal velocity of water drops in air at three temperatures (adapted from Yao (1980)). The right-hand ordinate gives the fire size for which $u_0(max)$ (Equation (4.52)) is equal to the terminal velocity given on the left-hand ordinate (see also Yao (1997))

then be reduced by the evaporative loss of the smallest droplets as they pass through the fire plume. Although this will tend to cool the flame gases, it will contribute little to the control of a fast-growing fire.

It is outside the scope of the present text to explore this subject further, but the above comments indicate some of the problems that must be considered in sprinkler design. Although development of the sprinkler has been largely empirical, there is now a much sounder theoretical base on which to progress (Rasbash, 1985; Yao, 1997). This is being enhanced by recent studies of the interaction of water droplets with sprinkler sprays using CFD modelling (e.g., Hoffmann *et al.*, 1989; Kumar *et al.*, 1997; see also McGrattan, 2006). Moreover, advanced experimental techniques have been brought to bear on the problem (e.g., Jackman *et al.*, 1992b).

4.4.4 The Removal of Smoke

A large proportion of fire injuries and fatalities can be attributed to the inhalation of smoke and toxic gases (Chapter 11). One technique that may be used in large buildings to protect the occupants from exposure to smoke while they are making way to a place of safety is to provide extract fans in the roof. However, it is necessary to know what rate of extraction will be required to prevent the space 'filling up with smoke', or more specifically, what rate will be sufficient to prevent those escaping from being exposed to untenable conditions due to smoke.

If we consider a theatre, with a ceiling 20 m high, a large fire at the front of the stalls could produce a buoyant plume that would carry smoke and noxious gases up to the ceiling, there to form a smoke layer which would progressively deepen as the fire continued to burn. This would place the people in the upper balcony at risk, particularly if they had to ascend into the smoke layer to escape. If 'head height' is 5 m from the ceiling of the theatre, then ideally, the smoke layer should not descend below this level. To achieve this, the extraction fans would have to remove smoke at a rate equivalent to the mass flow of smoke into this layer – which is the total mass flowrate at a height 15 m above the floor of the theatre. This can be calculated using Equation (4.49), provided that the rate of heat release from the fire is known. This has to be estimated, identifying likely materials that may become involved, using whatever information is available (see Chapter 9): it is commonly referred to as the 'design fire'. For example, it may be shown that a block of 16 seats at the front of the theatre could become involved and burn simultaneously, each with a maximum rate of heat release of 0.4 MW, giving a total rate of heat release of 6.4 MW. Taking this as the design fire, then the mass flowrate 15 m above the floor will be:

$$\dot{m} = 0.071 \dot{Q}_c^{1/3} z^{5/3} \left\lfloor 1 + 0.026 \dot{Q}_c^{2/3} z^{-5/3} \right\rfloor$$

$$= 0.071 (6400)^{1/3} 15^{5/3} \left\lfloor 1 + 0.026(6400)^{2/3} 15^{-5/3} \right\rfloor$$

$$= 132.1 \text{ kg/s} \tag{4.49}$$

A correction could be made for the virtual origin (Figure 4.11(b)), but for tall spaces it is a relatively minor correction. Thus, replacing z by $(z - z_0)$, where z_0 is obtained from Equation (4.26), and taking $D = 3$ m (the value for the area of seating involved), $z_0 = -0.3$, and the flowrate becomes 136.1 kg/s.

The temperature of the layer (T_{layer}) may be estimated from the rate of heat release ($\dot{Q}_{design}$, kW), the mass flowrate ($\dot{m}$, kg/s) and the heat capacity of air (c_p, kJ/kg) (Table 2.1), assuming that there are no heat losses (the system is *adiabatic*):[14]

$$\Delta T_{layer} = T_0 + \Delta T$$

where T_0 is the ambient temperature. The temperature rise, ΔT, is give by:

$$\Delta T = \frac{\dot{Q}_{design}}{\dot{m} c_p} = \frac{6400}{136.1 \times 1.0} = 47.0 \text{ K}$$

i.e., if the ambient temperature is 20°C, the smoke layer will be at 67°C (340 K).

The volumetric flowrate of the smoke ($\dot{V}_{smoke}$) can be estimated if the density of air at this temperature is known. Welty *et al.* (2008) give $\rho_{air} = 1.038$ kg/m³ at 340 K (see also Table 11.7), thus:

$$\dot{V}_{smoke} = \frac{\dot{m}}{\rho_{air}} = 131 \text{ m}^3/\text{s}$$

---

[14] The calculation which follows is similar to that used for the calculation of adiabatic flame temperature in Section 1.2.5.

In practice, calculations of this kind could be used to determine the sizes and temperature ratings of smoke extraction fans. In this example a steady fire has been used for illustrative purposes, but real fires grow with time (see Chapter 9). A more general calculation might take this into account coupled with evacuation calculations and analysis of the development of untenable conditions on escape routes to arrive at an appropriate smoke extraction rate.

The above calculation relates to a free-standing fire, producing an axisymmetric plume (Figures 4.9(a) and 4.26(a)). Fires burning against a wall or in a corner (Figure 4.26(b) and 4.26(c)) will produce lower flowrates of smoke than predicted by Equation (4.49), but the temperatures will be higher. In principle, using Equation (4.49) will give conservative values for the mass flowrates generated by these scenarios, but there are other scenarios that require a different approach – the most important of these being the 'spill plume'. This occurs when there is a fire in a compartment that is contiguous with an atrium or multi-level shopping centre and the fire gases 'spill' out into the larger space. The process is complex as the compartment fire develops in the normal manner, but when the smoke layer enters the larger space through the connecting opening, it turns to produce a vertical flow, carrying smoke upwards – entraining air as it does so (Butcher and Parnell, 1979; Milke, 2008b). Calculating the amount of smoke that would have to be extracted to maintain the smoke layer at a safe height is difficult as there are so many factors that must be taken into account. For example, if the flow emerging from the compartment has to flow under a balcony before it can 'turn' to produce an upward-flowing plume, the presence of screens that channel the flow – thus preventing sideways spread before 'turning' – will restrict the width of the plume in the turning region. This has the effect of keeping the flow in the vertical plume to a minimum. Another factor that has to be considered is whether or not the vertical plume 'attaches' to the wall above the turning region. Empirical formulae have been derived to enable the mass flowrate to be calculated as a function of height (cf. Equation (4.49) for the asymmetric plume), but there is considerable uncertainty regarding their validity (e.g., see McCartney *et al.*, 2008). Milke (2008b) quotes the following expression:

$$\dot{m} = 0.36(\dot{Q}L^2)^{1/3}(z + 0.25H_{\mathrm{b}}) \tag{4.77}$$

where L is the width of the balcony spill plume (m), z is the clear air height between the fuel surface and the smoke layer in the compartment (m), and H_{b} is the height of (the underside of) the balcony above the fuel surface (m). This is based on Law's (1986) interpretation of the results from small-scale experiments carried out by Morgan and Marshall (1979). The key to the problem is understanding the physics of the entrainment process in the turning region (e.g., see Thomas *et al.*, 1998; Kumar *et al.*, 2010) so that a model can be developed and tested against reliable data. There has been a paucity of suitable experimental data, but recent work published by Harrison and Spearpoint (2006, 2007) may lead the way to resolving the situation.

4.4.5 Modelling

The term 'modelling' has two connotations, physical and mathematical. Although 'mathematical' modelling of fire has become predominant in the last two decades, 'physical' modelling has provided the basis for our understanding of the fundamentals of fire dynamics. Many problems in other branches of engineering have been resolved by the same approach, applying procedures which permit full-scale behaviour to be predicted from the results of small-scale laboratory experiments. The prerequisite is that the physical model is 'similar' to the prototype in the sense that there is a direct correlation between the responses of the two systems to equivalent stimuli or events (Hottel, 1961). The procedure has been developed through the application of dimensional analysis (e.g., Quintiere, 1989a, 2006). Scaling the model is achieved by identifying the important parameters of the system and expressing these in the form of relevant dimensionless groups (Table 2.4).[15] For exact similarity, these must have the same values for the prototype and the model, but in fact it is not possible for all the groups to be preserved. Thus, in ship hull design, small-scale models are used for which the ratio L/u^2 (L = length scale and u = rate of flow of water past the hull) is identical to the full-scale ship, so that the Froude number (u^2/gL) is preserved, although inevitably the Reynolds number ($uL\rho/u$) will vary. Corrections based on separate experiments can be made to enable the drag on the full-scale prototype to be calculated from results obtained with the low Reynolds number model (Friedman, 1971).

There are obvious advantages to be gained if the same approach could be applied to the study of fire. The number of dimensionless groups that should be preserved is quite large as the forces relating to buoyancy, inertia and viscous effects are all involved. However, there are two methods that are available, namely Froude modelling and pressure modelling. Froude modelling is possible for situations in which viscous forces are relatively unimportant and only the group $u_\infty^2 \rho/lg \Delta \rho$ need be preserved. This requires that velocities are scaled with the square root of the principal dimension, i.e., $u/l^{1/2}$ is maintained constant. In natural fires, when turbulent conditions prevail, behaviour is determined by the relative importance of momentum and buoyancy: viscous forces can be ignored. In the introduction to this chapter, it was pointed out that the relevant dimensional group is the Froude number, $Fr = u^2/gD$ (Table 2.4), which can be expressed in terms of the rate of heat release. The non-dimensional group $\dot{Q}^*$ is the square root of a Froude number:

$$\dot{Q}^* = \frac{\dot{Q}_c}{\rho_\infty c_p T_\infty \sqrt{gD} \cdot D^2} \tag{4.3}$$

The significance of this group has already been demonstrated in correlations of flame height, etc. It is consistent with an early dimensional analysis that revealed that the rate of heat release (or rate of burning ($\dot{m}$, g/s)) of a fire had to be scaled with the five-halves power of the principal dimension, i.e., $\dot{m}/l^{5/2}$ (or $\dot{m}^2/l^5$) must be preserved. The quotient $\dot{m}^2/l^5$ appears in early correlations of flame height (Thomas *et al*., 1961), but is replaced by $\dot{Q}_c^2/D^5$ (or $\dot{Q}^*$) in more recent flame height correlations (see Figure 4.18, Zukoski *et al*. (1981a)). It also appears in correlations of temperatures and velocities in the fire plume (Figures 4.8(b) and 4.20) (McCaffrey, 1979) and ceiling temperatures directly above a fire at floor level (Alpert, 1972). However, limitations on Froude modelling are found when viscous effects become important, e.g., in laminar flow situations. For this reason, the physical model must be large enough to ensure turbulent flows (Thomas *et al*., 1963). Difficulties also arise when transient processes such as flame spread are being modelled

---

[15] This is demonstrated in Figure 2.16.

as the response times associated with transient heating of solids follow different scaling laws (de Ris, 1973).

Pressure modelling has the advantage of being able to cope with both laminar and turbulent flow (Alpert, 1975; Quintiere, 1989a). The Grashof number may be preserved in a small-scale model if the pressure is increased in such a way as to keep the product $\rho^2 l^3$ constant. This can be seen by rearranging the Grashof number thus:

$$\text{Gr} = \frac{gl^3 \Delta\rho}{\rho v^2} = \frac{g\rho^2 l^3 \Delta\rho}{\mu^2} = \frac{g}{\mu^2}\left(\frac{\rho}{\Delta\rho}\right)^{-1}\rho^2 l^3 \tag{4.78}$$

where μ (the dynamic viscosity), g and $(\Delta\rho/\rho)$ are all independent of pressure. Thus, an object 1 m high at atmospheric pressure could be modelled by an object 0.1 m high if the pressure was increased to 31.6 atm. In experiments carried out under these conditions, it is also possible to preserve the Reynolds number for any forced flow in the system:

$$\text{Re} = \rho u_\infty l/\mu \tag{4.79}$$

In designing the experiment, l has been scaled with $\rho^{-2/3}$, so that u (the velocity of any imposed air flow) must be scaled with $\rho^{-1/3}$ to maintain a constant Reynolds number. The Froude number is then automatically conserved, as $\text{Fr} = \text{Re}^2/\text{Gr}$. The validity of this modelling technique has been explored by de Ris (1973).

It might appear that Froude modelling or pressure modelling could be used as the basis for designing physical models of fire, but different non-dimensional groups have to be conserved for transient processes such as ignition and flame spread. Perhaps more significant is the fact that it is not possible to scale radiation as it is such a highly non-linear function of temperature. In effect, this means that small-scale physical models of fire cannot be used directly, either as a means of improving our understanding of full-scale fire behaviour, or in assessing the likely performance of combustible materials in fire situations.

Mathematical modelling, on the other hand, has been developed to a stage where it is now possible to gain valuable insight into certain types of fire phenomena, as well as being able to carry out an examination of the consequences of change – answering the 'what if' questions. The approach is based on the fundamental physics of fluid dynamics and heat transfer. Some of the simplest models are considered elsewhere: examples include the transfer of heat to a sprinkler head (Section 4.4.2) and the ignition of combustible materials exposed to a radiant heat source (Section 6.3.1). Modelling of flame spread has received considerable attention, starting with attempts to derive analytical solutions to the appropriate conservation equations (e.g., de Ris, 1969; Quintiere, 1981). In general, many simplifying assumptions have to be made to achieve this. Latterly, attention has been turned to developing numerical models which can be solved computationally. There has been much impetus to use this type of model to assist in the assessment of the fire hazards associated with combustible wall lining materials, using data obtained from small-scale tests, such as the cone calorimeter (Magnusson and Sundstrom, 1985; Karlsson, 1993; Grant and Drysdale, 1995). This is addressed briefly in Section 7.3.

Very significant advances have been made in modelling fires and fire phenomena using 'zone' and 'field' models. Zone models derive from work carried out at Harvard University in the late 1970s and early 1980s by Emmons and Mitler (Mitler and Emmons, 1981;

Mitler, 1985). The original model was developed for a fire in a single compartment, or enclosure, which is divided into a small number of control volumes – e.g., the upper smoke layer, the lower layer of clear air, the burning fuel and the fire plume (below the smoke layer). The basic conservation equations (for mass, energy and chemical species) are solved iteratively as the fire develops, entraining air which enters through the lower part of the ventilation opening and expelling hot smoke through the upper part (see Chapter 10). They generally rely on empirical or semi-empirical correlations to enable various features to be incorporated, such as entrainment into the fire plume and the rate of supply of air into the compartment. The effect of the transfer of radiant heat from the smoke layer to the surface of the burning fuel can be included, thus allowing the development of the fire to be modelled. Several zone models have been developed, many of which have been summarized by Walton *et al.* (2008). One of the most importance, CFAST, can deal with multiple interlinked compartments with multiple openings (Walton *et al.*, 2008; Peacock *et al.*, 2008). However, in this text, only the fundamental principles on which the zone models are based can be covered: readers may wish to consult relevant review articles such as those by Quintiere (1989b), Cox (1995) and Novozhilov (2001) specifically devoted to aspects of the subject.

The term 'field model' is used in the fire community as a synonym for computational fluid dynamics (CFD). Instead of a small number of zones, the relevant space is divided into a very large number of control volumes – from 10^4 (minimum) to 10^6–10^7 or more – and the partial differential conservation equations (the Navier–Stokes equations) are solved iteratively for every control volume, stepping forward in time (McGrattan and Miles, 2008). This is computationally intensive, but modern computers are fast enough to make this type of modelling feasible. In 1988, the first 3-D simulation of a major fire spread scenario using CFD had to be run on one of the largest available computers, the Cray 'supercomputer' at Harwell (Simcox *et al.*, 1992). In the year 2010, it is possible to run complex CFD models on a laptop. Indeed, it might be said that modern computer technology has allowed fire modelling to develop too rapidly, outstripping our understanding of fire dynamics and our ability to use the models in a safe and constructive manner. This is chiefly due to the fact that insufficient experimental data are available to enable verification of the models. This problem was highlighted by Emmons (1984) in his remarkable paper 'The further history of fire science', written from the perspective of the year AD2280. The point has been addressed by others (e.g., Beard, 2000; Novozhilov, 2001) and problems have been identified in attempts to predict *a priori* the outcome of fully instrumented fire tests (e.g., Rein *et al.*, 2009). Very few suitable databases exist against which models may be thoroughly tested, but with large capacity data-logging systems now available (Luo and Beck, 1994), this problem can be resolved.

The advantage of CFD is that it involves solving the fundamental equations of fluid dynamics, and coupled with empirical flame chemistry and radiation models it can give an adequate description of a variety of fire phenomena, including smoke movement in large spaces (Cox *et al.*, 1990), flame spread (coupling the gas phase and condensed phase processes (di Blasi *et al.*, 1988)), the structure of the flames of pool fires (Crauford *et al.*, 1985) and the development of a fire in a compartment (Cox, 1983). However, turbulence cannot be modelled using present computational power and approximations have to be made, either the Reynolds-averaged Navier–Stokes (RANS) equations, or large eddy simulation (LES) (Novozhilov, 2001; McGrattan and Miles, 2008).[16] Cox (1995) has provided an excellent review of the application of field models to fire problems, but it should be borne in mind that CFD models have their greatest value as research tools, guiding the research scientist towards a fuller understanding of the fundamentals of fire dynamics. Nevertheless, it is clear that they have great potential as tools to assist with the fire safety engineering design of complex buildings, but at the present time they must be applied with great care.

Problems

4.1   Calculate the length of the turbulent diffusion flame formed when pure methane is released at high pressure through a nozzle 0.1 m in diameter. Assume that the flame temperature for methane is 1875°C (Lewis and von Elbe, 1987) and the ambient temperature is 20°C. What would the length be if the nozzle fluid consisted of 50% methane in air?

4.2   Calculate the rate of heat release for the following fires: (a) natural gas (assume methane) released at a flowrate of 20×10^{-3} m^3/s (measured at 25°C and normal atmospheric pressure) through a sand bed burner 1.0 m in diameter (assume complete combustion); (b) propane, released under the same conditions as (a).

4.3   Calculate the value of $\dot{Q}^*$ for fires (a) and (b) in Problem 4.2.

4.4   Calculate the heights of the flames (a) and (b) in Problem 4.2. Check your answers to Problem 4.3 by calculating the flame heights from Equation (4.41). Estimate the flame volumes.

4.5   Calculate the frequency of oscillation of the flames (a) and (b) in Problem 4.2.

4.6   Using data from the previous questions, calculate the temperature on the centreline of the fire plume at a height equal to four times the flame height for fires (a) and (b). What difference does it make if allowance is made for the virtual source?

4.7   Given a 1.5 MW fire at floor level in a 4 m high enclosure which has an extensive flat ceiling, calculate the gas temperature under the ceiling (a) directly above a fire; and (b) 4 m and 8 m from the plume axis. Assume an ambient temperature of 20°C and steady state conditions.

4.8   Using the example given in Problem 4.7, calculate the maximum velocity of the gases in the ceiling jet 4 m and 8 m from the plume axis. How long will it take the sensing element of a sprinkler head to activate in these two positions if it is rated at 60°C and the RTI is $100 \, \text{m}^{1/2}\text{s}^{1/2}$? What will be the effect of the RTI being reduced to $25 \, \text{m}^{1/2}\text{s}^{1/2}$? (Assume that the air temperature in the plunge test is 200°C.)

4.9   What is the minimum size of fire at floor level capable of activating fixed temperature heat detectors (rated at 70°C) in a large enclosure 8 m high? Assume that the

---

[16] An alternative method, direct numerical simulation (DNS), does not require approximate methods to deal with turbulence, but as the required spatial and temporal resolution is so small it cannot be used for fire problems. So far it has only been applied successfully to small laminar flames and small turbulent jets (McGrattan and Miles, 2008).

ceiling is flat and that the detectors are spaced at 5 m centres. Ambient temperature is 20°C. Consider three situations in which the fire is (a) at the centre of the enclosure; (b) close to one wall; and (c) in a corner.

4.10 The space described in Problem 4.9 is to be protected by smoke detectors spaced at 5 m centres. On the assumption that these will activate when the temperature at the detector head has increased by 15 K, calculate the minimum fire sizes required to activate the heads when the fire is (a) at the centre of the enclosure; (b) close to one wall; and (c) in a corner if the detectors are at 5 m centres.

4.11 If the height of the space described in the previous question is reduced to 2.5 m, calculate the minimum fire sizes that would activate a smoke detector for cases (a), (b) and (c) with the detectors at 5 m centres and 2.5 m centres.

4.12 In Section 4.4.4, the rate of flow of smoke into a smoke layer 15 m above the floor of the front stalls of a theatre was calculated for a design fire of 6.4 MW. Calculate (a) the rate of extraction that must be provided at roof level to prevent smoke descending below 10 m from the floor for the same design fire (6.4 MW); and (b) the rate of extraction required to prevent the smoke layer descending below 15 m if the seats are replaced by ones that individually gave a maximum rate of heat release of 0.25 MW. Assume that the same number of seats (16) is involved in the design fire.

4.13 Calculate the temperatures of the smoke layers formed in parts (a) and (b) of the previous question, assuming no heat losses (adiabatic). Compare these with the value obtained for smoke layer 15 m above floor for the design fire (6.4 MW) in Section 4.4.4.

4.14 A pool fire involving a roughly circular area 25 m in diameter where a substantial depth of gasoline has accumulated is exposing surrounding structures and materials to radiant heat. Estimate the magnitude of the flux at ground level at (a) 25 m and (b) 50 m from the edge of the pool using the empirical formula derived by Shokri and Beyler (Equation (4.65)). Compare these figures with the values of the flux at the same positions using Equation (4.62), with the accompanying data. Finally, consider the consequence of the assumption made in Equation (4.62) regarding the radiation factor (χ) when compared with Figure 4.36.

5

Steady Burning of Liquids and Solids

In the previous chapter, it was shown that the size of a fire as perceived by flame height depends on the diameter of the fuel bed and the rate of heat release due to the combustion of the fuel vapour. The latter may be expressed in terms of the primary variable, the mass flowrate of the fuel vapours ($\dot{m}$), thus:

$$\dot{Q}_c = \dot{m} \chi \Delta H_c \tag{5.1}$$

where ΔH_c is the heat of combustion of the volatiles and χ is an efficiency factor that takes into account incomplete combustion (Tewarson, 1980, 2008). It was necessary to rely on the mass flowrate (i.e., the rate of mass loss, commonly associated with the term 'burning rate') until it became possible to measure $\dot{Q}_c$ directly, either on a small scale with the cone calorimeter (Babrauskas, 2008b), or for full-scale items in the furniture calorimeter, the room calorimeter, or one of many variants (Babrauskas, 1992a) using the technique of oxygen consumption calorimetry (OCC) (Section 1.2.3).

Information on the rate of heat release of liquid fuels and combustible solids is required not only to evaluate flame size (Section 4.3.2) but also to assess likely flame behaviour in practical situations such as interaction with compartment boundaries (Section 4.3.3), and to estimate the contribution that individual combustible items may make towards fire development in a compartment (Section 9.2.2, Figure 9.14).

In this chapter, the steady burning of combustible solids and liquids is considered in detail. Given that the rate of mass loss is the major factor in determining the rate of heat release (Equation (5.1)), the parameters that determine how rapidly volatiles are produced under fire conditions will be identified. However, it is important to remember that steady burning will only be achieved after an initial transient period following ignition (Chapter 6) and spread of flame over the combustible surface (Chapter 7). The transient stage will continue until a quasi-equilibrium is established (an illustration of this is provided in Section 5.1.1, Figure 5.5).

An Introduction to Fire Dynamics, Third Edition. Dougal Drysdale.
© 2011 John Wiley & Sons, Ltd. Published 2011 by John Wiley & Sons, Ltd.

5.1 Burning of Liquids

In this section, only those liquids that are in the liquid state under normal conditions of temperature and pressure (20°C and 101.3 kPa) are considered in detail, although cryogenic liquids such as LNG (liquefied natural gas) and pressurized liquids such as LPG (liquefied petroleum gas) are discussed briefly in Section 5.1.4.

The term 'pool' is used to describe a liquid with a free surface contained within a tray, an open tank or any similar confining configuration such as a bund, or a depression in the ground where liquid has been allowed to accumulate. Its depth may be anything from several metres (as in an oil storage tank) to a few millimetres. If the fuel depth is less than 5–10 mm, heat losses to the ground (or the substrate on which it lies) will reduce the burning rate to a lower steady state: depending on the flashpoint of the fuel (Section 6.2), extinction may occur if the heat losses become too great as the depth decreases. This is demonstrated most clearly by Garo et al. (1994) for layers of heating oil floating on water.[1] They found that if the layer of oil was less than c. 2 mm thick, heat losses to the water layer were too great for burning to be sustained (see Equation (5.3) et seq.).

The term 'pool' is normally applied to a liquid of substantial depth – certainly more than 10 mm, when heat losses to the ground have only a minimal effect on the rate of burning. However, the release of liquid on to flat ground (or on to water) will produce a shallow spill which, if ignited, will continue to spread fire and burn for as long as the release continues. The distinction between a pool fire and a spill fire is important, as will become apparent in Section 5.1.2 (Gottuk and White, 2008).

5.1.1 Pool Fires

The early Russian work on liquid pool fires remains the most extensive single study. Blinov and Khudiakov (1957) (Hottel, 1959; Hall, 1973) studied the rates of burning of pools of hydrocarbon liquids with diameters ranging from 3.7×10^{-3} to 22.9 m. A constant head device was used with all the smaller 'pools' to maintain the liquid surface level with the rim of the container. This point of detail is important in experimental work: if there is an exposed rim above the liquid surface, the characteristics of the flame are altered (Corlett, 1968; Hall, 1973; Orloff and de Ris, 1982; Brosmer and Tien, 1987; Bouhafid et al., 1988) due to the turbulence induced by the entrainment of air around the perimeter of the container. This causes an increase in the rate of convective heat transfer to the fuel surface, which in turn affects the rate of burning significantly (de Ris, 1979).

Blinov and Khudiakov found that the rate of burning expressed as a 'regression rate' R (mm/min) (equivalent to the volumetric loss of liquid per unit surface area of the pool in unit time) was high for small-scale laboratory 'pools' (0.01 m diameter and less), and exhibited a minimum at around 0.1 m (Figure 5.1). While the regression rate (R, mm/min) is convenient for some purposes, the mass flux (kg/m².s) is a more logical measure of the rate of burning. The conversion is straightforward:

$$\dot{m}'' = \rho_l \cdot R \cdot 10^{-3}/60 \quad \text{kg/m}^2 \cdot \text{s} \tag{5.2}$$

---

[1] The heating oil had a boiling point range of 250–350°C. This work was carried out to investigate the phenomenon of boilover, which is discussed below.

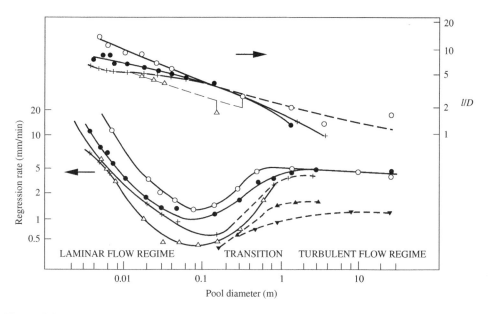

Figure 5.1 Regression rates and flame heights for liquid pool fires with diameters in the range 3.7×10^{-3} to 22.9 m. o, Gasoline; •, tractor kerosene; △, solar oil; +, diesel oil; ▲, petroleum oil; ▼, mazut oil (Blinov and Khudiakov, 1957, 1961; Hottel, 1959). By permission

where ρ_l is the density of the liquid (kg/m^3). This may be used to estimate a rate of heat release using Equation (1.4).

Three regimes can be distinguished. If the diameter is *less* than 0.03 m, the flames are laminar and the rate of burning, R, falls with increase in diameter, while for large diameters ($D > 1$ m), the flames are fully turbulent and R becomes independent of diameter. In the range $0.03 < D < 1.0$ m, 'transitional' behaviour, between laminar and turbulent, is observed. The dependence on pool diameter can be explained in terms of changes in the relative importance of the mechanisms by which heat is transferred to the fuel surface from the flame. The rate of steady burning can be expressed as

$$\dot{m}'' = \frac{\dot{Q}''_F - \dot{Q}''_L}{L_v} \text{ g/m}^2.\text{s} \tag{5.3}$$

(Equation (1.3)): this is written as a quasi-equilibrium, in which the net rate of heat transfer ($\dot{Q}''_F - \dot{Q}''_L$) exactly balances the energy required for the vaporization of the fuel ($\dot{m}''L_v$). The heat flux from the flame ($\dot{Q}''_F$) can be expressed as the sum of three terms, namely:

$$\dot{Q}''_F = \dot{Q}''_{\text{conduction}} + \dot{Q}''_{\text{convection}} + \dot{Q}''_{\text{radiation}} \tag{5.4}$$

where $\dot{Q}''_{\text{radiation}}$ takes into account surface re-radiation, which would normally be considered as part of $\dot{Q}''_L$ (see Equation (5.3)).

The conduction term refers to heat transfer through the rim of the container, thus

$$\dot{Q}''_{\text{conduction}} = k_1 \pi D (T_F - T_l) \tag{5.5}$$

where T_F and T_l are the flame and liquid temperatures, respectively, and k_1 is a constant which incorporates additional heat transfer terms and an effective container height (below the rim) through which heat is transferred to the liquid. For convection direct to the fuel surface:

$$\dot{Q}''_{convection} = k_2 \frac{\pi D^2}{4}(T_F - T_l) \tag{5.6}$$

where k_2 is the convective heat transfer coefficient. Finally, the radiation term is given by:

$$\dot{Q}''_{radiation} = k_3 \frac{\pi D^2}{4}\left(T_F^4 - T_l^4\right)(1 - \exp(-k_4 D)) \tag{5.7}$$

where k_3 contains the Stefan–Boltzmann constant (σ) and the configuration factor for heat transfer from the flame to the fuel surface (Section 2.4.1), while $(1 - \exp(-k_4 D))$ is the effective emissivity of the flame (Equation (2.83)). On inspection, it can be seen that k_4 must contain not only some factor of proportionality relating the mean beam length to the pool diameter (see Table 2.9), but also concentrations and emission coefficients of the radiating species in the flame.[2] Dividing Equations (5.5)–(5.7) by the pool surface area $\pi D^2/4$ and substituting the results into Equation (5.4):

$$\dot{Q}''_F = \frac{4\Sigma\dot{Q}}{\pi D^2} = 4\frac{k_1(T_F - T_l)}{D} + k_2(T_F - T_l) + k_3\left(T_F^4 - T_l^4\right)(1 - \exp(-k_4 D)) \tag{5.8}$$

The regression rate would then be given by:

$$R = \frac{\dot{Q}''_F - \dot{Q}''_L}{\rho L_v} \tag{5.9}$$

where ρ is the density of the liquid, L_v is the latent heat of evaporation and $\dot{Q}''_F$ is given by Equation (5.8). This has the correct mathematical form to account for the shape of the curves in Figure 5.1. When D is very small, conductive heat transfer determines the rate of burning while the radiative term predominates if D is large, provided that k_4 is of sufficient magnitude.

The increasing dominance of radiation in large diameter ($D > 1$ m) hydrocarbon fires may be deduced from results obtained by Burgess *et al.* (1961).[3] They correlated their data on the rates of burning of hydrocarbon liquids in open trays (diameters up to 1.5 m) using the expression:

$$R = R_\infty(1 - \exp(-k_4' D)) \tag{5.10}$$

where R_∞ is the limiting, radiation-dominated regression rate (compare Equations (5.7) and (5.10) and Figures 5.1 and 5.2). Some values of R_∞ obtained by Burgess *et al.* are given in Table 5.1.

These data apply to still air conditions. If there is an imposed airflow (e.g., wind; Lois and Swithenbank (1979)) or forced ventilation as in a tunnel (Ingason, 2005), or perhaps

---

[2] k_4 must also incorporate a factor for 'radiation blocking' when the layer of fuel vapours above the fuel becomes sufficiently thick to attenuate the flux falling on the surface (see, e.g., de Ris (1979) and Brosmer and Tien (1987)).
[3] Subsequently, Yumoto (1971) reported experimental data that showed clearly the increasing dominance of radiation over convection for hydrocarbon fires as the pool diameter was increased from 0.6 to 3.0 m.

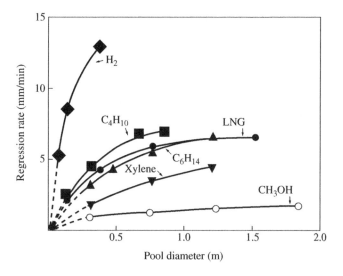

Figure 5.2   Regression rates of burning liquids in open trays (Burgess *et al.*, 1961)

Table 5.1   Limiting regression rates for liquid pool fires (Burgess *et al.*, 1961)

Liquid fuel	Limiting regression rate (R_∞, mm/min)[a]
LNG	6.6
n-Butane	7.9
n-Hexane	7.3
Xylene	5.8
Methanol	1.7

[a]From Equation (5.10) and Figure 5.2.

an induced flow by virtue of restricted ventilation (e.g., in a room with an open door; Steckler *et al.* (1984)), changes to the burning rates are to be expected. However, the data available are limited and do not allow generalizations to be made, as observed by Lam *et al.* (2004). Within a given data set, some order may be found – for example, Lois and Swithenbank (1979) report the burning rate of a 0.92 m pool of hexane to increase monotonically from c. 3 mm/min to about 4.5 mm/min as the wind velocity was increased from c. 2.5 to c. 5.5 m/s. However, in tunnel fire experiments, Apte *et al.* (1991) found that the burning rate of aviation fuel in a 1 m diameter tray decreased monotonically from 5.6 to 4.8 mm/min as the air velocity increased from 0.5 to 2 m/s. Different behaviour is reported for acetone pool fires, which appear to be independent of air velocity up to c. 1 m/s, and then decrease (to 1.4 m/s) for diameters in the range 0.3–0.6 m (Welker and Sliepcevich, 1966; Lam *et al.*, 2004). On the other hand, the regression rates of large pools (e.g., 20 m diameter) of hydrocarbon fuels (JP4 and JP8) have been found to increase with increasing air velocity (Lam *et al.*, 2004). It would appear that the variation of regression rate with imposed wind speed is influenced by many factors that have yet to be resolved. Moreover, although there are several ways of measuring regression rate, the accuracy is generally poor, particularly for larger pool diameters.

Zabetakis and Burgess (1961) recommended that the following expression be used to predict the burning rate (kg/m^2.s) of liquid pools of diameters greater than 0.2 m in still air conditions (cf. Equation (5.10)):

$$\dot{m}'' = \dot{m}''_\infty(1 - \exp(-k\beta D)) \tag{5.11}$$

where the product $k\beta$ is equivalent to k'_4 in Equation (5.10), and consists of an extinction coefficient (k, m^{-1}) and a 'mean beam length corrector' (β). From his survey of contemporary published data on a range of liquids, Babrauskas (1983b) proposed values of $\dot{m}''_\infty$ and $k\beta$ which are given in Table 5.2. What is surprising is the relatively small range of values of $\dot{m}''_\infty$ found: with the exception of the petroleum products quoted (gasoline, kerosene and crude oil), the hydrocarbon fuels lie between 0.06 and 0.10 kg/m^2.s, including two cryogenic liquids – methane and propane (see Section 5.1.4). Table 5.2 shows that the limiting burning rates for the simple alcohols methanol and ethanol are much less than that of the hydrocarbons. This is partly due to the greater values of L_v (see Equation (5.3)) for these liquids (Table 5.3), but is also a result of the much lower emissivity of the alcohol flames, which is associated with a relatively low radiative heat flux to the surface (see Section 2.4.2). The significance of the latter was first demonstrated by

Table 5.2 Data for estimating the burning rate of large pools (Babrauskas, 1983b)

Liquid	Density (kg/m^3)	$\dot{m}''_\infty$ (kg/m^2 · s)	$k\beta$ (m^{-1})
Cryogenics			
Liquid methane	415	0.078	1.1
Liquid propane	585	0.099	1.4
Alcohols			
Methanol	796	0.017	–
Ethanol	794	0.015	–
Simple organic fuels			
Butane	573	0.078	2.7
Benzene	874	0.085	2.7
Hexane	650	0.074	1.9
Heptane	675	0.101	1.1
Acetone	791	0.041	1.9
Petroleum products			
Gasoline	740	0.055[a]	2.1
Kerosene	820	0.039	3.5
Crude oil	830–880	0.022–0.045	2.8

[a]Chatris *et al.* (2001) quote 0.077 kg/m^2·s for gasoline burning in a 4 m diameter tray.

Table 5.3   Latent heats of evaporation of some liquids
(Lide, 1993/94)

Liquid	Boiling point (°C)	L_{v} (kJ/g)[a]
Water	100	2.258
Methanol	64.6	1.100
Ethanol	78.3	0.838
Methane	−161.5	0.512
Propane	−42.1	0.433
Butane	−0.5	0.387
Benzene	80.09	0.394
Hexane	66.73	0.335
Heptane	98.5	0.318
Decane	174.15	0.273

[a]The latent heat of evaporation refers to the boiling point
at normal atmospheric pressure.

Rasbash *et al.* (1956), who made a detailed study of the flames above 30 cm diameter
pools of alcohol, benzene, kerosene and petrol. Their apparatus is shown in Figure 5.3.
They measured the burning rates of these liquids and estimated the emission coefficients
of the flames (K) from measurements of flame shape, temperature and radiant heat loss,
assuming that emissivity could be expressed as:

$$\varepsilon = 1 - \exp(-KL) \tag{5.12}$$

where L is the mean beam length (see Table 2.9 and Equation (2.83) *et seq.*: an accurate
value can only be calculated if the flame shape is known). Their results are summarized
in Table 5.4. One significant feature of these data is that the temperature of the non-
luminous alcohol flame is much higher than that of the hydrocarbon flames, which lose

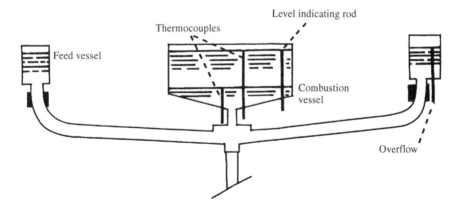

Figure 5.3   Details of the apparatus used by Rasbash *et al.* (1956) to study liquid pool fires

Table 5.4 Radiation properties of flames above 0.3 m diameter pool fires (Rasbash *et al*., 1956). The liquid level was maintained 20 mm below the rim of the vessel

	Flame temperature ($^\circ$C)[a]	Flame width (m)	K (m^{-1})	Emissivity (ε)
Alcohol	1218	0.18	0.37	0.066
Petrol	1026	0.22	2.0	0.36
Kerosene	990	0.18	2.6	0.37
Benzene	921			
after 2 min		0.22	3.9	0.59
after 5 min		0.29	4.1	0.70
after 8 min		0.30	4.2	0.72

[a]Time-averaged flame temperatures measured by the Schmidt method (Gaydon and Wolfhard, 1979).

a considerable proportion of heat by radiation from the soot particles within the flame (Section 2.4.3). The amount of heat radiated to the surface of the pool was calculated using Equation (5.12) with the appropriate configuration factor and compared with the rate of heat transfer required to produce the observed rate of burning (see Equation (5.3) *et seq*.).

The results of these calculations are given in Table 5.5 and show that the radiative flux to the surface in the case of the alcohol falls far short of that required to maintain the flow of volatiles.[4] Careful observation of the pale blue alcohol flame reveals that it burns very close to the surface, apparently touching it – as shown in Figure 5.4(a). This is consistent with both the low burning rate of alcohol and the apparent enhanced rate of convective heat transfer, compared to the hydrocarbon fuels studied by Rasbash *et al*. (1956). For these, there is a discernible vapour zone immediately above the liquid surface. This was particularly apparent for benzene, which eventually adopted the shape illustrated in Figure 5.4(d), thereafter oscillating occasionally between 5.4(d) and 5.4(e). For the three hydrocarbon fuels, the estimated radiant heat flux was greater than the heat

Table 5.5 Radiative heat transfer rates to the surface of burning liquids compared with the net heat transfer rates (Rasbash *et al*., 1956)

Liquid	Heat required to maintain steady burning rate (kW)	Estimated radiant heat transfer from flames to surface (kW)
Alcohol	1.22	0.21
Benzene	2.23	2.51
Petrol	0.94	1.50
Kerosene	1.05	1.08

---

[4] These calculations refer to the gross radiant heat transfer from the flame to the surface. Radiative losses from the surface were not taken into account, but in the case of liquids would be expected to be small.

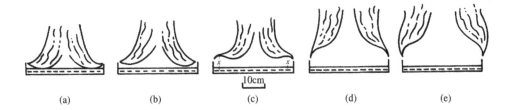

Figure 5.4 Shapes of flames immediately above the surface of burning liquids (Rasbash *et al.*, 1956). Reproduced from *Fuel*, **31** (1956) 94–107, by permission of the publishers, Butterworth & Co. (Publishers) Ltd. ©

flux required to produce the flow of vapours, suggesting that the vapour zone above the surface might be attenuating the radiation reaching the surface. However, it is not possible to draw any conclusions regarding the relative magnitudes of the effect for the three fuels as the flame shapes and thicknesses of the vapour zones (at x in Figure 5.4(c)) were different (values of x were 50 mm, 40–50 mm and 25–30 mm for benzene, petrol and kerosene, respectively).

The attenuation, or blocking effect has been studied by a number of authors. Brosmer and Tien (1987) calculated the radiant heat feedback to the surface of PMMA 'pool fires'[5] ($\dot{Q}''_F$ in Equation (5.3)), comparing their results with measurements taken by Modak and Croce (1977) on PMMA fires of diameter 0.31, 0.61 and 1.22 m. They made two important observations: (i) it was necessary to model the flame shape accurately to allow a realistic mean beam length to be calculated; and (ii) the predicted feedback was too high unless absorption of radiation by the cool vapour layer above the surface was included in the model. Moreover, for pool diameters greater than c. 0.5 m, it was predicted that the maximum radiant intensity would not lie at the centre of the PMMA 'pool', but some way towards the perimeter as a consequence of the 'blockage' effect.[6]

The rate of burning is unlikely to be constant across any horizontal fuel surface. In their study of *small* liquid pool fires in concentric vessels, 10–30 cm in diameter, Akita and Yumoto (1965) found that the rate of evaporation was greater near the perimeter than at the centre, an effect that was most pronounced with methanol. This is consistent with the fact that convection tends to dominate the heat transfer process from flames above small pools (see Figures 5.1 and 5.4(a)); but as the size of the pool is increased, radiative heat transfer becomes dominant, resulting in more rapid burning towards the centre, but moderated by absorption by the cool vapours (de Ris, 1979; Brosmer and Tien, 1987).

Only the surface layers of a deep pool of a pure liquid fuel will be heated during steady burning. A temperature distribution similar to that shown in Figure 5.5 will become established below the surface. This takes some time to develop, accounting for the characteristic 'growth period' between ignition and fully developed burning (Rasbash

---

[5] The term 'pool fire' in this context refers to the horizontal configuration.

[6] Wakatsuki *et al.* (2007) have shown that radiation blocking is significant for methanol pool fires and that modelling the effect requires reliable data on the absorption characteristics of the vapour at elevated temperatures.

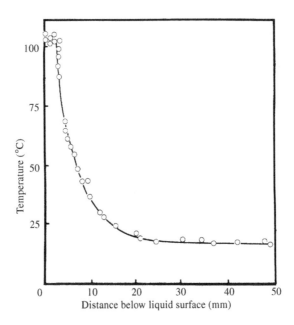

Figure 5.5   Temperature distribution below the surface of *n*-butanol during steady burning ('pool' diameter 36 mm). From Blinov and Khudiakov (1961), by permission

et al., 1956; Chatris *et al.*, 2001). Khudiakov showed that the distribution in Figure 5.5 can be described by the empirical equation:

$$\frac{T - T_{\infty}}{T_S - T_{\infty}} = \exp(-\kappa x) \tag{5.13}$$

where x is the depth and T_S and T_{∞} are the temperatures at $x = 0$ and $x = \infty$, respectively. Equation (5.13) is the solution of Equation (2.15), in which t is replaced by x/R_{∞}, where R_{∞} is the regression rate, i.e.

$$\frac{d^2T}{dx^2} = \frac{R_{\infty}}{\alpha}\left(\frac{dT}{dx}\right) \tag{5.14}$$

Accordingly, the constant κ in Equation (5.13) should be equal to $R_{\infty}\rho c/k$, but the agreement is poor for the results shown in Figure 5.5, possibly as a result of interactions with the rim and the walls of the container which are not taken into account in the simple heat balance described by Equation (5.14). Certainly, the shallow surface layer at constant temperature is not consistent with this model and may indicate in-depth absorption of radiation (e.g., Inamura *et al.*, 1992).

As a pool of liquid burns away and its depth decreases, there will come a time when heat losses to the base of the container will become increasingly important. If this represents a substantial heat sink, the rate will diminish and burning will eventually cease if the heat losses are sufficient to lower the surface temperature to below the firepoint (Section 6.2). This effect is encountered when attempts are made to burn oil slicks floating on water. Once the slick has 'weathered' or otherwise lost its light ends, e.g., by burning, the residue, although combustible, cannot burn if its thickness is less than a few millimetres (e.g., Petty, 1983; Garo *et al.*, 1994). The associated issue of 'thin film boilover' is discussed below.

The surface temperature of a freely burning liquid is close to, but slightly below, its boiling point. Liquid mixtures, such as petrol, kerosene and fuel oil, do not have a fixed boiling point and the lighter volatiles will tend to burn off first. The surface temperature will therefore increase with time as the residual liquid becomes less volatile.

A hazard associated with some hydrocarbon liquid blends (particularly crude oils) is that of hot zone formation (Burgoyne and Katan, 1947). In such cases, a steady state temperature distribution similar to that shown in Figure 5.5 does not form. Instead, a 'hot zone' propagates into the fuel at a rate significantly greater than the surface regression rate. This is illustrated in Figure 5.6. The danger arises with fires involving large storage tanks containing these liquids if the temperature of the hot zone is significantly greater than 100°C. If the hot zone reaches the foot of the tank and encounters a layer of water (which is commonly present), then explosive vaporization of the water can occur when the vapour pressure becomes sufficient to overcome the head of liquid above, thus ejecting hot, burning oil. This is known as 'boilover': the likely consequences need not be elaborated upon (Vervalin, 1973; Koseki, 1993/94; Mannan, 2005). Table 5.6 compares the regression rates with the rate of descent of the hot zone (or 'heat wave') for a number of crude oils.

The precise mechanism of hot zone formation has not been established, but as it is a phenomenon associated exclusively with fuel mixtures – particularly crude and other heavy oils (Hall, 1973) – it is likely to involve selective evaporation of the light ends. In his original review, Hall (1973) was unable to choose between the mechanism proposed by Hall (1925) involving the continuous migration of light ends to the surface, followed by distillation, and that of Burgoyne and Katan (1947), who suggested that light ends volatilize at the interface of the hot oil and the cool liquid below, and then rise to the surface. Certainly, bubbles are produced within the hot zone and rise to the surface

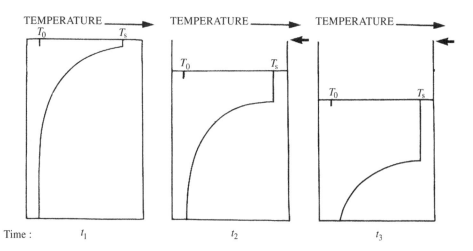

Figure 5.6 Three stages in the formation and propagation of a hot zone during a storage tank fire ($t_1 < t_2 < t_3$). The position of the original surface is marked by arrows

Table 5.6 Comparison between rates of propagation of hot zones and regression rates of liquid fuels (Burgoyne and Katan, 1947)[a]

Oil type	Rate of descent of hot zone (mm/min)	Regression rate (mm/min)
Light crude oil		
<0.3% water	7–15	1.7–7.5
>0.3% water	7.5–20	1.7–7.5
Heavy crude and fuel oils		
<0.3% water	up to 8	1.3–2.2
>0.3% water	3–20	1.3–2.3
Tops (light fraction of crude oil)	4.2–5.8	2.5–4.2

[a]Discussed by Hall (1973).

(Hasegawa, 1989), apparently enhancing the mixing process which may account for the uniform properties of the hot zone. The mechanism of downward propagation (descent of the hot zone) is unclear, but may involve slow vertical oscillations at the interface, which have been observed by Hasegawa (1989) and others. Boilover apparently occurs when the hot layer reaches the bottom of the tank. This has not been studied on a large scale, but some insights may be gleaned from work that has been carried out on the burning of shallow layers of fuel (less than 15 mm thick) floating on water. The term 'thin layer boilover' has been coined to describe the sudden increase in the intensity of the fire, which Garo *et al.* (1994, 1996, 2006) have shown to occur when the temperature of the fuel/water interface reaches some critical value (Garo *et al.* (1994) originally stated it to be 150°C, but in later work they revised it to 120°C (Garo *et al.*, 2006)). The intensity of boilover increases rapidly with increase in thickness[7] and with increasing difference between the boiling points of fuel and water (Garo *et al.*, 1996). The key seems to be the fact that the water becomes superheated and will undergo 'boiling nucleation' at a critical temperature, causing agitation of the fuel layer above and promoting vigorous burning. It seems likely that similar processes are involved when a hot zone forms during a tank fire and that the nucleate boiling of the water at the foot of the tank (present either as a layer or as a water/oil emulsion) is responsible for the expulsion of significant quantities of burning fuel.

In a full-scale tank fire, the onset of boilover is impossible to predict. In a series of laboratory-scale oil tank fires (typically 10 cm diameter), Fan *et al.* (1995) have shown that some warning of the onset of boilover may be obtained by monitoring the noise emitted at the oil/water interface. This increases dramatically just before boilover and is probably associated with the start of nucleate boiling of the water at the interface. The time to boilover increases with depth of liquid, as observed by Koseki (1993/94): this also applies to the 'thin layer boilover' phenomenon (Garo *et al.*, 2006).

---

[7] Koseki *et al.* (1991) showed this to be true for fuel layers 20–100 mm thick.

5.1.2   Spill Fires

Spill fires may be regarded as a subset of pool fires and will occur following the release of a liquid fuel in a location where there is no horizontal confinement. The liquid will continue to spread, thereby increasing the area of burning surface until a dynamic equilibrium is established. The source of the liquid may be a leak or a discharge from a pipe or a damaged container. Alternatively, it may be 'instantaneous', following the sudden failure of a tank or other vessel. Depending on the topography of the surface, the liquid may spread uniformly over the ground or be channelled in discrete directions – most obviously, downhill if the ground is sloping. The depth of liquid will depend on the properties of the liquid and the nature of the surface, but it will tend to be of the order of a few millimetres. If it accumulates at a low point, the depth will increase and a pool fire will develop.[8]

The size of the spill fire from a continuous leak will be determined by the rate of discharge of the liquid ($\dot{m}_{discharge}$) and its rate of burning ($\dot{m}_f'' A_f$) – which depends on the area of the free liquid surface that is involved (A_f). Assuming a flat horizontal surface with no obstructions, the burning liquid will spread radially away from the point of release, increasing the area of the free surface. If it is burning and the rate of discharge remains constant, a steady state will be reached when:

$$A_f = \frac{\dot{m}_{discharge}}{\dot{m}_f''} \qquad (5.15)$$

(Fay, 2003). However, as the layer of liquid will be very shallow, $\dot{m}_f''$ will be affected by heat losses to the surface on which the spill has formed (see Equation (5.3)). For example, Putorti *et al*. (2001) report a burning rate for a 1 litre gasoline spill fire on wood parquet flooring of c. 0.011 $kg/m^2.s$, which should be compared with the figure of 0.055 $kg/m^2.s$ quoted for a gasoline pool fire by Babrauskas (1983b) (see Table 5.2). The actual depth will depend on how easily the liquid can flow away from the source, under the influence of gravity. On a smooth horizontal surface, this will largely be determined by its viscosity (which decreases with temperature), but surface tension-driven flows (see Section 7.1) may also be important where there is a temperature gradient at the perimeter of the burning liquid spill. If a flammable liquid has been used to start a fire, this mechanism will cause hot liquid to be driven into cooler locations, shielded from the flames: consequently, in a fire investigation, evidence of the use of a liquid accelerant may be found in areas shielded from radiant heat, such as underneath low items of furniture.

If the release is sudden, the liquid will flow away from the point of release, creating a large area of liquid surface of shallow depth.[9] If ignited, the fire would rapidly achieve its maximum size, but then subsequently diminish as the shallow layer burns off. Surprisingly little work has been carried out on such incidents, although there has been some that focused on the behaviour of spillages onto water (e.g., Brzustowski and Twardus, 1982), given the hazards associated with leakages and spillages from tankers

---

[8] In this discussion, it is tacitly assumed that the liquid has been ignited at the point of discharge. Alternatively, the spill may develop before ignition occurs.

[9] This observation applies only to liquids that are stable at ambient temperatures and pressures. Cryogenic liquids (e.g., LNG) and pressurized liquids (e.g., LPG) behave quite differently and are discussed in Section 5.1.4.

at sea. This is discussed in some detail by Fay (2003), while Thyer (2003) reviews the available information on the behaviour of spills of cryogenic liquids (see Section 5.1.4).

An associated type of fire is the so-called 'running liquid fire', which may be encountered in process plant when there is a leak of flammable liquid from a high point on an item of equipment (Stark, 1972). The fire that ensues – involving liquid running down vertical surfaces – is a very difficult one to extinguish and can present a serious threat to the structural stability of the plant. Very high rates of heat transfer to structural members can occur, leading to loss of structural integrity (see Section 10.4 and Figure 10.28).

5.1.3 Burning of Liquid Droplets

If a combustible liquid is dispersed as a suspension of droplets in air, ignition can result in very rapid burning, regardless of how high its firepoint temperature might be (see Section 6.2). This is an extremely efficient method by which liquid fuels may be burnt and is used widely in industrial furnaces and other devices (e.g., the combustion chamber of the diesel engine). Similarly, accidental formation of a flammable mist or spray – for example, as a result of a small leak in a high-pressure hydraulic system – will present a significant fire and/or explosion hazard. Such is the concern about spray fires that much effort has been expended in developing test methods for assessing this particular hazard for various hydraulic fluids (Holmstedt and Persson, 1985; Yule and Moodie, 1992). A pinhole leak in a high-pressure hydraulic system can produce a discharge which, if ignited, will sustain a very large flame. The rate of discharge ($\dot{m}$, kg/s) is given by the expression (see, e.g., Wells (1997)):

$$\dot{m} = C_D A \sqrt{2\rho(p - p_0)} \tag{5.16}$$

where C_D is a discharge coefficient (normally taken to be 0.61), A (m^2) is the cross-sectional area of the 'pinhole', ρ (kg/m^3) is the density of the liquid and p (Pa) is the pressure in the system (p_0 is the ambient pressure, 101 300 Pa). Thus, with a system operating at 10 bar (1013 kPa), the discharge rate through a 1 mm hole for a typical hydrocarbon fluid (density 750 kg/m^3) will be 0.0177 kg/s. Taking the heat of combustion as 45 MJ/kg (Table 1.13), this will correspond to a maximum possible rate of heat release of 0.8 MW.

As indicated earlier (Section 3.1.3), flammability limits of mists exist and can be measured: for hydrocarbon liquids, the lower limit in air corresponds to 45–50 g/m^3, provided that the droplet diameter is less than 10–20 µm. In the experiments carried out by Burgoyne and Cohen (1954), the limit appeared to decrease as the droplet size increased (Figure 3.7) (see also Cook et al., 1977). Explosions involving flammable mists are recognized as a serious risk in certain well-characterized situations – e.g., crankcases of marine engines (Burgoyne et al., 1954) – and exhibit the same attributes as those involving flammable vapour/air mixtures (Section 3.5). A review of hydrocarbon mist explosions has been published by Bowen and Cameron (1999).

A significant amount of research has been carried out into the burning of mists and sprays (see Zabetakis, 1965; Kanury, 1975; Holmstedt and Persson, 1985). The propagation of flame through an aerosol is similar in principle to flame propagation though a flammable vapour/air mixture, but it involves evaporation of the droplets ahead of the flame front. This is a complex process that is difficult to study and there remains some uncertainty over the question of whether or not an aerosol explosion would be more violent than an equivalent vapour/air explosion (Bowen and Cameron, 1999).

One aspect of research in this area has involved the study of the burning of single droplets (Williams, 1973), a problem that first highlighted the interrelationship between heat and mass transfer at the burning surface. It is appropriate to discuss this work here as it provides an introduction to Spalding's mass transfer number, commonly referred to as the 'B-number' (Spalding, 1955; Kanury, 1975). It was derived originally in an analysis of droplet vaporization in which the latent heat of evaporation is supplied to a droplet at uniform temperature by convection from the surrounding free gas stream. The derivation, which is discussed in full by (inter alia) Spalding (1955), Kanury (1975), Glassman (1977) and Kuo (2005), hinges on the fact that the mass flux $(\dot{m}_s'')$ from the droplet surface can be expressed in two ways, either in terms of heat transfer

$$\dot{m}_s'' \cdot L_v = k_g \left(\frac{dT}{dr}\right)_s \tag{5.17}$$

(where L_v is the latent heat of vaporization, k_g is the thermal conductivity of air, $(dT/dr)_s$ is the gas phase temperature gradient at the surface and r is the radial distance from the centre of the droplet (using spherical coordinates)); or in terms of mass transfer:

$$\dot{m}_s'' \cdot Y_{fR} = \dot{m}_s'' Y_{fs} + \left(-\rho_g D_f \left(\frac{dY_f}{dr}\right)_s\right) \tag{5.18}$$

where Y_{fR} is the mass fraction of the fuel within the droplet, Y_{fs} is the mass fraction of the fuel in the gas phase at the surface, $\rho_g (dY_f/dr)_s$ is the concentration gradient of fuel vapour at the surface, ρ_g is the density of air and D_f is the diffusivity of the fuel vapour in air. Equation (5.18) equates the total rate of fuel (vapour) production to the sum of the rate of fuel vapour removal from the surface by convection and diffusion and can be rearranged to give:

$$\dot{m}_s'' = \frac{\rho_g D_f \left(\dfrac{dY_f}{dr}\right)_s}{Y_{fs} - Y_{fR}} \tag{5.19}$$

or, as $(Y_{fs} - Y_{fR})$ is a constant:

$$\dot{m}_s'' = \rho_g D_f \frac{d}{dr}\left(\frac{Y_f - Y_{f\infty}}{Y_{fs} - Y_{fR}}\right) = \rho_g D_f \left(\frac{db_D}{dr}\right) \tag{5.20}$$

where the variable $b_D = (Y_f - Y_{f\infty})/(Y_{fs} - Y_{fR})$ is introduced for convenience. Starting with Equation (5.17), a similar rearrangement leads to:

$$\dot{m}_s'' = \rho_g \alpha_g \frac{d}{dr}\left(\frac{c_g(T - T_\infty)}{L_v}\right) = \rho_g \alpha_g \left(\frac{db_T}{dr}\right) \tag{5.21}$$

where α_g is the thermal diffusivity of the gas $(k_g/\rho_g c_g)$, T_∞ is the value of T at $r = \infty$ and $b_T = c_g(T - T_\infty)/L_v$. Equations (5.20) and (5.21) are identical if the ratio $Le = \alpha_g/D_f = 1$, where Le is the Lewis number: this is a common approximation in combustion problems (Lewis and von Elbe, 1987). Equations (5.17b) and (5.17c) indicate that b_T and b_D are conserved variables which determine the direction and magnitude of the mass flux. The mass transfer number (B) is defined as the difference between b_∞ (i.e., at $r = \infty$) and b_s, the value at $r = R$, where R is the droplet radius, thus:

$$B = b_\infty - b_s \equiv \frac{c_g(T_\infty - T_s)}{L_v} \equiv \frac{(Y_{f\infty} - Y_{fs})}{(Y_{fs} - Y_{fR})} \quad (5.22)$$

Application of the steady state conservation of energy to the evaporating droplet leads to the expression:

$$\dot{m}_s'' = \frac{h}{c_g}\ln(1 + B) \quad (5.23)$$

where $\dot{m}_s''$ is the rate of mass loss from the surface, h is the convective heat transfer coefficient averaged over the entire surface of the droplet and c_g is the thermal capacity of air (Spalding, 1955; see also Kanury (1975) and Kuo (2005)). As $h = \text{Nu} \cdot k/2R$ (Section 2.3), it can be seen that the rate of evaporation is inversely proportional to droplet diameter, a factor of significance when rapid evaporation is required, as in a diesel engine. (It also accounts for the effectiveness of water mist as a flame suppressant (Grant *et al.*, 1999).)

If evaporation is accompanied by combustion of the vapour, some of the heat released in the flame will contribute to the volatilization process. Analysis of the conservation equations (energy, fuel, oxygen and products) permits identification of a series of conserved variables, similar to 'b' above, which are equivalent, provided that Le = 1 and that the diffusion flame can be assumed to be of the Burke–Schuman type (Section 4.1) – i.e., the reaction rate is infinite and burning is stoichiometric in the flame zone, which implies that there is no oxygen on the fuel side of the flame. The resulting mass transfer number is normally quoted as:

$$B \approx \frac{\Delta H_c(Y_{O_2\infty}/r_{ox}) + c_g(T_\infty - T_s)}{L_v} \quad (5.24)$$

where r_{ox} is the mass stoichiometric ratio (gm oxygen/gm fuel). The first term in the numerator is the heat of combustion per unit mass of air consumed, i.e., $\sim$3000 J/g (Section 1.2.3). The second term is small and may be neglected, so that B reduces to:

$$B \approx \frac{3000}{L_v} \quad (5.25)$$

From Equations (5.23) and (5.24), it is seen that combustible liquids with low heats of evaporation (and hence high values of B) will tend to burn more rapidly. The B-numbers of a range of fuels are compared in Table 5.7: inspection reveals the difference between methanol (which burns relatively slowly) and the alkanes, as discussed earlier (Figure 5.2).

The original concept of the B-number was developed by Spalding (1955) for evaporation of single droplets, but it can be applied to the burning of single droplets as the associated flame is non-luminous (very low emissivity) and convective heat transfer will dominate. As it stands, Equation (5.23) cannot be used under conditions where there is

Table 5.7 B-numbers of various fuels in air at 20°C (Friedman, 1971)

Fuel	B^a
n-Pentane	8.1
n-Hexane	6.7
n-Heptane	5.8
n-Octane	5.2
n-Decane	4.3
Benzene	6.1
Toluene	6.1
Xylene	5.8
Methanol	2.7
Ethanol	3.3
Acetone	5.1
Kerosene	3.9
Diesel oil	3.9

^{a}These refer to evaporation at ambient temperature.

significant radiative heat transfer, but it can be modified to take radiation into account using the expression:

$$B_R = \frac{(Y_{og}\Delta H_c/r)(1 - (\chi_R/\chi_A)) + c(T_g - T_s)}{L_v(1 - E)} \tag{5.26}$$

where χ_R/χ_A is the fraction of heat released in the flame that is radiated, and

$$E = (\dot{Q}_E'' + \dot{Q}_{FR}'' - \dot{Q}_L'')/\dot{m}'' \cdot L_v \tag{5.27}$$

This has been applied (*inter alia*) by Tewarson *et al.* (1981) (*vide infra*).

5.1.4 Pressurized and Cryogenic Liquids

In addition to 'stable' liquids that have been discussed in the previous sections, it is necessary to consider the behaviour of gaseous fuels that are transported and stored in the liquid state. A gaseous fuel (such as propane) may be liquefied at ambient pressure by cooling it to below its normal boiling point, or by pressurization, provided that the ambient temperature is below the critical temperature of the vapour (see Table 5.8). The so-called 'permanent gases' (which include hydrogen, helium and methane) cannot be liquefied by pressurization alone and are commonly stored as gases at high pressure (e.g., 140 bar, or 14 MPa).[10] Gaseous propane and n-butane, on the other hand, will liquefy if compressed to pressures of 849 and 196 kPa, respectively, *viz.* the vapour pressures that liquid propane and n-butane exhibit at ambient temperatures (e.g., 20°C). The critical temperatures of these fuels are above ambient and the cylinders in which they are stored need not be as strong as those required for the

---

[10] The permanent gases may be stored as cryogenic liquids if they are cooled to below their respective critical temperatures. Natural gas is commonly stored in this way as 'liquefied natural gas' (LNG).

Table 5.8 Boiling points and critical points

	Boiling point (°C)	Vapour pressure at 20°C (bar/kPa)	Critical temperature (°C)
Hydrogen (H_2)	−252.9	− *	−239.8
Oxygen (O_2)	−218.8	− *	−118.2
Methane(CH_4)	−164	− *	−81.9
Propane (C_3H_8)	−42	8.38/849	97
Butane (C_4H_{10})	−0.5	1.93/196	152

*These are 'permanent gases' which cannot be liquefied at 20°C. See Section 5.1.4.

permanent gases: the critical temperatures of H_2, O_2, CH_4, C_3H_8 and n-C_4H_{10} are shown in Table 5.8, along with the vapour pressures of propane and n-butane at 20°C.[11]

Pressurized and cryogenic liquids behave very differently when released from containment. If the pressure on liquid propane (for example) is suddenly released, a fraction of the liquid will vaporize almost instantaneously throughout its volume, drawing heat from the rest of the liquid, which in turn will be cooled to the atmospheric boiling point (−42°C). This process is known as 'flashing': if the 'theoretical flashing fraction' is more than c. 30% (i.e., the energy to vaporize 30% of the liquid is drawn from the remaining 70%), a sudden release of pressure caused by catastrophic failure of the container will produce a 'BLEVE' (boiling liquid expanding vapour explosion). This term was first coined for steam boilers which failed at high pressures due to overheating, but is now common parlance in the fire engineering community to describe an event involving a pressurized liquid which is flammable. It is associated with failure of a pressure vessel (either storage or transportation) during fire exposure: ignition occurs on release and a fireball is formed, the diameter of which has been shown to be proportional to the one-third power of the mass of fuel released (Roberts, 1981/82):

$$D = 5.8M^{1/3} \qquad\qquad (5.28)$$

where D is the diameter of the fireball (m) and M is the mass of the fuel released (kg). It applies to a wide range of fuels (Dorofeev *et al.*, 1995; Mannan, 2005; Abbasi and Abbasi, 2007; Zalosh, 2008).

Methane is the main constituent of LNG, the most ubiquitous cryogenic liquid fuel, and boils at −164°C. If it is released suddenly onto the ground, vigorous boiling will occur until the surface has cooled to −164°C, after which evaporation will level off, albeit at a relatively high rate (Clancey, 1974; Thyer, 2003): otherwise, after the initial rapid boil-off, it will behave as a stable liquid at normal atmospheric pressure. A large flash fire would be anticipated if ignition occurred immediately after the release, but it is found that the steady rate of burning of the pool is similar to that of any other hydrocarbon liquid, such as hexane (Table 5.1) (Burgess *et al.*, 1961).

---

[11] Cryogenic storage of LPG (a blend of mainly propane and butane) becomes economic when large quantities are involved.

5.2 Burning of Solids

It was pointed out in Chapter 1 that the burning of a solid fuel almost invariably requires chemical decomposition (pyrolysis) to produce fuel vapours ('the volatiles'), which can escape from the surface to burn in the flame. Pyrolysis is known to be enhanced by the presence of oxygen (e.g., Kashiwagi and Ohlemiller, 1982), but a detailed discussion of these chemical processes is beyond the scope of the present text. However, it is important to emphasize the complexity involved, and the wide variety of products that are formed in polymer degradation, whether or not oxygen is present (Madorsky, 1964; Cullis and Hirschler, 1981; Hirschler and Morgan, 2008). While the composition of the fuel vapours has direct relevance to the combustion process and product formation, the fire safety engineer normally bypasses this complexity by relying on the results of small-scale tests to provide relatively simple data which may be used in the assessment of the fire hazard of a given material. The best known example is perhaps the cone calorimeter (Section 1.2.3; Babrauskas, 1992a, 2008b), but the results (i.e., the performance of the material in the test) must be interpreted correctly if they are to be of any value. This requires a thorough understanding of fire processes and careful analysis of the test results – whether obtained using 'new generation' test procedures (e.g., the cone calorimeter or the Fire Propagation Apparatus[12] (ASTM, 2009)), or the older, more 'traditional' tests such as BS 476 Part 7 (British Standards Institution, 1997) and ASTM E84 (ASTM, 2009). For this reason, the remainder of this chapter is focused on fundamental issues relating to the steady burning of simple combustible solids, usually in the form of plane slabs, although in reality fires involve items of complex geometries that are composed of a variety of different materials (e.g., an armchair, or packaged goods in high-rack storage). The fundamentals are central to our understanding of many fire processes, including ignition (Chapter 6) and flame spread (Chapter 7), and are essential for the interpretation of fire test data.

The behaviour of synthetic polymers will be considered separately from that of wood, which merits special treatment (Section 5.2.2). The fire behaviour of finely divided solids is discussed briefly in Section 5.2.3.

5.2.1 Burning of Synthetic Polymers

Unlike liquids, solids can be burnt in any orientation – although thermoplastics will tend to melt and flow under fire conditions (Section 1.1.2) (see Sherratt and Drysdale, 2001). The important factors that determine rate of burning have already been identified in the equation:

$$\dot{m}'' = \frac{\dot{Q}_F'' - \dot{Q}_L''}{L_v} \tag{5.3}$$

Surface temperatures of burning solids tend to be high (typically $>350°C$ under steady burning conditions), so that radiative heat loss from the surface is significant. The heat required to produce the volatiles or 'heat of gasification' (L_v) is considerably greater for solids than for liquids as chemical decomposition is involved (compare $L_v = 1.76$ kJ/g

---

[12] This must be distinguished from BS476 Part 6 (BSI, 1989) which is commonly known as the 'Fire Propagation Test'.

Table 5.9 'Flammability parameters' determined by Tewarson and Pion (1976)

Combustibles	L_v (kJ/g)	$\dot{Q}_F''$ (kW/m^2)	$\dot{Q}_L''$ (kW/m^2)	$\dot{m}_{ideal}''$ (g/m^2·s)
FR phenolic foam (rigid)	3.74	25.1	98.7	11[b]
FR polyisocyanurate foam (rigid, with glass fibres)	3.67	33.1	28.4	9[b]
Polyoxymethylene (solid)	2.43	38.5	13.8	16
Polyethylene (solid)	2.32	32.6	26.3	14
Polycarbonate (solid)	2.07	51.9	74.1	25
Polypropylene (solid)	2.03	28.0	18.8	14
Wood (Douglas fir)	1.82	23.8	23.8	13[b]
Polystyrene (solid)	1.76	61.5	50.2	35
FR polyester (glass fibre reinforced)	1.75	29.3	21.3	17
Phenolic (solid)	1.64	21.8	16.3	13
Polymethylmethacrylate (solid)	1.62	38.5	21.3	24
FR polyisocyanurate foam (rigid)	1.52	50.2	58.5	33
Polyurethane foam (rigid)	1.52	68.1	57.7	45
Polyester (glass fibre reinforced)	1.39	24.7	16.3	18
FR polystyrene foam (rigid)	1.36	34.3	23.4	25
Polyurethane foam (flexible)	1.22	51.2	24.3	32
Methyl alcohol (liquid)	1.20[a]	38.1	22.2	32
FR polyurethane foam (rigid)	1.19	31.4	21.3	26
Ethyl alcohol (liquid)	0.97	38.9	24.7	40
FR plywood	0.95	9.6	18.4	10[b]
Styrene (liquid)	0.64[a]	72.8	43.5	114
Methylmethacrylate (liquid)	0.52	20.9	25.5	76
Benzene (liquid)	0.49[a]	72.8	42.2	149
Heptane (liquid)	0.48[a]	44.3	30.5	93

[a] Weast (1974/75).
[b] Charring materials. $\dot{m}_{ideal}''$ taken as the peak burning rate.

for solid polystyrene with 0.64 kJ/g for liquid styrene monomer, Table 5.9; Tewarson and Pion (1976)). However, it should be emphasized that Equation (5.3) (and its derivatives) refers to the quasi-steady state. In particular, the heat loss term is transient as it includes conductive losses through the solid which will gradually diminish with time as the solid heats up. Materials that char on heating (e.g., wood (Section 5.2.2), polyvinyl chloride, certain thermosetting resins, etc. (Table 1.2)) build up a layer of char on the surface that will tend to shield the unaffected fuel beneath (cf. Figure 5.14). Even higher surface temperatures are achieved and the burning behaviour is modified accordingly.

The apparatus developed at Factory Mutual Research Corporation (now FMGlobal) to examine parameters that determine 'flammability' (Tewarson and Pion, 1976) is illustrated in Figure 5.7.[13] It permits a small sample of solid material ($\sim$0.007 m^2 in area) to be weighed continuously as it burns in a horizontal configuration: the oxygen concentration

---

[13] This apparatus was the prototype for the Fire Propagation Apparatus (FPA) (ASTM, 2009).

in the surrounding atmosphere and the intensity of an external radiant heat flux ($\dot{Q}''_E$) can be varied as required. With an external heat flux, Equation (5.3) is modified to:

$$\dot{m}'' = \frac{\dot{Q}''_F + \dot{Q}''_E - \dot{Q}''_L}{L_v} \tag{5.29}$$

where $\dot{Q}''_F$ and $\dot{Q}''_E$ refer to the heat fluxes to the surface from the flame and from the external radiant heaters, respectively. This allows the various quantities implicit in Equation 5.29 to be determined. As the rate of burning is strongly dependent on the oxygen concentration, it was assumed that $\dot{Q}''_F = \xi\eta_{O_2}^{\alpha'}$, where ξ and α' are constants, and the relationship between $\dot{m}''$ and η_{O_2}, the mole fraction of oxygen in the surrounding atmosphere, examined. It was found that when $\dot{Q}''_E$ was held constant, $\dot{m}''$ is a linear function of η_{O_2} (i.e., $\alpha' = 1$) over the range of oxygen concentrations studied (Figure 5.8). The slope of the line in Figure 5.8 gives a value for ξ/L_v provided that $(\dot{Q}''_E - \dot{Q}''_L)/L_v$ is constant: this appears to be the case.

$$\dot{m}'' = \frac{\xi\eta_{O_2}}{L_v} + \frac{\dot{Q}''_E - \dot{Q}''_L}{L_v} \tag{5.30}$$

Similarly, if η_{O_2} is held constant, then a plot of $\dot{m}''$ against $\dot{Q}''_E$ will give a straight line of slope $1/L_v$ (Figure 5.9). Values of L_v, the heat required to produce the volatiles, for a number of polymeric materials are given in Table 5.9. These compare favourably with values obtained by other methods, such as differential scanning calorimetry (Tewarson and Pion, 1976). As both ξ/L_v and L_v can be derived by this method, the constant ξ is known so that the product $\xi\eta_{O_2}$ can be calculated for air ($\eta_{O_2} = 0.21$). This is the heat transferred from the flame to the surface of the fuel, i.e., $\dot{Q}''_F$. In Table 5.9, values of $\dot{Q}''_F$ are compared with those of $\dot{Q}''_L$ which have been calculated directly from Equation (5.29). This identifies clearly materials that will not burn unless an external heat flux is applied to render the numerator of Equation (5.29) positive (e.g., flame retarded (FR) phenolic foam).

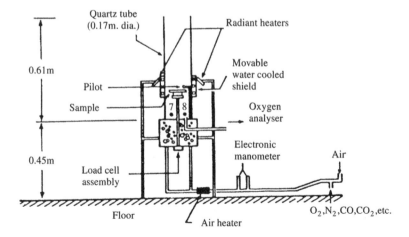

Figure 5.7 The apparatus developed at Factory Mutual Research Corporation for determining 'flammability' parameters (Tewarson and Pion, 1976). By permission

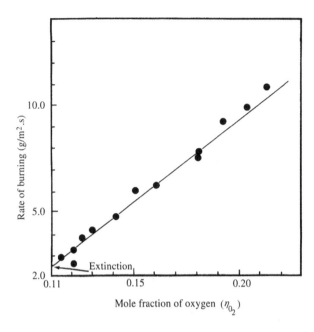

Figure 5.8 Mass burning rate of polyoxymethylene as a function of mole fraction of oxygen (η_{O_2}) with no external heat flux ($\dot{Q}_E'' = 0$). Adapted from Tewarson and Pion (1976), by permission of the Combustion Institute

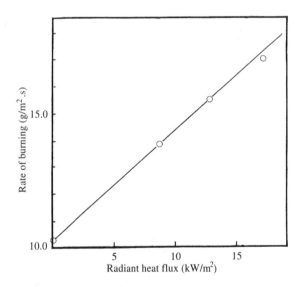

Figure 5.9 Mass burning rate of polyoxymethylene as a function of external heat flux ($\dot{Q}_E''$) in air ($\eta_{O_2} = 0.21$). Reproduced by permission of the Combustion Institute from Tewarson and Pion (1976)

Tewarson proposed that the quantity:

$$\dot{m}''_{\text{ideal}} = \frac{\dot{Q}''_{\text{F}}}{L_{\text{v}}} \tag{5.31}$$

be used as a measure of the 'burning intensity' of a material (Tewarson and Pion, 1976) (Table 5.9) – i.e., the maximum burning rate that a material could achieve if all heat losses were reduced to zero or exactly compensated by an imposed heat flux $\dot{Q}''_{\text{E}} = \dot{Q}''_{\text{L}}$ (see Equation (5.29)). While this gives results that appear to correlate reasonably well with data from existing fire tests, it would be more logical if heat loss by surface re-radiation was included in $\dot{m}''_{\text{ideal}}$. This would seem to provide a means of calculating burning rates under different heat gain and loss regimes, thus:

$$\dot{m}'' = \dot{m}''_{\text{ideal}} + (\dot{Q}''_{\text{E}} - \dot{Q}''_{\text{L}})/L_{\text{v}} \tag{5.32}$$

However, the values of $\dot{m}''_{\text{ideal}}$ shown in Table 5.9 were obtained with a small-scale apparatus in which the sample area was no more than 0.007 m^2 (0.047 m in diameter). It has already been shown (Section 5.1.1) that as the diameter of a burning pool of liquid is increased to 0.3 m and beyond, radiation becomes the dominant mode of heat transfer, except for fuels that burn with non-luminous flames, such as methanol (Figure 5.2). The same is true for solids. Regardless of the nature of fuels in Tewarson's original experiments, the transfer of heat from the flame to the surface ($\dot{Q}''_{\text{F}}$) would not have been dominated by radiation. Such data cannot be used directly to predict large-scale behaviour, or to make a hazard assessment. The problem of enhancing the radiation component at the small scale was addressed by Tewarson and his co-workers by carrying out the experiments at increased oxygen concentrations (Tewarson *et al.*, 1981). They showed that radiation becomes the dominant mode of heat transfer even on this scale if the oxygen concentration in the surrounding atmosphere is increased (Figure 5.7). This effect is achieved because elevated oxygen concentrations produce hotter, sootier and more emissive flames which radiate a greater proportion of the net heat of combustion back to the surface. This is consistent with the observation that the fraction of the heat of combustion that is radiated increases asymptotically as the oxygen concentration is increased. The increase in rate is self-limiting partly because the flow of volatiles will absorb a significant proportion of the radiative flux as well as tending to block convective transfer to the surface (Section 5.1.1; Brosmer and Tien (1987)).

There is strong evidence that radiation is the dominant mode of heat transfer for large fuel beds. Markstein (1979) observed that the emissivity of the flames above polymethylmethacrylate (PMMA) increased approximately three-fold as the diameter of the fuel bed was increased from 0.31 to 0.73 m. This was accompanied by an increase in $\dot{m}''$ from 10 to 20 g/m^2.s, which is consistent with the conclusion of Modak and Croce (1977) that radiation becomes increasingly important as the mode of heat transfer to the surface as the diameter of the fire increases above 0.2–0.3 m: indeed, they determined that over 80% of the heat transferred to the surface of a burning PMMA slab, 1.22 m square, was by radiation. Subsequent analysis of these data by Iqbal and Quintiere (1994) confirmed this conclusion and drew attention to the fact that the higher mass transfer rates associated with large fire diameters cause a reduction in the rate of heat transfer by convection. Similar results were obtained by Tewarson *et al.* (1981) by extrapolating their

Table 5.10 Convective and radiative components of $\dot{Q}_F''$ (Tewarson *et al.*, 1981)

Fuel[a]	$m_{O_2}^b$	$\dot{Q}_{F,c}''$ (kW/m^2)	$\dot{Q}_{F,r}''$ (kW/m^2)	$\dot{Q}_{F,c}'' + \dot{Q}_{F,r}''$ (kW/m^2)	$\dot{Q}_{F,r}''/\dot{Q}_{F,c}''$
PMMA	0.183	17	4	21	0.23
	0.195	16	7	23	0.44
	0.207	17	7	24	0.41
	0.233	15	15	30	1.0
	0.318	13	26	39	2.0
	0.404	12	38	50	3.2
	0.490	13	43	56	3.3
	0.513	12	44	56	3.7
PP	0.196	20	3	23	0.15
	0.208	15	14	28	0.93
	0.233	17	14	31	0.82
	0.266	15	23	38	1.5
	0.310	12	37	49	3.1
	0.370	20	41	61	2.1
	0.427	18	44	62	2.4
	0.507	13	53	66	4.1

[a]Fuel bed area = 0.0068 m^2.
[b]Mass fraction of oxygen in air is $m_{O_2} = 0.232$.

data for small samples to high oxygen concentrations. They used a version of Spalding's B-number (Equation (5.24)), corrected for radiation (Equation 5.26), to deduce the contributions to the heat flux to the surface by radiation and convection, and showed convincingly for a number of fuels that as the oxygen concentration was increased, radiation became predominant.

The comparison is given in Table 5.10 for PMMA and polypropylene (PP). Both fuels show the increasing importance of radiative heat transfer to the surface as the oxygen concentration is increased. Indeed, $\dot{Q}_{F,r}''$ increases while $\dot{Q}_{F,c}''$ decreases. The latter is a consequence of the so-called 'blowing effect' brought about by the fact that the increasing flow of fuel vapours *from* the surface inhibits convective heat transfer in the opposite direction (*to* the surface). This accounts for the increase of $\dot{Q}_{F,r}''/\dot{Q}_{F,c}''$ from significantly less than 1.0 in air to more than 3.0 in O_2/N_2 mixtures containing more than 50% oxygen. Iqbal and Quintiere (1994) have applied a one-dimensional analytical model, also using the modified Spalding B-number (Equation (5.26)), which showed the same pattern of behaviour (i.e., increasing $\dot{Q}_{F,r}''/\dot{Q}_{F,c}''$) with increased size of fire. This is entirely consistent with experimental work described above, but it should be borne in mind that the radiative heat transfer to an extended surface will not be uniform. This has been discussed by Brosmer and Tien (1987) (see p. 189).

Clearly, in steady burning of an isolated fuel bed, flame emissivity and the heat required to produce the volatiles are important properties which can be assigned to the material itself, rather than to interactions with its environment. Markstein (1979) has compared the radiative output of flames above 0.31 m square slabs of PMMA, PP, polystyrene (PS), polyoxymethylene (POM) and polyurethane foam (PUF), and found the emissivities to decrease in the order

$$PS > PP > PMMA > PUF > POM$$

This agrees closely but not exactly with the ranking of these plastics according to their rates of burning (Table 5.11), namely:

$$PS > PMMA > PP > PUF > POM$$

Only PP and PMMA are out of sequence, which can be explained at least in part by the differences in the heats required to produce the volatiles: that for PP is 25% larger than that for PMMA (Table 5.9).

If the material is burning in an enclosure fire, in which the heat flux to the surface comes from general burning within the space (Chapter 10), the rate at which it will contribute heat to the compartment will be calculated from Equation (5.1), i.e.

$$\dot{Q}_c = \dot{m}'' \chi \Delta H_c A_F \tag{5.1a}$$

where A_F is the fuel surface area. Writing $\dot{Q}''_{net}$ as the net heat flux entering the surface, Equation (5.1a) may be rewritten:

$$\dot{Q}_c = \frac{\dot{Q}''_{net}}{L_v} \chi \Delta H_c A_F \tag{5.1b}$$

or

$$\frac{\dot{Q}_c}{A_F} = \dot{Q}''_{net} \chi \left(\frac{\Delta H_c}{L_v} \right) \tag{5.1c}$$

Given that χ lies within a relatively narrow range (0.4–0.7, according to Tewarson (1980)), it can be seen that the rate of heat release from a burning material is strongly dependent on $\Delta H_c / L_v$, which Rasbash (1976) referred to as the 'combustibility ratio'. Values calculated from Tewarson's data (but using the heat of combustion of the solid) are given in Table 5.12 (Tewarson, 1980). This shows that combustible solids have values in the range 3 (for red oak) to 30 (for a particular rigid polystyrene foam),

Table 5.11   Burning rates of plastics fires (Markstein, 1979)

Fuel[a]	Emissivity[b]	$\dot{m}''$ (g/m^2·s)
Polystyrene	0.83	14.1 ± 0.8
Polypropylene	0.4	8.4 ± 0.6
Polymethylmethacrylate	0.25	10.0 ± 0.7^c
Polyurethane foam	0.17	8.2 ± 1.8
Polyoxymethylene	0.05	6.4 ± 0.5

[a]Except for polyurethane foam, the fuels were burnt as pools, 0.31×0.31 m^2. Data for PUF were deduced from a spreading fire.
[b]As measured 0.051 m above the fuel bed.
[c]0.73 m diameter pool of PMMA gave $\dot{m}'' = 20.0 \pm 1.4$ g/m^2 · s.

Table 5.12 $\Delta H_c / L_v$ values for fuels (Tewarson, 1980)

Fuel[a]	$\Delta H_c / L_v^b$
Red oak (solid)	2.96
Rigid PU foam (43)	5.14
Polyoxymethylene (granular)	6.37
Rigid PU foam (37)	6.54
Flexible PU foam (1-A)	6.63
PVC (granular)	6.66
Polyethylene 48% Cl (granular)	6.72
Rigid PU foam (29)	8.37
Flexible PU foam (27)	12.26
Nylon (granular)	13.10
Flexible PU foam (21)	13.34
Epoxy/FR/glass fibre (solid)	13.38
PMMA (granular)	15.46
Methanol (liquid)	16.50
Flexible PU foam (25)	20.03
Rigid polystyrene foam (47)	20.51
Polypropylene (granular)	21.37
Polystyrene (granular)	23.04
Polyethylene (granular)	24.84
Rigid polyethylene foam (4)	27.23
Rigid polystyrene foam (53)	30.02
Styrene (liquid)	63.30
Heptane (liquid)	92.83

[a] Numbers in parentheses are PRC sample numbers (Products Research Committee, 1980).

[b] ΔH_c measured in an oxygen bomb calorimeter and corrected for water as a vapour for fuels for which data are not available: L_v is obtained by measuring the mass loss rate of the fuel in pyrolysis in N_2 environment as a function of external heat flux for fuels for which data are not available. *Note:* If ΔH_c is replaced by the heat of combustion of the volatiles, $(\Delta H_c + L_v)$, then all the 'combustibility ratios' are increased by 1.00 and the ranking order is unchanged.

and places materials in a ranking order which in its broad outline matches the consensus based on common knowledge of the steady burning behaviour of these materials. Liquid fuels tend to have much larger values of $\Delta H_c / L_v$, ranging up to 93 for heptane, with methanol having a low value in line with its high latent heat of evaporation and relatively low ΔH_c (Table 1.13) (see Section 5.1). As hydrocarbon polymers (e.g., polyethylene) tend to have much higher heats of combustion than their oxygenated derivatives (e.g., polymethylmethacrylate), their 'combustibility ratios' tend to be greater.

However, while these figures are likely to give a reasonable indication of the ranking order for different materials, logically they should be calculated from the heat of combustion of the volatiles, rather than the net heat of combustion of the solid. The latter is determined by oxygen bomb calorimetry and, for char-forming materials (e.g., wood), will include the energy released in oxidation of the char which would normally burn very slowly in a real fire, much of it after flaming combustion has ceased. The result of this is that the combustibility ratio for charring fuels will tend to be overestimated (see Table 5.12).

Flame retardants can influence the 'combustibility ratio' by altering ΔH_c and/or L_v. This can be achieved by changing the pyrolysis mechanism (see Section 5.2.2) or effectively 'diluting' the fuel by means of an inert filler such as alumina trihydrate (Lyons, 1970). However, the rate of heat release (Equation (5.1a)) is influenced by χ, the combustion 'efficiency', which for some fire-retarded species may be as low as 0.4. Tewarson (1980) suggests that χ may vary from 0.7 to 0.4, decreasing in the following order:

Aliphatic > Aliphatic/Aromatic > Aromatic > Highly halogenated species

Some values obtained using the Factory Mutual Flammability Apparatus are given in Table 5.13 (Tewarson, 1982).

If the surface of a combustible solid is vertical, the interaction between the flame and the fuel is quite different. The flame clings to the surface, entraining air from one side only (Figure 5.10(a)), effectively filling the boundary layer and providing convective heating as the stream of burning gas flows over the surface. The surface 'sees' a flame whose thickness is a minimum at the base of the vertical surface where the flow is laminar, but increases with height as fresh volatiles mix with the rising plume to yield turbulent flaming above ~0.2 m. Measurements on thick, vertical slabs of PMMA, 1.57 and 3.56 m high, have shown that the local steady burning rate exhibits a minimum at approximately 0.2 m from the lower edge, thereafter increasing with height and reaching a maximum at the top (Figure 5.10(b)) (Orloff *et al.*, 1974, 1976). Calculations based on measurements of the emissive power of the flame as a function of height indicate that this can be attributed to radiation. It was estimated that 75–87% of the total heat transferred to the

Table 5.13 Fraction of heat of combustion released during burning in the Factory Mutual Flammability Apparatus (Figure 5.7) (Tewarson, 1982)

Fuel	$\dot{Q}''_E$ (kW/m^2)	χ	χ_{conv}	χ_{rad}
Methanol (*l*)	0	0.993	0.853	0.141
Heptane (*l*)	0	0.690	0.374	0.316
Cellulose	52.4	0.716	0.351	0.365
Polyoxymethylene	0	0.755	0.607	0.148
Polymethylmethacrylate	0	0.867	0.622	0.245
	39.7	0.710	0.340	0.360
	52.4	0.710	0.410	0.300
Polypropylene	0	0.752	0.548	0.204
	39.7	0.593	0.233	0.360
	52.4	0.679	0.267	0.413
Styrene (*l*)	0	0.550	0.180	0.370
Polystyrene	0	0.607	0.385	0.222
	32.5	0.392	0.090	0.302
	39.7	0.464	0.130	0.334
Polyvinylchloride	52.4	0.357	0.148	0.209

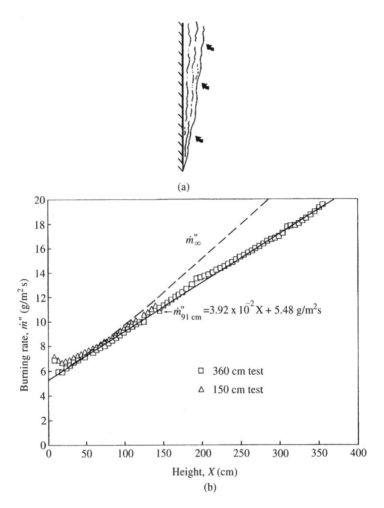

Figure 5.10 (a) Illustration of burning at a vertical surface. (b) Variation of local steady burning rate per unit area with distance from the bottom of vertical PMMA slabs 0.91 m wide, 3.6 m high (□) and 1.5 m high (△): _ _ _ _ predicted burning rate for an infinitely wide slab. Reproduced by permission of the Combustion Institute from Orloff *et al.* (1976)

surface was by radiation (Orloff *et al.*, 1976). While these results refer specifically to PMMA, it is likely that this conclusion will apply generally to steady burning of vertical surfaces. However, many synthetic materials (i.e., most thermoplastics) will melt and flow while burning. Not only will this lead to the establishment of a pool fire at the base of the wall (e.g., Zhang *et al.*, 1997), but it will also affect the burning behaviour of the vertical surface. In experiments in which plastic products were allowed to burn to completion, it was found that c. 80% of the mass burned as liquid pool fires (Sherratt and Drysdale, 2001). This is in agreement with Zhang *et al.* (1997), who estimated that 20% or less of a thermoplastic wall lining burned while still adhering to the wall, the remainder forming a pool fire underneath. The combination of a pool fire and a vertical combustible surface will produce vigorous burning and create special problems in confined spaces and enclosures (Chapter 10). Indeed, the use of large-scale tests as a means of assessing the fire hazard of wall lining materials reflects the awareness of potentially dangerous situations of this type (e.g., British Standards Institution, 2010), which may not be apparent from small-scale laboratory tests. The scenario is complex, however, and it has been shown that the rate at which the pool fire will develop depends on the nature and thermal properties of the surface on which it forms (Sherratt and Drysdale, 2001), consistent with our understanding of the behaviour of liquid spill fires (Section 5.1.2).

Burning of horizontal, downward-facing combustible surfaces tends not to occur in isolation and consequently has received limited attention. Combustible ceiling linings may become involved during the growth, or pre-flashover, period of a compartment fire (Section 9.2) and will contribute to the extension of flames under the ceiling (Hinkley and Wraight, 1969) (Section 4.3.4), but will rarely ignite and burn without a substantial input of heat from the primary fire at floor level or elsewhere. It has been shown that flames on the underside of small slabs of polymethylmethacrylate tend to be very thin and weak, providing a relatively low heat flux to the surface ($\dot{Q}_F''$) compared with burning in the pool configuration. Thus, Ohtani et $al.$ (1981) estimated $\dot{Q}_F''$ to be 8 kW/m^2 and 22 kW/m^2 for downward- and upward-facing burning surfaces, respectively, from data obtained with 50 mm square slabs of PMMA. The appearance and behaviour of flames in this configuration have been investigated by Orloff and de Ris (1972) using downward-facing porous gas burners to allow fuel flowrate to be independent of the rate of heat transfer. Flames with a cellular structure are produced, their size and behaviour depending on the flowrate of gaseous fuel. The 'cells' were small and quite distinctive, growing in size with increasing fuel flowrate, but always present even at the highest flows that could be achieved in their apparatus. Cellular-like flame structures have been observed occasionally on the underside of combustible ceilings during the later stages of compartment fires.

Steady burning of surfaces at other inclinations – i.e., neither horizontal nor vertical – has not been studied systematically, although work has been carried out on the spread of flame on sloping surfaces (Section 7.2.1). However, the discussion so far has referred to plane surfaces burning in isolation or in an experimental situation with an imposed heat flux. In 'real fires', isolated burning will only occur in the early stages before fire spreads beyond the item first ignited. Once the area of fire has increased, cross-radiation from flames and between different burning surfaces (which may be at any inclination) will enhance both the rate of burning and the rate of spread (Section 7.2.5). Indeed, wherever there is opportunity for heat to build up in one location, increased rates of burning will result (Section 9.1). This can be expected in any confined space in which combustible surfaces are in close proximity (Section 2.4.1). The most hazardous configurations in buildings are ducts, voids and cavities which, if lined with combustible materials, provide optimal configurations for rapid fire spread and intense burning (Section 10.7). Such conditions must be avoided or adequately protected.

5.2.2 Burning of Wood

Unlike synthetic polymers, wood is an inhomogeneous material which is also non-isotropic – i.e., many of its properties vary with the direction in which the measurement is made. It is a complex mixture of natural polymers of high molecular weight, the

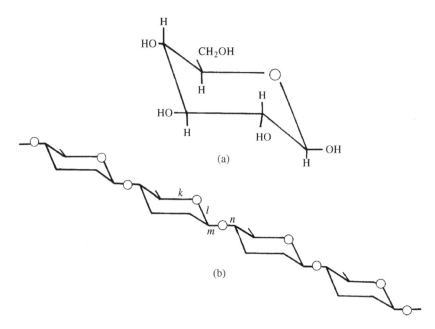

Figure 5.11   (a) β-D-Glucopyranose (the stable configuration of D-glucose); (b) part of a cellulose molecule. H and OH groups are shown in (a) but have been omitted from (b), for clarity

most important of which are cellulose ($\sim$50%), hemicellulose ($\sim$25%) and lignin ($\sim$25%) (Madorsky, 1964), although these proportions vary from species to species. For softwoods, lignin constitutes 23–33% of the wood substance, while the range reported for hardwoods is 16–25% (Miller, 1999): an apparent consequence is that softwoods tend to give higher char yields than hardwoods, as discussed by Di Blasi (2009). Cellulose, which is the principal constituent of all higher plants, is a condensation polymer of the hexose sugar, D-glucose (Figure 5.11(a)), and adopts the linear structure shown in Figure 5.11(b). This configuration allows the molecules to align themselves into bundles (microfibrils) which provide the structural strength and rigidity of the cell wall. The microfibrils are bound together during the process of lignification, when the hemicellulose and lignin are laid down in the growing plant. Wood normally contains absorbed moisture, some of which will be bound by weak hydrogen bonds to hydroxyl (OH) groups of the main constituents (e.g., see Figure 5.11) and – if the relative humidity is high enough – some will be present as free water contained in natural voids within the wood. The latter is held only by weak capillary forces and will be the first to be driven off when the temperature is increased towards 100°C: this is discussed briefly by Moghtaderi (2006).

Unlike cellulose, hemicellulose has a branched structure based on pentose and hexose sugars and its molecular weight is low in comparison. The structure of lignin (described as a three-dimensional phenylpropanol polymer) is vastly more complex (Miller, 1999). Thermogravimetric analysis of the degradation of wood, cellulose and lignin (Figure 5.12) shows that the constituents decompose to release volatiles at different temperatures, typically:

Hemicellulose $200-260°C$

Cellulose $240-350°C$

Lignin $280-500°C$

(Roberts, 1970). If lignin is heated to temperatures in excess of $400-450°C$, only about 50% volatilizes, the balance of the mass remaining as a char residue. On the other hand, pure 'α-cellulose' – the material extracted from cotton and washed thoroughly to leach out any soluble inorganic impurities – leaves only $\sim$5% char after prolonged heating at $300°C$. However, if inorganic impurities (e.g., sodium salts, etc.) are present, much higher yields are found: for example, viscose rayon (a fibre consisting of regenerated cellulose and having a relatively high residual inorganic content) can give over 40% char (Madorsky, 1964).[14] When wood is burnt, or heated above $450°C$ in air, $15-25\%$ normally remains as char, much of this coming from the lignin content (up to $10-12\%$ of the original mass of the wood). In addition to the presence of inorganic impurities (Lyons, 1970), the yield of char also depends on the temperature at which the conversion takes place and the rate of heating (Madorsky, 1964; Di Blasi, 2009), which in a fire will be influenced by the level of imposed heat flux and the oxygen concentration (Kashiwagi et al., 1987). This is significant as the nature and composition of the volatiles must change if the yield of char is altered:[15] a consequence is that the fire behaviour (particularly the ignition characteristics) will be altered.

Much of our understanding comes from studies of the decomposition of wood made under non-flaming conditions. The effect of inorganic impurities is illustrated very well by Brenden (1967). He illustrated this by comparing the yields of 'char', 'tar', water and 'gas' (mainly CO and CO_2) from samples of Ponderosa pine which had been treated with a number of salts capable of imparting some degree of flame retardancy (Table 5.14) (Lyons, 1970). The fraction designated 'tar' contains the combustible volatiles and consists of products of low volatility, the most important of which is believed to be levoglucosan (Figure 5.13).

It appears that there are two competing mechanisms of cellulose degradation. Referring to Figure 5.11(b) (Madorsky, 1964), if any of the bonds of the type marked k or l break, a six-membered ring will open but the continuity of the polymer chain remains intact. It is suggested that under these circumstances the products are char, with CO, CO_2 and H_2O as the principal volatiles. If, on the other hand, bonds m or n break, the polymer chain 'backbone' is broken, leaving exposed reactive ends from which levoglucosan molecules can break away and volatilize from the high temperature zone. Low rates of heating, or relatively low temperatures, appear to favour the char-forming reaction. Similarly, the range of flame retardants commonly used to improve the response of wood to fire (e.g., phosphates and borates) act by promoting the char-forming process at the expense of 'tar' formation. Table 5.14 shows how the char yields can be more than doubled by treating pine with phosphates and borates, while at the same time the composition of the volatiles changes in favour of a lower proportion of the flammable 'tar' constituent (Brenden, 1967). As a result, the heat of combustion of the volatiles is decreased, which will lower the

---

[14] Formation of char is a prerequisite for smouldering combustion: see Section 8.2.

[15] A comprehensive review of the decomposition of wood and other related materials has been published recently by di Blasi (2009).

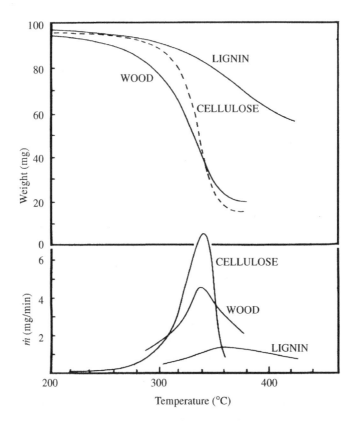

Figure 5.12 (a) Thermogravimetric analysis of 90–100 mg samples of wood (Ponderosa pine), cellulose powder (Whatman) and lignin, heated under vacuum at 3°C rise in temperature per minute. (b) Derivative TGA curves from (a). From Browne and Brenden (1964). Reproduced by permission of Forest Products Laboratory, Forest Service, USDA, Madison, WI

Table 5.14 Pyrolysis of Ponderosa pine (Brenden, 1967)[a]

	Concentration of applied solution	Treatment level	Char	Tar	Water	Gas[b]
Untreated wood	–	–	19.8	54.9	20.9	4.4
+ $Na_2B_4O_7$	5%	4.28%	48.4	11.8	30.4	9.4
+ $(NH_4)_2.HPO_4$	5%	6.69%	45.5	16.8	32.0	5.7
+ ammonium polyphosphate	5%	5.0%	43.8	19.0	34.6	2.6
+ H_3BO_3	5%	3.9%	46.2	10.7	33.9	9.2
+ ammonium sulphamate[c]	5%	6.3%	49.8	2.6	33.4	14.2
+ H_3PO_4	5%	6.8%	54.1	2.5	37.3	6.1

[a] Browne and Brendan (1964) showed that the heat of combustion of the volatiles was less for a flame retarded wood than for the parent wood. This is consistent with the suppressed yields of 'tar' observed for the retarded samples.
[b] 'Non-condensable gases': CO, CO_2, H_2, CH_4.
[c] $NH_4 . NH_2 \cdot SO_3$.

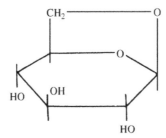

Figure 5.13 Structure of levoglucosan

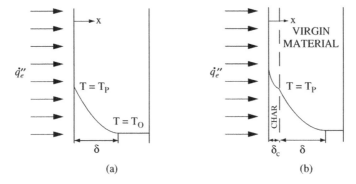

Figure 5.14 Temperature profiles in a slab of wood exposed to a radiant heat flux. (a) Temperature profile before a significant char layer has formed. (b) Temperature profile after the char layer has developed (after Moghtaderi, 2005)

amount of heat that can be transferred to the surface from the flame ($\dot{Q}_F''$ in Equation (5.29)) so that $\dot{m}_{ideal}''$ will be reduced (Equation (5.31)). Consequently, a higher imposed heat flux ($\dot{Q}_E''$ in Equation (5.29)) may be necessary to allow sustained burning (see Table 5.9). As it accumulates, the layer of char will protect the unaffected wood below, and higher temperatures at the surface of the char will be required to provide the necessary heat flow to produce the flow of volatiles This is illustrated schematically in Figure 5.14, which is adapted from Moghtaderi (2006), and is consistent with the results obtained in the cone calorimeter that show the burning rate of wood to increase rapidly to a peak value before decreasing, following a $t^{-1/2}$ relationship (Figure 5.15(a); Spearpoint (1999), Quintiere (2006)). The higher surface temperatures will mean greater radiative heat losses (included in $\dot{Q}_L''$), but against this there will be some surface (heterogeneous) oxidation of the char that will contribute positively to the heat balance. This was observed by Kashiwagi *et al.* (1987) in a study of the effect of oxygen on the rate of decomposition of samples of white pine. It should be noted that the rate of mass loss from non-charring materials as measured in the cone calorimeter is very different to that of wood: the burning rate reaches a steady value, as shown for polyethylene in Figure 5.15(b) (Hopkins and Quintiere, 1996). The increase in the rate of burning after 800 s is a consequence of a change in the boundary condition (reduced heat losses) as the rear face of the sample is insulated.

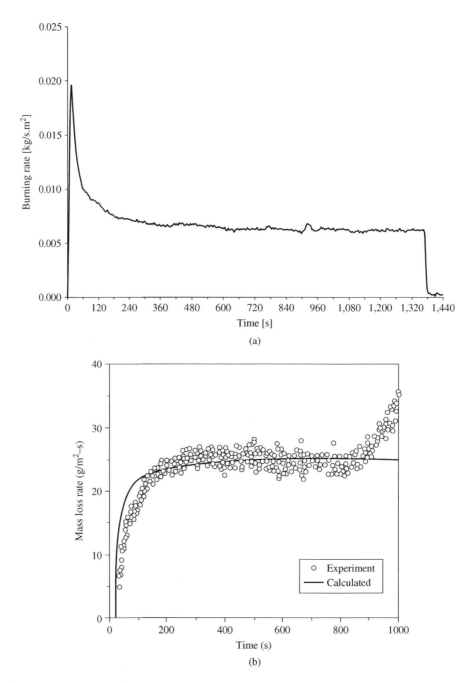

Figure 5.15 (a) Mass loss rate of red oak exposed to 75 kW/m^2 in the cone calorimeter (Spearpoint, 1999). (b) Mass loss rate of polyethylene exposed to 70 kW/m^2 in the cone calorimeter (Hopkins and Quintiere, 1996; reprinted with permission from Elsevier)

As a slab of wood burns away and a layer of char accumulates, it seems likely that the composition of the volatiles will change. Roberts (1964a,b) found no evidence for this, although his samples may have been too small to show the effect. He carried out combustion bomb calorimetry on small samples of wood (dry European beech), partially decomposed wood samples and 'char', which enabled him to deduce the 'heat of combustion of the volatiles'. His results are given in Table 5.15. In these experiments the char yield was 16–17% of the original wood, indicating on the basis of data in Table 5.15 that it accounted for ~30% of the total heat production and consumed ~33% of the total air requirement of the wood. Browne and Brenden (1964) carried out similar experiments with dry Ponderosa pine and found evidence that the composition of the volatiles did change, apparently becoming *more* combustible as the degradation proceeded. Their results were as follows:

$$\text{at 10\% weight loss} \quad \Delta H_c(\text{volatiles}) = 11.0 \text{ kJ/g}$$

$$\text{at 60\% weight loss} \quad \Delta H_c(\text{volatiles}) = 14.2 \text{ kJ/g}$$

$$\text{Parent wood} \quad \Delta H_c(\text{wood}) = 19.4 \text{ kJ/g}$$

Using the cone calorimeter, it is now possible to determine instantaneous values of the effective heat of combustion of wood as burning progresses (see Section 1.2.3). This is shown in Figure 5.16 for Western red cedar, exposed to a radiant flux of 65 kW/m^2 (Babrauskas, 2008b). The value of $\Delta H_{c,\text{effective}}$ is approximately constant from 120–480 s, but increases rapidly to over 35 kJ/g at 600 s, consistent with the combustion of char. This should be compared with Roberts' value of 34.3 kJ/g for the heat of combustion of char (Table 5.15). The initial peak of 17 kJ/g at 30 s has not been explained, but *may* be evidence for a change in the composition of the volatiles as the char layer builds up on the surface. The very low values of $\Delta H_{c,\text{effective}}$ before 30 s are consistent with the evaporation of water as the surface layers are heated.

(a) *Burning of wooden slabs and sticks.* The complexity of wood makes it difficult to interpret the burning behaviour in terms of Equation (5.3). Because of the grain structure, properties vary with direction: thus the thermal conductivity parallel to the grain is about twice that perpendicular to the grain, and there is an even greater difference in gas permeability (of the order of 10^3; Roberts (1971a)). Volatiles generated just below the surface of the unaffected wood can escape more easily along the grain than at right angles towards the surface. The appearance of jets of volatiles and flame from the end of a burning stick or log, or from a knot, is evidence for this.

Wood discolours and chars at temperatures above 200–250°C, although prolonged heating at lower temperatures (≥ 120°C) will have the same effect. The physical structure begins to break down rapidly at temperatures above 300°C. This is first apparent on the surface when small cracks appear in the char, perpendicular to the direction of the grain. This permits volatiles to escape easily through the surface from the affected layer (Figure 5.17) (Roberts, 1971a). The cracks will gradually widen as the depth of char increases, leading to the characteristic 'crazed' pattern that is frequently referred to as 'crocodiling' or 'alligatoring'. The appearance of such patterns in a fire-damaged building was once widely believed to give an indication of the rate of fire development (e.g., Brannigan, 1980), but there has been no systematic investigation of this and the method

Table 5.15 Combustion of wood and its degradation products (Roberts, 1964a)

	Wood[a]	Volatiles	Char
Gross heat of combustion (kJ/g)[b]	19.5	16.6	34.3
Mean molecular formula	$CH_{1.5}O_{0.7}$	CH_2O	$CH_{0.2}O_{0.02}$
Theoretical air requirements (g air/g fuel)[c]	5.7	4.6	11.2

[a] European beech.
[b] By combustion bomb calorimetry.
[c] Section 1.2.3.

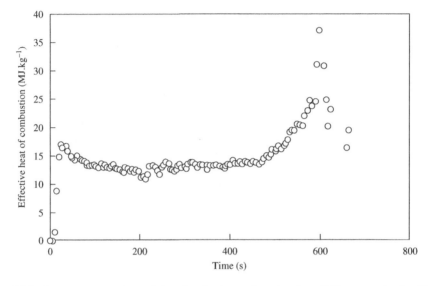

Figure 5.16 Instantaneous values of the effective heat of combustion of Western red cedar (17 mm thick samples) at an imposed radiant heat flux of 65 kW/m^2 in the cone calorimeter (Babrauskas, 2008b). Reproduced by permission of the Society of Fire Protection Engineers

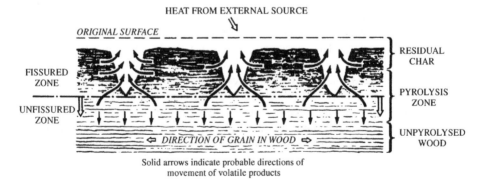

Figure 5.17 Representation of a cross-section through a slab of burning, or pyrolysing, wood. Reproduced by permission of the Combustion Institute from Roberts (1971a)

must be regarded as highly questionable, as illustrated by De Haan (2007) (see also Cooke and Ide (1985)).

Clearly, the burning of wood is a much more complex process than that of synthetic polymers, charring or non-charring. Any theoretical analysis must take into account not only the terms in Equation (5.3) – which will be complicated by the presence of the layer of char – but also the interactions within the hot char. Even during active burning, small quantities of oxygen may diffuse to the surface and react heterogeneously, releasing heat that would contribute to the decomposition of the virgin wood under the layer of char. This might be interpreted as a reduction in the apparent heat of gasification (see Equation (5.3)).

Indeed, there has been a lack of consensus in the literature regarding the value of L_v for wood, with values from 1.8 (and less) to 7 kJ/g having been reported for a range of species, including both hardwoods and softwoods (Thomas *et al.*, 1967b; Tewarson and Pion, 1976; Petrella, 1979). There is, of course, a wide variation in the composition and structure between woods of different species. Thomas *et al.* (1967b) proposed that there might be a correlation between L_v and permeability, while Hadvig and Paulsen (1976) suggested a link with the lignin content. Janssens (1993) carried out a very thorough study of six solid woods using the cone calorimeter to obtain experimental data, which were then analysed by means of an integral heat transfer model. He found that L_v was not constant, but varied as the depth of char increased: for example, Victorian ash showed an initial value of about 3 kJ/g, increasing slightly (to 3.5–4 kJ/g) then decreasing slowly to about 1 kJ/g when the char depth was c. 14 mm. The average value was 2.57 kJ/g. The averages for six species are shown in Table 5.16. There is an apparent difference between the softwoods (L_v; 3.2 kJ/g) and the hardwoods (L_v; 2.6 kJ/g), with one exception: L_v for Douglas fir, a softwood with a high resin content, is 2.64 kJ/g.

Results such as these still require detailed interpretation. Janssens' work shows where some of the variability in reported values of L_v may lie. Thus, Thomas *et al.* (1967b) found the values of L_v to increase with mass loss (comparing measurements at 10% and 30% loss), but the absolute values that they reported are much higher than Janssens' (e.g., 5.1 kJ/g at 10% mass loss for Douglas fir). It should also be noted that Tewarson and Pion (1976) and Petrella (1979) studied horizontal samples, while Thomas *et al.*

Table 5.16 Average values of L_v for various woods (Janssens, 1993): S = softwood, H = hardwood

Material	L_v (kJ/g)
Western red cedar (S)	3.27
Redwood (S)	3.14
Radiata pine (S)	3.22
Douglas fir (S)	2.64
Victorian ash (H)	2.57
Blackbutt (H)	2.54

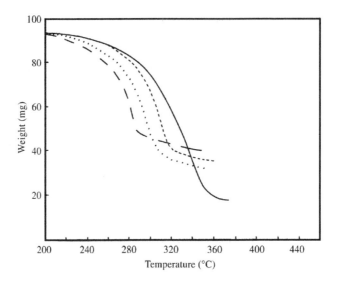

Figure 5.18 Thermogravimetric analysis of samples of wood (Ponderosa pine), untreated _____ and treated with various inorganic salts: _ _ _ _, 2% $Na_2B_4O_7.10H_2O$; . . . ., 2% NaCl; _ _ _ _, 2% $NH_4.H_2.PO_4$. From Brenden (1967). Reproduced by permission of Forest Products Laboratory, Forest Service, USDA, Madison, WI

(1967b) and Janssens (1993) held their samples vertically. The relevance of this is not clear, but other studies have revealed significant apparatus dependency (Rath *et al.*, 2003).[16]

It is known that the rate of decomposition of wood, or cellulose in particular, is very sensitive to the presence of inorganic impurities, such as flame retardants (Figure 5.18). Thus the difference between L_v for Douglas fir and fire-retarded plywood reported by Tewarson and Pion (1976) (1800 J/g and 950 J/g, respectively) is at least consistent with the catalytic action of the retardant chemicals on the char-forming reaction referred to above (see also Table 5.9). While the variation in the relative proportions of the three main constituents of wood from one species to another is likely to have an effect, variations in the content of inorganic constituents may predominate. However, this has not been investigated on a systematic basis.

It is common experience that a thick slab of wood will not burn unless supported by radiative (or convective) heat transfer from another source (e.g., flames from a nearby fire or burning surface). This is in agreement with Tewarson's observation that $\dot{Q}''_F \approx \dot{Q}''_L$ for Douglas fir (Table 5.9) (Tewarson and Pion, 1976) – i.e., the heat transfer from the flame was theoretically just sufficient to match the heat losses from the sample under burning conditions. Results obtained by Petrella (1979) with an apparatus similar to that of Tewarson indicate that generally $\dot{Q}''_F < \dot{Q}''_L$ for several species of wood. Clearly, the ability of wood to burn depends on there being an imposed heat flux. In a log fire, this 'imposed

---

[16] Their study of the pyrolysis of samples of wood used differential scanning calorimetry under a nitrogen atmosphere. It is not clear if their values of L_v are relevant to the burning of wood (given the complete absence of oxygen), but their data will be of value in resolving this complex problem.

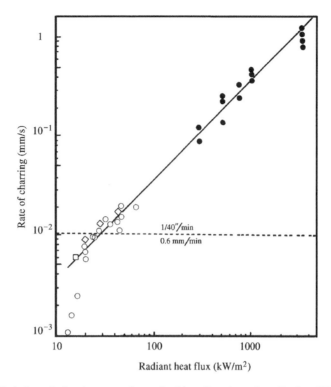

Figure 5.19   Variation of charring rate of wood with radiant heat flux (Butler, 1971). Reproduced by permission of The Controller, HMSO. © Crown copyright

heat flux' is provided by mutual cross-radiation between the internal burning surfaces: this mechanism also applies to the burning of wood cribs (see below, Figure 5.20).

The rate of burning of wood is frequently reported as the 'rate of charring' (mm/min): it is similar to the concept of 'regression rate' used for the burning of liquids (Figure 5.1), the difference being that a layer of char of increasing thickness forms over the 'regressing wood'. Determining the depth of char is a relatively easy measurement to make. In Figure 5.19, data on the rate of charring (R_W) as a function of imposed radiant heat flux I (from 12 to 3000 kW/m²)[17] is shown on a log–log plot (Butler, 1971). These data were obtained using slabs of wood which were sufficiently thick to behave as semi-infinite solids for the duration of burning. The linear correlation is described by the expression:

$$R_W = 2.2 \times 10^{-2} I \text{ mm/min} \tag{5.33}$$

where I is in kW/m². In a compartment fire, localized temperatures as high as 1100°C may be achieved, corresponding to black-body radiation of 200 kW/m². This would result in 'rates of burning of wood' as high as 4.4 mm/min. It is interesting to consider Tewarson's

---

[17] Data at the upper end of this range were sought to gain an understanding of the performance of wood exposed to high levels of thermal radiation associated with a nuclear explosion. These high intensities were achieved in the laboratory using arc lamps.

'ideal burning rate' ($\dot{m}''_{ideal}$) for Douglas fir of 13 g/m^2.s (Tewarson and Pion, 1976) in the context of Figure 5.19: assuming a density of 640 kg/m^3, this corresponds to a rate of charring of the order of 1 mm/min.

In the early fire investigation literature, it was stated that the burning rate of wood (R_W) was constant, and quoted as 1/40 inch per minute (0.6 mm/min). This figure was used to estimate how long a fire had been burning, simply on the basis of depth-of-char measurements. It is clear from the above arguments (particularly Figure 5.19) that this method is deeply flawed and should never be used. The figure came from observations of the depth of char on wooden beams and columns that had been exposed to a standard fire test (e.g., BS 476 Part 21 (British Standards Institution, 1987b)) (Butler, 1971).

Higher rates of burning will be observed for samples that are thermally thin, unless the heat losses from the rear face (included with $\dot{Q}''_L$ of Equation (5.3)) are high. This may be compared with earlier discussions about burning of liquid fuel spills (Section 5.1.2).

While a thick slab of wood cannot burn in isolation, 'kindling' and thin pieces (e.g., wood shavings and matchsticks) can be ignited relatively easily and will continue to burn, although flaming may have to be established on all sides. This is possible because thin samples will behave as systems with low Biot number (Section 2.2.2), so that once ignition has occurred and the ignition source has been removed, the rate of heat loss from the surface into the body of the sample will be minimal (Section 6.3.2). It is possible to estimate the maximum thickness of wood that can still be regarded as 'thin' from the point of view of ignition. It will depend mainly on the duration of contact with the ignition source (assumed to be a flame). The depth of the heated layer is of the order $(\alpha t)^{1/2}$, where t is the duration in seconds (Section 2.2.2). Thus, with a 10 s application time, the maximum thickness of a splint of oak ($\alpha = 8.9 \times 10^{-8}$ m^2/s) that may be ignited will be of the order of $2 \times (\alpha t)^{1/2} \approx 2 \times 10^{-3}$ m (assuming immersion of the splint in the flame). In principle, thicker samples could be ignited after longer exposure but then other factors, such as depletion of volatiles from the surface layers and direction of subsequent flame spread (Section 7.2.1), become important. The duration of burning of a 'thin' stick of wood varies roughly in proportion to D^n, where D is the diameter and $n \approx 1.6 \pm 0.2$ (e.g., Thomas, 1974a).

(b) *Burning of wood cribs.* In the earlier sections, burning at plane surfaces has been discussed as this is directly relevant to materials burning in isolation, although in real fires complex geometries and configurations can cause interactions that will influence behaviour strongly (Chapters 9 and 10). One type of fuel bed in which such complexities dominate behaviour is the wood crib, which comprises crossed layers of sticks as shown in Figure 5.20. The confinement of heat within the crib, including cross-radiation between the burning surfaces, allows sticks of substantial cross-section to burn efficiently. (This is the mechanism by which logs burn in a fire.) They are still used as a means of producing reproducible fire sources for research and testing purposes (e.g., British Standards Institute, 2006), although they are largely superseded by the sand-bed gas burner, particularly as the ignition source in large-scale fire tests (Babrauskas, 1992b, 2008a). Several parameters may be controlled independently to produce a fuel bed which will burn at a known rate and for a specific duration: these include stick thickness (b), number of layers (N), separation of sticks in each layer (s) and length of the sticks (Gross, 1962; Block, 1971; Babrauskas, 2008a). Moisture content is also controlled.

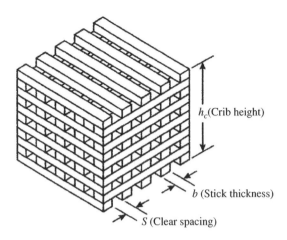

h_c(Crib height)

b (Stick thickness)

S (Clear spacing)

Figure 5.20 The structure of a wood crib (Babrauskas, 2008a). Reproduced by permission of the Society of Fire Protection Engineers

Gross (1962) identified two regimes of burning corresponding to 'under-ventilated' and 'well-ventilated' cribs. In the former, corresponding to densely packed cribs, the rate of burning is dependent on the ratio A_v/A_s, where A_s is the total exposed surface area of the sticks and A_v is the open area of the vertical shafts. He scaled the rate of burning with stick thickness, as $Rb^{1.6}$, where R is the rate in percentage mass loss per second, and compared it with a 'porosity factor' $\Phi = N^{0.5}b^{1.1}(A_v/A_s)$. This parameter is derived from the ratio $\dot{m}_{ac}/\dot{m}$, where $\dot{m}_{ac}$ is the mass flowrate of air through the vertical shafts and $\dot{m}$ is the total rate of production of volatiles. As $\dot{m}'' \propto b^{-0.6}$ (Gross, 1962) and assuming that $\dot{m}_{ac} \propto h_c^{1/2} \cdot A_v$, where h_c ($= Nb$) is the height of the crib, then:

$$\frac{\dot{m}_{ac}}{\dot{m}} \propto \frac{(Nb)^{1/2} \cdot A_v}{A_s \cdot b^{-0.6}} = N^{1/2}b^{1.1}\frac{A_v}{A_s} \tag{5.34}$$

Gross's plot of $Rb^{1.6}$ versus Φ is shown in Figure 5.21. For $\Phi < 0.08$, a linear relationship exists between $Rb^{1.6}$ and Φ, but $Rb^{1.6}$ is approximately constant when $\Phi > 0.1$. The latter case corresponds to good ventilation and substantial flaming within the crib, the rate of burning being controlled by the thickness of the individual sticks. (Sustained burning is not possible if $\Phi > 0.4$.)

5.2.3 Burning of Dusts and Powders

While finely divided combustible materials can behave in fire as simple fuel beds as described in the previous sections, two additional modes of behaviour must be considered, namely the ability to give rise to smouldering combustion and the potential to create an explosible dust cloud. Smouldering can only occur with porous char-forming materials such as sawdust or wood flour. The phenomenon of smouldering will be discussed at greater length in Section 8.2: the mechanism involves heterogeneous oxidation of rigid char, which in turn generates enough heat to cause pyrolysis of unaffected fuel adjacent

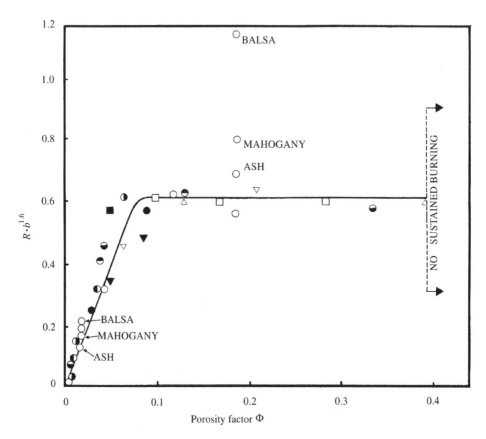

Figure 5.21 The effect of porosity on the scaled rate of burning of wood cribs (Douglas fir). Symbols refer to a range of stick thicknesses (b) and spacings (s) (Gross, 1962)

to the combustion zone. This forms fresh char, which in turn begins to oxidize. A non-charring thermoplastic powder cannot smoulder, as when it is exposed to heat it will simply melt and form a liquid pool.

Many, but not all, combustible dusts are capable of burning rapidly in air if they are thrown into suspension as a dust cloud. This is equivalent to the burning of a flammable mist or droplet suspension, to which reference has already been made (Section 5.1.3). The hazard associated with the formation and ignition of dust clouds in industrial and agricultural environments has long been recognized: extremely violent and damaging explosions are possible (e.g., coal dust explosions in mines, dust explosions in grain elevators, etc.). As with flammable gases and vapours, it is possible to identify limits of flammability (or more correctly, *explosibility*), minimum ignition energies, auto-ignition temperatures, etc. The subject has been studied in great depth but is considered to be outside the scope of the present text. There are several excellent reviews and monographs that cover the subject at all levels (Palmer, 1973; Bartknecht, 1981; Field, 1982; Eckhoff, 1997; Mannan, 2005).

Problems

5.1 Calculate the rate of heat release from a fire involving a circular pool of *n*-hexane of diameter 2 m, assuming that the efficiency of combustion is 0.85. Use the limiting regression rate given in Table 5.1. Compare your result with that given for *n*-heptane in Table 4.3.

5.2 Carry out the same calculation as in Problem 5.1 for a 2 m diameter pool of methanol, but assume 100% combustion efficiency. Consider what information you would need to calculate the radiant flux at a point distant from this flame.

5.3 An explosion occurs in a crude oil tank (15 m in diameter), leaving behind a fully developed pool fire. Massive attack with firefighting foam at 4 hours has the fire extinguished within 15 minutes. By this time, the hydrocarbon liquid level has fallen by 0.5 m. What was the average rate of burning and the average rate of heat release, if the density of the oil is 800 kg/m^3? Assume 85% combustion efficiency.

5.4 Considering the fire described in Problem 5.3, use the formula by Shokri and Beyler to calculate the maximum radiant heat flux falling on adjacent, identical tanks within the same bund. The minimum distance between tanks is 10 m.

5.5 Compare Tewarson's 'ideal burning rates' (Table 5.9) for *n*-heptane and methanol with the limiting burning rates for large pools of these liquids (see Tables 5.1 and 5.2). (Take R_∞ (*n*-heptane) = 7.3 mm/min.) What can you deduce from the results?

5.6 Using the data in Table 5.2, calculate the rates of burning of 0.5 m diameter trays of the following liquid fuels: (a) ethanol; (b) hexane; and (c) benzene. (Note the differences between these and the respective limiting values.)

5.7 You are required to 'design' a fire that will burn for about 10 minutes with a heat output of approximately 750 kW. What diameter of tray, or pan, would you require to contain (a) ethanol and (b) heptane to achieve a fire with these characteristics? Assume that combustion is 100% efficient, and take the heat of combustion of heptane to be 45 MJ/kg. (Hint: first try estimating the tray diameter assuming the limiting rates of burning.)

5.8 A horizontal sheet of black polymethylmethacrylate is allowed to burn at its upper surface while its lower surface is maintained at 20°C. Treating the sheet as an infinite slab, calculate the rate of burning if the burning surface is at 350°C, and the sheet is (a) 4 mm thick and (b) 2 mm thick. (Assume $\varepsilon = 1$ and the thermal conductivity is constant.)

6

Ignition: The Initiation of Flaming Combustion

Ignition may be defined as that process by which a rapid, exothermic reaction is initiated, which then propagates and causes the material involved to undergo change, producing temperatures greatly in excess of ambient. Thus, ignition of a stoichiometric propane/air mixture triggers the oxidation reaction which propagates as a flame through the mixture, converting the hydrocarbon to carbon dioxide and water vapour at temperatures typically in the range 2000–2500 K (Chapter 1). It is convenient to distinguish two types of ignition, namely piloted – in which flaming is initiated in a flammable vapour/air mixture by a 'pilot', such as an electrical spark or an independent flame – and spontaneous – in which flaming develops spontaneously within the mixture. To achieve flaming combustion of liquids and solids, external heating is required, except in the case of piloted ignition of flammable liquids that have firepoints below ambient temperature (see Section 6.2.1). The phenomenon of spontaneous ignition within bulk solids, which leads to smouldering combustion, will be discussed separately in Chapter 8.

The objectives of this chapter are to gain an understanding of the processes involved in ignition and to examine ways in which the 'ignitability' or 'ease of ignition' of combustible solids might be quantified. The subject is covered comprehensively in Babrauskas' *Ignition Handbook* (2003). The phenomenon of extinction has many features in common with ignition and is discussed briefly in Section 6.6. However, as initiation of flaming necessarily involves reactions of the volatiles in air, it is appropriate to start with a review of ignition of flammable vapour/air mixtures.

6.1 Ignition of Flammable Vapour/Air Mixtures

It has been shown elsewhere (Section 1.2.3) that the reaction between a flammable vapour and air is capable of releasing a substantial amount of energy, but it is the rate of energy release that will determine whether or not the reaction will be self-sustaining and propagate as a flame through a flammable mixture (Section 3.3). To illustrate this point, it can be assumed that the rate of the oxidation processes obeys an Arrhenius-type temperature

An Introduction to Fire Dynamics, Third Edition. Dougal Drysdale.
© 2011 John Wiley & Sons, Ltd. Published 2011 by John Wiley & Sons, Ltd.

dependence (Equation (1.2)). The rate of heat release within a small volume (V) would then be given by:

$$\dot{Q}_c = \Delta H_c V C_i^n A \exp(-E_A/RT) \qquad (6.1)$$

where A is the 'pre-exponential factor' whose units will depend on n, the order of the reaction, C_i is the concentration (mole/m^3) and ΔH_c is the heat of combustion (kJ/mole). Such is the nature of the exponential term that no temperature limit can be identified below which $\dot{Q}_c = 0$: oxidation occurs even at ordinary ambient temperatures, although in most cases at a negligible rate. The heat generated is lost to the surroundings and consequently there is no significant rise in temperature and $\dot{Q}_c$ remains negligible. This is illustrated schematically in Figure 6.1 as a plot of $\dot{Q}_c$ against temperature, superimposed on a similar plot of the rate of heat loss, $\dot{\mathscr{L}}$. The latter is assumed to be directly proportional to the temperature difference, ΔT, between the reaction volume and the surroundings, i.e.

$$\dot{\mathscr{L}}_1 = hS\Delta T \qquad (6.2)$$

where h is a heat transfer coefficient and S is the surface area of the reaction volume through which heat is lost. The intersection at p_1 represents a point of equilibrium ($\dot{Q}_c = \dot{\mathscr{L}}$), limiting the temperature rise to $\Delta T = (T_{p_1} - T_{a_1})$. (This is exaggerated in Figure 6.1 for clarity.) Slight perturbations about this point are stable and the system will return to its equilibrium position. This cannot be said for p_2: although $\dot{Q}_c = \dot{\mathscr{L}}$ at this point, perturbations lead to instability. For example, if the temperature is reduced infinitesimally, then $\dot{\mathscr{L}} > \dot{Q}_c$ and the system will cool and move to p_1. Alternatively, at a temperature slightly higher than T_{p_2}, $\dot{Q}_c > \dot{\mathscr{L}}$, and the system will rapidly increase in temperature to a new point of stability at p_3. This corresponds to a stable, high-temperature combustion reaction that can propagate as a premixed flame. Although Figure 6.1 is schematic, arguments based on it are valid in a qualitative sense. However, it should be noted that it does not indicate that there is a limit to the temperature that the reacting mixture can achieve because of the thermal capacity of the products (Section 1.2.3), nor that the heat loss function will change at higher temperatures, especially when radiative losses become significant (Section 2.4.2).

Referring to Figure 6.1, it can be seen that to ignite a flammable vapour/air mixture at an ambient temperature T_{a_1}, sufficient energy must be available to transfer the system from its stable state (p_1) at a low temperature (T_{p_1}) to an unstable condition at a temperature greater than T_{p_2}. The concept of a minimum ignition energy for a given flammable vapour/air mixture (Figure 3.3) is quite consistent with this, although when the ignition source is an electrical spark, Figure 6.1 is not entirely satisfactory. Given that an electrical discharge generates a transient plasma, rich in atoms, free radicals and ions, free radical initiation must contribute significantly in spark ignition. The energy dissipated in the weakest spark capable of igniting a stoichiometric propane/air mixture (0.3 mJ) is capable of raising the temperature of a spherical volume of diameter equal to the quenching distance (2 mm) by only a few tens of degrees. Without free radical initiation, a rise of several hundred degrees is necessary to promote rapid ignition (see below).

The minimum ignition energies quoted in Table 3.1 refer to electrical sparks generated between two electrodes whose separation cannot be less than the minimum quenching distance (Table 3.1), otherwise heat losses to the electrodes will cause the reaction zone to cool and prevent flame becoming established (Section 3.3). If the electrodes are free,

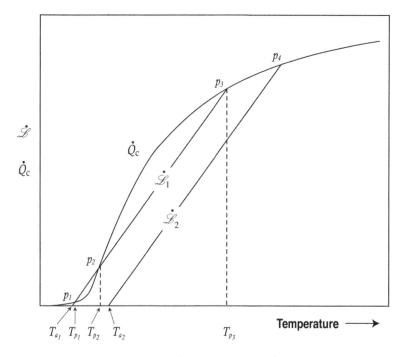

Figure 6.1 Rates of heat production ($\dot{Q}_c$) and heat loss ($\dot{\mathscr{L}}$) as functions of temperature

as defined in Figure 6.2, then ignition can be achieved when their separation is less than the quenching distance (d_q) simply by increasing the spark energy to overcome heat losses to the electrodes. However, if the electrodes are flush mounted through glass discs (Figure 6.2) and their separation is less than d_q, ignition is not possible because the flame will be quenched as it propagates away from the spark.

Other ignition sources include flames, mechanical sparks, hot surfaces and glowing wires.[1] These involve convective heat transfer from the solid surface to the gas, and ignition occurs spontaneously in the hot boundary layer. Figure 6.1 may be used qualitatively to illustrate the mechanism. Imagine a small volume of flammable vapour/air mixture within the boundary layer which, for simplicity, is assumed to be at a uniform temperature T_{a_2}. Under these conditions, $\dot{Q}_c > \dot{\mathscr{L}}$, and the temperature of the element of volume will rise rapidly. In this particular case, the rate of heat loss is unable to prevent a runaway reaction, and ignition will occur as the system transfers to the intersection p_4, corresponding to the high-temperature combustion process (see above). In reality, the temperature in the boundary layer is not uniform (Figure 2.15) and the rate of heat loss will be influenced strongly by any air movement or turbulence. Consequently, whether or not flame will develop depends on the extent of the surface, its geometry and temperature, as well as the ambient conditions. The minimum temperatures for ignition of stoichiometric

---

[1] Smouldering cigarettes cannot ignite common flammable gases and vapours such as methane, propane and petrol (gasoline) vapour. However, there is evidence that they can cause ignition of hydrogen, carbon disulphide, diethyl ether and other highly reactive species (Hollyhead, 1996).

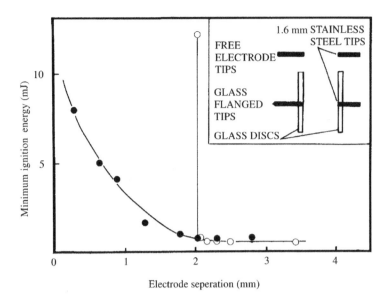

Figure 6.2 Minimum ignition energies for free (●) and glass-flanged (○) electrode tips as a function of electrode distance (stoichiometric natural gas/air mixture). Reproduced by permission of Academic Press from Lewis and von Elbe (1987)

vapour/air mixtures which are quoted in the literature (British Standards Institution, 2002; National Fire Protection Association, 2008) refer to uniform heating of a substantial volume of mixture enclosed in a spherical glass vessel (>0.2 litre). Under these conditions, the mixture is static and there will be a measurable delay, or 'induction period' (τ) before ignition occurs, particularly at temperatures close to the minimum 'auto-ignition temperature' (Figure 6.3) when τ may be found to be of the order of 1 s or more.

The existence of a critical ignition temperature for flammable mixtures led to the development of thermal explosion theory, based on Equations (6.1) and (6.2) (Semenov, 1928). Semenov assumed that the temperature within the reacting gas mixture remained uniform (Figure 6.4(a)) and that heat losses were described accurately by Equation (6.2), where ΔT is the temperature difference between the gas and the walls of the enclosing vessel. It was also assumed that reactant consumption was negligible and that the rate followed the Arrhenius temperature dependence (Equation (6.1)). Figure 6.5 shows $\dot{Q}_c$ and $\mathscr{L}$ plotted against temperature for three values of the ambient (i.e., wall) temperature.

The critical ambient temperature ($T_1 = T_{a,cr}$) is identified as that giving a heat loss curve which intersects the heat production curve, $\dot{Q}_c$, tangentially. This may be expressed mathematically as:

$$\dot{Q}_c = \dot{\mathscr{L}} \tag{6.3a}$$

and

$$\frac{d\dot{Q}_c}{dT} = \frac{d\dot{\mathscr{L}}}{dT} \tag{6.3b}$$

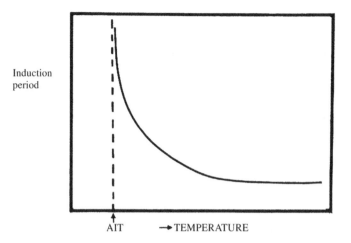

Figure 6.3 Variation of induction period with temperature for stoichiometric fuel/air mixtures (schematic) (AIT = auto-ignition temperature)

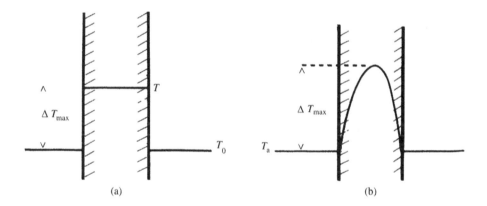

Figure 6.4 Temperature profiles inside spontaneously heating systems according to the models of (a) Semenov and (b) Frank-Kamenetskii (schematic)

which, stated in full, give

$$\Delta H_c V C_i^n A \exp(-E_A/RT) = h S(T - T_{a,cr}) \tag{6.4}$$

and

$$\frac{E_A}{RT^2} \Delta H_c V C_i^n A \exp(-E_A/RT) = h S \tag{6.5}$$

Dividing Equation (6.4) by Equation (6.5) gives:

$$\frac{RT^2}{E_A} = T - T_{a,cr} \tag{6.6}$$

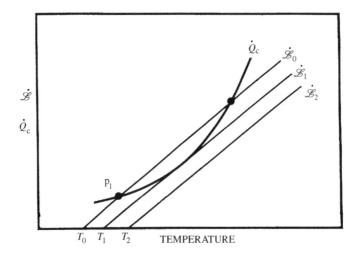

Figure 6.5 Illustrating Semenov's model for spontaneous ignition ('thermal explosion theory') (Equations (6.3)–(6.7))

where $T_{a,cr}$ is the critical ambient temperature and T is the corresponding (equilibrium) gas temperature. The difference $(T - T_{a,cr})$ is the maximum temperature rise that can occur spontaneously within this system without ignition taking place. Provided that $E_A \gg RT$, Equation (6.6) can be solved by binomial expansion to give

$$\Delta T_{crit} = T - T_{a,cr} \approx \frac{R T_{a,cr}^2}{E_A} \tag{6.7}$$

For typical values of $T_{a,cr}$ and E_A (700 K and 200 kJ/mol, respectively) $\Delta T_{crit} \approx 20$ K.

However, the Semenov model is unrealistic for most circumstances in that it ignores temperature differences within the reacting system: it is a 'lumped thermal capacity', or low Biot number model, as defined in Section 2.2.2. Frank-Kamenetskii (1939) developed a high Biot number model (Bi > 10: see Figure 6.4(b)) based on Equation (2.14), i.e.

$$\nabla^2 T + \frac{\dot{Q}_c'''}{k} = \frac{1}{\alpha} \frac{\partial T}{\partial t} \tag{6.8}$$

To simplify the solution (e.g., see Gray and Lee (1967)), this can be reduced to the one-dimensional case with uniform, symmetrical heating, thus:

$$\frac{\partial^2 T}{\partial r^2} + \frac{\kappa}{r} \frac{\partial T}{\partial r} + \frac{\dot{Q}_c'''}{k} = \frac{1}{\alpha} \frac{\partial T}{\partial t} \tag{6.9}$$

where κ takes values of 0, 1 and 2 for an infinite slab (thickness $2r_0$), an infinite cylinder (radius r_0) and a sphere (radius r_0), respectively. Further simplification was achieved by assuming that (i) the reaction rate can be described by a single Arrhenius expression (Equation (6.1)); (ii) there is no reactant consumption (cf. Semenov, 1928); (iii) the Biot number is sufficiently large for conduction within the reacting volume to determine the rate of heat loss; and (iv) the thermal properties of the system are constant (independent of temperature). Thus, the boundary conditions for Equation (6.9) are:

$$T_{r_0} = T_0 \text{ at } t \geq 0 \text{ (surface)} \tag{6.9a}$$

(i.e., the surface behaves as an isothermal heat sink)

$$\frac{\partial T}{\partial r} = 0 \quad \text{at } r = 0 \text{ and } t \geq 0 \text{ (centre)} \tag{6.9b}$$

$$\text{Rate of heat flow at surface} = -k \left(\frac{\partial T}{\partial r} \right)_{r=r_0} \tag{6.9c}$$

(see Figure 6.4(b)). If the reacting system is capable of achieving a stable steady state, analogous to the intersection p_1 in Figures 6.1 and 6.5, then Equation (6.9) will have a solution when $(\partial T / \partial t) = 0$.

Conventionally (e.g. Gray and Lee, 1967), the following dimensionless variables are introduced:

$$\theta = \frac{T - T_a}{RT_a / E_A}$$

$$z = \frac{r}{r_0}$$

which allow Equation (6.9) to be rewritten, with $\partial T / \partial t = 0$:

$$\frac{k}{r_0^2} \left(\frac{RT_a^2}{E_A} \right) \left(\frac{\partial^2 \theta}{\partial z^2} + \frac{\kappa}{z} \frac{\partial \theta}{\partial z} \right) = -\Delta H_c C_i^n A \exp \left[-\left(\frac{E_A}{RT_a} - \frac{\theta}{1 - \varepsilon \theta} \right) \right] \tag{6.10}$$

where $\varepsilon = RT_a / E_A$. Provided that $\varepsilon \ll 1$ (i.e., the activation energy is high), then Equation (6.10) can be approximated by:

$$\frac{\partial^2 \theta}{\partial z^2} + \frac{\kappa}{z} \frac{\partial \theta}{\partial z} = \frac{r_0^2 E_A \Delta H_c A C_i^n}{k R T_a^2} \exp(-E_A / RT_a) \cdot \exp(\theta) \tag{6.11}$$

or

$$\nabla^2 \theta = -\delta \exp(\theta) \tag{6.12}$$

where:

$$\delta = \frac{r_0^2 E_A \Delta H_c A C_i^n}{k R T_a^2} \exp(-E_A / RT_a) \tag{6.13}$$

Solutions to Equation (6.12) exist only for a certain range of values of δ corresponding to various degrees of self-heating. It may be assumed that conditions lying outside this range, i.e., when $\delta > \delta_{cr}$, correspond to ignition. The phenomenon of criticality is illustrated very well by results of Fine *et al.* (1969) on the thermal decomposition of gaseous diethyl peroxide (Figure 6.6), although in this case the exothermic reaction is the decomposition of an unstable compound rather than an oxidation process.

Mathematically it is possible to identify values for δ_{cr} for a number of different shapes (Table 6.1) (Gray and Lee, 1967; Boddington *et al.*, 1971). Equation (6.13) may then be used to investigate the relationship between the characteristic dimension of the system

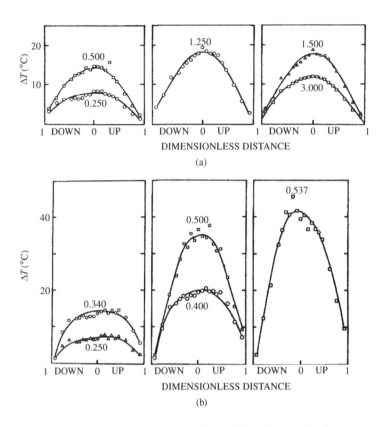

Figure 6.6 Spontaneous (exothermic) decomposition of diethyl peroxide. (a) Instantaneous temperature distributions across the diameter of a reaction vessel for subcritical system (1.1 torr of peroxide at 203.7°C); (b) as (a), but supercritical (1.4 torr at 203.7°C). Times given in seconds. Reproduced by permission of the Combustion Institute from Fine *et al.* (1969)

Table 6.1 Critical values of Frank-Kamenetskii's δ (Equation (6.13))

Shape	κ	δ_{cr}
Slab, thickness $2r_0$	0	0.88
Cylinder, radius r_0	1	2.00
Sphere, radius r_0	2	3.32
Cube, side $2r_0$	3.28	2.52

(r_0) and the critical ambient temperature ($T_{a,cr}$) above which it will ignite. Rearranging Equation (6.13), and taking Napierian logarithms:

$$\ln\left(\frac{\delta_{cr} T_{a,cr}^2}{r_0^2}\right) = \ln\left(\frac{E_A \Delta H_c C_i^n A}{kR}\right) - \frac{E_A}{RT_{a,cr}} \tag{6.14}$$

Table 6.2 Comparison of the minimum auto-ignition temperatures ($^\circ$C) of combustible liquids in spherical vessels of different sizes (Setchkin, 1954)[a]

	Volume of vessel ($m^3 \times 10^6$)				
	8	35	200	1000	1200
Diethylether	212	197	180	170	160
Kerosene	283	248	233	227	210
Benzene	668	619	579	559	–
Methanol	498	473	441	428	386
n-Pentane	295	273	–	258	–
n-Heptane	255	248	–	223	–

[a]The test involves introducing a small sample of liquid into the vessel. It is, of course, the vapour/air mixture that ignites.

shows that provided the assumptions are valid (i.e., the first term on the right-hand side of Equation (6.14) is constant), $\ln(\delta_{cr} T^2_{a,cr}/r^2_0)$ should be a linear function of $1/T_{a,cr}$. This has been found to hold for many systems and is widely used to investigate the spontaneous combustion characteristics of bulk solids (Section 8.1 and Figure 8.2). It indicates the strong inverse relationship that exists between r_0 and the critical ambient temperature $T_{a,cr}$, which is apparent when auto-ignition temperatures are measured in reaction vessels of different sizes (Table 6.2). Such critical temperature data, many of which are quoted in the literature without qualification, should be regarded as indicative as they refer to particular experimental conditions. Typical values are given in Table 6.3.

For spontaneous ignition to occur in the boundary layer close to a heated surface, the surface must be hot enough to produce temperatures sufficient for auto-ignition at a distance greater than the quenching distance, d_q. For a surface of limited extent, the temperature necessary for ignition increases as the surface area is decreased (Powell, 1969): this is shown clearly in Figure 6.7 (Rae *et al*., 1964; Laurendeau, 1982).[2] With mechanical sparks that comprise very small incandescent particles (<0.1 mm) generated by frictional impact between two solid surfaces, even higher temperatures must be achieved. The temperature of impact sparks is limited by the melting points of the materials involved (Powell, 1969; Laurendeau, 1982), and ignition must be rapid as the particles will cool quickly (this is akin to 'hot spot ignition', discussed in Section 8.1.2). Pyrophoric sparks – in which the particles (e.g., aluminium and magnesium) oxidize vigorously in air – are capable of achieving very high temperatures (>2000°C) and can ignite the most stubborn mixtures. The thermite reaction between aluminium and ferric oxide (rust) can be initiated by impact, e.g., aluminium paint on rusty iron struck by any rigid object, producing sparks burning at 3000°C. These are highly incendive.[3]

---

[2] This has also been observed for the hot surface ignition of liquids (Colwell and Reza, 2005) (Section 6.2.3).

[3] It is assumed that the meaning of terms such as 'incendive' ('able to cause ignition') is either known or is obvious, but a useful chapter on relevant terminology may be found in Babrauskas (2003).

Table 6.3 Typical values of the minimum auto-ignition
temperature for flammable gases and vapours

	Minimum auto-ignition[a] temperature $(^\circ C)^a$
Hydrogen	400
Carbon disulphide	90
Carbon monoxide	609
Methane	600
Propane	450
n-Butane	405[b]
iso-Butane	460[b]
n-Octane	220[b]
iso-Octane (2,2,4-trimethylpentane)	418[b]
Ethene	450
Acetylene (ethyne)	305
Methanol	464
Ethanol	423
Acetone	538
Benzene	562

[a] Data taken from Yaws (1999).
[b] Note that branched alkanes have much higher auto-ignition temperatures than their straight-chain isomers (compare iso-butane
and n-butane).

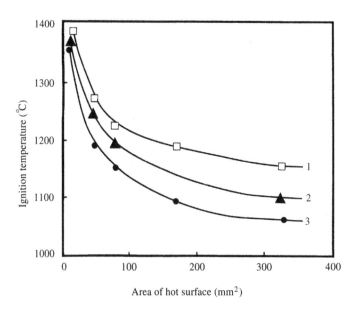

Figure 6.7 Minimum temperatures for ignition of 6% methane/air mixtures by hot surfaces of
different areas and locations within an explosion chamber: □, hot surface on wall; ▲, on ceiling;
● on floor. Rae *et al*. (1964), quoted by Laurendeau (1982). Reproduced by permission of the
Combustion Institute

6.2 Ignition of Liquids

Combustible liquids are classified according to their flashpoints, i.e., the lowest temperature at which a flammable vapour/air mixture exists at the surface at normal atmospheric pressure, 101.3 kPa (Burgoyne and Williams-Leir, 1949). This is normally determined using the Pensky–Martens Closed Cup Apparatus (ASTM, 1994b). The liquid is heated slowly (5–6°C per minute) in an enclosed vessel (Figure 6.8(a)) and a small, non-luminous pilot flame is introduced into the vapour space at frequent intervals through a port which is opened and closed automatically by a shutter. The 'closed cup' flashpoint is taken as the lowest temperature of the liquid at which the vapour/air mixture ignites producing a 'flash' of light blue flame, characteristic of premixed burning (see Section 3.2): some examples are given in Table 6.4. The proportion of vapour in air at the flashpoint can be calculated from the equilibrium vapour pressure of the liquid (Equation (1.14)), and while it is in reasonably good agreement with the published flammability limit data (Table 3.1), the value predicted tends to be slightly lower. Taking as an example pure liquid n-decane (which was discussed in Section 3.1.3), the vapour pressure at the flashpoint (44°C) can be calculated from Equation (1.14) as 5.33 mm Hg, corresponding to $5.33/760 = 7.0 \times 10^{-3}$ atm, i.e., 0.7% by volume at normal atmospheric pressure. The accepted figure for the lower flammability limit of n-decane vapour at 25°C is 0.75% (Table 3.1). Working backwards from this figure (LFL = 0.75% = 5.70 mm Hg), the closed cup flashpoint is predicted to be 45.2°C, over one degree higher than measured. The difference cannot be explained by the fact that the measured LFL refers to a temperature of 25°C (almost 20 degrees lower than the flashpoint) (see Equation (3.3d)), but is likely to be a result of the way in which the LFL is measured (Section 3.1.1). It is taken to be the lowest vapour concentration at which a flame will propagate 750 mm vertically upwards in a 50 mm diameter tube (Figure 3.1). Limited, localized propagation in the immediate vicinity of the ignition source is observed at lower concentrations, which may be what is observed in the closed cup apparatus (Drysdale, 2008).

Flashpoints of mixtures of flammable liquids can be estimated if the vapour pressures of the components can be calculated. For 'ideal solutions', to which hydrocarbon mixtures approximate, Raoult's law can be used (Equation (1.15)). As an example, consider the problem of deciding whether or not n-undecane ($C_{11}H_{24}$) containing 3% of n-hexane (by volume) should be classified as a 'highly flammable liquid' as defined in the 1972 UK

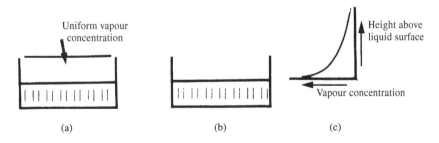

Figure 6.8 Determination of flashpoint: (a) closed cup, (b) open cup, (c) partial vapour pressure gradient above the surface of the fuel in the open cup

Table 6.4 Boiling points, flashpoints and firepoints of liquids[a]

	Formula	Boiling point (°C)	Closed cup flashpoint (°C)	Open cup flashpoint (°C)	Firepoint (°C)
n–Hexane	n–C_6H_{14}	69	−22	–	–
Cyclohexane	c-C_6H_{12}	81	−20	–	–
n-Heptane	n-C_7H_{16}	98	−4	–	
n-Octane	n-C_8H_{18}	125	−13	–	20^b
iso-Octane	iso-C_8H_{18}		−12	–	
n-Decane	n-$C_{10}H_{22}$	174	44	52^c	61.5^c
n-Dodecane	n-$C_{12}H_{26}$	216	72	–	103
Benzene	C_6H_6	80	−11	–	–
Toluene	$C_6H_5.CH_3$	110	4	7	–
p-Xylene	$C_6H_4.(CH_3)_2$	137	25	31^c	44^c
Methanol	CH_3OH	64	12	1^c $(13.5)^d$	1^c $(13.5)^d$
Ethanol	C_2H_5OH	78	13	6^c $(18.0)^d$	6^c $(18.0)^d$
n-Propanol	n-C_3H_7OH	97	15	16.5^c $(26.0)^d$	16.5^c $(26.0)^d$
iso-Propanol	iso-C_3H_7OH	82	12	–	–
n-Butanol	n-C_4H_9OH	117	29	36^c $(40.0)^d$	36^c $(40.0)^d$
n-Hexanol	n-$C_6H_{13}OH$	157	45	74	–
Acetone	$(CH_3)_2CO$	56	−14	−9	–
Diethyl ketone	$(C_2H_5)_2CO$	101	–	13	–

[a] A more comprehensive list of firepoints may be found in Drysdale (2008).
[b] Quoted by Ross (1994).
[c] Glassman and Dryer (1980/81).
[d] Figures in brackets refer to ignition by a spark (see last part of Section 6.2; also Glassman and Dryer (1980/81)).

Regulations[4] (Table 6.5). This can be reduced to determining if the mixture of liquids has a flammable vapour/air mixture above its surface at 32°C. The molar concentrations of n-hexane and n-undecane are calculated from the formula $V\rho/M_w$, where V, ρ and M_w are the percentage volume and the density of the liquid and the molecular weight of each component. The molar concentrations are thus:

$$n_{n\text{-hex}} = \frac{0.03 \times 660}{86} = 0.230 \qquad n_{n\text{-undec}} = \frac{0.97 \times 740}{156} = 4.601$$

which allow the mole fractions of n-hexane and n-undecane to be calculated (Equation (1.16)):

$$x_{n\text{-hex}} = \frac{0.230}{0.230 + 4.601} = 0.048 \qquad x_{n\text{-undec}} = \frac{4.601}{0.230 + 4.601} = 0.952$$

The vapour pressures of the pure liquids at 32°C are calculated (Equation (1.14) and Table 1.12) to be:

$$p^o_{n\text{-hex}} = 179.08 \, \text{mm Hg} \quad \text{and} \quad p^o_{n\text{-undec}} = 1.08 \, \text{mm Hg}$$

---

[4] These regulations were superseded in 2002 by European regulations (DSEAR – see Table 6.6).

Table 6.5 Classification of liquids in the UK (1972 Regulations) and the USA, compared with the system adopted by the UN – the 'Globally Harmonized System'. GHS classifies liquids with flashpoints <0°C and boiling points <35°C as 'extremely flammable' (cf. Table 6.6)

UK	Flashpoint	USA	Flashpoint	GHS
Highly flammable liquids	<32°C	Class IA, IB	<22.8°C	Highly flammable liquids (<23°C)
		Class IC	22.8–37.8°C	
Flammable liquids	32–60°C	Class II	37.8–60°C	Flammable liquids (23–60°C)
Combustible liquids	>60°C	Class IIIA	60–94.3°C	Combustible liquids (>60°C)
		Class IIIB	>94.3°C	

and their partial pressures above the liquid mixture are obtained by applying Raoult's law (Equation (1.15)), thus:

$$p^o_{n\text{-hex}} = 8.60 \, \text{mm Hg} \quad \text{and} \quad p^o_{n\text{-undec}} = 1.03 \, \text{mm Hg}$$

Whether or not this mixture would be flammable in a normal atmosphere can be established by applying Le Chatelier's law in the form given in Equation (3.2b). The lower flammability limits of n-undecane and n-hexane are 0.68% and 1.2%, respectively (at 25°C), thus Equation (3.2b) gives:

$$\frac{100 \times 8.60/760}{1.2} + \frac{100 \times 1.03/760}{0.68} = 1.14 > 1$$

indicating the lower limit has been exceeded and therefore the liquid mixture has a flashpoint below 32°C.

The reliability of this calculation depends on a number of assumptions which are known to be approximate, in particular that the liquid mixture behaves according to Raoult's law. For non-ideal mixtures, the activities of the components of the mixture must be known and applied as described briefly in Section 1.2.2 (Equation (1.15) *et seq*.) (see Babrauskas, 2003; Drysdale, 2008).

In a container of flammable liquid, it is possible for the vapour/air mixture in the headspace (i.e., the enclosed volume above the liquid surface) to be flammable. This is the case with methanol and ethanol at temperatures between c. 11 and 23°C. Indeed, there have been several serious incidents in restaurants when attempts have been made to refill flambé lamps without first ensuring that the flame had been extinguished (e.g., Mundwiler, 1990). Flame propagates into the headspace, causing an internal explosion which may be capable of scattering burning liquid over a wide area. This cannot occur with petrol (gasoline) tanks at normal ambient temperatures as the vapour pressure of the gasoline lies well above the upper flammability limit at normal temperatures and the mixture in the headspace is too rich to burn. Many common volatile liquids fall into this category, but it should be noted that if the temperature is reduced sufficiently, the vapour concentration will fall below the upper flammability limit. The temperature at which the vapour pressure corresponds to the upper flammability limit has been called the 'upper flashpoint'. Few measurements have been reported in the literature (Hasegawa and Kashuki, 1991; Babrauskas, 2003), although clearly it is an important concept as it defines the range of temperatures within which a flammable mixture will exist in the headspace of a container. An estimate of the upper flashpoint can be made if the upper flammability limit of the vapour is known. Taking n-hexane as an example, its upper flammability limit is 7.4% (Table 3.1), corresponding to a vapour pressure of 0.074×760 mm Hg. Using Equation (1.14) and substituting data from Table 1.12 (Weast, 1974/75), the 'upper flashpoint' is calculated as $+6°C$, indicating that the vapour/air mixture in the headspace of a container of n-hexane will become flammable if the temperature falls below 6°C. This scenario was discussed in Section 3.1.3 in the context of aircraft fuel tanks: on landing, the vapour/air mixture in the headspace of a tank containing JP-4 (a highly volatile fuel) can be flammable as the fuel temperature will have fallen during exposure to the low temperatures experienced at high altitude.

In Section 3.1.3, Figure 3.8(a), it was shown that the lower flammability limit is remarkably insensitive to a reduction in pressure, at least until the pressure falls below c. 0.2 atm. This has a significant consequence regarding flashpoint: if the temperature of a liquid fuel is held constant but the pressure is reduced, the ratio of fuel vapour to air will increase so that a 'flammable' liquid such as n-decane (normal flashpoint 44°C) will become 'highly flammable' (using the terminology of the UK 1972 Regulations, Table 6.5) if the atmospheric pressure is reduced sufficiently. The large pressure reductions that are required to create this situation are encountered as a matter of course during the ascent of an aircraft. The implications for the interpretation of 'flashpoint' are clear: it will vary with pressure. Indeed, it is recommended in the standard that a correction is made for atmospheric pressure. At sea level, this is a minor adjustment, but at the altitude of Denver, CO, which is famously 1 mile high (1609 m), atmospheric pressure is 631 mm Hg. Taking the lower flammability limit for n-decane as 0.75%, and using Equation (1.14) and substituting data from Table 1.12, it can be shown that its flashpoint will be 41.8°C, about 2 degrees less than the value quoted in Table 6.4 (44°C) and about 3.5 degrees less than the theoretical value calculated above (45.2°C).[5] The difference is even greater at higher altitudes: thus in Mexico City (2240 m) and Lhasa (3650 m) the flashpoint of n-decane would be approximately 39.4°C and 35.9°C, respectively.

Flashpoint measurements can be made in an open cup (Figure 6.8(b)), but in this case vapour is free to diffuse away from the surface, producing a vapour concentration gradient which decreases monotonically with height (this is illustrated schematically in Figure 6.8(c)). As ignition can only occur when the vapour/air mixture is above the lower flammability limit at the location of the pilot flame, it is found that the open cup flashpoint is dependent on the height of the ignition source above the liquid surface (Burgoyne et al., 1967): results that demonstrate this are shown in Figure 6.9 (Glassman and Dryer, 1980/81). In the standard open cup test (ASTM, 1990a), the size of the pilot flame and its height above the liquid surface are strictly specified. In general, open cup flashpoints are greater than those measured in the closed cup. However, ignition of the vapour in the open cup test will not lead to sustained burning of the liquid unless its temperature is increased further, to the 'firepoint'. This is found to be significantly greater than the flashpoint for hydrocarbons, although there are surprisingly few data quoted in the literature (see Table 6.4). Values that do exist indicate that the concentration of vapour at the surface must be greater than stoichiometric for a diffusion flame to

---

[5] In view of the difference between the measured and 'theoretical' values, it is more logical to compare the calculated values of the flashpoint at 760 and 631 mm Hg.

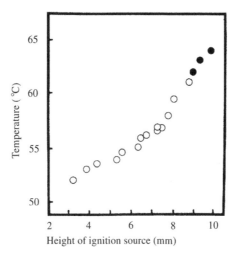

Figure 6.9 Variation of the measured open cup flashpoint with height of ignition source above the liquid surface (*n*-decane). ○, flash only; ●, flash followed by sustained burning. From Glassman and Dryer (1980/81), by permission

become established. Roberts and Quince (1973) reported values of between 1.33× and 1.92× the stoichiometric concentration, i.e., the mixture close to the surface is fuel rich, but still within the flammability limits (see Table 3.1, column 8). (Alcohols appear to behave differently and will be discussed below.) Ignition of the vapour/air mixture above the liquid gives a transient, fuel-rich premixed flame in which some of the fuel vapour will not be completely consumed. A diffusion flame will then remain only if the rate of supply of vapour is sufficient to support it. If the rate is too low (i.e., the liquid is below its firepoint, but above the flashpoint), such a flame cannot survive as it will be quenched (extinguished) as a consequence of heat losses to the surface (Section 3.3). Roberts and Quince (1973), Rasbash (1975) and others, have argued that the nascent flame will self-extinguish if it loses more than ~30% of the heat of combustion to the surface.[6] However, at and above the firepoint, the flame will stabilize and the continuing heat loss to the surface becomes available to heat the surface layer and increase the rate of supply of vapour. Consequently, the diffusion flame will strengthen and grow in size, culminating in steady burning when the surface temperature achieves a value close to the boiling point and a stable temperature profile has been established below the surface (see Figure 5.5).

The firepoint of a liquid is defined as the lowest temperature at which ignition of the vapours in an open cup is followed by sustained burning. In the standard test (ASTM, 2005), the temperature is maintained by external heating. Under steady state (non-fire) conditions, the rate of evaporation is given by:

$$\dot{m}''_{evap} = \frac{\dot{Q}''_E - \dot{Q}''_L}{L_v} \tag{6.15}$$

---

[6] As the amount of heat radiated by the nascent flame will be very small, the heat loss to the surface is assumed to be by convection.

where $\dot{m}''_{\mathrm{evap}}$ is the mass flux leaving the surface, L_{v} is the latent heat of evaporation of the liquid, and $\dot{Q}''_{\mathrm{E}}$ and $\dot{Q}''_{L}$ are the rates of external heating and heat loss, respectively, expressed in terms of unit area of exposed liquid surface. If flame is established at the surface, it provides an additional source of heat and the rate of evaporation (now the 'rate of burning') increases:

$$\dot{m}''_{\mathrm{burn}} = \frac{f \Delta H_{\mathrm{c}} \dot{m}''_{\mathrm{burn}} + \dot{Q}''_{\mathrm{E}} - \dot{Q}''_{L}}{L_{\mathrm{v}}} \qquad (6.16)$$

where f is the fraction of the heat of combustion of the vapour (ΔH_{c}) that is transferred back to the surface and is made up of radiative and convective components, indicated by f_{r} and f_{c}, respectively. Equation (6.16) may be rewritten:

$$((f_{\mathrm{r}} + f_{\mathrm{c}}) \Delta H_{\mathrm{c}} - L_{\mathrm{v}}) \dot{m}''_{\mathrm{burn}} + \dot{Q}''_{\mathrm{E}} - \dot{Q}''_{L} = S \qquad (6.17)$$

where $S = 0$ for 'steady burning'.[7] The proportions f_{r} and f_{c} vary with the size of the fire. At the firepoint, f_{r} is small as the flame is non-luminous and may be assumed to be zero for the present argument. However, f_{c} (which becomes small for steady burning of large fuel beds (Section 5.1)) approaches a maximum value (ϕ) at the firepoint corresponding to the stability limit of flame at the surface. Setting $f_{\mathrm{r}} = 0$ and $f_{\mathrm{c}} = \phi$ and writing $\dot{m}''_{\mathrm{burn}} = \dot{m}''_{\mathrm{cr}}$, the limiting or critical flowrate of the volatiles at the firepoint (Section 6.3.2), Equation (6.17) can be used to identify the limiting conditions under which ignition of the volatiles can lead to sustained burning:

$$(\phi \Delta H_{\mathrm{c}} - L_{\mathrm{v}}) \dot{m}''_{\mathrm{cr}} + \dot{Q}''_{\mathrm{E}} - \dot{Q}''_{L} = S \qquad (6.18)$$

This equation, first proposed by Rasbash (1975), will be discussed in more detail in relation to the ignition of solids (Section 6.3.2). Steady burning will develop after ignition of the volatiles, only if there is sufficient excess heat available from the flame to cause the surface temperature to increase (i.e., $S > 0$), thereby increasing the rate of volatilization and allowing the flame to strengthen. The value of S will subsequently diminish as steady state burning is approached. Equation (6.18) identifies formally those factors which determine whether or not sustained burning can develop.

Glassman and Dryer (1980/81) found an apparent anomaly in the relationship between open and closed cup flashpoints and firepoints for alcohols. With a pilot flame as the ignition source, it is possible to ignite an alcohol in an open cup at a temperature significantly lower than the closed cup flashpoint (Table 6.4). This may be explained if the pilot flame is contributing heat to the surface, raising the temperature locally to the firepoint; the anomaly disappears if a spark ignition source is used instead, although the (open cup) flashpoint and firepoint remain coincident. This behaviour is quite different from that of hydrocarbon liquids, but while the explanation may lie in the different infra-red absorption characteristics of these fuels, the effect has not been fully investigated.

This is not simply an academic point, as it suggests that the closed cup flashpoint may not be a valid method of assessing the fire risks associated with all liquid fuels. Indeed, for some liquids of reduced flammability, such as blends containing certain chlorinated

---

[7] S may be regarded as the rate of transfer of 'sensible heat', i.e. heat transfer that manifests itself in a temperature change.

hydrocarbons, it can give a false result (Babrauskas, 2003). One such blend (a commercial cleaning fluid) did not give a flashpoint when tested according to the procedure prescribed for the standard closed cup apparatus (Tyler, 2008), although it had been implicated in one or two fires and explosions. The principal component was methyl chloroform (1,1,1 trichloroethane, $CCl_3 \cdot CH_3$). When the flashpoint of the blend was measured in larger (non-standard) vessels (of diameter greater than 124 mm), a value of 12°C was obtained (James, 1991), suggesting that flame quenching was dominant in the standard closed cup apparatus. Another situation in which incorrect or misleading results may be obtained is in the case of fuel blends which contain highly volatile components (such as 'live' crude oil). If the liquid has been exposed to the atmosphere at ambient temperature for a sufficient period of time, some of the 'light ends' may be lost and the measured flashpoint will be too high and unrepresentative of the original fuel. If this is seen to be a possibility, then the liquid and the apparatus should be chilled prior to carrying out the test (Drysdale, 2008).

6.2.1 Ignition of Low Flashpoint Liquids

Bearing in mind the above caveats, classifying combustible liquids according to their flashpoints is a convenient way of indicating their relative fire hazards. Liquids with 'low' flashpoints present a risk at ambient temperatures as their vapours may be ignited by a spark or flame. If such a vapour/air mixture is confined, overpressures that are capable of causing structural damage may be generated (see Section 1.2.5), although fire will only result if the liquid is above its firepoint (e.g., gasoline). The closed cup flashpoint is always used to indicate the hazard as this will err on the side of safety if the risk is only that of fire. Various classification systems exist. In the UK, the Highly Flammable Liquids and Liquefied Petroleum Gases Regulations, 1972, defined liquids with flashpoints less than 32°C[8] as 'highly flammable'. Liquids with flashpoints between 32°C and 60°C were termed 'flammable', while those with flashpoints above 60°C were classified as 'combustible'. A similar scheme in use in the USA is summarized in Table 6.5, where it can be seen that the low flashpoint liquids are also divided into two groups. It is worth noting that the upper bound of the 'Group I'-type flammable liquid (37.8°C) is higher in the USA than in the UK, reflecting the higher ambient temperatures that are encountered in the States. In still warmer climates, it would be appropriate to set an even higher upper boundary.

New classification systems have been developed under the auspices of the European Union and the United Nations to facilitate international trade. In the UK, the 1972 Regulations have been replaced under a European Directive by the Dangerous Substances and Explosive Atmospheres Regulations (DSEAR) 2002, which uses the classification system defined in the Chemical Hazard Information and Packaging for Supply (CHIPS) Regulations. In these, the boundary between 'Highly Flammable' and 'Flammable' has been lowered to 21°C and a new category 'Extremely Flammable' introduced, as shown in Table 6.6. The United Nations have published a slightly different classification for international transportation, identifying four categories (1–4) as shown in Table 6.7. This is known as the UN Globally Harmonized System (GHS) and is compared with the UK 1972 Regulations and NFPA 30 (National Fire Protection Association, 2008) in Table 6.5.

---

[8] The regulation specifies that this is determined using the Abel closed cup apparatus (British Standards Institution, 1982).

Table 6.6 The classification system used in DSEAR 2002

Classification	Flashpoint	Boiling point
Extremely flammable	<0°C	≤ 35°C
Highly flammable	<21°C	>35°C
Flammable	≥ 21 and <55°C	

Table 6.7 Classification developed by the UN for transport of hazardous chemicals, known as the Globally Harmonised System (GHS)

Category	Criteria
1	Flash point <23°C and initial boiling point ≤ 35°C
2	Flash point <23°C and initial boiling point >35°C
3	Flash point ≥ 23°C and ≤ 60°C
4	Flash point >60°C and ≤ 93°C

In the open, a large pool of a highly flammable liquid (e.g., petrol or gasoline) will produce a significant volume of flammable vapour/air mixture which – given time – can extend beyond the limits of the pool boundary. Introduction of an ignition source to this volume will cause flame to propagate back to the pool, burning any vapour near the surface which was initially above the upper flammability limit, producing a large, transient diffusion flame before steady burning is established. This is sometimes referred to as a 'flash fire'. The flammable zone, or volume, will extend downwind of the pool and its extent will depend on the vapour pressure of the liquid (which determines its rate of evaporation) and the windspeed and degree of atmospheric turbulence. As the latter increases, the rate of dispersion increases and the horizontal extent of the flammable zone will decrease. Further discussion of this topic is beyond the scope of this text (see Wade, 1942; Sutton, 1953; Clancey, 1974), but it should be pointed out that under quiescent, or still air conditions associated with an atmospheric inversion, a heavier-than-air vapour formed by the evaporation of a volatile liquid will tend to 'slump' and spread as a gravity current, with very limited dilution. An extreme example of this occurred at the Buncefield Oil Storage and Transfer Depot on 11 December 2005, when an unleaded gasoline tank was overfilled (Newton, 2008). A pancake-shaped vapour cloud formed which was over 300 m in diameter by the time ignition occurred, producing a devastating explosion (see Section 3.6).

6.2.2 Ignition of High Flashpoint Liquids

Sustained burning of a high flashpoint liquid can only be achieved if it has been heated to above its firepoint. In the standard open cup test (ASTM, 2005), the bulk of the liquid is heated uniformly, although in principle only the surface layers need be heated. While this may be achieved by applying a heat flux to the entire surface (e.g., a pool of liquid exposed to radiation from a nearby fire), it is more common to encounter local application of heat, such as a flame impinging on, or burning close to the surface. Ignition and sustained burning can take some time as heat is dissipated rapidly from the affected area by a convective mechanism that is maintained by surface tension-driven flows (Sirignano and Glassman, 1970). Surface tension – defined as the force per unit length acting at the surface of a liquid – is temperature-dependent, decreasing significantly with increase in temperature. The result of this is that there is a net force acting at the surface, drawing hot liquid away from the heated area, causing fresh, cool liquid to take its place from below the surface: the convection currents so induced are illustrated in Figure 6.10. In a pool of limited extent, full involvement will eventually be achieved following flame spread across the surface, but only after a substantial amount of heat has been transferred convectively to the bulk of the liquid and the temperature at the surface has increased to the firepoint (Burgoyne and Roberts, 1968). Flame spreads over the liquid surface as an advancing ignition front in which the surface temperature adjacent to the leading edge of the flame reaches the firepoint. This general mechanism is common to flame spread over both liquids and solids and is discussed in Chapter 7.

However, a high firepoint liquid can be ignited very easily if it is absorbed onto a wick, i.e., a porous medium of low thermal conductivity such as is used in kerosene lamps and candles. Application of a flame to the fuel-soaked wick causes a rapid local increase in temperature, not only because the layer of liquid is too thin for convective dissipation of heat to occur, but also because the wick is an effective thermal insulator (low $k\rho c$). Ignition will be achieved at the point of application and will be followed by flame spread over the wick surface (Section 7.1). One mechanism by which pools of high firepoint liquids may be ignited without recourse to bulk heating is to ignite the liquid absorbed on a wick (a cloth or any other porous material) that is lying in the pool. Flame will establish itself in a position in which it can begin to heat the surface layers of the free liquid. This

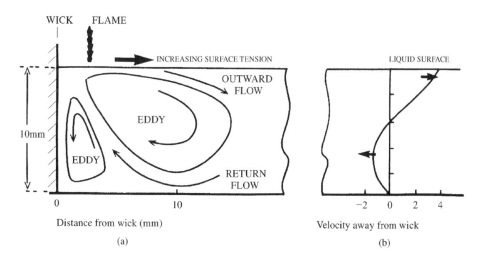

Figure 6.10 (a) Surface tension-driven flows and convective motion in a liquid subjected to a localized ignition source. (b) Velocity profile 10 mm from wick. After Burgoyne *et al.* (1968), by permission

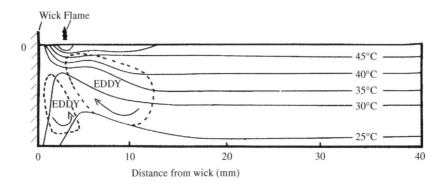

Figure 6.11 Wick ignition of a pool of high flashpoint liquid. Temperature distribution immediately before flame spread over the surface. From Burgoyne *et al*. (1968), by permission

type of ignition was studied by Burgoyne *et al*. (1968) using three alcohols (butanol, iso-pentanol and hexanol). The liquid was contained in a long trough and ignited on a wick located at one end (Figure 6.11). By monitoring the temperature at various points within the liquid, it was found that flame began to spread away from the wick only when the temperature at the surface exceeded the flashpoint. The associated induction period is illustrated in Figure 6.12(a). It was shown that the amount of heat transferred from the flame at the wick to the liquid was in agreement with the amount of heat stored in the liquid at the end of the 'induction period'.

In view of this, Burgoyne *et al*. (1968) examined the effect of reducing the depth of the liquid on the induction period. They found that the latter exhibited a minimum at about 2 mm, no ignition being possible for the fuels used at depths less than 1 mm (Figure 6.12(b)). The minimum exists because less heat needs to be transferred from the flame on the wick to raise the surface temperature of a shallow pool to the firepoint. However, if the pool is too shallow, then in addition to convective flow being restricted, heat losses to the supporting surface become too great and the firepoint will not be attained. This is consistent with the work on oil slick ignition by Brzustowski and Twardus (1982) and applies in general to hydrocarbon liquids floating on water: reference was made to this effect in Section 5.1.1 and there will be further discussion in Section 7.1 in relation to flame spread over liquids (Mackinven *et al*., 1970).

With wick ignition, the source of heat is also acting as the pilot for ignition of the volatiles, but other situations can be envisaged in which the heat source is quite independent of the agency that initiates flaming (as in the determination of firepoint using the Cleveland open cup (ASTM, 2005)). If the temperature of the environment is increased to above the firepoint of the liquid, then it will eventually behave as a highly flammable liquid as it becomes heated (Section 6.2.1). If heating is confined to the liquid and its container, then additional evaporation will take place, giving a localized flammable zone, although the vapour may condense on nearby cold surfaces. Over a long period of time the liquid may evaporate completely if sufficiently volatile. On the other hand, a relatively involatile combustible liquid may be heated sufficiently for spontaneous ignition to occur. This can happen with most cooking oils and fats and can be demonstrated easily using the

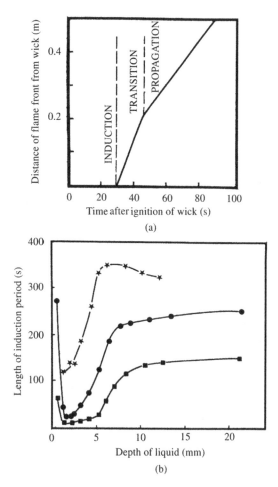

Figure 6.12 (a) Induction period for wick ignition of high firepoint fuels. (b) Effect of liquid depth on the duration of the induction period. ✶, Hexanol; ●, isopentanol; ■, butanol. From Burgoyne *et al*. (1968), by permission

Cleveland open cup. Careful observation shows that the flame starts in the plume of hot volatile products, well clear of the liquid surface, then flashes back to give immediately an intense fire, as the liquid is well above its firepoint and may already be close to its boiling point (see Figure 6.28). The liquid temperature at which auto-ignition occurs will depend on the surface area of the liquid and be very sensitive to any air movement which would tend to disturb and cool the plume. This point is addressed in the next section.

6.2.3 Auto-ignition of Liquid Fuels

In Section 6.1, the auto-ignition of flammable vapour/air mixtures in contact with hot surfaces was considered. The ignition of liquid fuels in this manner is a recognized problem in the automotive and aviation industries, where oils and hydraulic fluids may come into contact with hot surfaces. The overall process must involve evaporation of the liquid under circumstances in which a flammable vapour/air mixture can form and undergo spontaneous ignition. A standard test has been developed to determine the auto-ignition temperature of 'liquid chemicals', as they are described in the ASTM Standard (ASTM, 2005). The test is similar to that used by Setchkin (1954), but involves dropping a small sample (100 μl) into a uniformly heated 0.5 l glass flask containing air at a predetermined temperature and observing whether or not ignition occurs within 10 minutes. This is the technique used to determine the auto-ignition temperatures of the liquids that are included in Table 6.3.

However, these values refer to a closed system in which the flammable mixture is held within a uniformly heated spherical vessel. As with flammable vapour/air mixtures, liquids may come into contact with open surfaces at high temperature (e.g., hot exhausts) and can ignite after evaporation if the conditions are suitable. Colwell and Reza (2005) studied the ignition of single droplets of combustible liquids falling onto a flat heated plate (exposed area 57.9 cm by 10.8 cm) under still air conditions. There was no well-defined auto-ignition temperature and they had to carry out 200 ignition tests (with each of 14 fluids) over a range of temperatures to enable ignition probability distributions to be developed. Examples of these are shown in Figure 6.13 for aviation fluids ranging from turbine engine lubricants to aviation gasoline.[9] Taking the turbine engine lubricant specified as MIL-L-7808 as an example, there was a 5% probability of ignition at 570°C which increased to 95% at 625°C. The probability was 50% at 595°C, considerably higher than the

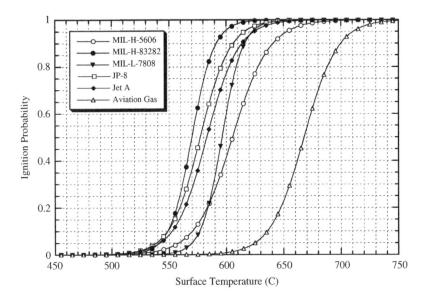

Figure 6.13 Ignition probability as a function of the surface temperature of a flat plate for aviation fluids. From Colwell and Reza (2005): *Fire Technology*, 41 (2005) 105–123, 'Hot surface ignition of automotive and aviation fluids', J.D. Colwell and A. Reza, Figure 8, with kind permission from Springer Science+Business Media B.V

[9] A similar study of the ignition of diesel fuels has been reported recently by Shaw and Weckman (2010).

auto-ignition temperature of approximately 360°C as determined in the standard test (ASTM, 2005; Colwell and Reza, 2005). The distributions reported by Colwell and Reza (2005) refer specifically to the geometry of their apparatus. For any degree of confinement around the hot surface (or indeed if the hot surface was profiled in a significant manner), the temperatures are likely to be reduced – the limiting (minimum) value corresponding to the total confinement of the standard test. On the other hand, higher surface temperatures would be expected with smaller surface areas, as observed by Colwell and Reza (2005) and reported for methane/air mixtures by Laurendau 1982 (see Figure 6.7), or indeed if there was any imposed air movement that would dilute (and cool) the fuel vapour.

6.3 Piloted Ignition of Solids

The phenomena of 'flashpoint' and 'firepoint' can be observed with solids under conditions of surface heating (e.g., Deepak and Drysdale, 1983; Thomson and Drysdale, 1987; Rich *et al.*, 2007), but cannot be defined in terms of a bulk temperature. The generation of flammable volatiles involves chemical decomposition of the solid (pyrolysis), which is an irreversible process: there is no equivalent to the equilibrium vapour pressure that may be used to estimate the flashpoint of a liquid fuel (Section 6.2).

However, it is reasonable to assume that the same principles apply, namely that the flashpoint is associated with the minimum conditions under which pyrolysis products achieve the lower flammability limit close to the surface, and that the firepoint corresponds to a fuel-rich, near-stoichiometric mixture at the surface. As the system is open (cf. the open cup flashpoint test), these concentrations will be associated with specific rates of pyrolysis, or mass fluxes (Rasbash *et al.*, 1986; Janssens, 1991a), which should be capable of measurement. A number have been measured by Tewarson (1995) and Drysdale and Thomson (1989), but under significantly different conditions: these will be discussed in Section 6.3.2. If it is assumed that sustained, piloted ignition can only occur if a critical mass flux of fuel vapours is exceeded, then the process of piloted ignition can be described in terms of Figure 6.14 (Drysdale, 1985a), where 'sufficient flow of volatiles' corresponds to a mass flux greater than the critical value, and 'suitable conditions' implies that the environmental conditions remain favourable for the flame to become established.

Continuing the analogy with liquids, it might be supposed that the critical condition corresponding to the firepoint of a solid could be identified in terms of a surface temperature. This can be measured, albeit with difficulty. Attaching a thermocouple to a surface in such a way that it records the surface temperature accurately during the heating period and up to the point of ignition is not easy. Figure 6.15 illustrates how the surface temperature of a combustible solid increases with time as it is exposed to a heat flux. After t_1 seconds, a temperature is reached at which the rate of pyrolysis is just sufficient for the first visible appearance of 'smoke' (air-borne pyrolysis products) from the surface. After a further delay (at t_2 s) the rate of production of the pyrolysis products is sufficient to form a flammable vapour/air mixture near the surface, at which time ignition of the vapours by an independent 'pilot' would produce a flash of flame – the flashpoint – while at t_3 the rate of release of vapours is sufficient to sustain a flame at the surface if a pilot ignition source is present. In the absence of a pilot source, continued heating will lead to

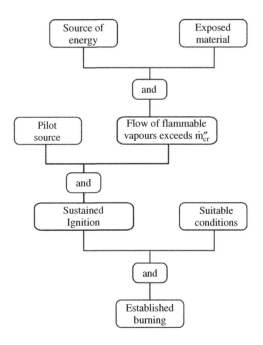

Figure 6.14 Scheme for piloted ignition of a combustible solid (Drysdale, 1985a)

spontaneous appearance of flame in the vapour/air mixture above the surface at t_4. This is spontaneous ignition (see Section 6.4).

Despite the experimental difficulties, there have been numerous studies in which fire-point temperatures have been measured directly for a range of materials, including wood (e.g., Atreya *et al.*, 1986; Atreya and Abu-Zaid, 1991; Janssens, 1991c) and various thermoplastics (Thomson and Drysdale, 1987; Long *et al.*, 1999; Cordova *et al.*, 2001, and many others). It is also possible to deduce values of T_{ig} from measurements of time to ignition as a function of the imposed heat flux (e.g., from plots of $1/t_{ig}$ vs. $\dot{Q}_R^{\prime\prime 2}$ – see Equation (6.32)), but the effective thermal inertia ($k\rho c$) must be known. Inevitably this introduces uncertainties. Tables of values of T_{ig} are available (e.g., Tewarson, 2008), but the individual values should be used with caution if their pedigree is not given.

Indeed, recent work has shown that the concept of a firepoint temperature (T_{ig}) needs careful interpretation. In experiments with vertically oriented slabs of PMMA exposed to a radiant heat flux, Cordova *et al.* (2001) found that under conditions of forced convective flows, the measured value of T_{ig} is greater than that under conditions of natural convection. This can be explained in terms of dilution of the flammable pyrolysis products, which must achieve a flammable concentration at the location of the pilot source before they will ignite.[10] Under conditions of natural convection and heat fluxes above 20 kW/m^2,

---

[10] This should be compared with the observation that the firepoint of a combustible liquid increases with the height of the ignition source above the surface of the liquids (see Figure 6.10).

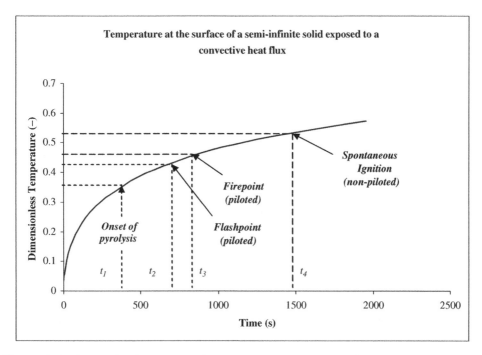

Figure 6.15 Schematic diagram showing the surface temperature of a thermally thick solid exposed to a heat flux, showing the times to the onset of pyrolysis (t_1), flashpoint (t_2), firepoint (t_3) and spontaneous ignition (t_4) (when no pilot is present)

they found that T_{ig} was approximately constant (310°C), but decreased if the heat flux was reduced, falling below 290°C at 11 kW/m². A similar observation was made by Thomson and Drysdale (1987), but no satisfactory explanation has been put forward. Clearly, a more robust definition is required to describe the ignition process properly. The key issue is in defining the conditions under which a diffusion flame can stabilize at the surface: this can be understood in terms of a heat balance using Equation (6.18). This is discussed below.

Most of the early studies of ignition concentrated on the response of an inert solid to an imposed heat flux and determining the time to ignition, tacitly assuming that a critical ignition temperature is a meaningful concept. Indeed, this has been the basis for most of the standard 'ignition tests' that simply measure the time to (piloted) ignition under a defined heat flux (e.g., the ISO Ignitability Test (ISO, 1984), the BS Ignitability Test (British Standards Institution, 1991), and – at their simplest functionalities – the cone calorimeter (ASTM, 2009c) and the Fire Propagation Apparatus (ASTM, 2009d)). In modelling the response of the solid to the heat flux, it is usually assumed that material is inert and there is no decomposition before the ignition temperature (firepoint) is reached. Section 6.3.1 discusses this aspect of the ignition process in more detail as there are a number of issues that should be considered when assessing ease of ignition. Here it will be assumed that the heat flux is constant. In Section 6.3.2, the details of flame stabilization will be discussed under the circumstances in which the heat flux ceases, or substantially decays after piloted ignition has occurred.

For the sake of clarity, the following sections deal with the piloted ignition scenario which is illustrated in Figure 6.14, i.e., the source of energy and the 'pilot' are distinct. However, when there is direct flame impingement (Section 6.5), the flame acts both as the source of energy and as the ignition source for the vapours. This is more difficult to analyse from first principles as the heat flux imposed by the flame depends on the size of the flame, its radiation characteristics and the geometry of the adjacent surfaces, which can influence the flow dynamics, and thus the magnitude of the convective component (Hasemi, 1984; Kokkala, 1993; Back *et al.*, 1994; Foley and Drysdale, 1995).

6.3.1 Ignition during a Constant Heat Flux

If the heat flux at the surface is constant,[11] the firepoint may be defined as the minimum surface temperature (T_{ig}) at which the flow of volatiles is sufficient to allow flame to persist at the surface (this corresponds to 'sustained ignition' (Janssens, 1991)). In this context, the time taken to achieve the firepoint is the most important parameter and provided that T_{ig} is known, it may be calculated from first principles if the problem is reduced to one of heat transfer to the surface of an inert solid (Simms, 1963; Kanury, 1972; Mikkola and Wichman, 1989). However, there are a number of factors that are inconsistent with the 'inert solid' concept arising particularly from physical changes and chemical reactions that occur at and below the surface. These include evaporation of moisture, the endothermic pyrolysis reactions that are responsible for the formation of the fuel volatiles (e.g., Dakka *et al.*, 2002) and heterogeneous oxidative reactions which promote the pyrolysis process (evidence for this has been reported for PMMA by Kashiwagi and Ohlemiller (1982) and Dakka *et al.* (2002); and for wood (Kashiwagi *et al.* (1987)). Nevertheless, to a first approximation the assumption that the material is inert until the firepoint temperature is reached is a satisfactory one, particularly if the activation energy of the pyrolysis process is high (Table 1.5).

Kanury (1972) considered various solutions to the one-dimensional heat conduction equation (Equation (2.15)) in which the boundary conditions were chosen to represent a number of configurations, including both the 'infinite slab' and the 'semi-infinite solid' discussed in Section 2.2.2. In all cases, the solid is assumed to be opaque and inert, with uniform thermal properties that are independent of temperature. This allows the main underlying principles which influence the ignition behaviour of solids to be examined. Most theoretical and experimental investigations have concentrated on ignition brought about by radiative heat transfer. The original impetus for this followed the realization that thermal radiation levels from a nuclear explosion would be sufficient to ignite combustible materials at great distances from the blast centre. However, it has become apparent that radiation is of fundamental importance in the growth and spread of fire in many diverse situations, such as open fuel beds (Section 7.4) and compartments (Chapter 9). The high level of interest in radiative ignition has been maintained, although ignition by convection

---

[11] It should also be considerably greater than the minimum heat flux required for (piloted) ignition, a concept that is discussed below.

cannot be neglected. In the following sections, relevant solutions for the one-dimensional heat conduction equation in an inert slab:

$$\frac{\partial^2 T}{\partial x^2} = \frac{1}{\alpha}\frac{\partial t}{\partial t} \tag{6.19}$$

are discussed.

6.3.1.1 The Thin Slab

It was shown in Section 2.2.2 that the temperature (T) of a thin 'slab' exposed to convective heating at both faces varies with time according to the expression (Equation (2.21)):

$$\frac{T_\infty - T}{T_\infty - T_0} = \exp(-2ht/\tau\rho c) \tag{6.20}$$

where T_0 and T_∞ refer to the initial and final (i.e., gas stream) temperatures, respectively, and τ is the thickness. For example, this would be relevant to a curtain fabric exposed to a rising current of hot gases. If the 'firepoint' of the material can be identified as a specific temperature T_{ig}, then assuming ignition of the volatiles, the time to ignition (t_{ig}) would be given by:

$$t_{ig} = \frac{\tau\rho c}{2h}\ln\left(\frac{T_\infty - T_0}{T_\infty - T_{ig}}\right) \tag{6.21}$$

Assuming that both T_{ig} and c are constant, this indicates that the time to ignition is directly proportional to the mass per unit surface area $(\tau\rho)$, and inversely proportional to h, the convective heat transfer coefficient. The same general conclusions may be deduced if the positive heat transfer is on one side only. Note that for a single material (e.g., paper), t_{ig} is directly proportional to the thickness of the sample, at least up to the thin (low Biot number) limit. This is best illustrated in terms of flame *spread* over thin fuels (see Section 7.2.2 and Figure 7.10).

The problem of an infinite slab exposed to radiant heating on one face $(x = +l)$ with convective cooling at both faces $(x = \pm l)$ (Figure 6.16) can be solved with the following boundary conditions applied to Equation (6.19):

$$T = T_0 \text{ for all } x \text{ at } t = 0 \tag{6.22a}$$

$$a\dot{Q}_R'' = h\theta - k\frac{d\theta}{dx} \text{ for } x = +l \text{ and } t > 0 \tag{6.22b}$$

and

$$h\theta = -k\frac{d\theta}{dx} \text{ at } x = -l \text{ and } t > 0 \tag{6.22c}$$

where $\theta = T - T_0$, a is the absorptivity (assumed constant) and $\dot{Q}_R''$ is the radiant heat flux falling on the surface. Radiative heat loss is neglected here, although it will become significant at temperatures approaching T_{ig} when it cannot be ignored (Mikkola and Wichman, 1989).[12] The full analytical solution is available (Carslaw and Jaeger, 1959),

---

[12] Radiation may be included by replacing h by $h_c + h_r$, where h_r is a linearized coefficient that approximates the radiative loss term (Torero, 2008).

but Simms (1963) adopted the 'lumped thermal capacity' (low Biot number) approach (Section 2.2.2) to obtain the following equation for a thermally thin material within which the temperature is uniform at all times:

$$a\dot{Q}_R'' = \tau\rho c\frac{d\theta}{dt} + 2h\theta \tag{6.23}$$

where τ, the thickness of the material, is equivalent to $2l$ (Figure 6.16). As θ is now independent of x, Equation (6.23) may be integrated to give:

$$\theta = T - T_0 = \frac{a\dot{Q}_R''}{2h}(1 - \exp(-2ht/\rho c\tau)) \tag{6.24}$$

If $T = T_{ig}$, this can be rearranged to give the time to ignition as:

$$t_{ig} = \frac{\tau\rho c}{2h}\ln\left(\frac{a\dot{Q}_R''}{a\dot{Q}_R'' - 2h(T_{ig} - T_0)}\right) \tag{6.25}$$

Equations (6.21) and (6.25) are similar in form, showing that regardless of the mode of heat transfer, the time to ignition for thin materials is directly proportional to the mass per unit area $(\tau\rho)$ (or, more properly, the thermal capacity per unit area $(\tau\rho c)$). They also define the limiting conditions for ignition, i.e., $T_\infty > T_{ig}$ for convective heating and $a\dot{Q}_R'' > 2h(T_{ig} - T_0)$ for radiative heating. (Recall that radiative losses are neglected in this derivation.)

Analysis of ignition of sheets or slabs which cannot be regarded as thermally thin (e.g., Bi > 0.1) requires more complex solutions (e.g., Equation (2.18) for convective heat transfer to an infinite slab). Time to ignition increases asymptotically with τ to a limiting value that corresponds to the ignition of a semi-infinite solid, for which the mathematics is simpler.

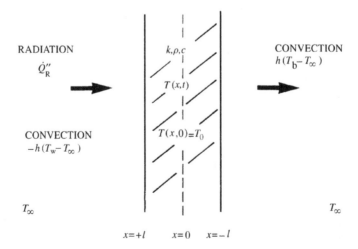

Figure 6.16 Radiative heat transfer to one face of a vertical, infinite slab, with convective heat losses at $x = \pm l$

6.3.1.2 The Semi-infinite Solid

The surface temperature of a semi-infinite solid exposed to convective heating (or cooling) varies with time according to Equation (2.26),[13] i.e.

$$\frac{\theta_s}{\theta_\infty} = \frac{T_s - T_0}{T_\infty - T_0} = 1 - \exp(\beta^2) \cdot \text{erfc}(\beta) \tag{6.26}$$

where T_s is the surface temperature and:

$$\beta = (h(\alpha t)^{1/2}/k) = (ht^{1/2}/(k\rho c)^{1/2}) = \text{Bi} \cdot \text{Fo}^{1/2}$$

The complimentary error function (erfc(β)) does not have an analytical solution, but numerical values are available (see Table 2.2). Thus, θ_s/θ_∞ may be plotted as a function of β according to Equation (6.26) (see Figure 6.17). Given that sustained piloted ignition will be possible if $T_s \geq T_{ig}$, then an estimate of the minimum time to ignition under convective heating may be deduced from the value of β corresponding to $\theta_s/\theta_\infty = \theta_{ig}/\theta_\infty$ (from Figure 6.17), if the thermal inertia ($k\rho c$) and the convective heat transfer coefficient (h) are known. This of course assumes that T_{ig} is also known, which was not the case when this work was carried out by Simms. However, he was able to deduce a value of T_{ig} from experimental data on the ignition of wood exposed to a radiant heat flux.

For radiative heating, the appropriate boundary conditions for Equation (6.19) are:

$$a\dot{Q}_R'' - h\theta = -k\frac{d\theta}{dx} \text{ at } x = 0 \text{ for } t > 0 \tag{6.27a}$$

$$T = T_0 \text{ at } t = 0 \text{ for all } x \tag{6.27b}$$

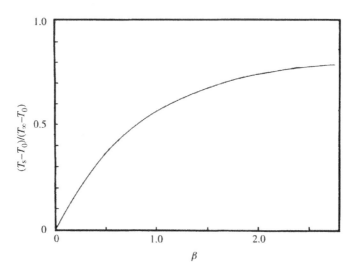

Figure 6.17 Variation of $(T_s - T_0)/(T_\infty - T_0)$ with β (Equation (6.26))

---

[13] As stated in Section 2.2.2, a thick slab will approximate to 'semi-infinite' behaviour provided that $\tau > 2(\alpha t)^{1/2}$ where t is the duration of heating.

An approximate solution to Equation (6.26) may be obtained by substituting $(T_\infty - T_0) = a\dot{Q}''_R/h$, a condition that is valid at the steady state $(t = \infty)$ *if heat loss by radiation is ignored*, i.e.

$$a\dot{Q}''_R = h(T_\infty - T_0) \qquad (6.28)$$

Thus Equation (6.26) becomes

$$\theta_s = \frac{a\dot{Q}''_R}{h}(1 - \exp(\beta^2) \cdot \mathrm{erfc}(\beta)) \qquad (6.29)$$

Simms (1963) referred to β as the 'cooling modulus' and cast Equation (6.29) in a different form by rearranging and multiplying both sides by β, giving:

$$\gamma = \frac{\beta}{1 - \exp(\beta^2) \cdot \mathrm{erfc}(\beta)} \qquad (6.30)$$

where:

$$\gamma = \frac{a\dot{Q}''_R t}{\theta_s \rho c (\alpha t)^{1/2}} \qquad (6.31)$$

and may be called the 'energy modulus'. Simms (1963) used Equation (6.30) to correlate data of Lawson and Simms (1952) on the piloted ignition of wood, in which vertical samples (5 cm square) of several species were exposed to radiant heat fluxes in the range 6.3–63 kW/m². A pilot flame was held in the plume of fuel vapours rising from the surface, as shown in Figure 6.18(a), and the time to ignition recorded. Simms plotted γ vs. β, and selected a single value of $\theta_s = \theta_{ig}$ that gave the most satisfactory correlation for all the data. This is shown in Figure 6.19, and while there is a non-random scatter which reflects density differences, the correlation is reasonable for a value of $\theta_{ig} = 340°C$ (i.e., $T_{ig} \approx 320–325°C$).[14] Since then, there have been a number of direct

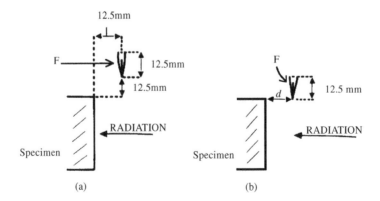

Figure 6.18 Piloted ignition of wood exposed to a radiant heat flux by means of a flame burning downwards (F) (Simms, 1963). Reproduced by permission of The Controller, HMSO. © Crown Copyright

---

[14] The exception is fibre insulating board: this could be due to a number of factors, including the onset of smouldering after prolonged exposure times (see Section 6.4 for a discussion of this phenomenon).

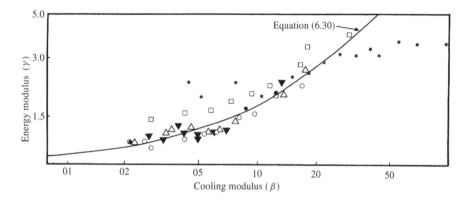

Figure 6.19 Correlation between the energy modulus (γ) and the cooling modulus (β) for woods of different density (Equation (6.30)). Data obtained from configuration shown in Figure 6.18(a). $\theta_{ig} = 340°C$, $h = 33 \, \mathrm{W/m^2 \cdot K}$. $\star$, Fibre insulation board; $\square$, cedar; $\triangle$, freijo; o, mahogany; $\bullet$, oak; $\blacktriangledown$, iroko (Simms, 1963). Reproduced by permission of The Controller, HMSO. © Crown Copyright

experimental measurements of the surface temperature corresponding to the firepoint of different species of wood that tend to support this figure, although there is a spread of data reflecting an apparent species dependency (Table 6.8; Janssens, 1991c). This is discussed below.

Simms (1963) found that if the location of the pilot flame was changed, the value of θ_{ig} had to be altered to maintain a satisfactory correlation. Thus, with the tip of the pilot flame level with the top edge of the sample, as shown in Figure 6.18(b), θ_{ig} was found to be 300°C, 380°C and 410°C for $d = 6.2$, 12.5 and 19 mm, respectively, the time to ignition also increasing in this order (Figure 6.20). This emphasizes that the pilot source must be within the zone of flammability of the volatiles, a point that was made earlier regarding the measurement of the flashpoint and the firepoint of a liquid fuel (see Figure 6.10). When $d > 20$ mm (see Figure 6.18(b)), no ignition was possible, even at the highest heat fluxes, as the ignition source lay outside the stream of volatiles. While care must be taken in interpreting results of this type of experiment, they do not negate the underlying concept of a critical surface temperature acting as a limiting ignition criterion, at least under a constant heat flux (see Section 6.3.2).

A critical radiant heat flux is sometimes quoted as the limiting criterion for piloted ignition. This is illustrated in Figure 6.20, but its value will be sensitive to changes in heat loss from the surface and hence the orientation and geometry of that surface. Lawson and Simms (1952) obtained an estimate of the limiting flux for vertical samples of wood by extrapolating a plot of $\dot{Q}''_R$ versus $\dot{Q}''_R/t_{ig}^{1/2}$ to $t_{ig} = \infty$, where t_{ig} is the time to ignition under a radiant heat flux $\dot{Q}''_R$ (Figure 6.21): however, it should be remembered that in the above model, radiative heat losses are neglected. From these and other data, a minimum flux for piloted ignition of wood was deduced as approximately 12 kW/m^2 (0.3 cal/cm^2·s). This figure was incorporated into the Scottish Building Regulations in 1971 as a basis for determining building separation (Section 2.4.1) (Law, 1963). It is believed that this was the first application of quantitative 'fire safety engineering'.

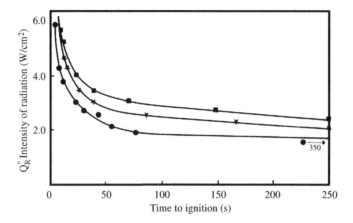

Figure 6.20 Effect of position of the pilot flame on the time to ignition of Columbian pine (Figure 6.18(b)). •, $d = 6.2\,\text{mm}$; ✶, $d = 12.5\,\text{mm}$; ▪, $d = 19.0\,\text{mm}$ (Simms, 1963).[15] Reproduced by permission of The Controller, HMSO. © Crown Copyright

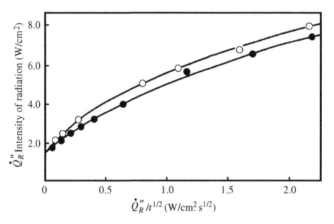

Figure 6.21 Determination of the critical radiant heat flux for piloted ignition of oak (Figure 6.18(b)). •, $d = 6.2\,\text{mm}$: ○, $d = 19.0\,\text{mm}$ (Simms, 1963). Reproduced by permission of The Controller, HMSO. © Crown Copyright

Care must be taken in the interpretation of these minimum heat fluxes. A considerable amount of data has been published on the 'time to ignition' (t_{ig}) as a function of the imposed (incident) heat flux in the cone calorimeter (e.g., Babrauskas and Parker, 1987), in the ISO ignitability test (e.g., Bluhme, 1987) and in other experimental apparatuses (Simms, 1963; Thomson *et al.*, 1988; Dakka *et al.*, 2002), including the FMRC flammability apparatus and its derivatives (e.g., Tewarson and Ogden, 1992; Beaulieu and Dembsey, 2008). Such data may be used to estimate the minimum heat

---

[15] Note that Quintiere (1981) and others have plotted similar data, but with the radiant heat flux as the abscissa (*x*-axis).

flux capable of producing the conditions for piloted ignition defining some arbitrary (but experimentally valid) time interval beyond which ignition would be deemed not to have occurred (e.g., 15 minutes (ISO, 1997a)).[16] The technique involves 'bracketting', as described for the determination of the flammability limits of gaseous fuel/air mixtures (see Section 3.1.1). Janssens (1991) refers to this as 'the minimum heat flux for ignition' to distinguish it from the 'critical heat flux' for ignition which is obtained by extrapolating $\dot{Q}_R''$ to $t_{ig} = \infty$. This may be done by plotting $\dot{Q}_R''$ vs. t_{ig} and extrapolating the asymptotic value of $\dot{Q}_R''$ (Simms, 1963 (Figure 6.20); Quintiere, 1981; Boonmee and Quintiere, 2002; Dakka *et al.*, 2002 (Figure 6.22)).

However, it is more common to use a correlation based on a simplification of the heat transfer models discussed in Chapter 2. Thus, taking Equation (2.30) (in which $\dot{Q}_R''$ is constant and *heat losses are ignored*) and setting $\theta_{ig} = T_{ig} - T_0$, we obtain the time to ignition for a thermally thick solid:

$$t_{ig} = \frac{\pi}{4}k\rho c \frac{(T_{ig} - T_0)^2}{\dot{Q}_R''^2} \tag{6.32}$$

The following expression for thin fuels may be derived from Equation (6.23) if $a = 1.0$ and heat losses are assumed to be negligible (i.e., the heat loss term $(h\theta)$ is set equal to 0):

$$t_{ig} = \rho c \tau \frac{(T_{ig} - T_0)}{\dot{Q}_R''} \tag{6.33}$$

These suggest that data for thick fuels should be examined by plotting $1/\sqrt{t_{ig}}$ vs. $\dot{Q}_R''$, and for thin fuels, $1/t_{ig}$ vs. $\dot{Q}_R''$. These have been widely used, but sometimes with scant regard

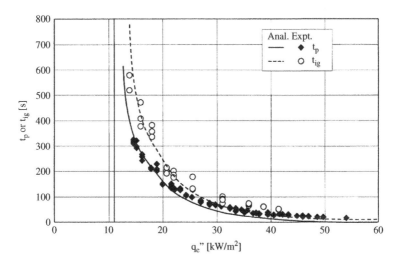

Figure 6.22 Characteristic ignition delay times (t_{ig}) and times to the onset of pyrolysis (t_p) for PMMA for a wide range of heat fluxes (Dakka *et al.*, 2002). Reproduced by permission of the Combustion Institute

---

[16] This is purely arbitrary. Ignition of wood samples in the cone calorimeter has been observed at low heat fluxes ($\sim 10\,kW/m^2$) after extended periods of 1–2 hours. This is discussed in Section 6.4.

to their origins (Mikkola and Wichman, 1989). Deceptively simple correlations can be obtained which lend themselves to linear extrapolation, yielding an apparent value for the critical radiant flux, and a straight line whose slope (for a thermally thick sample) is related to the thermal inertia ($k\rho c$) of the material (Equation (6.32)). However, non-linearities are found if data sets are extended to low heat fluxes, corresponding to long ignition times (>5–10 min) when it is impossible to ignore the effects of radiative and convective heat losses (Mikkola and Wichman, 1989). It was shown in Section 2.2.2 that the characteristic thermal conduction length ($\sqrt{\alpha t}$) could be used as an indicator of the depth of the heated layer of a thick material, and that heat losses from the rear face of a material would be negligible if $L > 4 \times \sqrt{\alpha t}$, indicating 'semi-infinite behaviour' (Figure 2.9). A thermally thin material could be defined as one with $L < \sqrt{\alpha t}$. 'Thermal thickness' increases with $\sqrt{t}$, and for a sufficiently long exposure time a physically thick material will no longer behave as a semi-infinite solid, and will begin to show behaviour which is neither 'thick' nor 'thin'. The consequences of this are normally lost in the scatter unless the data set includes ignition times significantly greater than c. 5 min. This is clearly shown in data presented by Toal *et al.* (1989) and Tewarson and Ogden (1992) (Figure 6.23).[17] A related observation can be made when the insulation of the rear face of a physically thick sample is altered: the material will behave as a semi-infinite solid ('thermally thick') while undergoing piloted

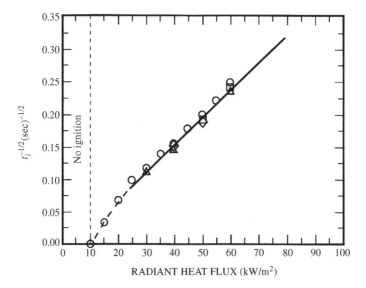

Figure 6.23   Time to ignition data of Tewarson and Ogden (1992) for 25 mm thick black PMMA, obtained in the FMRC flammability apparatus and plotted according to Equation (6.31) ($t_i^{-1/2}$ vs. $\dot{Q}_R''$). The surface was coated with a thin layer of fine graphite powder to ensure a high, reproducible absorptivity. Symbols refer to different flows of air past the sample (0–0.18 m/s). By permission of the Combustion Institute

---

[17] Recently, Beaulieu and Dembsey (2008) reported deviation from linearity at high heat fluxes (>60 kW/m^2) for several materials (PMMA, POM, PVC, wood (pine), plywood and asphalt shingle). The reason for this is not yet understood.

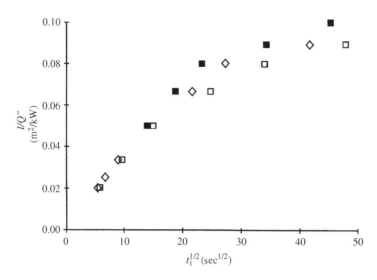

Figure 6.24   Time to ignition data of Thomson *et al.* (1988) for 6 mm thick PMMA measured in the ISO ignitability apparatus (ISO, 1997a) and plotted according to Equation (6.32) (specifically, $1/\dot{Q}_R''$ vs. $t_i^{1/2}$ to reveal the differences in the data at long exposure times): $\diamond$, clear PMMA; $\square$, black PMMA; ■, black PMMA backed with the insulating material Kaowool rather than the standard ISO backing material. Reprinted from Thomson *et al.* (1988), by permission

ignition under a high radiative heat flux and the results are independent of the conditions at the rear face. However, at low heat fluxes, the differences in the heat losses through the rear face show up as differences in the times to ignition (Figure 6.24) (Thomson *et al.*, 1988).

Delichatsios *et al.* (1991) have examined the interpretation of such data and have shown how a linear extrapolation of the plot of $1/\sqrt{t_{ig}}$ vs. $\dot{Q}_R''$ is likely to give a critical heat flux which is only 70%[18] of the 'true value'. Their work emphasizes the importance of matching the correlation to the physical characteristics of the fuel (i.e., in the limits, thick or thin).

Clearly, the critical value is not a material property *per se*. It has been shown that it is strongly influenced by the convective heat transfer boundary conditions at the surface (natural or forced convection) (Cordova and Fernandez-Pello, 2000), which may also be affected by the configuration (e.g., horizontal or vertical) and shape (e.g., flat or curved) of the sample. Moreover, the heat losses to the apparatus in which the measurements of t_{ig} are being made will inevitably affect the value of $\dot{q}''_{crit}$. In the test based on the LIFT apparatus (Quintiere, 1981), the values of $\dot{q}''_{crit}$ deduced from measurements of time to ignition are consistently higher than values obtained by other authors (e.g., Simms and Tewarson (see Table 6.8)). This may reflect higher heat losses in this apparatus compared with (e.g.) the FMRC apparatus. Structural changes at the surface can also have a profound effect: for example, if delamination occurs and the surface layer separates from the bulk of the material, it can become an easily ignitable 'thin fuel' (Rasbash and Drysdale, 1983).

---

[18] This is supported by the data of Tewarson and Ogden, as shown in Figure 6.23, but the figure has been questioned by Babrauskas (2003).

Table 6.8 Criteria for ignition (various sources)

Material	Critical radiant heat flux (kW/m^2)		Critical surface temperature (°C)	
	Pilot	Spontaneous	Pilot	Spontaneous
'Wood'	12[a]	28[a]	350[b]	600[c]
Western red cedar	13.3[d]	–	354[d]	–
Redwood	14.0[d]	–	364[d]	–
Radiata pine	12.9[d]	–	349[d]	–
Douglas fir	13[d]	–	350[d]	–
Victorian ash	10.4[d]	–	311[d]	–
Blackbutt	9.7[d]	–	300[d]	–
Polymethylmethacrylate	21[e]	–	–	–
Polymethylmethacrylate	11[f]	–	310 ± 3[h]	–
Polyoxymethylene	13[g]	–	281 ± 5[h]	–
Polyethylene	15[g]	–	363 ± 3[h]	–
Polypropylene	15[g]	–	334 ± 4[h]	–
Polystyrene	13[g]	–	366 ± 4[h]	–

[a]General value for wood, vertical samples, Lawson and Simms (1952). The piloted value is consistent with the range of values found by Mikkola and Wichman (1989).
[b]Deduced from flame spread under conditions of radiant heating (Atreya et al., 1986). Value for wood compatible with Simms (1963).
[c]Deduced by Simms (1963) for radiative heating. Lower value observed for convective heating (Section 6.4).
[d]Janssens (1991c). Note that the hardwoods (ash and Blackbutt) have significantly lower 'firepoint' temperatures than the softwoods quoted in this table.
[e]Quintiere (1981). Comparatively, these values are very high (see Section 7.2.5(c)).
[f]Thomson et al. (1988). Horizontal samples.
[g]Tewarson (1995). Horizontal samples, Factory Mutual Flammability Apparatus.
[h]Thomson and Drysdale (1987). Horizontal samples.

Multiple layers of gloss (oil-based) paint on a non-combustible surface (e.g., plaster) can behave in this manner. When heated, the upper layers may separate from the lower layers, forming 'blisters' that will ignite easily as they are no longer attached to the heat sink afforded by the plaster. This has serious implications for upward flame spread (see Section 7.2) (Murrell and Rawlins, 1996).

Tewarson and Ogden (1992) (Tewarson, 2008) have introduced the concept of 'thermal response parameter' (TRP), given by an expression which has its origin in Equation (6.32):

$$TRP = (T_{ig} - T_0)\sqrt{k\rho c} \qquad (6.34)$$

It is derived from the linear part of the plot of $1/\sqrt{t_{ig}}$ vs. $\dot{Q}''_R$ (see Figure 6.23), and has been suggested as a means of assessing the ignition resistance of materials. However, it is valid only while the material is behaving as a thermally thick solid. Tewarson (2008) tabulates values obtained in the FMRC flammability apparatus and in the cone calorimeter, but in most cases where direct comparison is possible it appears that they are apparatus-dependent. Its value is primarily for ranking different materials: applying the TRP to 'real' fire scenarios would require a detailed knowledge of the likely heat transfer boundary conditions in the fire.

Despite these caveats, useful information has been gleaned from such data sets. One that is historically interesting, and still perfectly valid, is the original work by Lawson and Simms (1952) on the ignition of a range of samples of wood in which they found a relationship between time to ignition and the 'thermal inertia' of the solid, which is revealed in an excellent correlation between $(\dot{Q}''_R - \dot{Q}''_{R,0})t_{ig}^{2/3}$ and $k\rho c$, where $\dot{Q}''_{R,0}$ is the minimum radiant intensity for piloted ignition of a vertical sample and t_{ig} is the time to ignition under an imposed heat flux $\dot{Q}''_R$. This is shown in Figure 6.25, in which the straight line is given by:

$$(\dot{Q}''_R - \dot{Q}''_{R,0})t_{ig}^{2/3} = 0.6(k\rho c + 11.9 \times 10^4) \tag{6.35}$$

The importance of $k\rho c$ is well illustrated in a more recent study by Janssens (1991). He developed a model from the conservation equation (Equation (6.19), with appropriate boundary conditions) and derived the expression:

$$\frac{\dot{q}''}{\dot{q}''_{cr}} = 1 + 0.73 \left(\frac{k\rho c}{h_{ig}^2 t_{ig}} \right)^{0.547} \tag{6.36}$$

where h_{ig} is the total heat transfer coefficient, as defined in the steady state expression:

$$\varepsilon \dot{q}''_{cr} = h_c(T_{ig} - T_\infty) + \varepsilon\sigma(T_{ig}^4 - T_\infty^4) = h_{ig}(T_{ig} - T_\infty) \tag{6.37}$$

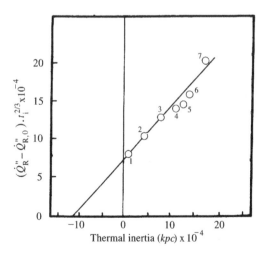

Figure 6.25 Correlation between time to ignition and thermal inertia for a number of woods, plus fibre insulation board (Equation (6.35)) (Lawson and Simms, 1952). 1. Fibre insulation; 2. cedar; 3. whitewood; 4. mahogany; 5. freija; 6. oak; 7. iroko. The samples were 50 mm square and 19 mm thick. Reproduced by permission of The Controller, HMSO. © Crown Copyright

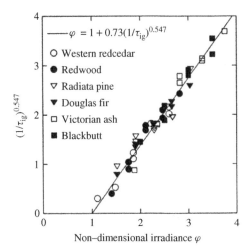

Figure 6.26 Correlation of data on time to ignition of various species of wood using Equation (6.35), where $\tau_{ig} = (h_{ig}^2 t_{ig}/k\rho c)$, $\varphi = (\dot{Q}''_R/\dot{Q}''_{cr})$, h_{ig} is the total heat transfer coefficient at ignition (kW/m^2.K) and $\dot{Q}''_{cr}$ is the critical heat flux (Janssens, 1991c). Values of T_{ig} for these species are shown in Table 6.8. Reproduced by permission of the International Association for Fire Safety Science

Using Equation (6.36), he correlated data on t_{ig} of six species of wood over a range of heat fluxes (15–45 kW/m^2). The correlation, which is shown in Figure 6.26 (Janssens, 1991c), is very good considering that the ambient temperature values of $k\rho c$ (W^2·s/m^4·K^2) ranged from 8.7×10^5 (Western redwood) to 39.3×10^5 (Blackbutt), and the values of T_{ig} fell in the range 364°C (redwood) to 300°C (Blackbutt) (see Table 6.8).

As wood is the most common combustible material in general use, it is not surprising that there have been numerous studies relating to its ignition (see reviews by Janssens (1991a) and Babrauskas (2001)). Invariably, its complex chemical and physical nature (see Section 5.2.2) influences its behaviour. Thus, at low heat fluxes (close to the minimum necessary for piloted ignition) the surface layers are heated slowly, creating a situation in which temperatures remain below 300°C for a prolonged period, promoting the reactions that favour the formation of char rather than 'tar' (see Section 5.2.2). Prolonged exposure at these relatively low heat fluxes[19] can lead to the onset of glowing combustion on the surface, involving an exothermic, heterogeneous reaction between the char and oxygen from the air. This is independent of whether or not a pilot ignition source is present and may properly be considered to be an 'auto-ignition', but no flaming is involved. At considerably higher heat fluxes (say, >25 kW/m^2), ignition of the volatiles occurs at the pilot source when the char surface is of the order of 350°C (see Table 6.8), and glowing combustion of the char surface has not had time to develop. However, at an intermediate heat flux – low, but above the minimum necessary for piloted ignition – glowing combustion may commence before the appearance of the flame. The question of whether or not the

---

[19] Swann *et al*. (2008) report a minimum heat flux of 7.5–8.0 kW/m^2 for the onset of smouldering of maple plywood.

glowing combustion can act as the pilot source for the onset of flaming is a moot one and is raised by Boonmee and Quintiere (2002) (see Sections 6.4, 8.2 and 8.3). It should also be noted that its surface absorptivity (*a* in Equation (6.27a) *et seq.*) is unlikely to remain constant as a layer of char will begin to form when the surface temperatures exceeds 150–200°C. This behaviour is common to all materials which char on heating, but unlike synthetic char-forming materials (such as the polyisocyanurates), wood exhibits unique behaviour due to its anisotropy and its ability to absorb water.

Regarding anisotropy, it takes longer to ignite a piece of wood at a cut end than on its surface, as the thermal conductivity (hence $k\rho c$) is greater along the grain than across it (Vytenis and Welker, 1975; Janssens, 1991a; Spearpoint and Quintiere, 2001; Boonmee and Quintiere, 2002). Exposed knots are difficult to ignite for the same reason, although their greater density is also a factor. There is considerable resistance to the flow of volatiles perpendicular to the grain. Only when the structure of the wood begins to break down at temperatures of the order of 250–300°C will volatiles tend to move directly to the surface, across the grain (cf. Figure 5.13). Before this happens, volatiles can be observed to issue from a cut end or from around a knot where the resistance to flow is much less.

Moisture contained in the wood affects the ignition process in two ways: physically, by increasing the effective thermal capacity of a sample (including the latent heat requirement); and chemically, as the water vapour dilutes the pyrolysis products, effectively reducing the heat of combustion of the evolved vapours (see Section 6.3.2). Thus, the ignition time is found to increase with increasing moisture content (Atreya and Abu-Zaid, 1991 (Table 6.9); Moghtaderi *et al.*, 1997). This is considered in more detail by Mikkola (1992). However, the 'critical heat flux', measured asymptotically as in Figures 6.20 and 6.22, is unaffected by the initial moisture content for the simple reason that for long ignition times, exposed samples will have had time to dry out. This has been observed by Khan *et al.* (2008) in a study of the ignition of corrugated cardboard.

6.3.2 Ignition Involving a 'Discontinuous' Heat Flux

While a critical surface temperature may be a satisfactory method of characterizing the firepoint of a solid exposed to a constant heat flux and may be suitable for engineering calculations (Thomson *et al.*, 1988), it is not always suitable, particularly if the flux is removed after piloted ignition has occurred. Bamford *et al.* (1946) studied the ignition of slabs of wood (deal) by subjecting both sides to the flames from a pair of 'batswing' burners and determining how long it took to reach a stage where flaming would persist when

Table 6.9 Effect of moisture on time to ignition (horizontal samples, Douglas fir, 26.5 kW/m²) (from Atreya and Abu-Zaid, 1991)

Moisture content (%)	Time to ignition (s)
0	55
11	100
17	145
27	215

the burners were removed. By comparing the results of a numerical analysis of transient heat conduction within the slabs, they concluded that a critical flowrate of volatiles from the surface, $\dot{m}_{cr}'' \geq 2.5\,\text{g/m}^2{\cdot}\text{s}$, was necessary for sustained ignition to occur. However, this is not a sufficient condition. It remained for Martin (1965), Weatherford and Sheppard (1965) and others (see Kanury, 1972) to emphasize the significance of the heating history and temperature gradient within the solid (at the moment of 'ignition'). This can be illustrated directly using the firepoint equation (Equation (6.18)) that was introduced in Section 6.2:

$$(\phi \Delta H_c - L_v)\dot{m}_{cr}'' + \dot{Q}_E'' - \dot{Q}_L'' = S \tag{6.18}$$

This describes 'ignition' if $S \geq 0$ but 'extinction' if $S < 0$. It is possible for self-extinction to occur following ignition if the imposed heat flux $\dot{Q}_E''$, which was initially responsible for the surface achieving the firepoint temperature, is reduced or removed completely. Consider Equation (6.18): values of L_v are available from the literature (Table 5.8), and while there is some disagreement over measured values of $\dot{m}_{cr}''$, it is a meaningful concept (see below). Rasbash (1975) has argued that the value of ϕ may be taken as 0.3 for many combustible materials, although it will be less than 0.2 for those that are fire retarded, and may be as high as 0.4 for certain oxygenated polymers (Table 6.10). Only the terms $\dot{Q}_E''$ and $\dot{Q}_L''$ need to be calculated for the relevant fuel configuration before Equation (6.18) can be applied.

As an example, consider a thick slab of PMMA (assumed to act as a semi-infinite solid) exposed to a radiant heat flux of $50\,\text{kW/m}^2$ in the presence of a pilot ignition source. Given that the firepoint temperature of PMMA is $310°\text{C}$ (Thomson and Drysdale, 1987) and the thermal inertia $(k\rho c)$ is $3.2 \times 10^5\,\text{W}^2{\cdot}\text{s/m}^4{\cdot}\text{K}^2$ (Table 2.1), Equation (6.32) can be used to show that it will take approximately 8.5 s to reach the firepoint. After this

Table 6.10 Parameters in the firepoint equation (Equation (6.18)

Sample	Forced convection[a]		Natural convection[b]		Thomson & Drysdale (1988)[c]	
	$\dot{m}_{cr}''$ (g/m²s)	ϕ	$\dot{m}_{cr}''$ (g/m²s)	ϕ	T_{ig} (°C)	$\dot{m}_{cr}''$ (g/m²s)
Polyoxymethylene	4.4	0.43	3.9	0.45	281	1.8
Polymethylmethacrylate	4.4	0.28	3.2	0.27	310	2.0
Polyethylene	2.5	0.27	1.9	0.27	363	1.31
Polypropylene	2.7	0.24	2.2	0.26	334	1.1
Polystyrene	4.0	0.21	3.0	0.21	366	1.0

[a]Tewarson and Pion (1978). $\dot{m}_{cr}''$ was determined experimentally and ϕ calculated from Equation (6.40), using $h/c_p = 13\,\text{g/m}^2\text{s}$ for forced convection.
[b]As[a], but using $h/c_p = 10\,\text{g/m}^2\text{s}$ for natural convection.
[c]These data were obtained in an apparatus of design based on the ISO Ignitability Test (ISO 5657), quite different from the Factory Mutual Flammability Apparatus of Tewarson. The values of $\dot{m}_{cr}''$ are about one-half of those determined by Tewarson and Pion. This has not been resolved, but it seems more likely that it reveals a strong sensitivity to the value of the heat transfer coefficient h rather than a fundamental flaw in the concept of a critical mass flux at the firepoint (see Drysdale and Thomson, 1989). This is discussed further in the text.

period of exposure, the depth of the heated layer will be approximately $\sqrt{\alpha t} = 10^{-3}\,\text{m}$ (see Table 2.1 and Section 2.2.2, after Equation (2.24b)). $\dot{Q}''_L$ can then be estimated as:

$$\dot{Q}''_L = \varepsilon \sigma T_{ig}^4 + h(T_{ig} - T_0) + k \left(\frac{\mathrm{d}T}{\mathrm{d}x}\right)_{\text{surface}} \tag{6.38}$$

where $T_{ig} = 583\,\text{K}$, $T_0 = 293\,\text{K}$ and $k = 0.19\,\text{W/mK}$. For this argument, ε and h are taken as 0.8 (dimensionless) and 15 (W/m$^2\cdot$K), respectively, and it is assumed that the temperature gradient at the surface $(\mathrm{d}T/\mathrm{d}x)_{\text{surface}}$ can be approximated by $(583 - 293)/(\alpha t)^{1/2}$.

Converting from W to kW, the heat loss term in Equation (6.18) is estimated as:

$$\dot{Q}''_L = 5.2 + 4.4 + 55.1 = 64.7\,\text{kW/m}^2$$

Taking Equation (6.18) with $\phi = 0.3$, $\Delta H_c = 26\,\text{kJ/g}$, $L_v = 1.62\,\text{kJ/g}$ (Tewarson and Pion, 1976) and $\dot{m}''_{cr} = 2.5\,\text{g/m}^2.\text{s}$ (Thomson and Drysdale, 1988), it is possible to test whether or not a flame will remain stabilized at the surface if the supporting radiation is removed immediately piloted ignition takes place (i.e., $\dot{Q}''_R \to 0$ at $t = 8.5\,\text{s}$):

$$S = (0.3 \times 26.0 - 1.62) \times 2.5 - 64.7$$
$$= -49.25\,\text{kW/m}^2 < 0$$

i.e., flaming will not be sustained.

If exactly the same calculation is carried out for a slab of low thermal inertia material, such as polyurethane foam, the same conclusion is drawn. However, while theoretically correct, the exercise is unrealistic, as the firepoint temperature (assumed for convenience to be 310°C (Drysdale and Thomson, 1990)) is achieved after only 0.025 s, when the depth of the heated layer will be only 0.17 mm. (The conductive heat loss then will be 57 kW/m^2.) Such a short exposure time is physically impossible for the types of heat source that are relevant to this discussion: 0.5 s might be more realistic, after which the heated layer would be 0.8 mm thick and the surface temperature would be much higher than 310°C. The latter has a dominant effect on subsequent behaviour. The rate of production of fuel vapours will now greatly exceed the critical value ($\dot{m}'' > \dot{m}''_{cr}$), and the associated flame will be capable of providing a much greater heat flux to the surface (by radiation and convection), thus sustaining the burning process.

Similarly, to achieve sustained burning on a thick slab of PMMA, the surface temperature must be increased to well above 310°C before the imposed heat flux is removed. A relevant example of this effect is to be found in the original British Standard Test for the ignitability of materials (BS 476 Part 4, now withdrawn): this involved exposing a vertical sheet of material to a small impinging diffusion flame for 10 s. Most dense materials greater than 6 mm thick will 'pass' the test, as flaming will not be sustained when the igniting flame has been removed. Typically, to achieve sustained flaming following the application of a 'BS476 Part 4' flame, an exposure duration of 30 s or more would be required.

Equation (6.18) can only be used at the firepoint, when $\dot{m}'' = \dot{m}''_{cr}$. When $\dot{m}''$ is greater than $\dot{m}''_{cr}$, the *proportion* of the heat of combustion that is transferred from the flame to the surface (by convection and radiation) will actually reduce, while the heat flux to the surface will increase as the flame will be strengthening. Values of f_c and f_r (see Equations

(6.17) and (6.18)) can no longer be assumed. These factors must be borne in mind when attempting to interpret the ignition process in terms of this equation.[20]

An alternative to this detailed argument provides a more general overview, which can also be used to explain directly why low density materials, once ignited, progress to give an intense fire so quickly. As burning continues, the depth of the heated layer ($\sim\sqrt{\alpha t}$) increases as heat is conducted into the body of the solid. It is possible to calculate the effective 'thermal capacity' of this layer as a function of time for different materials. This is shown in Table 6.11 as the product $\rho c\sqrt{(\alpha t)}(=\sqrt{(k\rho ct)})$, which has the units J/m^2·K, i.e., the amount of energy required to raise the *average* temperature of a unit area of the heated layer of material by one degree. The amount required for the polyurethane foam is only 6% of that for polymethylmethacrylate. Thus, even if the flames above these two materials had the same heat transfer properties with respect to the surface, the polyurethane foam would achieve fully developed burning in only a fraction of the time taken by the PMMA. These results show that thermal inertia ($k\rho c$) is an important factor in determining rate of fire development as well as ease of ignition.

Of the other factors in Equation (6.18) that influence the ignition process (specifically T_{ig}, ΔH_c, $\dot{m}''_{cr}$ and ϕ), there are few values available in the literature. Although simple in concept, the firepoint temperature (T_{ig}) is very difficult to measure experimentally and tabulated values may not be reliable (see Section 6.3.1). Values of ΔH_c can be obtained by combustion bomb calorimetry (Section 1.2.3), but the results refer to complete combustion (to CO_2 and H_2O) of a small sample which may not accurately represent the material of interest. Moreover, for char-forming materials, it is the heat of combustion of the volatiles that should be used.

The critical mass flux ($\dot{m}''_{cr}$) can be determined experimentally by monitoring the mass of a sample of material continuously while it is exposed to a constant radiant heat flux and subjected at regular intervals to a small pilot flame (cf. the open cup flashpoint test (ASTM, 2005)), a spark (as used in the cone calorimeter) or other ignition device (e.g., an electrically heated coil (Cordova *et al.*, 2001)). The rate of mass loss at the firepoint is very small and a load cell of very high resolution is required.

Figure 6.27 shows a typical record of sample weight as a function of time, the gradient changing when the surface ignites and sustains flame (Deepak and Drysdale, 1983).

Table 6.11 Effective 'thermal capacities' of surfaces

	Time of heating (s)	Thermal diffusivity[a] (m^2/s)	Depth of heated layer[b] (m)	'Effective thermal capacity' (J/m^2.K)
PMMA	10	1.1×10^{-7}	1×10^{-3}	1690
Polypropylene	10	1.3×10^{-7}	1.1×10^{-3}	1965
Polystyrene	10	8.3×10^{-8}	0.9×10^{-3}	1188
Polyurethane foam	10	1.2×10^{-6}	3.5×10^{-3}	98

[a]Data taken from Tables 1.2 and 2.1.
[b]Assumed to be equal to $\sqrt{\alpha t}$.

[20] It is important to emphasize that at the present time it is not possible to determine some of the terms in the firepoint equation and more research is required to enable us to resolve the associated problems. Nevertheless, the underlying concept appears to be valid.

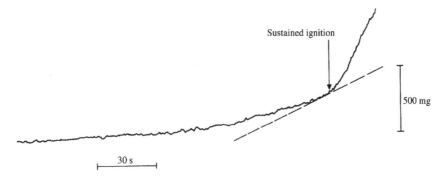

Figure 6.27 Determination of the critical mass flowrate ($\dot{m}_{cr}$) at the firepoint. Deepak and Drysdale (1983), by permission

'Flashing' is normally observed before this point is reached: although this is detectable in experiments in which the surface temperature is measured (Atreya et al., 1986; Thomson and Drysdale, 1988), it does not show on the mass vs. time curve. $\dot{m}''_{cr}$ is taken from the gradient of the first section of the curve at the discontinuity, as shown.

The quantity ϕ in Equation (6.18) is more elusive, but may be derived from $\dot{m}''_{cr}$ via Spalding's mass transfer number, B (see Section 5.1.2). Rasbash (1976) suggested that $\dot{m}''_{cr}$ can be related to a critical value of Spalding's mass transfer number (Spalding, 1955), thus:

$$\dot{m}''_{cr} = \frac{h}{c_p}\ln(1 + B_{cr}) \qquad (6.39)$$

where h is the coefficient that applies to convective heat transfer between the flame and surface (kW/m^2·K), c_p is the thermal capacity of air (kJ/K·g), and $B_{cr} = A/\phi\Delta H_c$, where $A \approx 3000$ kJ/g (see Equation (5.23)). Equation (6.39) applies to the 'steady state' firepoint condition (Equation (6.18), with $S = 0$) and assumes that $L_v = \phi\Delta H_c$: it offers the only means by which ϕ may be estimated (Rasbash, 1975; Tewarson 1980).

A number of measurements of critical mass flux ($\dot{m}''_{cr}$) have been made. Individual papers report values for a number of materials that are internally consistent (e.g., Thomson and Drysdale, 1988; Tewarson, 2008), but comparisons of results obtained from different apparatuses reveal differences that are too great to be explained by random scatter. For example, Thomson's values of firepoint ($\dot{m}''_{cr}$) are approximately half of those measured by Tewarson (Table 6.10). This may be due to the different flow conditions that exist at the surface of the samples in the two apparatuses. This has still to be resolved, but what is encouraging is that within each data set, there are common trends: thus the critical mass fluxes for the oxygenated polymers (POM and PMMA) are roughly twice those for the hydrocarbon polymers (PE, PP and PS), qualitatively consistent with the lower heats of combustion of the oxygenated polymers (cf. Equation (6.18)). Unfortunately, most other measurements that have been reported for the critical mass flux at the firepoint are limited to PMMA. Thus, a value of ~ 1.8 g/m^2·s has been estimated by Cordova and Fernandez-Pello (2000) with an ignition model in which time-to-ignition data for PMMA were used (Cordova et al., 2001). Values of $\dot{m}''_{cr}$ for PMMA have been obtained experimentally over a range of heat fluxes by Panagiotou and Quintiere 2004 and Rich et al. (2007) (~ 2 g/m^2·s and 1.3–2.3 g/m^2·s, respectively). These are comparable to Thomson's value of 2.0 g/m^2

for PMMA, but no general conclusion can be taken from this observation. Few values have been obtained for wood, but Moghtaderi *et al*. (1997) obtained a value of $\dot{m}''_{cr}$ c. 1.8 g/m^2·s for dry radiata pine, rising to about 4 g/m^2·s when the moisture content was 30%.

It should be noted that Rich *et al*. (2007) appear to have carried out the most extensive examination of the critical mass flux concept to date. Not only did they vary the imposed heat flux, but they also varied the oxygen concentration and the air flow velocity over the sample surface. In addition, they developed a theoretical model against which they compared their results.

If burning in the nascent flame immediately after ignition was stoichiometric, then a value of $\phi = 0.45$ would be anticipated on the basis that the flame must be quenched to a temperature of $\sim$1600 K (Section 3.1.2). The fact that it is found to be $\sim$0.3 for many materials (Tewarson, 1980) could be due to a number of reasons, including non-stoichiometric burning in the limiting flame and radiative heat loss from the flame (although this will be minor). The reactivity of the volatiles will certainly influence the magnitude of this factor. Fire retardants that reduce reactivity by inhibiting the flame reactions increase the limiting flame temperature and consequently reduce ϕ. The effect on the ignition properties of a material can be seen by examining Equation (6.18), although the value of h/c must be known accurately. Tewarson selected $h = 13$ W/m^2·K and 10 W/m^2·K for 'forced convection' and 'natural convection', respectively, and obtained a reasonable range of values of ϕ (Table 6.10). However, the lower values of $\dot{m}''_{cr}$ obtained by Thomson give impossibly high values (i.e., $\phi > 0.45$) with the same heat transfer coefficients (Thomson and Drysdale, 1988). Yet, as before, each data set gives an internally consistent group of values of $\dot{m}''_{cr}$, suggesting that further research would prove fruitful to our understanding of this aspect of the ignition process.

Indeed, from the above discussion, it is possible to identify several material properties which influence ease of ignition. A material will be difficult to ignite if L_v is large and ϕ and/or ΔH_c are small – or if $\dot{Q}''_L$ is large. Materials may be selected on the basis of these properties, or treated with fire retardants to alter these properties in an appropriate way. For example, retardants containing bromine and chlorine release the halogen into the gas phase along with the volatiles, rendering the latter less reactive, thus decreasing ϕ (Section 3.5.4). (However, they can be driven off under a sustained low-level heat flux, thus causing the polymer to lose its fire retardant properties (e.g., Drysdale and Thomson, 1989).) The use of alumina trihydrate as a filler for polyesters increases the thermal inertia ($k\rho c$) of the solid and effectively lowers ΔH_c as water vapour is released with the fuel volatiles. (The latter effect may be seen in the results of Moghtaderi *et al*. (1997): the value of $\dot{m}''_{cr}$ was found to increase as the moisture content of samples of radiata pine was increased.) Phosphates and borates, when added to cellulosic materials, promote a degradation reaction which leads to a greater yield of char, and an increase in the proportion of CO_2 and H_2O in the volatiles, which reduces ΔH_c (Table 5.10). Thermally stable materials which have high degradation temperatures will exhibit greater radiative heat losses at the firepoint (increased $\dot{Q}''_L$). Similarly, the formation of a layer of char insulates the fuel beneath, and higher temperatures will be required at the surface of the char to maintain the flow of volatiles. However, it is often overlooked that the thermal response of a thick combustible solid can dominate ignition behaviour through the effect of thermal inertia ($k\rho c$), as has been demonstrated above.

6.4 Spontaneous Ignition of Solids

If the surface of a combustible solid is exposed to a sufficiently high heat flux in the absence of a pilot source (Figure 6.15), the fuel vapours may ignite spontaneously if, somewhere within the plume, the volatile/air mixture is within the flammability limits and at a sufficiently high temperature (Section 6.1).[21] This is summarized in Figure 6.28 and the process described in detail by Torero (2008) in terms of a critical Damköhler number (see Section 6.6.2, Equation (6.41)). Spontaneous ignition requires a higher heat flux than piloted ignition because a higher surface temperature is required to produce a flow of volatiles that is hot enough to undergo this process. A value is quoted in Table 6.8 for 'wood', but it must be recognized that this is highly apparatus-dependent, and should not be accepted as a general value (see below).

The mechanism for spontaneous ignition ('auto-ignition') of a uniformly heated, homogeneous flammable vapour/air mixture has already been discussed in Section 6.1, but in the present case the mixture is neither homogeneous nor uniformly heated. Nevertheless, the same thermal mechanism is responsible for the instability that leads to the appearance of flame. Under an imposed radiative heat flux it is possible that absorption of radiation by the volatiles may contribute towards the onset of reaction. For example, Kashiwagi (1979) showed that the volatiles can attenuate the radiation reaching the surface quite strongly and it is known that the volatiles can be ignited by subjecting them to intense radiation from a laser. Others have found that attenuation is important and must be included in models of the ignition process (Zhou *et al.*, 2010), although Beaulieu and Dembsey (2009) could find no evidence for attenuation of a black-body radiant flux of $120 \, kW/m^2$ falling on the surface of PMMA after pyrolysis had started.

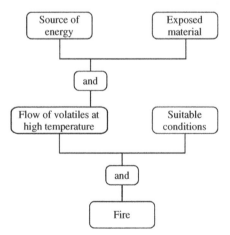

Figure 6.28   Scheme for spontaneous ignition of a combustible solid

---

[21] The term 'auto-ignition' is commonly used to describe the spontaneous appearance of flame in a homogeneous gaseous fuel/air mixture (see Table 6.2). However, some authors (e.g., Boonmee and Quintiere, 2002) have used it to describe what is referred to in this chapter as 'spontaneous ignition of solids'. The distinction is moot, but this author prefers to reserve 'auto-ignition' for premixtures of gaseous fuel and air.

Kanury (1972) has made an interesting observation on the surface temperatures required for piloted (PI) and spontaneous ignition (SI) of wood under radiative and convective heating, thus:

Mode of heat transfer	*Surface temperature of wood for:*	
	PI	SI
Radiation	300–410°C	600°C
Convection	450°C	490°C

These results were obtained using an experimental arrangement similar to that shown in Figure 6.18. They may be explained by the fact that the volatiles will be diluted signifi-cantly more by a forced convective flow and consequently a higher surface temperature will be required to produce a mixture that is above the lower flammability limit at the pilot source (PI). On the other hand, for spontaneous ignition to occur as a result of radiative heat transfer, the volatiles released from the surface must be hot enough to produce a flammable mixture above its auto-ignition temperature when it mixes with unheated air. With convective heating, as the volatiles are entering a stream of air that is already at a high temperature, they need not be so hot.

As the volatiles released from the surface under radiative heating must mix with the surrounding air that is entrained into the buoyant plume, spontaneous ignition is very sensitive to air movement, and indeed air flows associated with the orientation of the surface itself. This is well illustrated by results obtained by Shields *et al.* (1993), which clearly indicate that spontaneous ignition occurs more readily at a horizontal surface than a vertical one (Table 6.12). This may be explained qualitatively in terms of the differences in boundary layer flows in the two situations. Not only will the vertical surface be exposed to more effective convective cooling than the horizontal one, but the dilution of the volatiles will be more efficient in the vertical boundary layer. However, it should be noted that these experiments were terminated at 600 s. Boonmee and Quintiere (2002) found spontaneous flaming ignition to occur after 1000–1100 s at c. 27 kW/m^2, albeit with a different species of wood (redwood). Below 40 kW/m^2, glowing ignition could be observed at the surface before the onset of flaming. They also noted that the surface temperature corresponding to the occurrence of spontaneous flaming *decreased* from c. 700°C to c. 300°C as the radiant heat flux was *increased* from c. 40 kW/m^2 to c. 75 kW/m^2. They account for this by the fact that when a layer of char has built up on the surface (as will happen for the longer exposure times), a higher surface temperature is needed to deliver the required heat flux through the char to the as yet unaffected wood underneath (cf. Figure 5.14). It is perhaps surprising that a surface temperature as low as 300°C was observed for spontaneous flaming at the highest radiant fluxes.

In addition to the onset of spontaneous flaming, Boonmee and Quintiere (2002) also studied the onset of glowing combustion[22] on the surface of the layer of char that forms during extended periods of heating. With the exposed surface of samples of redwood (as 40 mm cubes) in the vertical orientation, glowing was recorded at heat fluxes above

---

[22] The paper by Boonmee and Quintiere is entitled 'Glowing and flaming autoignition of wood', but the onset of 'glowing' is referred to as 'radiant smouldering ignition' by Swann *et al.* (2008). This is discussed further in Chapter 8.

Table 6.12 Times to spontaneous ignition (seconds) at horizontal and vertical surfaces in a Cone Calorimeter (Shields *et al.* 1993)

Heat flux	15 mm Chipboard[a]		20 mm Sitka Spruce	
(kW/m^2)	Horizontal	Vertical	Horizontal	Vertical
20	$-$ [b]	$-$	$-$	$-$
30	123 ± 22	$-$	$-$	$-$
40	61 ± 15	$-$	74 ± 17	$-$
50	27 ± 5	$-$	25 ± 6	47 ± 5
60	19 ± 2	37 ± 5	17 ± 3	28 ± 6
70	14 ± 2	22 ± 6	9 ± 2	15 ± 2

[a] *aka* particleboard.
[b] No ignition within 600 seconds.

c. 18 kW/m^2, with a delay ('time to ignition') on the order of 200 s. In this particular set of experiments, when the heat flux was increased to c. 27 kW/m^2, the delay fell to approximately 100 s, followed $\sim$1000 s later by the spontaneous appearance of flame. Between 27 and 40 kW/m^2, glowing preceded flaming by a significant amount, but at 45 kW/m^2 (and above), they were almost coincident.

6.5 Surface Ignition by Flame Impingement

This mode of ignition has been referred to in Section 6.3.2 and describes the situation in which the pilot flame impinges on the surface of the material, with or without an imposed radiant flux. Ignition will occur after a delay, the duration of which will depend on $k\rho c$ (see Table 6.11) and the total heat flux to the surface, comprising heat transfer from the flame and any external source: thereafter, flame may spread across the surface. This can occur at heat fluxes much lower than those required for piloted ignition, in which the only source of heat is from an external source (Figure 6.14). With flame impingement, Simms and Hird 1958 found that the minimum external heat flux for igniting pinewood was only 4 kW/m^2, compared with 12 kW/m^2 for the mode of ignition illustrated in Figure 6.18. It is essentially a flame spread property, which has been investigated in detail by Quintiere (1981) (Section 7.2.5(c)).

It should be noted that the size and characteristics of the impinging flame are extremely important. The pilot flame used in the (now withdrawn) British Standard Ignitability Test (BS 476 Part 4) was much too weak to overcome the intrinsic ignition resistance of dense combustible solids 6 mm thick or more in the 10 seconds specified in the test (see Section 6.3.2). A laminar flame of this size will have low emissivity and will transfer heat mainly by convection. However, with increasing size and (in particular) thickness of the flame, radiation from the flame will come to dominate and the flow may become turbulent[23] (Section 4.3.3.1). Under these circumstances, the effective heat transfer to the combustible surface is dramatically increased, causing the surface in contact with the

---

[23] A set of seven standard ignition sources, of increasing intensity, was developed for testing upholstered furniture in the UK (BSI, 1990), although their heat transfer characteristics have not been quantified.

flame to burn vigorously, contributing vapours which will in turn enhance the size of the flame. Taking as an example a vertical surface, such as a combustible wall lining, exposed to a large flame (such as from a burning item of furniture (Williamson *et al.*, 1991)), the heat transfer may be sufficiently great to overcome any ignition resistance that the material may be judged to have on the basis of results from small-scale tests, such as the ISO ignitability test, or the cone calorimeter. Clearly, such data must be interpreted very carefully, bearing in mind the end-use scenario in which the material is to be used. For this reason, several large-scale tests have now been developed to assess the fire performance of wall lining materials in their most vulnerable configuration, viz. forming the corner of a compartment (e.g., ISO, 1981, 1999b, 2002b; British Standards Institution, 2010).

6.6 Extinction of Flame

Conceptually, extinction can be regarded as the obverse of piloted ignition and may be treated in a similar fashion, as a limiting condition or criticality. As with piloted ignition, there are two principal aspects to the phenomenon, namely: (i) extinction of the flame, (ii) reducing the supply of flammable vapours to below a critical value ($\dot{m}'' < \dot{m}''_{cr}$).

While the latter will cause the flame to go out, it is possible to extinguish the flame while ($\dot{m}'' > \dot{m}''_{cr}$), leaving the risk of re-ignition, which may occur spontaneously with combustible solids and liquids of high flashpoint (e.g., cooking oil) that have been burning for some time. The risk of re-ignition from a pilot will remain until the fuel cools to below its firepoint. However, for gas leaks and low flashpoint liquids, simply suppressing the flame will leave a continuing release of gaseous fuel, which in an enclosed space could lead to the formation of a flammable atmosphere. Under these circumstances, the rate of supply of fuel vapour must be stopped, or at the very least reduced to a non-hazardous level.

6.6.1 Extinction of Premixed Flames

The stability of premixed flames was discussed in Section 3.3 in relation to the existence of flammability limits. In a confined space, an explosion following the release of a flammable gas can be prevented by creating and maintaining an atmosphere that will not support flame propagation even under the most severe conditions (Section 3.1 and Figure 3.12). This is 'inerting' rather than extinction, and must exist before the ignition event occurs in order that flame does not become established. A premixed flame can be extinguished if a suitable chemical suppressant is released very rapidly ahead of the flame front. This is achieved in explosion suppression systems by early detection of the existence of flame, usually by monitoring a small pressure rise within the compartment, and rapid activation of the discharge of the chemical (Bartknecht, 1981; Ural and Garzia, 2008). Typical agents include the halons CF_2Br_2 and CF_2BrCl, as well as certain dry powders (see Section 1.2.4). The halons (Section 3.5.4) have been phased out since the early 1990s (Montreal Protocol, 1987), and have been replaced by 'ozone-friendly' alternatives, such as Inergen®, a proprietary mixture of nitrogen, argon and carbon dioxide, and FM 200 (HFC-227, or heptafluoropropane). However, none of the gaseous substitutes are as effective as the halons and there has been much interest in the development of water mist systems (see, e.g., Brenton *et al*. (1994) and Di Nenno and Taylor (2008)). Nitrogen and carbon dioxide are suitable only for pre-emptive inerting as the amounts required are too large to be released at the rates necessary for rapid suppression.

Premixed flames may also be extinguished by direct physical quenching. This involves (*inter alia*) cooling the reaction zone and is believed to be the principal mechanism by which flame arresters operate. This device consists of a multitude of narrow channels, each with an effective internal diameter less than the quenching distance, through which flame cannot propagate. The mechanism is described at some length in Section 3.3(a). Flame arresters are normally installed to prevent flame propagation into vent pipes and ducts in which flammable vapour/air mixtures may form (Health and Safety Executive, 1996; National Fire Protection Association, 2008).

6.6.2 Extinction of Diffusion Flames

In addition to cutting off the supply of fuel vapours (e.g., closing a valve to stop a gas leak or blanketing the surface of a flammable liquid with a suitable firefighting foam), diffusion flames may be extinguished by the same agents that are used for premixed flames. However, as there are already considerable heat losses from a diffusion flame, theoretically, less agent is required than for premixed flames: in practice, this distinction becomes academic as the fire size is increased. It is understood that the mechanism by which extinction occurs is essentially the same as in premixed flames. Thus, 'inert' diluents (e.g., N_2 and CO_2) cool the reaction zone by increasing the effective thermal capacity of the atmosphere (per mole of oxygen) (Sections 1.2.5 and 3.5.4) and chemical suppressants such as the halons inhibit the flame reactions (Sections 1.2.4 and 3.5.4).

These agents may be applied locally from hand-held appliances directed at the flame. Small, developing fires are easily extinguished in this way as the local concentration of agent can greatly exceed the minimum requirement. Greater skill is required as the fire size increases, particularly if the supply of agent is limited. All flame must be extinguished before the supply runs out otherwise the fire will simply re-establish itself. This problem can be overcome by 'total flooding', provided that the compartment in which the fire has occurred can be effectively sealed to maintain the necessary concentration of the agent. This is economic only in special circumstances – e.g., when the possibility of water damage by sprinklers is unacceptable, such as in the protection of works of art and valuable documents, and of marine engine rooms and ships' holds. The advantage of chemical suppressants in this role is that the protection system can be activated while personnel are still within the compartment, while prior evacuation is necessary in the case of carbon dioxide (and nitrogen) as the resultant atmosphere is non-habitable. In principle, a halon system can be activated sooner than a CO_2 system, but in addition to the environmental problem, the agent is much more expensive and can generate harmful and corrosive degradation products at unacceptable levels if the fire is already too large when the agent is released. It should be noted that while total flooding may be used to hold in check a deep-seated smouldering fire, it is unlikely to extinguish it completely as such fires can continue at very low oxygen concentrations, particularly if they are well established (see Chapter 8). Cooling of the smouldering mass by water (and physically removing the fuel) is the ultimate means of control.

The role of cooling in fire control must not be overlooked as this is the predominant method by which fires are extinguished. Water is particularly effective as it has a high latent heat of evaporation (2.4 kJ/g at 25°C). Indeed, it can extinguish a diffusion flame *per se* if it can be introduced into the flame in the form of a fine mist[24] or as steam. The suppression of fires by water, including the use of sprays, has been reviewed in detail by Grant *et al*. (1999). The most commonly used mode of suppression by water is in cooling the fuel surface: recalling Equation (6.17), an additional heat loss term, $\dot{Q}''_w$, may be introduced:

$$((f_r + f_c) \cdot \Delta H_c - L_v)\dot{m}''_{burn} + \dot{Q}''_E - \dot{Q}''_L - \dot{Q}''_w = S \tag{6.40}$$

When S becomes negative, the surface of the fuel will cool until ultimately $\dot{m}''_{burn} < \dot{m}''_{cr}$ and flame can no longer exist at the surface. These concepts are considered in some detail by Beyler (1992).

Water is ideal for fires involving solids and can be effective with high flashpoint hydrocarbon liquid fires provided it is introduced at the surface as a high velocity spray, which causes penetration of the droplets and cooling of the surface layers. If this is not effective, the water will sink to the bottom and may eventually displace burning liquid from its containment.

Diffusion flames may also be extinguished by the mechanism of 'blowout', familiar with the small flames of matches and candles. It is also the main method by which oilwell fires are tackled. The mechanism involves distortion of the reaction zone within the flame in such a way as to reduce its thickness, so that the fuel vapours have a much shorter period of time in which to react. If the reaction zone is too thin, then combustion will be incomplete and the flame is effectively cooled, ultimately to a level at which it can no longer be sustained ($T_f < 1600$ K). This can be interpreted in terms of a dimensionless group known as the Damkohler number (D):

$$D = \frac{\tau_r}{\tau_{ch}} \tag{6.41}$$

where τ_r is the 'residence time' (which refers to the length of time the fuel vapours remain in the reaction zone) and τ_{ch} is the chemical reaction time (i.e., the effective duration of the reaction at the temperature of the flame). A critical value of D may be identified, below which the flame will be extinguished. The residence time will depend on the fluid dynamics of the flame, but as the reaction time, τ_{ch}, is inversely proportional to the rate of the flame reaction, we can write:

$$D \propto \tau_r \exp(-E_A/RT_f) \tag{6.42}$$

Blowout will occur if sufficient airflow can be achieved to reduce τ_r and T_f, thus reducing D ultimately to below the critical value. This approach is quite compatible with the concept of a limiting flame temperature and has been explored by Williams (1974, 1982) as a means of interpreting many fire extinction problems. It can account for chemical suppression, which acts by increasing the effective chemical time by reducing the reaction rate (Section 1.2.4). As with local application of a limited quantity of extinguishant, blowout must be totally successful. This is particularly true with an oilwell fire, where the flow of fuel will continue unabated after extinction has been attempted.

---

[24] It has been shown that firefighters can influence the development of flashover in a compartment (Section 9.2) by application of water spray into the hot ceiling layer (e.g., Schnell, 1996).)

Problems

6.1 Using data given in Tables 1.12 and 3.1, calculate the closed cup flashpoint of n-octane. Compare your results with the value given in Table 6.4.

6.2 Would a mixture of 15% iso-octane + 85% n-dodecane by volume be classified as a 'highly flammable liquid' under the 1972 UK Regulations? Assume that the mixture behaves ideally, and that the densities of iso-octane and n-dodecane are 692 and 749 kg/m^3, respectively. Take the lower flammability limit of n-dodecane to be 0.6%.

6.3 Calculate the temperature at which the vapour pressure of n-decane corresponds to a stoichiometric vapour/air mixture. Compare your result with the value quoted for the firepoint of n-decane in Table 6.4.

6.4 n-Dodecane has a closed cup flashpoint of 74°C. What percentage by volume of n-hexane would be sufficient to give a mixture with a flashpoint of 32°C?

6.5 A vertical strip of cotton fabric, 1 m long, 0.2 m wide and 0.6 mm thick, is suspended by one short edge and exposed uniformly on one side to a radiant heat flux of 20 kW/m^2. Is it possible for the fabric to achieve its piloted ignition temperature of 300°C and if so, approximately how long will this take? Assume that the convective heat transfer coefficient $h = 12$ W/m^2·K, $\rho = 300$ kg/m^3 and $c = 1400$ J/kg K, and that the fabric surface has an emissivity of 0.9. The initial temperature is 20°C.

6.6 What difference would there be to the result of Problem 5 if the unexposed face of the fabric was insulated (cf. fabric over cushioning material)? (Assume the insulation to be perfect.)

6.7 A vertical slab of a 50 mm thick combustible solid is exposed to convective heating which takes the form of a plume of hot air flowing over the surface. If the firepoint of the solid is 320°C, how long will the surface take to reach this temperature if the air is at (a) 600°C; and (b) 800°C? (See Equation (2.26).) Assume that the convective heat transfer coefficient is 25 W/m^2·K, and that the material has the same properties as yellow pine (Table 2.1). Ignore radiative heat losses and assume that the properties of the wood remain unchanged during the heating process.

6.8 With reference to Problem 6.7, estimate the conductive heat losses from the surface into the body of the slab of wood at the moment the firepoint is reached. How does this compare with the radiative losses from the exposed face (at 320°C, assuming $\varepsilon = 1$)?

7

Spread of Flame

The rate at which a fire will develop depends on how rapidly flame can spread from the point of ignition to involve an increasingly large area of combustible material. In an enclosure, the attainment of fully developed burning requires growth of the fire beyond a certain critical size (Section 9.1) capable of producing high temperatures (typically >600°C) at ceiling level. Although enhanced radiation levels will increase the local rate of burning (Section 5.2), it is the increasing area of the fire that has the greater effect on flame size and rate of burning (Thomas, 1981). Thus the characteristics of flame spread over combustible materials must be examined as a basic component of fire growth.

Flame spread can be considered as an advancing ignition front in which the leading edge of the flame acts both as the source of heat (to raise the fuel ahead of the flame front to the firepoint) and as the source of pilot ignition. The flame front represents a formal boundary, referred to by Williams (1977) as the 'surface of fire inception', which lies between the two extreme states of unburnt and burning fuel. Movement of this boundary over the fuel can be regarded as the propagation of an ignition front and involves non-steady state heat transfer processes similar if not identical to those discussed in the context of pilot ignition of solids (Sections 6.3 and 6.5). Consequently, the rate of spread can depend as much on the physical properties of a material as on its chemical composition. The various factors which are known to be significant in determining the rate of spread over combustible solids are listed in Table 7.1 (Friedman, 1977). Several reviews of flame spread have been published in the intervening years (Fernandez-Pello and Hirano, 1983; Fernandez-Pello, 1984, 1995; Wichman, 1992; Ross, 1994).

Following the precedent set in earlier chapters, the behaviour of liquids will be reviewed before that of solids. Spread of flame through flammable vapour/air mixtures has already been discussed (Section 3.2).

7.1 Flame Spread Over Liquids

The rate at which flame will spread over a pool of liquid fuel depends strongly on its temperature and in particular whether or not this lies above or below its flashpoint or firepoint. The concentration of vapour above the surface of a highly flammable liquid will

An Introduction to Fire Dynamics, Third Edition. Dougal Drysdale.
© 2011 John Wiley & Sons, Ltd. Published 2011 by John Wiley & Sons, Ltd.

Table 7.1 Factors affecting rate of flame spread over combustible solids (after Friedman, 1977)

Material factors		Environmental factors
Chemical	Physical	
Composition of fuel	Initial temperature	Composition of atmosphere
Presence of retardants	Surface orientation	Pressure of atmosphere
	Direction of propagation	Temperature
	Thickness	Imposed heat flux
	Thermal capacity	Air velocity
	Thermal conductivity	
	Density	
	Geometry	
	Continuity	

be above the lower flammability limit at ambient temperature:[1] ignition will be followed by the propagation of a premixed flame through that part of the vapour/air mixture that is within the flammability limits. If the liquid is above the firepoint, this will develop into a diffusion flame and steady burning of the liquid will follow. On the other hand, if the temperature of a liquid is below its flashpoint, quite different behaviour is observed. The surface ahead of the flame front must be heated to allow the flame to advance and relatively low rates of spread are found. This is shown clearly by the results obtained by Akita (1972) on the rate of flame spread over methanol in the temperature range $-16°C$ to $+28°C$ (Figure 7.1). Several regimes of behaviour are noted, but above the flashpoint (noted as $11°C$ by Akita) the rate increases rapidly to a plateau beyond $20.5°C$. This is the temperature at which the vapour pressure of methanol corresponds to a stoichiometric vapour/air mixture. This will be discussed below, but the following section begins with flame spread at temperatures below the flashpoint – referred to by Ross and Miller (1999) as 'subflash pseudo-uniform spread'.

Glassman and Hansel (1968) were the first to propose that surface tension-driven flows were involved in the spread of flame over the surface of a pool of combustible liquid. This was demonstrated conclusively by Sirignano and Glassman (Section 6.2.2). Other physical properties of the liquid are important (e.g., viscosity (Glassman and Hansel, 1968)), but while they may be of more significance under microgravity, they will not be discussed further here (see Ross, 1994). The mechanism depends on the fact that surface tension decreases as the temperature is raised. Consequently, at the surface of the liquid, the decrease in temperature ahead of the flame front is directly responsible for a net force which causes hot fuel to be expelled from beneath the flame, thereby displacing the cooler surface layer (Mackinven *et al.*, 1970; Akita, 1972: Ross, 1994) (see Figure 6.11). This movement of hot liquid is accompanied by advancement of the flame. Some of the observations made by Mackinven *et al.* (1970) on hydrocarbon fuels contained in trays or channels (1.2–3.0 m in length) merit discussion. In these experiments, preheating of the fuel – which inevitably occurs with wick ignition as illustrated in Figures 6.11 and 6.12 (Burgoyne *et al.*, 1968) – was avoided by partitioning a short section at one end of

---

[1] Unless otherwise stated, it is assumed that the liquid will also be at ambient temperature

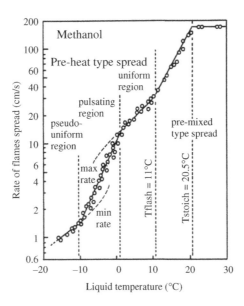

Figure 7.1 Relationship between liquid temperature and the rate of flame spread over the surface of methanol in a channel 2.6 cm wide, 1.0 cm deep and 100 cm long. From Akita (1972). Reproduced by permission of the Combustion Institute

the channel with a removable barrier and igniting only the enclosed liquid surface. The barrier was then taken away and flame allowed to spread over liquid whose surface was still at ambient temperature.

The diagram shown in Figure 7.2 represents flame spread over the surface of a liquid which is below its flashpoint. Behind the flame front, steady pool burning will develop (Section 5.1). Under quiescent conditions, a flow of air against the direction of spread will be established as a direct consequence of entrainment into the base of the developing fire (Miller and Ross, 1998) (see Section 4.3.1). This spread mechanism is described as 'counter-current' (or 'opposed flow'). The influence of an imposed airflow will be discussed below.

The leading edge of the spreading flame is blue, similar in appearance to a premixed flame, and pulsates or 'flashes' ahead of the main flame (Glassman and Dryer, 1980/81). This is the behaviour one would expect if the temperature of the region of surface just ahead of the main flame lay between the flashpoint and the firepoint. A flash of premixed flame would occur periodically whenever a flammable concentration of vapour in air had formed (cf. Section 6.2.2) (Ito *et al.*, 1991). Raising the bulk temperature of the liquid has the effect of reducing the pulsation period and increasing the rate of spread. If the temperature of the liquid is below its flashpoint, then it is found that for shallow pools, the rate will decrease as the depth is reduced (Mackinven *et al.*, 1970; Miller and Ross, 1992). This is due mainly to restriction of the internal convection currents which accompany the surface tension-driven flow (Figure 6.11). In the limit, these will be completely suppressed, as with a liquid absorbed onto a wick (Section 6.2.2): if the heat losses to the supporting material are too great, then flame spread will not be possible (Figure 7.3).

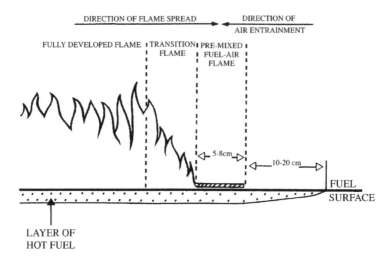

Figure 7.2 Spread of flame across a combustible liquid, initially at a temperature significantly less than its flashpoint, showing the net effect of the surface tension-driven flow. Reproduced by permission of Gordon and Breach from McKinven *et al*. (1970). The extent to which the flow and the flashes of premixed flame precede the main flame decrease if the initial temperature is increased towards the flashpoint (see Figure 7.1 (Akita, 1972))

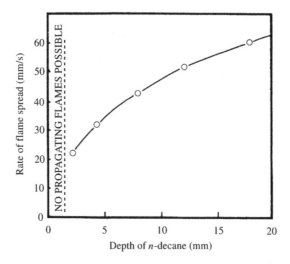

Figure 7.3 Variation of the rate of flame spread over *n*-decane floating on water as a function of depth of fuel. Container dimensions: 1.8 m ×0.195 m wide and 25 mm deep. Total depth of liquid (*n*-decane + water) 18 mm, initial temperature 23°C; flashpoint of *n*-decane 46°C. Reproduced by permission of Gordon and Breach from McKinven *et al*. (1970)

Another situation of interest is the spread of flame over fuel-soaked soil or sand. This has been studied by Ishida (1992) using glass beads of various diameters as a model for soil, with *n*-decane as the fuel. He examined flame spread along a 100 cm long, 5 cm wide and 2 cm deep tray completely filled with glass beads with sufficient liquid fuel added to soak the bed of beads. He found that the rate of spread increased if the diameter of the beads was reduced from 1.0 mm to 0.1 mm. Ishida suggests that this may be due to a combination of effects such as the suppression of surface tension-driven flows as the bead diameter is decreased and capillary effects which determine the rate of fuel supply to the surface. Similar conclusions were drawn from a study of radial flame spread from the point of ignition on a fuel-soaked bed of beads (Ishida, 1992). In this case, a central 'column' of flame developed behind the propagating front and heat transfer ahead of the spreading flame appeared to be dominated by radiation.

McKinven *et al.* (1970) also found that the rate of spread is independent of the width of the channel containing the liquid for widths between 15 and 20 cm (Figure 7.4). For narrower channels, heat losses to the sides are important, while for wider ones the established flame behind the advancing front becomes so large that radiative heat transfer to the unaffected fuel becomes significant.

If the liquid is above its firepoint, then the rate of flame spread is determined by propagation through the flammable vapour/air mixture above the surface and is no longer dependent on surface tension-driven flows. This has been demonstrated by Glassman and Hansel (1968) and more recently by White *et al.* (1997) by observing the motion of small beads of polystyrene foam floating on the surface of the liquid. When the temperature was less than the firepoint, the beads moved ahead and away from the spreading flame while if it was greater than the firepoint, they did not move and the flame passed over them. Glassman and Hansel (1968) proposed a simple diagram that distinguished 'surface tension-driven spread' and 'gas phase flame spread', the two regimes lying on either side of the firepoint temperature as shown in Figure 7.5.

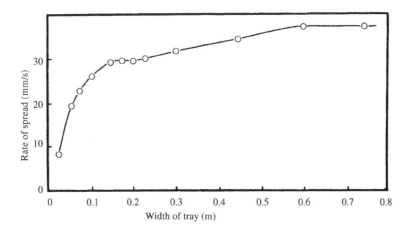

Figure 7.4 Rate of flame spread over *n*-decane as a function of tray width (4 mm of *n*-decane floating on water, other conditions as in Figure 7.3). Reproduced by permission of Gordon and Breach from McKinven *et al.* (1970)

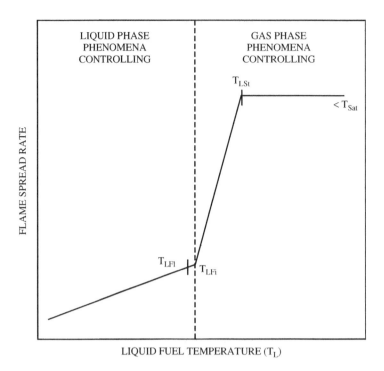

Figure 7.5 Schematic representation of the variation of flame spread rate as a function of liquid fuel temperature. $T_{L,Fl}$ and $T_{L,Fi}$ are the closed cup flashpoint and the firepoint temperatures, respectively and $T_{L,St}$ is the temperature at which the vapour pressure corresponds to a stoichiometric vapour/air mixture. After Glassman and Hansel (1968)

It is worth emphasizing that it is the firepoint that defines this boundary and not the closed cup flashpoint. White *et al.* (1997) measured the rate of flame spread over the aviation fuels JP-5 and JP-8 and plotted the results against $(T_{l,o} - T_{flash})$, i.e., the difference between the initial liquid temperature and the closed cup flashpoint. (This method of presentation was first used by Hillstrom (1975).) The data for both fuels are shown in Figure 7.6. Within experimental error, they fall on a single curve, despite their different flashpoints (63°C and 38°C for JP-5 and JP-8, respectively) and clearly show that the discontinuity associated with the switch from surface tension-driven spread to gas phase flame spread occurs ~15 K above the flashpoint (at $T_{l,o} - T_{flash} = 0$). The proposition is that $(T_{flash} + 15)°$C is approximately equal to the firepoint, although there are very few data on hydrocarbon liquids against which this can be tested (e.g., see Table 6.4).

It was argued by Glassman and Hansel (1968) that the maximum rate of spread would be determined by the fundamental burning velocity of the stoichiometric vapour/air mixture (as defined in Section 3.4). In fact, the limiting rate is four or five times this value, according to Burgoyne and Roberts (1968), Akita (1972) and others (Ross, 1994), reaching the maximum value at the temperature at which the vapour pressure of the liquid corresponds to the stoichiometric mixture at the surface (Figure 7.7). The effect is seen in the experimental data presented in Figures 7.1 and 7.6. This suggests that what is being observed is flame propagation in a partially confined system in which the unburnt gas

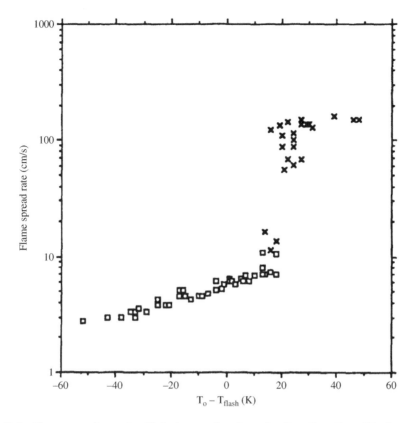

Figure 7.6 Flame spread rate for JP fuels as a function of $\Delta T = T_o - T_{flash}$ (K). Open circles represent surface tension-driven flow and crosses gas phase flame spread. The two fuels (JP-5 and JP-8) are not distinguished. White *et al.* (1997), reproduced with permission from Elsevier

is pushed ahead of the flame front (cf. Figure 3.21). A similar effect has been reported in flame spread through layers of methane/air mixtures trapped beneath the ceiling of an experimental mine gallery (e.g., Phillips, 1965).

Up to this point, the discussion has focused on flame spread in a quiescent atmosphere. The effect of a forced airflow on the rate of spread of flame over methanol has been examined in detail by Suzuki and Hirano (1982). They used a channel 4.2 cm wide, 3.3 cm deep and 100.8 cm long. If the air flow was in the opposite direction (counter-current), the rate of spread was found to decrease, ultimately to zero (and extinction). The critical air velocity required to achieve this increased with the temperature of the liquid, and thus with the flame spread velocity that would be found under quiescent conditions. For an imposed airflow in the same direction (concurrent), there was no effect until the velocity exceeded the quiescent flame spread velocity – and then it was found that the rate of flame spread increased, effectively matching the air flow rate. Under these circumstances, the flame is deflected forward, enhancing the rate of heat transfer ahead of the flame front and promoting the flame spread rate. Concurrent flow occurs naturally with upward flame spread on vertical combustible solids, as will be discussed in Section 7.2.1.

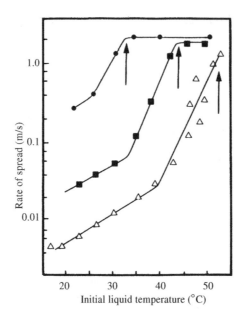

Figure 7.7 Dependence of the rate of spread of flame over flammable liquids on initial temperature: •, propanol; ■, butanol; △, isopentanol. Container 33 mm wide, liquid depth 2.5 mm. Burgoyne and Roberts (1968), by permission. The arrows indicate the temperatures at which a stoichiometric vapour/air mixture exists at the surface

7.2 Flame Spread Over Solids

Attention will now be focused on the surface spread of flame over combustible solids, examining the factors listed in Table 7.1 systematically. Unlike liquid pools, the surface of a solid can be at any orientation, which can have a dominating effect on fire behaviour. This is particularly true for flame spread, as surface geometry and inclination have a strong influence on the mechanism by which heat is transferred ahead of the burning zone.

As defined above (Figure 7.2), the term 'counter-current spread' describes the situation in which there is a flow of air opposed to the direction of spread, while concurrent spread is said to exist when the flow of air and the direction of spread are in the same direction. While these are clearly relevant to cases in which there is an imposed airflow, it also applies to flame spread where the air movement is generated naturally, by the dynamics of the flame. Naturally induced counter-current flow is developed in an otherwise quiescent atmosphere by a flame spreading along a horizontal surface (Figure 7.2), while naturally induced concurrent flow is observed when a flame is spreading upwards on a vertical surface (Section 7.2.1). The consequences of these effects will be discussed below.

7.2.1 Surface Orientation and Direction of Propagation

In general, solid surfaces can burn in any orientation, but flame spread is most rapid if it is directed upwards on a vertical surface. This can be illustrated with results of Magee and McAlevy (1971) on the upward propagation of flame over strips of filter paper at

Table 7.2 Rate of flame spread over strips of filter paper (dimensions not given) (Magee and McAlevy, 1971)

Orientation	Rate of flame spread (mm/s)
0° (horizontal)	3.6
+22.5°	6.3
+45°	11.2
+75°	29.2
+90° (vertically upwards)	46–74 (erratic)

inclinations between the horizontal and the vertical (Table 7.2). Propagation of flame was monitored by following the leading edge of the burning, or pyrolysis, zone: the results show more than a 10-fold increase in rate between these two orientations. Downward propagation is much slower, and the rate less sensitive to change in orientation. With computer cards as a typical 'thin' fuel, Hirano et al. (1974) found the rate of spread to be approximately constant (c. 1.3 mm/s) as the angle of orientation was changed from −90° (vertically downwards) to −30°, while increasing more than three-fold when the angle was changed from −30° to 0° (horizontal) (Figure 7.8(a)). These combined results suggest at least a 50-fold increase in the rate of spread between −90° and +90° for thin fuels.[2] (The effect of orientation on flame spread on thick fuels will be discussed below.)

The reason for this behaviour lies in the way in which the physical interaction between the flame and the unburnt fuel changes as the orientation is varied (Figure 7.9). For downward and horizontal spread, air entrainment into the flame leads to 'counter-current spread' (i.e., spread against the induced flow of air), but with upward spread on a vertical surface, the natural buoyancy of the flame generates 'concurrent spread'. This produces greatly enhanced rates of spread as the flame and hot gases rise in the same direction, filling the boundary layer and creating high rates of heat transfer ahead of the burning zone. The length of the flame becomes a critical factor as it defines the length of the heating zone (Section 4.3.3.1).

With physically thin fuels, such as paper or card, burning can occur simultaneously on both sides. This must be taken into account when interpreting flame spread behaviour. The increase in the rate of downward spread on computer cards as the inclination is changed from −30° to 0° (Hirano et al., 1974 (Figure 7.8(a))) is due to the flame on the underside of the card contributing to the forward heat transfer process. Kashiwagi and Newman (1976) have linked this to the onset of instability of the flame on the underside at low angles of inclination. Upward flame spread at inclinations greater than 0° increases monotonically (Drysdale and Macmillan, 1992 (Figure 7.8(b))), supported by the flow of the flame and hot gases on the underside of the card. Markstein and de Ris (1972) made similar observations in their study of flame spread on cotton fabrics. However, they also noted that while for inclinations between 45° and 90° (i.e., vertical) the flames on the upper and lower surfaces were of equal length, for inclinations between 20° and 45° the flames on the lower surface (the underside) extended further than the flames on the upper surface, which tend to lift away from the surface as a consequence of buoyancy

---

[2] In this book, angles are measured from the horizontal so that 90° corresponds to a vertical surface. In some publications, the angle is measured from the vertical (e.g., Quintiere, 2001)

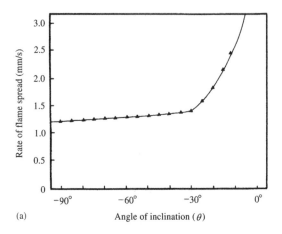

(a)

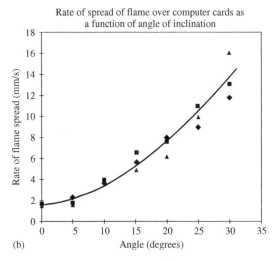

(b)

Figure 7.8   Variation of rate of flame spread over a thin fuel (computer card) as a function of angle of inclination (θ). (a) $\theta = -90°$ (vertically downwards) to $\theta = 0°$ (horizontal) (reproduced by permission of the Combustion Institute from Hirano *et al.* (1974)); (b) $\theta = 0°$ to $\theta = 30°$ (reproduced by permission of Elsevier Science from Drysdale and Macmillan (1992))

(this effect may be seen in Figure 7.11(c) and 7.11(d), albeit for a 'thick fuel'). They concluded that in these circumstances the flow on the underside exerted 'primary control' over the flame spread process. Drysdale and Macmillan (1992) found that if the flow on the underside was suppressed, upward flame spread on computer cards inclined at angles up to 30° was not possible. This was demonstrated by fixing a card on stenter pins fixed to a metal baseplate. If the gap between the card and the baseplate was ≤ 4 mm, the flame did not propagate and self-extinguished.[3]

---

[3] Quintiere (2001) has studied the effect of orientation on flame spread over thin fuels that were supported on an insulating substrate so that burning occurred only on one side.

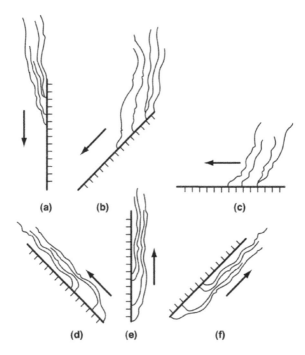

Figure 7.9 Interaction between a spreading flame and the surface of a (thick) combustible solid for different angles of inclination: (a) −90°; (b) −45°; (c) 0°; (d) +45°; (e) +90°. (a)–(c) involve counter-current spread, while (d)–(f) involve co-current spread. The switch from counter-current to co-current spread takes place at an angle of c. 15–20° (see Figure 7.10)

Quite different behaviour has been observed with physically thick fuels. Using PMMA as the fuel (>10 mm thick), Fernandez-Pello and Williams (1974) and Ito and Kashiwagi (1988) reported a small increase in the rate of spread as the angle of orientation was changed from −90° (vertically downwards) to horizontal, rather similar to the results shown in Figure 7.8(a) for a thin fuel. As the angle of inclination was increased to allow upward spread, Ito and Kashiwagi (1988) found that the rate continued to increase, achieving a maximum value at +90°. However, Drysdale and Macmillan (1992) found that significant enhancement of upward spread is observed only when the inclination of the surface is increased above +15°. The effect is quite pronounced if the flame is spreading up a plane incline (Figure 7.10, curve a), but is greatly enhanced if entrainment of air from the side is prevented by sidewalls (Figure 7.10, curve b). This was noted during the investigation into a fire that spread very rapidly on a wooden escalator at the King's Cross Underground Station in London (Fennell, 1988). The sides of the escalator completely suppressed side-entrainment.[4] It has become known as the 'trench effect', and has received much attention, both experimental (Smith, 1992; Atkinson *et al.*, 1995) and through CFD modelling (Cox *et al.*, 1989; Simcox *et al.*, 1992; Woodburn and Drysdale, 1997).

---

[4] The effect of sidewalls on flames was first noted by Markstein and de Ris (1972) in their study of upward flame spread on cotton fabrics. Significantly, extension of the flames was observed when there was no entrainment from the sides.

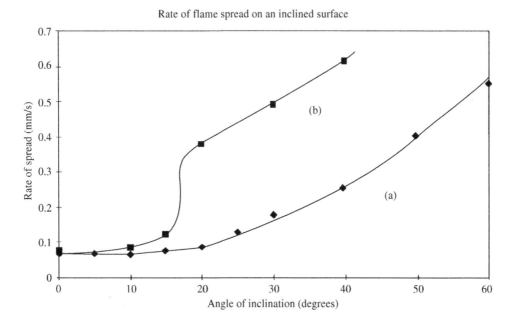

Figure 7.10   Rate of spread of flame on an inclined surface. 60 mm wide samples of PMMA: (a) without (♦) 'sidewalls' and (b) with (■) sidewalls (Drysdale and Macmillan, 1992)

There is a switch from counter-current spread that is dominant at inclinations below c. 15°, to concurrent spread at higher inclinations (certainly 25° and above). Figure 7.11 shows the effect of the slope on the shape of the flame during flame spread on an open, plane PMMA surface (6 mm thick), but it is impossible to establish precisely the critical angle at which the switch takes place from experiments of this kind. When there are sidewalls (i.e., there is no side-entrainment), an estimate of the critical angle can be made. Woodburn and Drysdale (1997) found that it depends on a number of factors including the geometry of the trench and the fire itself.[5] The same type of behaviour would be expected with thin fuels, but it tends to be masked by the flow of hot gases on the underside whenever there is the slightest upward inclination (Figure 7.8(b)). Upward spread on the underside of an incline (see Figure 7.9(f)) has not been studied, although as described above (see Markstein and de Ris, 1972) an argument can be made for more rapid upward spread on the underside of a thick fuel than on the topside (compare Figure 7.9(d) and (f)). Buoyancy keeps the flame much closer to the surface, enhancing the heat transfer efficiency and consequently the rate of spread: this is discussed below in the context of the behaviour of flames under ceilings. Results obtained by Quintiere (2001) on surface spread of flame over thin fuels (one being brown paper 'napkins') insulated on one side with fibreglass are consistent with this view.

---

[5] Using CFD, Woodburn found that critical angles do exist, depending on the boundary conditions (Woodburn and Drysdale, 1998). For example, in an inclined trench 0.27 m wide and 0.27 m deep, a 'switch' occurred when the inclination was increased from 19° to 20°, but a much lower angle (~10°) is predicted for 2-D flow, corresponding to an infinitely wide fuel bed.

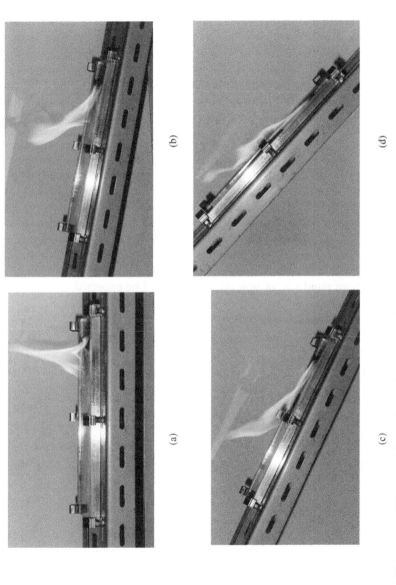

Figure 7.11 Flame spreading over the surface of PMMA slabs, 60 mm wide, 150 mm long and 6 mm thick: (a) $\theta = 0°$ (horizontal); (b) $\theta = 10°$; (c) $\theta = 20°$; and (d) $\theta = 45°$. Photographs by Stuart Kennedy and David Oliphant, University of Edinburgh

For thin fuels, downward spread ($-90°$) achieves a slow, steady rate of propagation almost immediately, but the rate of upward spread ($+90°$) increases rapidly until a quasi-steady state is achieved. This has been found for freely suspended strips of fabric (Thomas and Webster, 1960; Markstein and de Ris, 1972). Following ignition at the bottom edge there is a short period of laminar burning that quickly develops turbulence as the flame size increases. In their experiments, which involved 1.5 m lengths of fabric (maximum width 0.6 m), Markstein and de Ris (1972) showed that the instantaneous rate of flame spread was dependent on the length of the pyrolysing zone, i.e., the zone from which the volatiles were being released. As the rate of flow of volatiles will increase with area of burning, the rate of heat release will also increase, producing larger flames (Section 4.3.3.1). This will determine the extent of preheating of the unaffected fabric, which in turn determines how quickly fresh fuel is brought to the firepoint. Markstein and de Ris (1972) showed that after an initial transient period:

$$V_p \propto l_p^n$$

where V_p is the rate of vertical spread, l_p is the length of the pyrolysis zone (Figure 7.12) and n is a constant, approximately equal to 0.5, and developed a simple mathematical model to describe the behaviour. Information regarding forward heat transfer was obtained experimentally using a combination of flat sintered metal gas burners and water-cooled heat transfer plates. Their model shows that the growth of the fire depends *only* on the total forward heat transfer, which was limited by the onset of turbulence and the burnout time of the fabric. With a heavier fabric the burnout time would be longer (i.e., l_p would be longer) and a more rapid rate of upward spread would be expected.

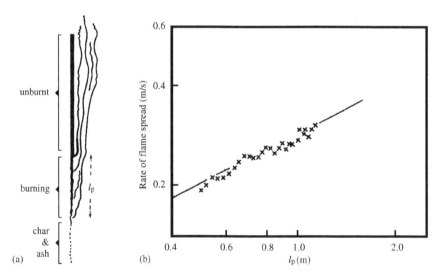

Figure 7.12 Upward spread of flame on a vertical strip of fabric: (a) showing the burning (pyrolysis) zone of length l_p; (b) increase of rate of upward spread with increasing length of the pyrolysis zone. Cotton broadcloth (103 g/m^2) of width 0.457 m and length 1.524 m (Markstein and de Ris, 1972)

Markstein and de Ris (1972) worked with 1.5 m lengths of fabric and extrapolated their data to give the limiting rates of spread of up to 0.45 m/s, which is likely to be one or two orders of magnitude greater than the rate of vertically downward spread, emphasizing the difference between concurrent and counter-current spread. In vertically downward spread, the flame gases flow away from the unburnt material so that convective transfer cannot occur and radiation is likely to be unimportant as the relevant configuration factor (i.e., the flame to the unburnt fuel) will be very small. Conduction through the gas phase is believed to be the dominant mechanism for thin fuel beds (Parker, 1972; Hirano *et al*., 1974), while for thick fuels conduction through the solid is a significant contributory factor (Fernandez-Pello and Williams, 1974; Fernandez-Pello and Santoro, 1980), particularly for vertically downward spread. On the other hand, its contribution to upward spread on a vertical slab is insignificant in comparison with the combined effects of convection and radiation (Ito and Kashiwagi, 1988) (see Section 7.3).

Observations have been made by Orloff *et al*. (1974) on upward flame spread on thick vertical PMMA slabs, 0.41 m wide and 1.57 m high, which revealed that the rate of spread was approximately constant (0.5–0.6 mm/s) over the first 0.1–0.15 m while the flame was laminar, but underwent a transition to increase exponentially as the flame became fully turbulent (Figure 7.13). Moreover, they found that radiation from the flame can ultimately be responsible for over 75% of the total heat transfer ahead of the pyrolysis front. In view of the dependence of the upward rate of spread on the length of the burning zone (l_p above), it follows that upward flame spread on a semi-infinite solid can never

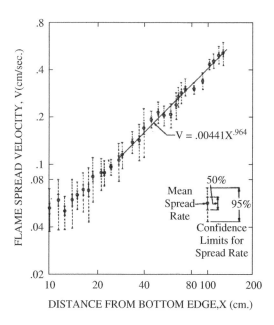

Figure 7.13 Flame spread velocity up a vertical sheet of PMMA as a function of the distance from the bottom edge. The results are averaged values from five experiments. The solid line is a least squares fit to the data above 30 cm (Orloff *et al*., 1974). By permission, the Combustion Institute

achieve a steady state. Indeed, Alpert and Ward (1984) proposed that the development of such a fire can be approximated by exponential growth (see Section 7.5).

The remaining orientation to be discussed is the horizontal ceiling. A combustible ceiling will burn and spread flame, but the process generally has to be driven by a 'ceiling jet' created by a fire at floor level, or by a wall fire (see Chapter 9).[6] Without this imposed flow, the rate of spread will be slow as the hot gases will tend to stagnate underneath the horizontal surface. It can only spread effectively as a co-current flow – as demonstrated in wind tunnel studies by Zhou and Fernandez-Pello (1993) with PMMA slabs in a ceiling configuration. They concluded that the same factors that determine the rate of upward flame spread on a wall also determine the rate of spread under a ceiling, given that a concurrent flow of air is required (see Fernandez-Pello, 1995). However, buoyancy effectively 'pushes' the ceiling flame close to the surface so that under equivalent experimental conditions, the rate of spread under a ceiling will be more rapid than spread over a floor configuration. The degree of turbulence of the airflow has a significant effect and for the ceiling configuration there will be some flame quenching close to the surface, leading to an increased yield of partially burned combustion products such as carbon monoxide: this was noted by Hasemi et al. (2001) (Section 4.3.3.2). As discussed earlier, it is anticipated that upward spread on a sloping ceiling would not require an imposed flow as it can develop its own concurrent flow (see Figure 7.9(f)).

7.2.2 Thickness of the Fuel

While flame spread can be treated theoretically as a quasi-steady state problem (e.g., de Ris, 1969), it involves transient heat transfer processes identical to those encountered in the previous chapter on ignition (Section 6.3), including heat transfer by conduction from the surface to the interior of the fuel. Thus, if the fuel is very thin and can be treated by the 'lumped thermal capacity model' in which there is no temperature gradient between the faces of the sample (Section 2.2.2), it can be shown theoretically that the rate of spread will be inversely proportional to the thickness (τ) of the material (Sections 7.2.3 and 7.3). There is ample evidence in the literature to support this conclusion. Magee and McAlevy (1971) quote data obtained by Royal (1970), which show that for downward spread on vertical specimens of thin cellulosic fuels, $V \propto \tau^{-1}$, for thicknesses less than 1.5 mm (Figure 7.14(a)). This has been confirmed by results of Suzuki et al. (1994) with samples prepared from filter paper, and similar relationships have been demonstrated for fabrics where the rate has been shown to be inversely proportional to the 'weight' of the fabric (see, for example, Hirschler et al. (2009)). Air permeability of the material has no significant effect (Moussa et al., 1973).

As the thickness is increased, the rate of spread ultimately becomes independent of thickness. This is seen in Figure 7.14(b). Suzuki et al. (1994) extended the range of thickness of paper to 10 mm and found that the dependence on thickness became less pronounced when $\tau > 1.5$ mm, becoming approximately constant in the range c. 5.0–7.5 mm. For thicker samples, the rate became erratic, and if $\tau > 8.4$ mm, continuous spread was not possible. However, with PMMA (Fernandez-Pello and Williams, 1974), which is a non-charring material, downward spread achieves a constant rate when the sample becomes

---

[6] The behaviour of flames from a burner flush-mounted to the underside of a ceiling has been studied by Hasemi et al. (2001) and Weng and Hasemi (2008). (See Page 209).

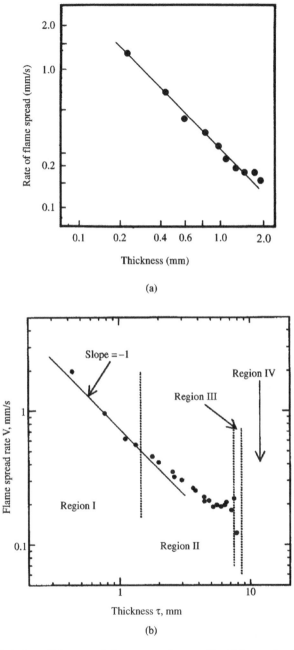

(a)

(b)

Figure 7.14 (a) Variation of downward flame spread rate with thickness for thin cellulosic specimens (Royal (1970), quoted by Magee and McAlevy (1971)). Reproduced by permission of Technomic Publishing Co. Inc., Lancaster, PA 17604, USA. (b) Dependence of rate of downward flame spread on thickness (paper samples). Regions I and II: stable spread. Region III: unstable spread. Region IV: no spread (thickness >8.4 mm). Suzuki *et al*. (1994), by permission of the Combustion Institute

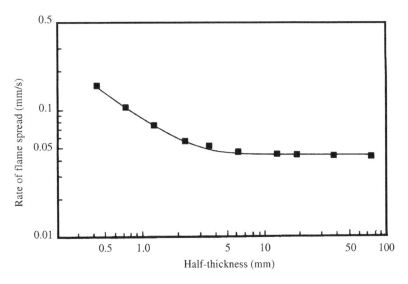

Figure 7.15 Variation of the rate of downward flame spread on vertical sheets of PMMA of different thickness. Reproduced by permission of the Combustion Institute from Fernandez-Pello and Williams (1974)

'thermally thick' (corresponding to $1/2\tau$ > c. 5 mm) in the sense described earlier (Section 2.2.2) (Figure 7.15). In going from thermally thin to thermally thick fuels, there is a change in the dominant mode by which heat is transferred ahead of the flame front, from heat conduction through the gas phase for thin fuels (see Parker (1972), Section 7.3) to heat conduction through the solid for thick fuels (Fernandez-Pello and Williams, 1974; Fernandez-Pello and Santoro, 1980; Ito and Kashiwagi, 1988).

It should be borne in mind that the above discussion relates to materials that do not change their physical form during the burning process. Processes such as melting (see, for example, Zhang *et al*. (1997), Sherratt and Drysdale (2001)) and delamination should be considered and taken into account. Delamination can turn a composite material which initially has the hallmarks of a 'thermally thick' fuel into one in which the surface lamina detaches and becomes a thin fuel, more easily ignited and with the propensity to spread flame rapidly. This is particularly relevant to upward flame spread: the most startling example involves delamination of multiple layers of gloss paint on a non-combustible substrate such as plaster. This is recognized as a serious fire hazard in old buildings such as hospitals which have been regularly maintained (Murrell and Rawlins, 1996).[7]

7.2.3 Density, Thermal Capacity and Thermal Conductivity

The distinction between 'thermally thick' and 'thermally thin' materials has already been made and the concept developed in Section 6.3.1. The 'depth of heating' is given

---

[7] There is now a UK Code of Practice addressing this issue: 'Refurbishment of Communal Buildings and the Fire Risk of Multilayer Paints', published by the Department of Trade and Industry in 2005 (PII Reference F-02-WFRC1).

approximately by $(\alpha t)^{1/2}$, where α is the thermal diffusivity $(k/\rho c)$ and t is the time in seconds during which the surface of the solid is exposed to a heat flux. For an advancing flame front, the exposure time for the unburnt fuel is l/V, where V is the rate of spread and l is the 'heating length' – i.e., the length of sample perpendicular to the advancing flame over which the temperature rises from T_0 (ambient) to the temperature corresponding to the firepoint. (This is marked clearly in Figure 7.24 (Section 7.3).) A critical thickness τ_{cr} for flame spread is then estimated from the expression:

$$\tau_{cr} = (\alpha l/V)^{0.5} \tag{7.1}$$

Thus, thin fuels may be defined as $\tau < \tau_{cr}$ and treated by the lumped thermal capacity model for which the time to achieve the firepoint (t_{ig}) under a given heat flux is directly proportional to the product $\rho c \tau$ (Equation (6.30)). As the rate of spread will be inversely proportional to t_{ig}, then

$$V \propto (\rho c \tau)^{-1} \tag{7.2}$$

which gives the dependence on τ that was noted above (Figure 7.14(a), Section 7.2.2). To discover how the thermal properties of a 'thick' fuel ($\tau > \tau_{cr}$) influence the rate of spread, τ must be replaced by an expression for the depth of the heated layer at the surface of the material (δ). Thus, following Equation (7.1):

$$\delta = (\alpha l/V)^{0.5} \tag{7.3}$$

If this is substituted in place of τ in Equation (7.2), rearrangement gives

$$V \propto \frac{1}{k\rho c} \tag{7.4}$$

provided that l is constant. The appearance of thermal inertia ($k\rho c$) is consistent with the fact that ease of ignition is strongly dependent on this factor. Examining it in more detail, as the thermal conductivity (k) of a solid is roughly proportional to its density (ρ), it can be seen from Equation (7.4) that the rate of flame spread is extremely sensitive to the density of the fuel bed (approximately, $V \propto \rho^{-2}$) (see Table 7.3). This is why foamed plastics and other combustible materials of low density spread flame and develop fire so rapidly: only a very small mass of material at the surface needs to be heated to allow

Table 7.3 Rate of spread of flame

Material	Density (kg/m^3)	Rate of spread (mm/s)	Reference
PMMA (thick)	1190	~0.07	Fernandez-Pello and Williams (1974)
Polyurethane[a]	15	3.7	Paul (1979)
Polyurethane[a]	22	2.5	Paul (1979)
Polyurethane[a]	32	1.6	Paul (1979)

[a]Formulation unknown.

the flame to spread (Table 6.8). Some authors have attributed this behaviour to surface roughness and porosity (e.g., Magee and McAlevy, 1971), but these have at most only second-order effects (Moussa *et al.*, 1973).

7.2.4 Geometry of the Sample

(a) *Width*. The width of a sample has little or no effect on the rate of vertically downward spread provided that edge effects do not dominate the behaviour (see (b) below). The situation is different for vertically upward burning. Thomas and Webster (1960) found that for freely suspended strips of cotton fabric:

$$V \propto (\text{width})^{0.5} \tag{7.5}$$

For widths ranging from 6–100 mm this result can be correlated with the increase in flame height arising from the increased area of burning. This illustrates one of the difficulties encountered in devising a small-scale test for assessing the fire behaviour of fabrics. In addition, the propensity of some synthetic materials to melt and drip while they are burning can lead to results that are difficult to interpret. The original fabric flammability tests used narrow strips of material, as specified in BS 3119 (British Standards Institution, 1959) and DOC FF-3-71 (US Department of Commerce, 1971a), but these have been replaced by ones that call for substantial widths (Holmes, 1975; British Standards Institution, 1976), and even full garments on instrumented manikins (e.g., ASTM, 2002).

Until recently, the effect of sample width on the rate of flame spread over 'thick' materials has not received the same amount of attention, but the same general features would be anticipated. The major difference lies in the fact that the flame behind the leading edge will be substantially greater as the area of burning (the 'pyrolysis zone' of Section 7.2.1) will be potentially much larger. This will affect the rate of horizontal spread if the width of the fuel bed is great enough. To illustrate this, recall that McKinven *et al.* (1970) found that radiative heat transfer ahead of the flame front became significant in the spread of flame over the surface of *n*-decane contained in trays more than 0.2 m wide (Figure 7.4). A similar effect would be expected for solids. Indeed, for fires spreading on fuel beds of substantial width, radiative heat transfer ahead of the flame front can dominate the fire spread mechanism, simply because of the size of the flames. This is particularly true for wildland fires, although for surface fires (across forest litter), Wotton *et al.* (1999) found that for flame spreading across horizontal beds the rate increased only as the width was increased from 0.5 to 2 m. In their experiments, there was no further increase if the width was increased to 10 m, indicating that the height of the flames behind the leading edge of the burning zone did not increase significantly.

Monitoring the front of the pyrolysis zone as a means of following flame spread is difficult unless char is formed (e.g., following the spread of flame over cellulosic fuels such as paper or cards). With PMMA, the leading edge of the pyrolysis zone corresponds to the first appearance of sub-surface bubbling: this has been widely used (e.g., Orloff *et al.*, 1974; Drysdale and Macmillan, 1992; Pizzo *et al.*, 2009). The experimental set-up developed by Pizzo *et al.* (2009) allowed a more accurate determination of the position of the pyrolysis front: they studied the rate of upward spread on vertical PMMA slabs 0.4 m high and 0.03 m thick, with widths in the range 0.025 to 0.2 m. For the narrower widths (0.025 and 0.05 m), the upward spread did not accelerate over the height of the slabs used (0.4 m), remaining constant at c. 0.4 mm/s and c. 0.65 mm/s, respectively (Figure 7.16). The wider samples (0.1–0.2 m wide) showed an initial flame spread rate of between 0.7 and 0.75 mm/s over the first 0.15–0.2 m but thereafter the flame began to accelerate. The latter results are in good agreement with those of Orloff *et al.* (1974) who used 0.41 m wide samples (Figure 7.13).

(b) *Presence of edges.* Flame spreads much more rapidly along an edge or in a corner than over a flat surface. This has been studied systematically by Markstein and de Ris (1972) using 'wedges' of PMMA, as shown in Figure 7.17(a). The rate of downward propagation[8] at the edge was measured as a function of the angle (θ), the following dependence being found for $20° \leq \theta \leq 180°$:

$$V \propto \theta^{4/3}$$

(Figure 7.17(b)). The narrower the angle θ, the closer the edge of the solid approaches thin-fuel behaviour, with flame spreading down both sides. Thus it is partly a thermal capacity effect and partly due to the fact that heat is being transferred to the fuel on both sides of the edge. The rate of downward spread is a minimum for $\theta = 180°$. If θ is greater than $180°$ (i.e., the angle becomes re-entrant), the rate of downward spread is enhanced by cross-radiation near the junction of the walls (Figure 7.17(c)). Markstein and de Ris (1972) estimated that $V \approx 0.56$ mm/s for $\theta = 270°$, this referring to the maximum, close to the corner. When $\theta > 270°$, cross-radiation effects will increasingly dominate, approaching the case of two inward-facing burning surfaces (see Section 2.4.1). This configuration has been adopted for upward spread by FM Global as a very versatile and severe test for materials that are to be used in clean room facilities in microprocessor manufacturing plant[9] (Wu, 1999; de Ris and Orloff, 2005).

7.2.5 Environmental Effects

(a) *Composition of the atmosphere.* Combustible materials will ignite more readily, spread flame more rapidly and burn more vigorously if the oxygen concentration is increased. This is of practical significance as there are many locations where oxygen enriched atmospheres (OEAs) may be produced accidentally (e.g., leakage from the oxygen supply system in a hospital or leakage from oxygen cylinders used for oxyacetylene welding) and several where OEAs are produced deliberately (e.g., 'oxygen tents' in intensive care units). An OEA is one in which the partial pressure of oxygen is greater than that of the normal atmosphere (160 mm Hg) (National Fire Protection Association, 2004), so that any environment in which the air pressure has been increased artificially (e.g., in a diving bell or for tunnel boring operations) must be considered potentially hazardous (see Section 7.2.5(d)).

Any increase in oxygen concentration in the air is accompanied by an increase in the rate of flame spread (see Figures 7.18 and 7.19). This is because the flame is hotter

---

[8] Downward flame spread has been used to study many of the factors that affect flame spread – such as thickness and the edge effect – simply because the experiments are easier to carry out and control. The conclusions are equally applicable to upward spread.

[9] See 'FM Approvals. Cleanroom Materials: Flammability Test Protocol', Class Number 4910, FM Approvals (FM Global, September 2007).

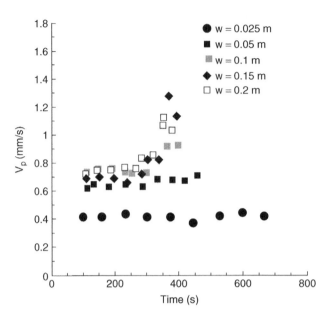

Figure 7.16 Development of the rate of vertical flame spread on slabs of PMMA of different widths. From Pizzo *et al*. (2009), by permission of Elsevier

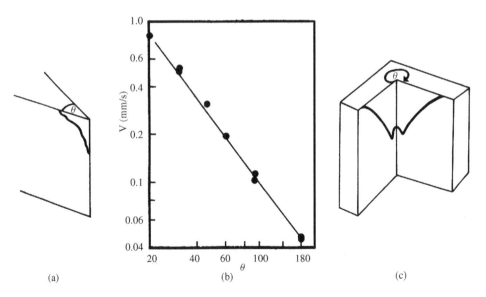

Figure 7.17 Rate of downward spread of flame on edges and in corners. (a) Definition of the angle (θ) at an edge. (b) Variation of V for $20° < \theta \leq 180°$. (c) Definition of θ for a corner. Figures (a) and (c) show typical shapes of the leading edge of the advancing flame (Markstein and de Ris, 1975). Reproduced by permission of Technomic Publishing Co. Inc., Lancaster, PA 17604, USA

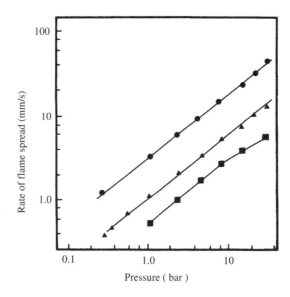

Figure 7.18   Variation of rate of flame spread over horizontal samples of PMMA with pressure of mixtures of oxygen and nitrogen: ■, 46% O_2; ▲, 62% O_2; ●, 100% O_2. Reproduced by permission of the Combustion Institute from McAlevy and Magee (1969)

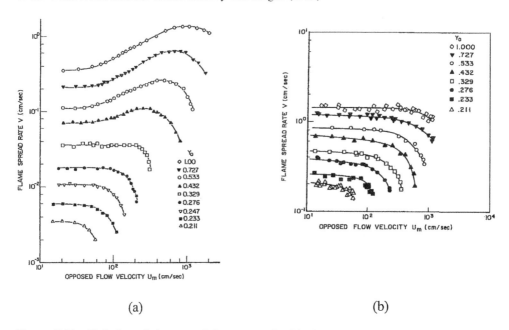

(a)   (b)

Figure 7.19   Variation of the rate of flame spread with the opposed forced flow velocity for several mass fractions of oxygen in O_2/N_2 mixtures. (a) Horizontal slabs of PMMA (23 cm long × 4 cm wide × 1.27 cm thick); (b) horizontal sheets of filter paper (23 cm long × 4 cm wide × 0.02 cm thick). The edges were inhibited by an insulating material to preferential flame spread (Fernandez-Pello *et al*., 1981). Reproduced by permission of the Combustion Institute

(see Chapter 1, Question 1.12 on premixed pentane/air and pentane/oxygen flames) and is capable of losing a larger fraction of the heat of combustion to the fuel (i.e., ϕ is greater, see Section 6.3.2). This allows it to lie closer to the fuel surface, thus producing a greater rate of heat transfer. Magee and McAlevy (1971) record that the dependence of propagation rate (V) on oxygen concentration is greater for 'thick' fuels than for 'thin', at least for vertically downward spread. Huggett et al. (1966) found that the rate of flame spread over wooden surfaces in O_2/N_2 and O_2/He mixtures depended on the ratio $[O_2]/c_g$, where $[O_2]$ is the concentration of oxygen and c_g is the thermal capacity of the atmosphere per mole of oxygen. Thus, flame spreads more rapidly in an O_2/He atmosphere than in an equivalent O_2/N_2 atmosphere because of the reduced c_g (Section 1.2.5). This allows the flame to move closer to the surface without being quenched.

A corollary to the work of Huggett is that it is possible to create an atmosphere that will support life but not flame. If the thermal capacity of the atmosphere per mole of oxygen is increased to more than c. 275 J/K (corresponding to 12% O_2 in N_2), flame cannot exist under normal ambient conditions. A level of oxygen as low as 12% will not support normal human activity but if this atmosphere is pressurized to 1.7 bar, the oxygen partial pressure will be increased to 160 mm Hg, equivalent to that in a normal atmosphere and therefore perfectly habitable – although incapable of supporting combustion (Huggett, 1973).

(b) *Fuel temperature*. Increasing the temperature of the fuel increases the rate of flame spread. Intuitively this is to be expected as the higher the initial fuel temperature the less heat is required to raise the unaffected fuel to the firepoint ahead of the flame. Magee and McAlevy (1971) report that for thin fuels:

$$V \propto \frac{1}{(T_p - T_0)} \quad \text{(approximately)} \tag{7.6}$$

while for thick fuels:

$$V \propto \frac{1}{(T_p - T_0)^2} \quad \text{(approximately)} \tag{7.7}$$

where T_p is the minimum temperature at which decomposition occurs and T_0 is the initial fuel temperature. These experimental results are consistent with the theoretical analysis carried out by de Ris (1969) (see Section 7.3).

(c) *Imposed radiant heat flux*. An imposed radiant heat flux will cause an increase in the rate of flame spread, primarily by preheating the fuel ahead of the flame front (Fernandez-Pello, 1977a,b; Hasemi et al., 1991) (cf. Equations (7.6) and (7.7)). However, the increased rate of burning behind the flame front will give stronger flames, which will provide additional forward heat transfer and thus enhance the process (Kashiwagi, 1976; Hirano and Tazawa, 1978).[10] These observations are quite general, although the relative importance of the effects will depend on orientation: thus, the effect is particularly marked if the spread is vertically upwards (Saito et al., 1989; Hasemi et al., 1991).

A relatively low flux is capable of producing a measurable effect. Alvares (1975) reports that a radiant heat flux equivalent to three or four times that of the summer sun in the UK (0.67 kW/m^2 (Table 2.9)) will increase the rate of flame spread over inclined

---

[10] Kashiwagi noted that forward heat transfer by radiation became significant for flame heights exceeding 0.15 m.

upholstered panels by 70%. It is even more pronounced with thin fuels such as paper (Hirano and Tazawa, 1978). Such an effect is significant during the early stages of a compartment fire when the levels of radiant heat flux from the compartment boundaries and the layer of hot smoky gases trapped below the ceiling are increasing (Section 9.1).

The response of a surface to the imposed flux is not instantaneous and the effect of transient heating must be considered if the flame begins to spread over the surface before thermal equilibrium has been reached (Figure 7.20) (Kashiwagi, 1976; Fernandez-Pello, 1977a; Quintiere, 1981). This is illustrated by the results of Kashiwagi, which are shown in Figure 7.20(a): he found the rate of flame spread over the surface of a thermally thick material to increase with the duration of exposure to a constant heat flux. The initial response to thermal radiation will depend on the thermal inertia ($k\rho c$) of the material (see Figure 2.10), while the steady state velocity (corresponding to a long preheating time) will be determined by the surface temperature achieved at equilibrium. This will depend on the heat losses from the surface. Kashiwagi's experiments, in which the 'fuel bed' was a carpet, showed that the steady state flame spread was faster if the carpet had an underlay to reduce heat losses to the floor (Figure 7.20(a)). It would be anticipated that if the equilibrated surface temperature is greater than the firepoint, the rate of spread would be very high as the flame would be propagating through a premixed vapour/air mixture (compare with Figure 7.7). This behaviour will not necessarily be observed for thermally thin materials. For example, although an imposed heat flux will promote flame spread over paper and similar materials, high levels will cause charring and rapid degradation, which can deplete the fuel bed of most of its 'volatiles' before the arrival of the flame front (Fernandez-Pello, 1977a). Ultimately, this will place a limit on the maximum rate of spread. (Note that thin thermoplastic films and sheets will soften and melt if exposed to the same conditions.)

When it was realized that radiation had such a profound influence on the rate of flame spread (and thus the rate of fire growth), laboratory tests involving exposure to radiant heat were developed to assess the flame spread potential of wall lining materials (e.g., British Standards Institution, 1997; ASTM, 2008c) and floor coverings (ISO, 2002c; ASTM, 2010). These were developed on an *ad hoc* basis to provide a means by which materials could be 'ranked' for regulatory purposes. However, the results of these tests are strongly apparatus-dependent and cannot yield quantitative data on material performance: indeed, different tests designed for the same purpose in different European countries gave widely different ranking orders for a given set of materials (Emmons, 1974).

This presented a barrier to international trade, particularly within Europe, that could only be resolved by developing a new generation of tests that would be acceptable to all member states. These were described collectively as 'reaction to fire' tests, and were designed and developed with the benefit of an improved understanding of fire science. These test methods – some of which are listed in Table 7.4 – have been subjected to detailed analyses with the intention of developing procedures for extracting fundamental material data that could be used by the regulator and in fire safety engineering design.

Quintiere (1981) analysed a surface spread of flame apparatus that has since been developed as an ISO fire test.[11] Flame is allowed to spread horizontally along a vertical sheet of material 155 mm high and 800 mm long, which is set with its long axis at an

---

[11] This apparatus became the Lateral Ignition and Flame Spread Test (LIFT) (ISO, 2006) in which the radiant panel is at an angle of 15° to the sample (cf. Figure 7.21).

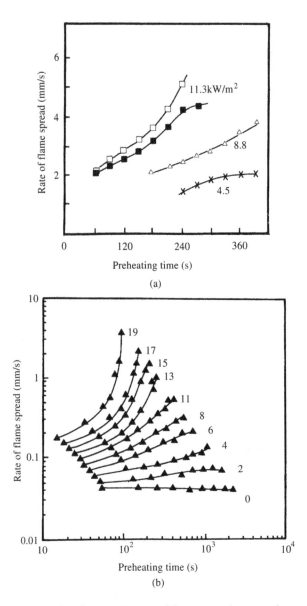

Figure 7.20   Effect of preheating time on the rate of flame spread over surfaces exposed to radiant heat. (a) Horizontal acrylic carpet: ×, △ and □ with underlay; ■ without underlay. Reproduced by permission of the Combustion Institute from Kashiwagi (1975). (b) Vertical PMMA sheets (downward spread). Figures refer to the imposed radiant heat flux (kW/m²). Reproduced by permission of Gordon and Breach from Fernandez-Pello (1977a)

Table 7.4 ISO reaction to fire tests (Vandevelde *et al.*, 1996)

Test method	Reference
Ignitability of building products	ISO 5657 (1997)
Lateral flame spread on building products	ISO 5658-2 (2006)
Rate of heat release from building products – cone calorimeter	ISO 5660-1 (2002a)
	BS 475 Part 15 (1993)
Smoke generated by building products – dual chamber test[a]	ISO 5924 (1989)
Full-scale room test for surface products	ISO 9705 (2002b)
	BS 476 Part 33 (1993)

[a]It appears that this has not progressed beyond the Technical Report stage.

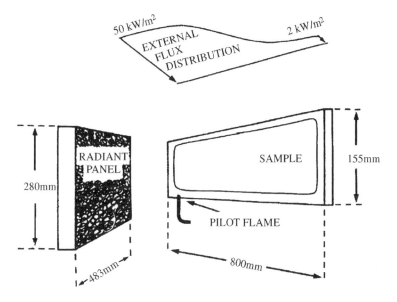

Figure 7.21 Details of the ISO Surface Spread of Flame Apparatus (Quintiere, 1981). Reproduced with permission

angle of 15° to a radiant panel as shown in Figure 7.21. Thus the radiant heat flux falling on the surface diminishes with distance from the panel, from 50 kW/m² at one end to 2 kW/m² at the other. The fuel vapours released close to the radiant heater are ignited by a pilot flame and subsequent flame spread along the sample is measured.

Quintiere's analysis of the heat transfer at the flame boundary was based on that developed earlier to analyse flame spread over carpets (Quintiere, 1975b; NFPA, 2006a). Several simplifying assumptions were required to enable the equations to be solved, including the following: (i) the solid is thermally thick; (ii) heat is transferred forward from the flame through the gas phase; and (iii) flame moves forward when the unburnt fuel reaches an 'ignition temperature', T_{ig} (equivalent to the firepoint) and below which the material is regarded as inert. Although there is some doubt about the compatibility of (i) and (ii), the equations reduce to

$$V^{-1/2} = \left[\frac{(\pi k\rho c)^{1/2}}{2hl^{1/2}\dot{q}_F''} \right] [h(T_{ig} - T_0) - \dot{q}_E''] \tag{7.8}$$

where V is the flame spread rate, $\dot{q}_F''$ is the heat flux from the flame to the surface across a distance l, $\dot{q}_E''$ is the externally imposed heat flux, T_0 is the initial temperature and h is an effective surface heat transfer coefficient. This equation can be rewritten:

$$V^{-1/2} = C(\dot{q}_{0,i}'' - \dot{q}_E'') \tag{7.9}$$

where C is referred to as a 'rate coefficient' and $\dot{q}_{0,i}''$ is the limiting heat flux necessary for pilot ignition (Section 6.3.1). These quantities contain the basic properties of the material and could reasonably be expected to be 'constants' in a given flame spread situation. If so, a plot of $V^{-1/2}$ versus $\dot{q}_E''$ would be a straight line of gradient $-C$ and intercept $C \cdot \dot{q}_{0,i}''$. Quintiere (1981) examined this hypothesis by using data on the variation of V and $\dot{q}_E''$ as a function of distance from the radiant panel in the ISO Surface Spread of Flame Apparatus. However, he found that $V^{-1/2}$ was a linear function of $\dot{q}_E''$ only for rapid rates of spread and only if the sample had been allowed a substantial preheat time before it was ignited (Figure 7.22), i.e., thermal equilibrium must be achieved. Quintiere's analysis provides a means by which C and $\dot{q}_{0,i}''$ can be evaluated, but in addition identifies

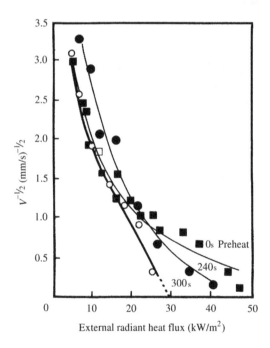

Figure 7.22 Flame spread correlation for hardboard according to Equation (7.9): ■, no preheat; ●, 240 s preheat; ○, 300 s preheat (Quintiere, 1981). Reproduced with permission

Table 7.5   Values of $C, \dot{q}_{0,s}''$ and $\dot{q}_{0,i}''$ from the Surface Spread of Flame Apparatus (Quintiere, 1981)

Material	$\dot{q}_{0,s}''$ (kW/m²)	$\dot{q}_{0,i}''$ (kW/m²)	C (s/mm)$^{1/2}$ (m²/ kW)
Coated fibreboard	12	19	0.30
Uncoated fibreboard	≤2	19	0.057
Chipboard	7	28	0.12
Hardboard	4	27	0.13
PMMA	≤2	21	0.16
Plywood panelling (vinyl side)	15	29	0.17
Plywood panelling (rear face)	8	29	0.08
Polyester	≤2	28	0.11

those materials for which there is a minimum external heat flux necessary to permit flame spread, $\dot{q}_{0,s}''$. At low values of $\dot{q}_E''$, Equation (7.9) does not fit the data: $V^{-1/2}$ is found to increase asymptotically (corresponding to $V \to 0$) as $\dot{q}_E''$ is reduced. The value of $\dot{q}_E''$ at the asymptote was taken to be $\dot{q}_{0,s}''$. Some of these figures are quoted in Table 7.5, although it must be emphasized that these relate to the heat transfer conditions pertaining to the configuration under study (cf. Section 6.3.1). Thus, these figures should not be accepted blindly for flame spread over any other orientation of surface, although in principle the *model* may be used for any orientation (e.g., Quintiere, 1975a).

The concept of a limiting heat flux for flame spread has already been incorporated into a standard test for assessing the fire properties of carpets and floor covering, etc. (NFPA, 2006a). In this test, the floor covering is lying horizontally with the radiant heater at an angle of 15°, as in the LIFT apparatus (ISO, 2006b) but turned through 90°. The heat flux is set at 10 kW/m² at the end closest to the panel, falling to 1 kW/m² at the other end of the 1070 mm long sample. The results that are given in Table 7.6 show clearly the insulation effect of an underlay or pad. (Note that if Equation (7.9) is rearranged in the form

$$V = \left\{ C(\dot{q}_{0,i}'' - \dot{q}_E'') \right\}^{-2} \tag{7.10}$$

then, as $C \propto (k\rho c)^{1/2}$, this predicts $V \propto 1/k\rho c$ as deduced in Section 7.2.3 for a semi-infinite solid.)

(d) *Atmospheric pressure.* Higher rates of flame spread are observed at increased atmospheric pressure as a result of two contributory factors: (i) there is effective oxygen enrichment (Section 7.2.5(a)) which enhances flame stability at the surface; and (ii) the convective heat transfer coefficient is increased due to its dependence on either the Reynolds number (forced convection) or the Grashof number (natural convection) (see Table 2.5), which both increase with pressure (see Table 2.4 and Section 4.4.5 (Equations (4.77) and (4.78))). The effect of pressure on flame spread rate over slabs of PMMA is illustrated in Figure 7.18 (McAlevy and Magee, 1969), although it should be noted that the dependence is much less for thin than for thick materials.

(e) *Imposed air movement (wind).* In general, confluent air movement will enhance the rate of spread of flame over a combustible surface (cf. co-current spread) and through

Table 7.6 Limiting radiant heat fluxes for flame spread over carpets

Carpet	Underlay	Critical radiant flux (kW/m²)	Reference
Nylon	–	2.4	Alderson and Breden, 1976[a]
Nylon	Hair jute	1.2	Alderson and Breden, 1976[a]
Wool	–	10.0	Alderson and Breden, 1976[a]
Wool	Hair jute	6.6	Alderson and Breden, 1976[a]
Acrylic	–	4.1	Alderson and Breden, 1976[a]
Acrylic	Hair jute	2.5	Alderson and Breden, 1976[a]
Polyester	–	3.2	Alderson and Breden, 1976[a]
Polyester	Hair jute	<1.0	Alderson and Breden, 1976[a]
100% wool	Hair jute	11	Moulen and Grubits, 1976[b]
80% wool, 20% nylon	Hair jute	12	Moulen and Grubits, 1979[b]

[a]Critical flux determined using the flooring radiant panel test (NFPA, 2006a).
[b]Critical flux determined using a modified ISO ignitability apparatus with central ignition (methenamine tablet). Underlay was hessian, 240 g/m².

open fuel beds, as encountered in bush and forest fires (Section 7.4). Friedman (1968) originally reported that the rate increases quasi-exponentially up to a critical level at which extinction, or blow-off, will occur (Section 6.6.2), but more recently Zhou and Fernandez-Pello (1990) have shown that the rate of flame spread on 12.7 mm thick strips of PMMA (76 mm wide and 300 mm long) increases linearly with the free stream velocity, at least if the air flow is turbulent. The mechanism involves flame deflection in the direction of spread which promotes radiative heating of the fuel ahead of the burning zone. When the degree of turbulence in the airflow was increased, it was found that the effect of the air flow velocity was reduced, quite significantly. This was attributed principally to a reduction in the length of the deflected flame as a consequence of the more efficient entrainment due to turbulent mixing, which in turn reduced the length of the 'preheat zone' (see Section 7.3). This is illustrated in Figure 7.23. Zhou and Fernandez-Pello observed that at air flow velocities greater than 4 m/s there were signs of extinction at the upstream leading edge.[12]

If the direction of air flow is opposed to the spread of flame, the net effect depends on the air velocity. At relatively low velocities (<0.4 m/s), Fernandez-Pello *et al*. (1981) found that the rate of spread over PMMA slabs is unaffected (Figure 7.19(a), $Y_O = 0.233$), although Ito *et al*. (2005) reported a slight increase in the spread rate compared to still air conditions at very low air flows (<0.2 m/s). At higher air velocities the rate of spread is found to decrease and ultimately the flame extinguishes at approximately 1 m/s (in these experiments). However, if the oxygen concentration is increased significantly ($Y_O > 0.4$), the rate of flame spread is found to increase with oxidizer flowrate, ultimately to extinguish at much higher velocities (>10 m/s in pure oxygen). These observations are consistent with the data of Lastrina (1970) that were reported by Magee and McAlevy (1971) for oxygen concentrations >45%: this work did not include measurements in normal air.

---

[12] It should be noted that these results apply to flame spread on a horizontal flat surface and should not be assumed to apply to the types of open fuel bed discussed in Section 7.4.

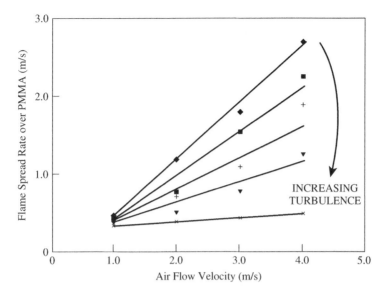

Figure 7.23   Variation of the rate of flame spread over PMMA slabs 76 mm wide and 300 mm long as a function of the concurrent air velocity at difference turbulence intensities. From Zhou and Fernandez-Pello (1990), by permission of the Combustion Institute

Fernandez-Pello *et al.* (1981) also examined the effect of a counter-current flow on the rate of flame spread over a thin fuel (filter paper) (Figure 7.19(b)). The results were very similar, although at elevated oxygen concentrations there was no increase in the spread rate, which either remained constant (as it did for $Y_O = 1$) or fell slightly until at high velocities the rate fell sharply towards extinction. It is suggested that the difference in behaviour between thick and thin fuels lies in the fact that for thin fuels heat transfer ahead of the flame can only occur through the gas phase and is therefore very sensitive to the opposed flow. On the other hand, conduction through the solid is the dominant mechanism for thick fuels and this is unaffected by the opposed flow.

An imposed counter-current flow will have two effects: (i) promoting mixing and combustion at the leading edge of the flame, generating higher temperatures and tending to increase the spread rate; and (ii) cooling the fuel ahead of the flame front (tending to decrease the rate). The latter will become increasingly important as the air velocity is increased.

Air movement is more common out of doors, and can have a dramatic effect on the development of forest and bush fires, i.e., in 'open' fuel beds. Experimental studies have made use of wood cribs and other open fuel bed configurations. This work is discussed briefly in Section 7.4, but is reviewed in depth by Pitts (1991).

7.3   Flame Spread Modelling

Early attempts at modelling flame spread were confined to analytical studies, casting the problem in terms of the heat transfer processes and then seeking solutions of the appropriate differential equations (de Ris, 1969; Quintiere, 1981). This gave valuable insight into the processes involved, but because the problem is so complex, some sweeping assumptions had to be made to allow the equations to be solved. However, these studies laid the groundwork for numerical simulations which became increasingly sophisticated as computational facilities became readily available. As an additional bonus, these numerical techniques are now providing the means by which data from tests such as the cone calorimeter may be used directly to model the fire spread process.

Williams (1977) provided an excellent review of contemporary knowledge of flame spread in which he attempted to unify all types of fire spread (premixed flames, smouldering, forest fires, etc.) by using the concept of the 'surface of fire inception'. This was first used by Fons (1946), Emmons (1965), Thomas *et al.* (1964) and others to describe mass fires, and defines the boundary between burning and non-burning fuel. The rate of heat transfer across this surface then determines the rate of fire spread. The 'fundamental equation of fire spread' is a simple energy conservation equation (Williams, 1977):

$$\rho V \Delta h = \dot{q} \tag{7.11}$$

where $\dot{q}$ is the rate of heat transfer across the surface, ρ is the fuel density, V is the rate of spread and Δh is the change in enthalpy as unit mass of fuel is raised from its initial temperature (T_0) to the temperature (T_i) corresponding to the 'firepoint'. If it is possible to identify $\dot{q}$ in a given fire spread situation, some insight can be gained into the factors that affect the rate of spread. Just such a model was developed by Parker (1972) to test the hypothesis that the rate of downward flame spread over a vertical piece of card (as a thermally thin, typical cellulosic fuel) was determined by the rate of conductive heat transfer through the gas phase from the leading edge of the flame to the unaffected fuel. He monitored the temperature at a point at the mid-plane of the card as the flame progressed downwards: the record of one temperature/time history is shown in Figure 7.24. Just below the advance edge of the visible flame the temperature was found to be only 110°C, thereafter rising steeply to 300°C, above which the cellulose begins to decompose,

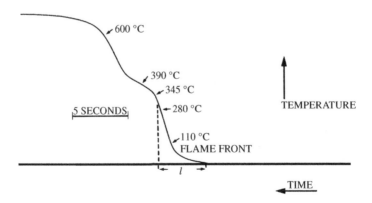

Figure 7.24 Temperature history of a point on a vertical card during downward flame propagation (Parker, 1972). Reproduced by permission of Technomic Publishing Co. Inc., Lancaster, PA 17 604, USA

producing volatiles. From the rate of temperature rise between 100 and 300°C, Parker calculated that the rate of heat transfer ahead of the flame was 20 kW/m².

Thus the flame appears to advance ahead of the pyrolysis zone: its leading edge is blue (cf. Figure 7.2) and stands about 1 mm off the surface. Parker (1972) derived the following simple expression for the rate of downward propagation (V) over white index cards, assuming that heat losses by radiation are negligible and that heat transfer to the fuel ahead of the burning zone is by gas phase conduction from the leading edge of the flame:

$$V = \frac{l}{t_p} = l \cdot \frac{k_g(T_f - T_0)}{d} \cdot \frac{2}{\rho c \tau (T_p - T_0)} \tag{7.12}$$

where:

t_p = time required to heat the surface to the pyrolysis temperature T_p;
l = length of the preheat zone (measured as 1.5 mm);
τ = thickness of the material (τ = 0.12 mm);
ρ = density of the material ($\rho\tau$ = 182 g/m²);
c = heat capacity of the material (c = 1.26 J/g.K);
T_p = threshold temperature for active pyrolysis (280°C);
T_f = temperature of the leading edge of the flame (1300°C);
T_0 = initial temperature (20°C);
k_g = thermal conductivity of air (2.5 × 10⁻² W/m.K at 20°C); and
d = 'standoff distance' (measured as 1 mm).

Although approached in a slightly different way, this equation has essentially the same form as Equation (7.11), where

$$\rho\Delta h = \rho c \tau (T_p - T_0) \tag{7.13}$$

and

$$\dot{q} = \frac{k_g(T_f - T_0)}{d} . l \tag{7.14}$$

where l is effectively the area per unit width of the material to which heat is transferred from the flame: the factor of two (in Equation (7.12)) takes account of the fact that flame spreads down both sides of the card simultaneously. Using the data given, the expression gives V = 1.6 mm/s, close to the observed value of 1.5 mm/s (Parker, 1972). Although this remarkable agreement must be regarded as fortuitous since several factors (k_g, ρ and c) are temperature-dependent, this result was taken as vindication of the original assumption regarding the mechanism of forward heat transfer. It has since been confirmed experimentally for downward spread over thin fuels generally (Hirano et al., 1974; Fernandez-Pello and Santoro, 1980). Note that Equation (7.12) predicts that $V \propto \tau^{-1}$ and $V \propto (T_p - T_0)^{-1}$, as found experimentally (Figure 7.14 and Equation (7.6)).

More sophisticated mathematical models of flame spread over horizontal surfaces have been developed (e.g., de Ris, 1969). Generally these give solutions for the rate of spread that are consistent with observation, but many assumptions need to be made to make the mathematics tractable and capable of yielding workable solutions. Thus, in his analysis of thermally thick fuels, de Ris (1969) found it necessary to ignore conduction through the solid as a means by which heat is transferred ahead of the flame front, although it has since been shown that this is the principal mechanism (Fernandez-Pello and Williams, 1974; Fernandez-Pello and Santoro, 1980). The difficulty arises because many of the factors which are important in determining the rate of spread are interdependent and cannot be uncoupled. If the gas phase flame kinetics are fast compared with the rate at which the surface responds to heating, then the flame spread mechanism is dominated by heat transfer from the flame to the surface ahead of the burning zone. However, if the flame reactions are slow – e.g., at low oxygen concentrations – the flame kinetics have to be taken into account. This has been shown clearly by Fernandez-Pello and co-workers for counter-current spread (Fernandez-Pello, 1995). Accurate prediction of flame spread rates cannot be obtained by analytical solutions to the governing equations, but considerable success has been achieved through numerical modelling. However, in all cases, significant assumptions still have to be made.

Involvement of combustible wall lining materials in a fire is recognized as a particularly hazardous scenario. Upward flame spread on a vertical surface is known to be rapid, and for this reason there has been great interest in developing a model which could be used to calculate the rate of fire development. Theoretical treatments have been explored by Markstein and de Ris (1972), Orloff *et al.* (1974) and Williams (1982), but while they assist in our understanding of the processes involved, they cannot be used directly in practical applications, *per se*. There are too many unknown factors and it is necessary to develop a semi-empirical approach which can incorporate relevant data relating to fire performance. A suitable numerical model involves analysing flame spread in terms of a series of discrete steps that are defined in terms of the rate at which material ahead of the burning zone is heated to a temperature at which it will support flame (i.e., the firepoint). A simplified illustration of the model used by Karlsson and others (Karlsson, 1993) is given in Figure 7.25. It has been shown that for vertical flame spread, the material above

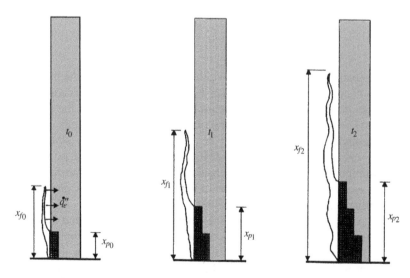

Figure 7.25 Simple model for upward flame spread, without burnout. The heat flux from the flame is assumed to be constant in the region $x_f < x < x_p$ (Karlsson, 1993). Grant and Drysdale (1995), by permission

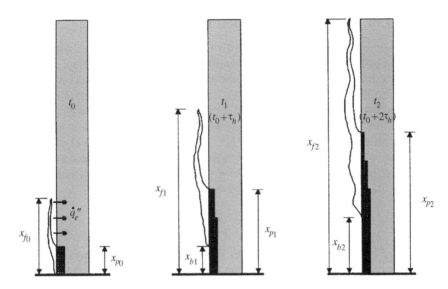

Figure 7.26 Simple model for upward flame spread, with burnout, where x_b is the burnout height (cf. Figure 7.25). Grant and Drysdale (1995), by permission

the burning zone is exposed to a heat flux of c. 25 kW/m² (Quintiere *et al.*, 1986), at least in the early stages of fire development when the flames are relatively thin. The flame height $((x_{fo} - x_{po})$ in Figure 7.25) is taken as a function of the rate of heat release per unit width ($\dot{Q}'_c$) over the height x_{po}:

$$x_{fo}(t) - x_{po}(t) = K(\dot{Q}'_c(t))^n \tag{7.15}$$

where K and n are empirical constants (Tu and Quintiere (1991) give $K = 0.067$ and $n = 2/3$ for x in m and $\dot{Q}'_c$ in kW/m – see Equation (4.50)). By incorporating data on time to ignition and rate of heat release from measurements in the cone calorimeter, the rate of development of a wall fire may be calculated (e.g., Magnusson and Sundstrom, 1985; Saito *et al.*, 1985; Hasemi *et al.*, 1991; Karlsson, 1993). The inclusion of burnout of the fuel is more difficult to incorporate (Figure 7.26), but has been addressed by Grant and Drysdale (1995) to model early fire growth on a vertical sheet of cardboard using data files from the cone calorimeter directly.

The value of this type of model is in providing a means by which the results of small-scale tests can be used to assess full-scale fire behaviour. At the present time, it is considered necessary to subject wall lining materials to large-scale 'standard' tests such as ISO 9705 (2002b), which are much more expensive than existing 'bench-scale' tests. For this reason, the models are being developed with a view to modelling large-scale test behaviour, particularly the rate of growth of the rate of heat release (e.g., Magnusson and Sundstrom, 1985; Karlsson, 1993). This approach is discussed in general terms by Cleary and Quintiere (1991) (see also Section 9.2.2).

7.4 Spread of Flame through Open Fuel Beds

Previous sections have dealt specifically with the spread of flame over continuous fuel surfaces, but the behaviour of 'open' or 'discontinuous' fuel beds is extremely important in the context of bush and forest fires. Most of the research into this topic has been carried out by the Forest Fire Research Community (Pitts, 1991; Albini, 1997), although experimental studies using wood cribs (Section 5.2.2(b)) provided many of the underlying principles. The mechanism of propagation has been shown to involve radiative heat transfer ahead of the flame front (Thomas *et al*. (1964, 1965) for wood cribs; Wotton *et al*. (1999) and Morandini *et al*. (2001) for forest litter). Under still air conditions, the advancing front adopts a vertical orientation and, in agreement with Equation (7.11), the rate of spread is inversely proportional to the bulk density of the fuel bed:

$$V = \frac{\dot{q}}{\rho \Delta h} \qquad (7.11)$$

where $\dot{q} = \varepsilon \sigma (T_F^4 - T_0^4)$. This equation does not incorporate all the factors (such as (*inter alia*) moisture content, relative humidity and orientation) that affect the rate of flame spread over (or through) natural fuels but it is relevant, particularly if the individual fuel elements (e.g., twigs, leaves, etc.) are thermally thin. If this is not the case, ρ must be replaced by an effective density which takes into account the fact that only the surface layer of each element (Equation (7.3)) will be heated before the arrival of the flame front (Thomas, 1974a,b; Williams, 1982) (cf. Table 6.7). General conclusions from these studies are compatible with many field observations. Thus, in a forest, fire will spread most rapidly through the brush at ground level, or through the tree crown where the bulk density is low and the fuel elements are thin (Emmons, 1965; Albini, 1997). In an extensive set of experiments in a large wind tunnel using layers of various types of forest litter, Catchpole *et al*. (1998) confirmed that flame spread most rapidly over lightly packed fine fuels (high surface to volume ratio). They demonstrated that the rate of spread was reduced by increased moisture content and was increased by an imposed airflow. There was an approximate linear dependence between the rate of spread and the windspeed. Very high rates are possible under drought conditions when moisture content will be low. For example, the catastrophic bush fires of 7 February 2009 in Victoria, Australia, occurred after a prolonged drought and were driven by strong northerly winds when the temperature was over 40°C.

Experience has shown that the topography of the land can have a profound effect on the way in which wildfire will spread, whether as a surface fire (grassland or forest litter) or through scrub or the forest crown. Such fires spread more rapidly if they are travelling uphill than they would on level ground, or downhill. In experiments using 1 m wide beds of forest litter, Morandini *et al*. (2001) found that the rate of spread up a slope of 10° was about 25% greater than on the horizontal and became more pronounced as the angle was increased. Similar observations were made by Macmillan and Drysdale (1992) on flame spread over a solid surface (Figure 7.10, curve a), but the samples were only 0.06 m wide and the influence of slope was not observed until the angle was greater than 20°. The effect was observed at significantly smaller angles when sidewalls were added to prevent lateral entrainment of air (Figure 7.10, curve b). This was investigated using CFD: Woodburn found that the flows associated with flame spread within a 'trench' switched from counter-current to concurrent at a critical angle of 19° (Woodburn and Drysdale, 1997) while for an infinitely wide 'trench' – effectively an infinitely wide flame front – the critical angle was only 10°.

When the terrain is complex, wildland fires occasionally propagate without warning at unusual speeds, trapping firefighters and causing multiple fatalities. This type of 'eruptive' propagation is well documented (Viegas, 2006): one of the worst incidents occurred in Colorado, USA, in 1994 and involved the deaths of 14 firefighters (Butler *et al.*, 1998). Although there is no general consensus on the mechanisms that lead to an eruptive fire, it appears that hillside canyon-like configurations are particularly hazardous (Viegas and Pita, 2004). However, other factors also need to be considered, such as humidity, ambient temperature, the direction and strength of the wind and the transportation of 'firebrands', which can cause secondary ignitions at some distance from the main fire (see, for example, Manzello *et al.* (2007, 2008)). Models are being developed to predict how and where such fires will spread on complex terrain so that they may be used for decision-making during firefighting operations.

7.5   Applications

As fire growth proceeds by flame spread from the point of ignition, it would be valuable to be able to analyse any potential fire spread situation and quantify the rate of development. In this context, much of the information in this chapter is essentially qualitative but two aspects merit closer examination, namely spread under the influence of radiation and upward spread on a vertical surface. A limited amount of quantification is possible with each.

7.5.1   Radiation-enhanced Flame Spread

The enhancement of flame spread by radiant heat is very significant and is shown clearly in Figure 7.18(b). Its relevance to fire development in compartments is discussed in Section 9.2 where it is argued that when the heat flux at floor level reaches 20 kW/m^2 there is an onset of a very rapid change in the character of the fire, from localized burning to full room involvement, when all combustible items are burning. This transition is known as flashover (Section 9.1) and requires that the fire has grown beyond a certain minimum size, mainly by flame spread (Thomas, 1981).

Unfortunately, there is little quantitative information available on the dependence of flame spread rate on radiant heat flux. A few data on the rate of spread over thermally thick materials in the absence of supporting radiation are given in Table 7.3 for horizontal fuel beds. Some materials will spread flame only if supported by a radiant flux: the values of the minimum flux necessary for flame spread ($\dot{q}''_{0,s}$) obtained by Quintiere (1981) using the ISO Surface Spread of Flame Apparatus (Figure 7.19) are recorded in Table 7.5. These refer to horizontal spread along a vertical surface and would not apply to thicknesses and orientations where different heat loss conditions hold. Similar data on the spread of flame over carpets obtained using NFPA 253 (National Fire Protection Association, 1995) are also given in Table 7.6. These show clearly that an underlay reduces $\dot{q}''_{0,s}$, which can be attributed to the reduced heat losses through the fuel bed. Information of this kind can be used to assess how far fire would spread away from a source of radiant heat over a combustible surface (e.g., carpets (Quintiere, 1975b) and wall lining materials in corridors).

On exposure to radiant (or convective) heat, the surface temperature of a combustible solid will increase, eventually to reach a steady state (equilibrium) value. This can be estimated on the basis of simple heat transfer calculations, assuming the material to be inert, and taking heat losses into account. Equations (7.6) and (7.7) could then be used to estimate the limiting rate of flame spread (V_2) under a given heat flux, provided that we know T_p (the pyrolysis temperature) and the flame spread rate V_1 at a single fuel temperature T_1. Thus, for a thick fuel:

$$V_2 = V_1 \frac{(T_p - T_1)^2}{(T_p - T_2)^2} \tag{7.16}$$

where T_2 – the equilibrium surface temperature reached when exposed to the heat flux – must first be calculated. The limits of applicability of this method are not known but it will break down unless the fuel surface temperature is considerably less than T_p.

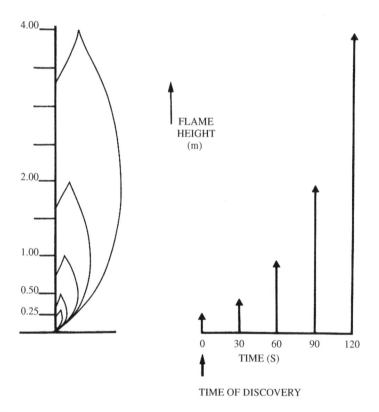

Figure 7.27 Illustration of the exponential growth in flame height for vertical spread. Doubling time 30 s. After Alpert and Ward (1984), with permission

7.5.2   Rate of Vertical Spread

Alpert and Ward (1984) have pointed out that as the spread of flame up a vertical surface accelerates exponentially, the rate of spread from an initial (small) fire can be estimated if the time it takes for the fire to double in size is known. This can be obtained simply from relatively small-scale laboratory tests, and provides a measure of the hazard of a vertical surface. If it is found that the doubling time for a particular material is 30 s and a fire is discovered at the base of a 4 m high vertical wall of the same material when the flame is 0.25 m high, the flame will be 0.50 m high at 1 min and will reach the top of the wall at $2\frac{1}{2}$ min (Figure 7.27). This type of information may be used to assess the risks associated with storage of goods in high bay warehouses and to develop fire protection strategies based on early detection of an outbreak of fire.

Problems

7.1   The rate of downward spread of flame on a vertical piece of card 0.4 mm thick is found to be 1.0 mm/s. What rate of spread would be expected for a card 0.7 mm thick if it is made from the same material and held in the same configuration?

7.2   Flame was found to spread down one face of a large vertical sheet of PMMA, 6 mm thick, at a rate of 0.05 mm/s when the initial temperature was 20°C. What would the equilibrium spread rate be if that face was exposed to a uniform radiant flux of 4 kW/m^2? Assume that the surface has an emissivity of 0.5 and that the convective heat transfer coefficient is $h = 12$ W/m^2.K. Ignore radiative heat loss for the calculation, but check subsequently if this is justified. Take the ambient air temperature to be 20°C and the 'firepoint' temperature to be 280°C.

7.3   Taking the example in Problem 7.2, what would the rate of downward spread be if the rear face was maintained constant at 20°C during exposure to the same radiant flux?

7.4   Timber flooring will not burn and spread flame unless it is exposed to (radiant) heating from elsewhere. The minimum level is of the order of 4 kW/m^2. Figure 9.4 shows the results of a series of experiments carried out at Fire Research Station, Borehamwood, which show the radiant heat flux at floor level produced by a crib fire at the end of a corridor – with and without the ceiling in place. Estimate roughly how far flame would spread along a wooden floor for the three situations illustrated in the diagram.

8

Spontaneous Ignition within Solids and Smouldering Combustion

The phenomenon of spontaneous ignition in vapour/air mixtures was discussed in some detail in Section 6.1 in relation to gaseous, liquid and solid fuels. However, certain combustible solids can ignite as a result of internal heating which arises spontaneously if there is an exothermic process liberating heat faster than it can be lost to the surroundings. There appear to be two main factors necessary for this type of spontaneous combustion to occur. First, the material must be sufficiently porous to allow air (oxygen) to permeate throughout the mass and second, it must yield a rigid char when undergoing thermal decomposition. The phenomenon is normally associated with relatively large masses of material and its principal characteristic is that combustion begins deep inside the material where the effects of self-heating are greatest (Figure 6.4(b)). It will give rise to smouldering combustion, which will slowly propagate outwards.

Smouldering involves surface oxidation of the char, which provides the heat necessary to cause thermal degradation of the contiguous combustible material. Successful propagation requires that volatiles are progressively driven out away from the zone of active combustion to expose fresh char which will then begin to burn. This is described in more detail in Section 8.2. However, the chapter begins with a brief discussion of spontaneous combustion and how it may be quantified by the application of thermal explosion theory. The subject is covered in depth by Gray and Lee (1967) and Bowes (1984), and is reviewed by Beever (1995) and Gray (2008).

8.1 Spontaneous Ignition in Bulk Solids

The principal characteristic of spontaneous ignition in a bulk solid is that combustion starts as a smouldering reaction within the material and propagates slowly outwards. The fire is initially deep-seated, and although it may lead to flaming combustion when it breaks through to the surface, it leaves behind evidence of prolonged burning at depth. The phenomenon can be described by thermal explosion theory in the form developed by Frank-Kamenetskii (1939) (Section 6.1). Heat must be generated within the mass by an exothermic process, in this context a heterogeneous surface oxidation within the interstices of a porous combustible material, although other sources are observed (e.g., decomposition of endothermic compounds (cf. Figure 6.6) and microbiological heating (Section 8.1.5)). If the energy is not removed as rapidly as it is released, then the material will self-heat, perhaps leading to thermal runaway and smouldering ignition if the conditions are right. One prerequisite is that the mass of material must have a low surface to volume ratio. A list of materials which are known to be subject to spontaneous heating is given in the *NFPA Handbook* (2008), from which Table 8.1 has been derived. However, although some precautions against spontaneous heating are given, there is very limited information on the conditions under which spontaneous ignition might occur. The list includes reactive oils, such as tung and linseed, which present a hazard when absorbed onto fibrous combustible materials such as cotton rags and waste. The oil is dispersed on a rigid porous structure, thereby presenting a large surface area at which oxidation can take place. A similar situation is found in chemical plant where thermal insulation, or lagging, is applied to pipes and vessels carrying hot process fluids. Leakage from a faulty flange or valve into the lagging can lead to a 'lagging fire', even if the insulation material is non-combustible. This phenomenon has been examined in some detail by Bowes (1974), Gugan (1976) and McIntosh *et al.* (1994) (Section 8.1.4).

The main difference between spontaneous ignition within a combustible solid and that of a flammable vapour/air mixture is that the latter is a premixed system. With a porous solid, oxygen must diffuse into the material to maintain the oxidation reactions on the surface of the individual particles or fibres wherever oxidation is occurring. The rate of development of self-heating which precedes ignition is slow and will be significant only if the insulation provided by the surrounding mass of material is sufficient. Consequently, the delays or 'induction periods' for spontaneous ignition of bulk solids tend to be many orders of magnitude greater than those associated with the auto-ignition of gases (Section 6.1 and Figure 6.3). On the other hand, in practical situations the risk of spontaneous ignition in bulk solids exists at temperatures much lower than typical auto-ignition temperatures quoted for gases and vapours (Table 3.1). If the accumulation of material is large enough, spontaneous combustion can occur at relatively low temperatures, perhaps as low as ambient (Bowes, 1984) (see Section 8.1.1). For example, there can be substantial losses of stocks of coal by this process if very large accumulations are permitted to lie undisturbed for long periods: the induction period is likely to be on the order of weeks. Practical aspects of this problem are discussed by Dungan (2008).

An excellent review of the methods that may be used to study the propensity of combustible solids to undergo self-heating and spontaneous combustion is given by Wang *et al.* (2006). However, only one of these methods will be discussed here – that based on the work of Frank-Kamenetskii, which was first applied to the spontaneous ignition of flammable vapour/air mixtures in 1969 (Section 6.1).

8.1.1 Application of the Frank-Kamenetskii Model

It is possible to use the results obtained from the Frank-Kamenetskii model (Equation (6.14)) to investigate the spontaneous heating characteristics of different materials:

$$\ln\left(\frac{\delta_{cr}T_{a,cr}^2}{r_0^2}\right) = \ln\left(\frac{E_A \Delta H_c A C_i^n}{k \cdot R}\right) - \frac{E_A}{R T_{a,cr}} \tag{8.1}$$

The relationship between $T_{a,cr}$ (the critical ignition temperature) and r_0 (the characteristic dimension of the sample) for a given geometry of material may be determined experimentally. Thus, cubical samples of the material can be prepared and exposed to constant elevated temperatures in a thermostatically controlled oven, monitoring the temperature at the centre of the sample by means of a thermocouple. In this way, the extent to which a sample of given size will tend to self-heat or ignite at different temperatures can be determined. The value of $T_{a,cr}$ is obtained for each cube size (side $= 2r_0$) by a process of trial-and-error and 'bracketing', an example of which is shown in Figure 8.1. Once $T_{a,cr}$ has been determined for several cube sizes, then using $\delta_{cr} = 2.52$ (Table 6.1), the data may be plotted in the form $\ln(\delta_{cr} T_{a,cr}^2/r_0^2)$ versus $1/T_{a,cr}$, as suggested by Equation (8.1). This has been done in Figure 8.2 with data from various sources for samples of wood fibre insulation board (FIB) in the form of cubes ($\delta_{cr} = 2.52$), slabs ($\delta_{cr} = 0.88$) and octagonal piles ($\delta_{cr} = 2.65$) (Thomas and Bowes, 1961a; Thomas, 1972a).

For experiments of this type, corrections to δ_{cr} may be necessary if the assumptions used in the original derivation of Equation (8.1) do not apply, for example, if the Biot number (Bi $= hr_0/k$) is less than about 10 (see Section 8.1.2). This will not be necessary for FIB unless $r_0 < 0.05\,$m as the thermal conductivity (k) of this material is very low ($0.041\,$W/m·K; Table 2.1). The linear correlation shown in Figure 8.2 suggests that the Frank-Kamenetskii model is a good approximation for this material within the range of temperatures covered. It may be used tentatively to predict spontaneous ignition behaviour outside this range, provided that the extrapolation is not too long. However, in addition to the assumptions introduced into the derivation of Equation (8.1) (Section 6.1), it is implicit that there is no change in the mechanism of heat production within the temperature range

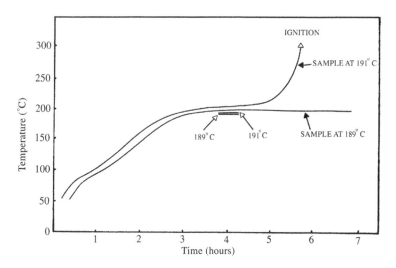

Figure 8.1 Determination of the critical temperature for 50 mm cubes of ground grain. Ignition occurred after 5 h at $T_a = 191°$C, but there was no ignition at $T_a = 189°$C. Therefore, $T_{a,cr} = 190°$C. (Curves displaced for clarity)

Table 8.1 Some materials subject to spontaneous heating[a]

Name	Tendency to spontaneous heating	Usual shipping container or storage method	Precautions against spontaneous heating	Remarks
Charcoal	High	Bulk, bags	Keep dry. Supply ventilation	Hardwood charcoal must be carefully prepared and aged. Avoid wetting and subsequent drying
Coal, bituminous	Moderate	Bulk	Store in small piles. Avoid high temperatures	Tendency to heat depends on origin and nature of coals. Highly volatile coals are particularly liable to heat
Distillers' dried grains with oil content (brewing grains)	Moderate	Bulk	Maintain moisture at 7–10%. Cool below 38°C before storage	Very dangerous if moisture content is 5% or lower
Fish meal	High	Bags, bulk	Keep moisture at 6–12%. Avoid exposure to heat	Dangerous if overdried or packaged over 38°C
Foam rubber in consumer products	Moderate		Where possible remove foam rubber pads, etc. from garments to be dried in dryers or over heaters. If garments containing foam rubber parts have been artificially dried, they should be thoroughly cooled before being piled, bundled or put away. Keep heating pads, hair dryers, other heat sources from contact with foam rubber pillows, etc.	Foam rubber may continue to heat spontaneously after being subjected to forced drying as in home or commercial dryers and after contact with heating pads and other heat sources. Natural drying does not cause spontaneous heating
Grain (various kinds)	Very slight	Bulk, bags	Avoid moisture extremes	Ground grains may heat if wet and warm

Material	Tendency	Form	Precautions	Remarks
Hay	Moderate	Bulk, bags	Keep dry and cool	Wet or improperly cured hay is almost certain to heat in hot weather. Baled hay seldom heats dangerously
Linseed oil	High	Tank cars, drums, cans, glass	Avoid contact of leakage from containers with rags, cotton or other fibrous combustible materials	Rags or fabric impregnated with this oil are extremely dangerous. Avoid piles, etc. Store in closed containers, preferably metal
Manure	Moderate	Bulk	Avoid extremes of low or high moisture contents. Ventilate the piles	Avoid storing or loading uncooled manures
Oiled rags	High	Bales	Avoid storing in bulk in open	Dangerous if wet with drying oil
Rags	Variable	Bales	Avoid contamination with drying oils. Avoid charring. Keep cool and dry	Tendency depends on previous use of rags. Partially burned or charred rags are dangerous
Sawdust	Possible	Bulk	Avoid contact with drying oils. Avoid hot, humid storage	Partially burned or charred sawdust may be dangerous
Wool wastes	Moderate	Bulk, bales, etc.	Keep cool and ventilated or store in closed containers. Avoid high moisture	Most wool wastes contain oil, etc. from the weaving and spinning and are liable to heat in storage. Wet wool wastes are very liable to spontaneous heating and possible ignition

[a]Extracted from Table 6.17.11 of the *NFPA Handbook*, 20th edn. Copyright © 2008 National Fire Protection Association, Quincy, MA. Reprinted with permission

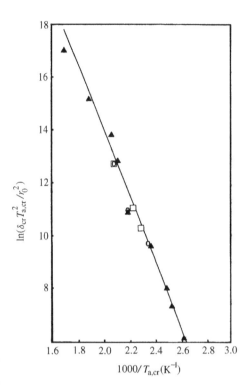

Figure 8.2 Correlation of spontaneous ignition data for wood fibre insulating board according to Equation (8.1) (Thomas and Bowes, 1961a). ▲, Mitchell (1951); o (cube) and □ (slab), Thomas and Bowes (1961a). Reproduced by permission of The Controller, HMSO. © Copyright

studied (see Section 8.1.4). If an extrapolation outside this range is to be attempted, one must have confidence in the assumption that the mechanism is unchanged.[1]

Spontaneous ignition is not instantaneous, but exhibits a delay, or induction period, τ' (Figure 6.3). However, whereas the induction period for auto-ignition of a flammable vapour/air mixture is seldom longer than 1 s (Section 6.1), that for a bulk solid may be hours, days or even weeks, depending on the mass of the stored material and the ambient temperature. The larger the mass of material, the lower the temperature that will lead to spontaneous ignition. This is illustrated by data of Bowes and Cameron (1971) on the spontaneous ignition of chemically active carbon (Table 8.2). This work was part of an investigation of fires involving substantial cargoes of active carbon (4–14 tonnes) stored in the holds of ships passing through the tropics. They established that spontaneous ignition was indeed the cause, and that the delay before the discovery of the fire (3–4 weeks) was compatible with, although slightly less than, an extrapolation of the small-scale experimental data to the ambient temperature relevant to the cargo spaces (29–38°C) (Figure 8.3). It was subsequently recommended that this material be baled

---

[1] An example in which this does not hold is the exothermic decomposition of hydrated calcium hypochlorite. Extrapolation from measurements made at high temperatures (>400 K (127°C)) will underestimate the risk at low temperatures (e.g., 50°C) by overestimating the value of r_0 (Gray and Halliburton, 2000).

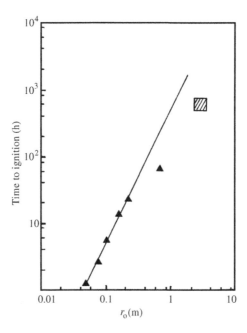

Figure 8.3   Time to spontaneous ignition of cubes of chemically activated carbon. ▲, Data from Table 8.2. Hatched area represents the times to discovery of fires on board ship involving this material (Bowes and Cameron, 1971). Reproduced by permission of The Controller, HMSO. © Crown Copyright

Table 8.2   Ignition of cubes of chemically activated carbon (Bowes and Cameron, 1971)

Linear dimension $(2r_0/mm)$	Critical temperature $(T_{a,cr}/°C)$	Time to ignition (τ/h)
51	125	1.3
76	113	2.7
102	110	5.6
152	99	14
204	90	24
601	60	68

for transportation, with a 400 gauge polyethylene liner incorporated into the sacking to prevent the ingress of oxygen (Board of Trade, 1966; Bowes and Cameron, 1971).

Another group of materials that is prone to self-heating and spontaneous combustion is coal, particularly sub-bituminous and low-grade coals (McIntosh and Tolputt, 1990; Carras and Young, 1994). It is accepted practice that when coal has to be stockpiled, large accumulations should be avoided and piles should not be left undisturbed for any significant length of time.[2] The important factor is the ratio of surface area to volume

---

[2] This practice may break down at times of industrial action when coal stocks are not being regularly moved.

(A_s/V): this should be maximized to ensure that heat losses are large enough to prevent self-heating developing into active combustion. Thus, instead of one large pile of coal, the practice is to stockpile the coal in several smaller piles.[3]

8.1.2   The Thomas Model

The Frank-Kamenetskii model assumes that heat loss from the bulk of the material is restricted only by conduction through the solid (Section 6.1) – i.e., that the temperature of the surface and the ambient atmosphere are equal, as illustrated in Figure 8.4(a) (see also Figure 6.4(b)). This will hold provided that the Biot number is large, i.e., Bi > 10. For very small values of Bi (i.e., Bi → 0), the Semenov model will be applicable (Section 6.1), but between these limits conduction through the solid and convective heat loss from the surface must both be considered. Thus the surface boundary conditions, Equations (6.9a) and (6.9c), are replaced by:

$$h(T_s - T_a) = k(dT/dr) \qquad (8.2)$$

This is illustrated schematically in Figure 8.4(b). Provided that the heat of combustion (ΔH_c) is large, δ_{cr} becomes a function of Bi, as shown in Figure 8.5 (Thomas, 1960, 1972a). In small-scale experiments, Bi may be sufficiently low to make it necessary to use corrected values of δ_{cr}, but otherwise application of the method is identical to that described in the previous section (Thomas and Bowes, 1961a).

Thomas (1973) developed an approximate theory of 'hot spot criticality' to analyse the behaviour of a large mass of material within which there is a small, hot region. Anthony and Greaney (1979) applied the method to the problem encountered when a large quantity

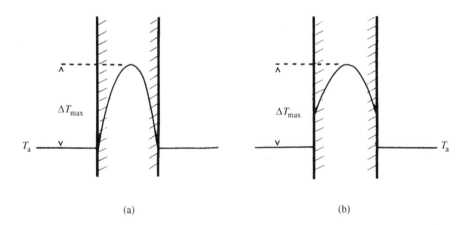

(a)   (b)

Figure 8.4   Temperature profiles inside spontaneously heating materials according to the models of (a) Frank-Kamenetskii (1939) and (b) Thomas (1960)

---

[3] For example: 27 cubic metres of material may be stored as a single cube, 3 m × 3 m × 3 m, with $A_s/V = 54/27 = 2$. If it is stored as 27 cubes, each 1 m × 1 m × 1 m, then $A_s/V = (27 \times 6)/27 = 6$, a three-fold increase in the surface to volume ratio.

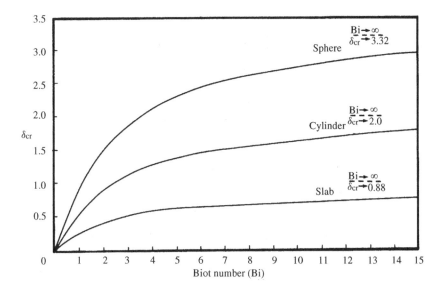

Figure 8.5 δ_{cr} as a function of Biot number for a sphere, a cylinder and a slab (Thomas, 1958). Reproduced by permission of The Royal Society of Chemistry

of hot material is stored in a cool environment (e.g., hot, dry sheets of fibre insulation board stored directly from the production line (Back, 1981/82) or hot linen removed from industrial tumble driers and placed in large baskets, as is common practice in hospital laundries (Perdue, 1956)). Fires resulting from this type of incident can be avoided by cooling the material before it is stored.

The concept of 'hot spot criticality' was mentioned in Section 6.1: given some mechanism by which a local temperature rise occurs, under what circumstances will ignition follow? In the case of a flammable vapour/air mixture, the energy source could be a mechanical spark, although for liquid explosives adiabatic heating of minute trapped air bubbles may be sufficient to cause ignition if the liquid is subjected to a shock. The 'hot spot' will cool by conductive losses to the surrounding material but if the rate of heat production within the affected volume exceeds the losses, then runaway reaction will occur. A good example of this is the situation in which a live electric light fitting is buried in a grain silo. It could act as the source of energy to initiate smouldering at depth and the outcome could be mistaken for 'true' spontaneous ignition on the basis of a cursory examination of the evidence. Another example of 'hot spot' ignition is encountered in wildland fires when hot embers are carried on thermal currents for some distance beyond the main fire perimeter. Many will cool harmlessly, but secondary fires can result: this is known as 'spotting' and represents a serious fire spread hazard (Albini, 1983; Manzello *et al.* 2007).

8.1.3 Ignition of Dust Layers

It is known that a layer of a combustible dust lying on a hot surface has the potential to ignite 'spontaneously'. Bowes and Townsend (1962) have examined this in detail and have shown that the Frank-Kamenetskii model may be modified for this situation. They carried out experiments in which layers of sawdust of various thicknesses were heated on a hot plate (Figure 8.6(a)) and determined the minimum plate temperatures which would cause ignition of each thickness by the process of 'bracketing'. Figure 8.6(b) shows the result of one of their tests in which the temperature of sample (beech sawdust) was monitored 10 mm above the hot surface by means of a thermocouple (cf. Figure 8.6(b) with Figure 8.1). They found that the tendency to ignite was insensitive to particle size in the range of sieve fractions 18–120 BS (~853–124 μm particle diameter) and that packing density had a significant effect only with the thinnest layers (~5 mm). A more recent study of the ignition of dust layers of coal, oil shale, corn starch, grain and brass powder is reported by Miron and Lazzara (1988). This was carried out as part on an investigation into the development of a standard 'hot plate test' for dust layers and although the study was limited, their observations are in general consistent with the work by Bowes and Townsend (1962). They found that the minimum hot surface temperature for ignition was higher for 'coarse' dusts than for 'fine' (e.g., the minimum temperatures for ignition of bituminous coal of mass median diameters of 330 μm and 61 μm were >380°C and 300°C, respectively).

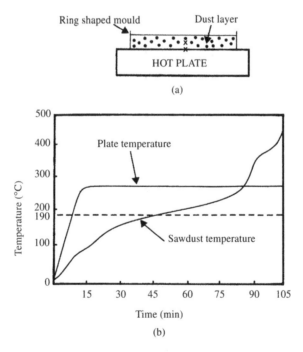

Figure 8.6 (a) Schematic diagram of the apparatus used to determine critical ignition temperatures for dust layers. Thermocouples located at ×: the mould was removed before heating the plate to a preselected temperature. (b) Ignition of a 25 mm thick layer of beech sawdust on a surface at 275°C. The sawdust temperature refers to a point 10 mm above the centre of the plate. If the sawdust was chemically inert, the equilibrium temperature at this location would be 190°C (Bowes and Townsend, 1962). Reproduced by permission of The Controller, HMSO. © Crown Copyright

The layer may be treated as an infinite slab, of thickness $2r_0$, resting on a hot surface, temperature T_{pl} at $r = 0$. The problem can be reduced to the same form as Equation (6.12), i.e.

$$\frac{d^2\theta}{dz^2} = -\delta \exp(\theta) \tag{8.3}$$

where $\theta = (T - T_{pl})E_A/RT_{pl}^2$ and $z = r/r_0$, but the boundary conditions are more complex as a result of non-symmetrical heating. These are:

$$\theta = 0 \text{ at } z = 0 \tag{8.3a}$$

and

$$-\frac{d\theta}{dz} = \frac{hr_0}{k}(\theta_s - \theta_0) \text{ at } z = 2 \tag{8.3b}$$

where $hr_0/k = \text{Bi}$. At the upper surface of the layer (i.e., at $r = 2r_0$), the temperature is $T = T_0$. As in the case of symmetrical heating, solutions to Equation (8.3) exist only for $\delta < \delta_{cr}$. Consequently, δ_{cr} must be determined to enable critical states to be identified.

Thomas and Bowes (1961b) showed δ_{cr} to be a function of both Bi and θ_0, as indicated in Figure 8.7, both of which must be evaluated. Bi can be obtained from the properties of the dust layer and the heat transfer conditions at the surface, but θ_0 can only be calculated if E_A (the activation energy) is known. This may be derived from the gradient of a plot of $\log(\delta_{cr} \cdot T_{pl}^2/r_0^2)$ vs. $1/T_{pl}$ (cf. Figure 8.2), but initially an assumed, trial value must be

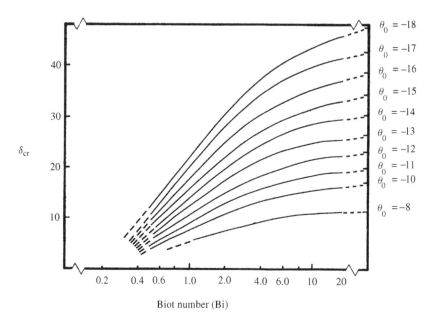

Figure 8.7 δ_{cr} as a function of Biot number and θ_0 for layers, or slabs heated on one face (Thomas and Bowes, 1961b). Reproduced by permission of The Royal Society of Chemistry

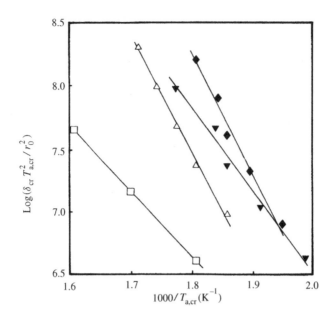

Figure 8.8 Correlation of spontaneous ignition data for dust layers on a hot surface according to Equation (8.1): ⊏, cork; △, provender; ◆, cocoa; ▼, cotton seed hull 'pepper' (Bowes and Townsend, 1962). Reproduced by permission of The Controller, HMSO. © Crown Copyright

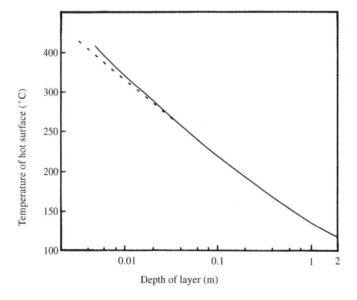

Figure 8.9 Extrapolation of results on spontaneous ignition of layers of sawdust on hot surfaces. _____, Mixed hardwood; _ _ _ _ _, beech (Bowes and Townsend, 1962). Reproduced by permission of The Controller, HMSO. © Crown Copyright

Table 8.3 Range of iodine values, taken from Bowes (1984)

Oil	Iodine number	Hazard
Tung, linseed	160–185	Most hazardous
Sunflower seed	127–136	
Soya bean	124–133	
Cottonseed	103–111	
Olive oil	80–85	
Coconut oil	8–10	Least hazardous

used. The procedure can then be repeated using the 'improved' value of E_A obtained from the plot based on the trial value. The success of this approach may be judged by the data shown in Figure 8.8. Bowes and Townsend (1962) use this method to extrapolate their small-scale data on wood sawdust to substantial thicknesses (Figure 8.9).

8.1.4 Ignition of Oil – Soaked Porous Substrates

It is possible for liquid fuels with high boiling points to undergo a form of smouldering if dispersed in a rigid, porous substrate. If it is dispersed as a thin film within the substrate, then provided that it is of low volatility it may undergo a slow exothermic oxidation that leaves behind a char-like residue on the internal surfaces. If there is sufficient insulation, self-heating will occur and a form of smouldering combustion will be initiated. One of the best-known examples involves the so-called 'drying oils', such as linseed and tung oils. These are unsaturated vegetable oils containing a high proportion of carbon–carbon double bonds which are reactive and prone to oxidation (Section 1.1.1). The proportion of these unsaturated linkages can be quantified in terms of the 'iodine number', which is proportional to the percentage of iodine absorption due to the reaction of iodine at the carbon–carbon double bonds. Oils with high iodine numbers have a greater propensity to self-heat, e.g., when dispersed on rags, etc. Typical ranges of values are given in Table 8.3.

The most hazardous oils can self-heat when dispersed onto quite small amounts of material (25 g has been reported). The less reactive oils will present problems if dispersed on much larger quantities of porous material, such as bales of wool (see Bowes, 1984). The chemical mechanism has not been studied in detail, but the consequence appears to be that char is produced on the fibres which can then oxidise more vigorously, producing a smoulder which has the potential to undergo transition to flaming (Section 8.2.2). The smoulder will involve the substrate if it is combustible (e.g., cotton rags), but it is not necessary for the substrate to be combustible.

Spontaneous combustion is also encountered with oil-soaked lagging (Bowes, 1974; McIntosh *et al.*, 1994). This type of fire involves the insulation ('lagging') which is used to reduce heat losses from pipes carrying hot process fluids in chemical plants and refineries. A model similar to that developed by Bowes for the ignition of dust layers has been applied, based on the geometry of the infinite cylinder. The lagging material is usually non-combustible, but a slow leak of combustible liquid will be absorbed by the insulation, and spontaneous heating and ignition will occur under certain conditions, which include the following:

(i) the liquid is insufficiently volatile to evaporate too rapidly;
(ii) the lagging is sufficiently porous to allow oxygen to diffuse to the surface of the absorbed liquid;
(iii) the leak is not so rapid as to fill the pores of the lagging material and thereby exclude oxygen from the high temperature region.

These conditions have been examined quantitatively by Brindley *et al.* (1998), who observed that very little oxygen is consumed during the early stages of self-heating, suggesting that even when the vapour pressure of the liquid is relatively high, thereby displacing some of the air within the porous lagging, there may still be enough oxygen present to support the self-heating process. These fires can develop undetected and remain so for a considerable period of time, perhaps becoming noticeable only when the leak increases and the whole mass bursts into flame. Even if detected before this stage they are difficult to deal with, since when the lagging is stripped away, extensive flaming can develop spontaneously. One precaution that can be taken to avoid this type of incident is to exclude oxygen by using non-porous lagging (e.g., foamed glass) or to seal the surface of the insulating material. One can also identify those parts of the system that are particularly susceptible to leaks (e.g., flanges, valves, etc.) and isolate these from the lagging material. Although it is technically possible to identify the maximum allowable thickness of insulation to avoid spontaneous ignition, this is not an acceptable method of prevention as the required thicknesses are generally insufficient to provide a satisfactory reduction in heat loss.

Lindner and Seibring (1967) and Britton (1990) have developed the following empirical relationship to aid the identification of fluids that have a propensity to undergo self-heating when dispersed on porous insulation materials:

$$\frac{\text{AIT}}{\text{AIT} - \text{FP}} > 1.55 \tag{8.4}$$

where AIT is the auto-ignition temperature and FP is the (closed cup) flashpoint. While it may seem illogical to incorporate AIT (which refers to auto-ignition in the gas phase) into a correlation that deals with a heterogeneous process, its significance is that it is a measure of the reactivity of the fuel. FP is also associated with a gas phase process, but in this expression it is a surrogate for a measure of the volatility of the fuel as it correlates with the boiling point (see, for example, Thorne (1976)). A 'reactive fuel' (low AIT) of low volatility (high FP) will therefore give a high value of the ratio AIT/(AIT − FP) – consistent with the above arguments although not providing a quantitative understanding.

8.1.5 Spontaneous Ignition in Haystacks

Spontaneous ignition of large piles of hay is one of the best known manifestations of this phenomenon. However, the mechanism is complex, involving at least three possible stages – the first of which involves spontaneous heating due to microbiological activity, a process that occurs in significant accumulations of vegetable matter. Gardeners are familiar with self-heating in compost heaps and in nature the Mallee fowl (*Leipoa ocellata*) – a resident of SE Australia – takes advantage of the process to incubate its eggs.[4] In large haystacks, the initial microbiological activity is capable of raising the temperature within the stack to over 70°C.[5] Chemical oxidation then takes over, although the reactions involved at these relatively low temperatures appear to be catalysed by moisture. The initial heating stage requires a relatively high moisture content for vigorous bacterial growth (~63–92% by weight), but higher water contents will enhance the effective thermal conductivity, increase the heat losses and thus limit the extent of self-heating (Rothbaum, 1963).

There is no information on the critical size for a haystack, although Bowes (1971) has estimated theoretically that it will be on the order of $2r_0 = 2$ m at ambient temperatures. This is not unreasonable and is consistent with the fact that there have been no reports of spontaneous ignition of rolled hay bales. The normal practice adopted to avoid self-heating and ignition of hay (and similar materials) is to make certain the material is not damp when it is stored (Walker, 1967).

8.2 Smouldering Combustion

The basic requirements for materials to be able to undergo self-sustained smouldering combustion are that they are porous and form a solid carbonaceous char when heated (Ohlemiller, 1985, 2008). Included are a wide range of materials of vegetable origin, such as paper, cellulosic fabrics, sawdust, fibreboard and latex rubber, as well as some thermosetting plastics in expanded form. In the natural environment, coal and peat are capable of smouldering (Rein, 2009). On the other hand, materials which can melt and shrink away from a heat source will not exhibit this mode of combustion. The reason for this will be clear once the mechanism of smouldering has been discussed.

Moussa *et al.* (1977) have examined the propagation of smouldering along horizontal, cylindrical cellulosic elements and have shown that the combustion wave has three distinct regions as shown in Figure 8.10, namely:

Zone 1 Pyrolysis zone in which there is a steep temperature rise and an outflow of visible airborne products from the parent material.

Zone 2 A charred zone where the temperature reaches a maximum, the evolution of the visible products stops and glowing occurs.

Zone 3 A zone of very porous residual char and/or ash which is no longer glowing and whose temperature is falling slowly.

Heat is released in Zone 2 where surface oxidation of the char occurs and it is here that the temperature reaches a maximum, typically in the region of 600°C for smouldering of cellulosic materials in still air. Heat is conducted from Zone 2 into the virgin fuel

---

[4] http://www.bbc.co.uk/nature/species/Malleefowl
[5] Thermophilic bacteria show maximum activity at 60°C and can survive up to 75°C in these circumstances (see Bowes (1984)).

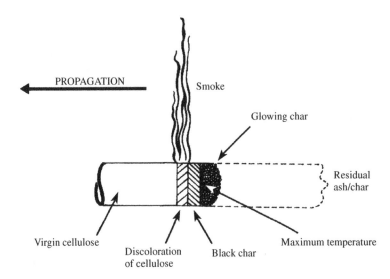

Figure 8.10   Representation of steady smouldering along a horizontal cellulose rod. Reproduced by permission of the Combustion Institute from Moussa *et al*. (1977)

(across the 'surface of fire inception', Section 7.3) and is responsible for the temperature rise observed in Zone 1. This causes thermal decomposition of the fuel, yielding the outflow of pyrolysis products and a residue of carbonaceous char. Temperatures greater than 250–300°C are required to accomplish this change in most organic materials. (If there is a significant moisture content, there will be a delay in reaching Zone 1 as the moisture is evaporated.) The three zones discussed by Moussa *et al*. (1977) appear to be common to all self-sustaining smouldering processes, although the reactions occurring in Zone 1 are complex (see Section 8.2.1). A distinction needs to be made between two types of propagation: forward smoulder, in which the flow of air is in the same direction as the propagation process and reverse smoulder, when the airflow opposes the direction of smoulder. These are discussed in the next section.

The volatile degradation products which are driven out from Zone 1 are not oxidized significantly. They represent the gaseous fuel which, in flaming combustion, would burn as a flame above the surface of the fuel. They are released ahead of the zone of active combustion and comprise a very complex mixture of products including high boiling point liquids and tars which condense to form an aerosol, quite different from smoke produced in flaming combustion (Section 11.1). These products are combustible and if allowed to accumulate in an enclosed space can, in principle, give rise to a flammable atmosphere. A smoke explosion arising from a smouldering fire involving latex rubber mattresses was reported in England in 1974 (Anon., 1975; Woolley and Ames, 1975). These products of smouldering combustion are rich in carbon monoxide and present a serious toxic hazard (see Section 11.1.4). Ohlemiller and Lucca (1983) took samples immediately behind the reacting zone in reverse smoulder through cellulose insulation and found that the ratio $[CO]/[CO_2]$ was approximately 0.8: it was slightly less (0.6) for forward smoulder. Similar ratios have been reported for other smouldering materials (e.g., peat (Rein *et al*., 2009)).

From the model of smouldering described above (Figure 8.10), it can be seen that if there is any tendency for the unaffected material to shrink away from the source of heat (Zone 2), there will be an effective reduction in the forward heat transfer. If this effect is sufficiently large, propagation of smouldering will not be possible. An interesting corollary is that it is possible for certain combustible liquids to exhibit smouldering provided that they are dispersed and absorbed on a rigid, but porous substrate. Lagging fires provide an example of such behaviour (Section 8.1.4).

8.2.1 Factors Affecting the Propagation of Smouldering

The mechanism by which smouldering propagates through a porous fuel bed has received increasing attention in recent years, promoted at least in part by the need to develop an understanding of its behaviour in microgravity (Ohlemiller and Lucca, 1983; Dosanjh et al., 1987; Torero and Fernandez-Pello, 1995, 1996; Rein et al., 2007; Rein, 2009). The process is modelled most simply by using the concept of the 'surface of fire inception' introduced in Section 7.3. This emphasizes the role of heat transfer ahead of the zone of active combustion, and the need to identify the rate controlling process. The basic equation is:

$$V = \dot{q}'' / \rho \Delta h \tag{8.5}$$

(Equation (7.11)), where V is the rate of spread, $\dot{q}''$ is the net energy flux across the surface of fire inception, ρ is the bulk density of the fuel and Δh is the thermal enthalpy change in raising unit mass of fuel from ambient temperature to the ignition temperature (Williams, 1977). This can be used to obtain an order of magnitude estimate of the rate of propagation of a smouldering combustion wave. If it is assumed that the ignition temperature is not too different from the maximum temperature in Zone 2 (i.e., T_{max}), then:

$$\Delta h = c(T_{max} - T_0) \tag{8.6}$$

where c is the heat capacity of the fuel and T_0 is the ambient temperature. If it is also assumed that heat transfer from Zone 2 to Zone 1 is by conduction and that a quasi-steady state exists, then:

$$V \approx \frac{k(T_{max} - T_0)}{x} \cdot \frac{1}{\rho c(T_{max} - T_0)} \tag{8.7}$$

i.e.

$$V \approx \frac{k}{\rho c} \cdot \frac{1}{x} = \frac{\alpha}{x} \tag{8.8}$$

where k and α are the thermal conductivity and the thermal diffusivity of the fuel, respectively; x is the distance over which heat is being transferred (Figure 8.11) and is found to be on the order of 0.01 m (Palmer, 1957). Taking fibre insulation board, for which $\alpha \approx 10^{-7}$ m^2/s (Table 2.1) as an example, the rate of propagation will be approximated by:

$$V = O(10^{-5}) \, \text{m/s}$$

or

$$= O(10^{-2}) \, \text{mm/s}$$

i.e., the rate of propagation is of the order of 10^{-2} mm/s, which is as observed. However, although this model gives the right order of magnitude for V, it is too crude: the

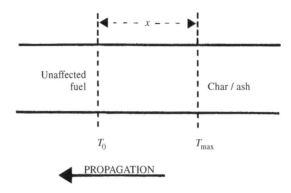

Figure 8.11 Simple heat transfer model for propagation of smouldering

calculations predict that the rate of propagation will be independent of both the initial temperature, which it is not (Krause and Schmidt, 2001), and the maximum temperature in Zone 2 although this is known to be incorrect. At elevated oxygen concentrations, Moussa *et al*. (1977) found a significant increase in the rate of spread which correlated with an increase in the temperature of Zone 2 (Figure 8.12). They examined a model in which the rate of heat generation in Zone 2 is determined only by the rate of diffusion of oxygen through a free convective boundary layer around the char zone. The steady rate of propagation and the maximum temperature (in Zone 2) are then determined for a given ambient condition by combining (i) the heat flux required to degrade the cellulose in Zone 1; (ii) the heat generated by the oxidation of the char in Zone 2; and (iii) the heat losses from the system. The agreement between two solutions of this model and the experimental data are shown in Figure 8.12. In view of the uncertainties regarding the heat required to degrade the cellulose, the agreement is remarkably good and provides support for the assumption that the rate of oxygen transport (by convection and diffusion) determines the rate of heat release. (It should be noted that much more sophisticated models for smouldering have now been developed – e.g., Ohlemiller (1985), Dosanjh *et al*. (1987), Schult *et al*. (1995), Rein *et al*. (2007).)

When a smoker draws on a cigarette, he is creating a flow of air moving in the same direction as the smoulder 'wave'. This mode of spread has come to be known as 'forward smoulder' (Ohlemiller and Lucca, 1983; Torero and Fernandez-Pello, 1996). Reverse smoulder[6] would be created by blowing through the cigarette – an action unlikely to satisfy the smoker. However, the concepts of forward and reverse smoulder are relevant to our understanding of the mechanism by which a smoulder wave propagates upwards or downwards – e.g., through a large accumulation of sawdust, under conditions of natural convection. Buoyancy will induce a flow of air into the smoulder wave: in the case of downward propagation (e.g., if a cigarette initiates smouldering on the top of a mattress or at the upper surface of a deep layer of sawdust), the spread will be against the induced airflow, and hence will be 'reverse smoulder'. 'Forward smoulder' occurs for upward propagation.

---

[6] Also referred to as 'opposed smoulder'.

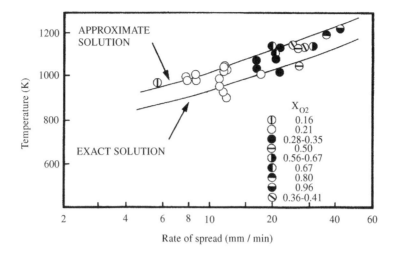

Figure 8.12 Data of Moussa *et al*. (1977) showing the correlation between rate of spread and the maximum temperature in Zone 2 for smouldering along horizontal cellulose rods. The various symbols refer to different proportions of oxygen in nitrogen (X_{O2}), except for the symbol ⊘, which refers to oxygen/helium mixtures. Reproduced by permission of the Combustion Institute from Moussa *et al*. (1977)

These two modes of spread have been studied experimentally as one-dimensional spread with an imposed airflow (usually of the order of 1 mm/s, but higher flows have been examined) (Ohlemiller and Lucca, 1983; Torero and Fernandez-Pello, 1996). There is a significant difference between the two modes. For reverse smoulder, the front edge of the smoulder wave (Moussa's Zone 1) encounters fresh, cool air with a full complement of oxygen. With forward smoulder, the flow of air reaching Zone 1 will have been significantly (if not completely) depleted of oxygen, and will be hot and contain combustion products. There is good evidence that two char-forming reactions will be in competition in this zone, one of which requires oxygen (Torero and Fernandez-Pello (1995) refer to this as the 'smoulder'), while the other is a true pyrolysis reaction and will proceed without oxygen, requiring only a sufficiently high temperature. In their experiments, Ohlemiller and Lucca (1983) observed that the velocity of reverse smoulder through cellulosic insulation was significantly greater than that of forward smoulder.

8.2.1.1 Experimental Studies

Very little research into smouldering combustion was carried out until there was an upsurge of interest in the fire properties of polyurethane foams. Early work by Palmer (1957) dealt mainly with the propagation of smouldering in layers of dust. Until the late 1970s, only isolated studies of smouldering in cellulosic materials (Kinbara *et al*., 1963; Moussa *et al*., 1977), string (Heskestad, 1976) and cigarettes (Egerton *et al*., 1963; Baker, 1977) had been published, but there is now interest in the behaviour of synthetic materials, including polyurethane foam and cellulose products used for insulation purposes. Only a brief review of this work can be given here (see Ohlemiller, 2008).

(a) *Dust layers.* The effect of the depth of a layer of dust on its ability to propagate smouldering horizontally through the fuel bed was studied by Palmer (1957) using wedge-shaped dust deposits in which smouldering was allowed to propagate from the thick towards the thin end of the wedge. The point at which propagation ceased was taken as a measure of the limiting depth. The minimum depths for sustained smouldering of cork dusts are shown in Table 8.4 as a function of particle size. For sawdust, the minimum depth was reduced substantially if the sample was subjected to an imposed airflow in a small wind tunnel. This is shown in Figure 8.13. The effect of an imposed airflow on the rate of propagation of forward and reverse smouldering in layers of cellulosic material has been studied quantitatively by Sato and Sega (1988) and Ohlemiller (1990). They used cellulosic powder (thickness 4–10 mm and bulk density 463 kg/m^3) and loose cellulosic insulation (thickness 107 mm and bulk density 80–100 kg/m^3), respectively. The rate of spread in the higher density material increased from 0.02 mm/s to 0.07 mm/s as the free-stream air velocity (u_0) was increased to 3 m/s (see Figure 8.14), but it then levelled out, with self-extinction occurring when $u_0 > 6$ m/s. It was observed that this was partly due to instability of the burning edge of the layer, which became increasingly steep as the airflow was increased. For reverse smouldering ($u_0 < 0 \ m/s$), the rate of spread was insensitive to u_0, until $u_0 < -0.8$ m/s, when it became unstable.

Ohlemiller also observed an increase in the rate of spread in loose cellulosic insulation,[7] but at values of u_0 above c. 2.5 m/s, transition to flaming occurred (Figure 8.15). In a limited set of experiments with reverse smoulder, he found the rate of spread to increase linearly from 0.03 mm/s to 0.05 mm/s as the airflow was increased to 5 m/s (the limit of the apparatus), but no transition to flaming was observed (Figure 8.15). He recorded the fact that localized blue flames could be seen in cracks and holes in the char at the higher flowrates (Ohlemiller, 1990).

There appears to be no upper limit to the depth of layer that can support smouldering. Palmer (1957) found that if smouldering is initiated at the bottom of a deep layer of sawdust, the time taken for the combustion wave to propagate vertically to the top surface (forward smoulder) is proportional to the square of the depth. Increasing the moisture content and the density of packing also increased the time of penetration. Packing density has this effect because the rate of propagation is dependent on the diffusion of oxygen to

Table 8.4 Minimum depth for sustained smouldering of cork dusts (Palmer, 1957)

Mean particle diameter (mm)	Minimum depth (mm)
0.5	~12
1.0	~36
2.0	~47
3.6	~36[a]

[a] Accompanied by glowing.

[7] Similar behaviour has been observed in layers of grain dust: Leisch *et al.* (1984) found the rate of propagation to increase by a factor of 2.5 when the airflow was increased from 0 to 4 m/s.

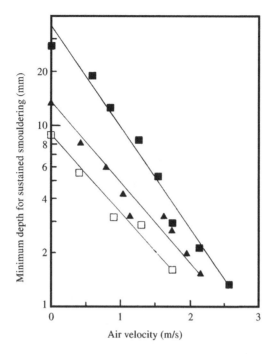

Figure 8.13 Variation with airflow of the minimum depth for sustained smouldering. ■, Deal sawdust (mean particle diameter, 1.0 mm); ▲, beech sawdust (0.48 mm); □, beech sawdust (0.19 mm). Reproduced by permission of the Combustion Institute from Palmer (1957)

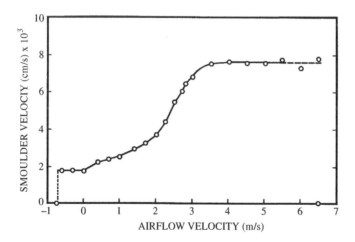

Figure 8.14 Effect of airflow on the rate of spread of smouldering in layers. Dependence of the rate of spread of smoulder along a bed of cellulose powder (density 463 kg/m³, thickness 8 mm) on the free-stream air velocity u_0 (Sato and Sega, 1988). Forward smoulder. Propagation was not stable for $u_0 > c$. 6 m/s: the smouldering self-extinguished. Transition to flaming was not observed. Reproduced with permission

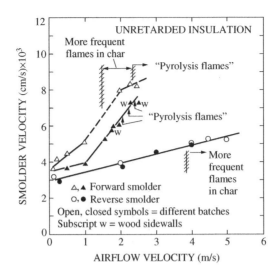

Figure 8.15 Effect of airflow on the rate of spread of smouldering in layers. Dependence of the rate of spread of smoulder along a bed of non-retarded loose cellulosic insulation (density $80-100\,kg/m^3$, thickness 107 mm) on the free-stream air velocity u_0 (Ohlemiller, 1990). Forward and reverse smoulder, as indicated. Open and closed symbols represent different batches of material. Forward smoulder gave transition to flaming with $u_0 > c.\ 2\,m/s$. There was no such transition for reverse smoulder (see text). By permission of the Combustion Institute

the surface of the reacting char. On the other hand, heat losses are low for deep-seated smouldering as the surrounding material provides insulation.

Upward propagation involves forward smouldering and tends to be more rapid than horizontal as hot combustion gases and degradation products rise into the unaffected fuel. The overlying layers absorb the degradation products and there may be no indication that smouldering is in progress within a heap of dust until the combustion zone approaches the surface. The first visible sign is dampness at the surface. As this dries out a musty smell develops, to be followed by the appearance of charring. During those later stages, substantial quantities of steam and acrid smoke are released. Palmer (1957) found that smouldering initiated at the foot of a pile of mixed wood sawdust 0.85 m deep took 10 days to penetrate to the surface. Ohlemiller and Lucca (1983) demonstrated that forward smoulder is more rapid than reverse smoulder in their study of propagation within a 17.8 cm deep layer of loose cellulosic insulation. It was held within a cylindrical, insulated drum 16.2 cm in diameter (Figure 8.16). They created either forward or reverse smoulder by imposing an upward flow of air through the fuel bed and initiating ignition either at the bottom or the top of the layer, respectively. At an air flow velocity of 0.3 cm/s, the rates of propagation for forward and reverse smoulder were approximately 1.8 and 1.0 mm/s, respectively. Their results also indicated that the rate of forward smoulder was much more sensitive to the air flow velocity than was the rate of reverse smoulder.

Most of Palmer's work dealt with the horizontal propagation of smouldering through dust layers (referred to as 'trains') that were formed on a solid surface: some of the many factors which affect the rate are summarized in Table 8.5 (Palmer, 1957). He observed

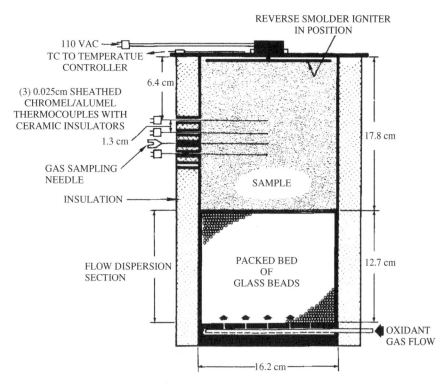

Figure 8.16 Cross-sectional view of the smoulder apparatus used by Ohlemiller and Lucca (1983). The flow of air is supplied through the bed of glass beads, and reverse smoulder is initiated by means of an electric coil heater located at the top, as shown. The ignition source is located at the foot of the sample to initiate forward smoulder. The array of thermocouples along the axis of the sample allows the spread of the smouldering front to be monitored. Reproduced by permission of the Combustion Institute

the onset of flaming at higher imposed airflows: this may be compared with Ohlemiller's experiments with low density cellulosic insulation in which the transition to flaming was observed for forward smoulder at airflows of c. 2.5 m/s (Figure 8.15; Ohlemiller (1990)). Leisch *et al*. (1984) studied the propagation of smouldering in 10 cm deep dust layers with the surface flush with the floor of a small wind tunnel, igniting them by an electrically heated platinum wire coil at various depths. They found that a cavity opened up behind the burning zone (effectively, behind Moussa's Zone 3) and that spontaneous flaming would occur at air velocities of approximately 4 m/s, provided that the cavity had become of 'substantial size'. This did not occur if the layer was ignited at 0.52 cm or 1.27 cm below the surface, but did when ignition was effected at a depth of 2.54 cm and 3.81 cm. The cavities created when ignition source was <1.27 cm were very shallow, suggesting that the geometry of the cavity was an important factor in determining whether or not flaming would occur.

(b) *Cellulose strips and boards.* The various studies of smouldering in cellulose rods and fibreboard tend to confirm the mechanism of propagation described above. Moussa *et al*. (1977) found the rate of propagation correlated well with the maximum temperature

Table 8.5 Factors affecting the horizontal rate of propagation of smouldering in dust layers (Drysdale, 1980).[a] Reproduced by permission of the Fire Protection Association

Factor	Dust	Effect on rate of propagation	Interpretation
Particle size	Beech	Rate increases very slightly ($\sim$15%) as mean particle diameter decreased from 0.5 to 0.1 mm (16.5 mm deep layer)	Not possible to resolve. Several conflicting factors including changes in ease of access of O_2 to surface, thermal diffusivity and thermal conductivity (improved insulation). Behaviour suggests latter is slightly dominant
Moisture content	Beech	Decreases slightly ($\sim$20%) as moisture content increased from 0.5% to 18.8%	Thermal capacity of unaffected fuel increased; more heat required to drive volatiles from Zone 1.
Depth of layer	Cork, grass	No effect on coarse dusts. With fine dusts (mean particle diameter <0.065 mm), rate decreases with depth.	Restricted access of oxygen to combustion zone
Imposed air velocity	Beech, deal, cork, grass	Increases; effect much more pronounced if airflow in same direction as propagation. Effect greater for coarse dusts. Glowing occurs at higher airflows (may lead to flaming)	Improved transfer of oxygen to combustion zone. When airflow in same direction as propagation, heat transfer to Zone 1 also improved (forced convection)

[a]Taken mostly from Palmer (1957).

in the combustion zone at different oxygen concentrations and partial pressures, as shown in Figure 8.12. Palmer's results show a similar correlation for dust layers when there was an imposed airflow blowing in the same direction as the smoulder. This will enhance the rate of supply of oxygen to Zone 2.

Propagation is more rapid in an upward direction (Palmer, 1957; Heskestad, 1976), as convection will contribute to the heat transfer yet not oppose access of oxygen to Zone 2. Most work has been carried out with horizontal and vertical, downward-burning samples, to simplify the heat transfer problem, but insufficient data have been obtained to generalize about the effect of sample size and shape.

(c) *Rubber latex foams.* Although rubber latex can be made to smoulder easily (Section 8.2.3), no fundamental study of this material has been published. The incident mentioned earlier regarding the production of a flammable smoke involved smouldering latex rubber mattresses stored in a large warehouse (Anon., 1975).

(d) *Leather.* Only a few commercially available leathers have been found to smoulder. Those which have been prepared by traditional vegetable tanning will not burn in this way, but there is evidence that chrome-tanned leathers will smoulder if they contain a high level of chrome (Prime, 1981, 1982).

(e) *Flexible polyurethane foams.* There is now a large variety of flexible 'polyurethane foams' available commercially, but apparently there has been no systematic comparative study of their smouldering potential. The original 'standard', PUF which was widely used in upholstered furniture in the 1970s, has virtually disappeared from the market, at least in the UK. This was in response to widespread concern about its behaviour in fire (Joint Fire Prevention Committee, 1978). Certain standard-grade foams are capable of smouldering in isolation, while others (e.g., some high-resilience foams), require contact with a substantial smouldering source, such as a heavy fabric, for sustained smouldering to be initiated (described as 'piloted smoulder' by Rogers and Ohlemiller (1978)). This is illustrated in Figure 8.17 (McCarter, 1976). The rate of propagation is likely to be slow as the maximum temperatures achieved in Zone 2 are not significantly greater than 400°C under conditions of natural convection (Ohlemiller and Rogers, 1978; Torero *et al.*, 1994), although this will depend on the thermal insulation afforded by the bulk of the foam. It is substantially less than the temperatures associated with smouldering cellulosics ($\geq$ 600°C), which may be one reason why most flexible polyurethane foams will only continue to smoulder (if at all) while there is a supporting heat flux (e.g., from a smouldering fabric cover). Ohlemiller and Rogers (1978) suggest that the propagation mechanism in these foams may involve radiative heat transfer across the open cell structure. Nevertheless, transport of oxygen is the dominant factor in determining the rate of spread under conditions of natural convection (e.g., Torero *et al.*, 1994).

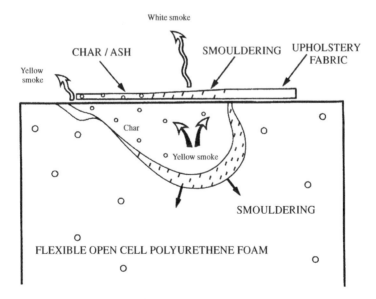

Figure 8.17 Representation of the interaction in a smouldering fabric/foam assembly (after McCarter, 1976). Reproduced by permission of Technomic Publishing Co. Inc., Lancaster, PA 17604, USA

If there is any tendency for the foam to soften or melt on exposure to heat (e.g., 'standard' polyurethane foam, and those 'combustion modified foams' formulated with melamine), it is unlikely to smoulder on its own. Many are very difficult to set smouldering and may self-extinguish unless supported by an independent source of heat. Torero and Fernandez-Pello (1995) showed that it was possible to study the upward propagation of smouldering in an open-celled polyurethane foam only if the sample was well insulated to minimize heat loss. (Their apparatus had similarities to that of Ohlemiller and Lucca (1983), shown in Figure 8.16.)

(f) *Rigid polyurethane and polyisocyanurate foams.* A limited amount of information regarding the behaviour of rigid foams is reported by Ohlemiller and Rogers (1978). These are used mainly for insulation and tend to have a closed cell structure, hence a very low, if not zero, air permeability. Although the cell walls rupture at relatively low temperatures (less than 275°C), the resulting permeability remains low until pyrolysis and charring commence. Consequently, smouldering of a rigid foam is a surface phenomenon and cannot occur at depth unless some mechanical breakdown has occurred to permit air access. Configuration is important as this will determine the heat loss from the reaction zone. Thus, smouldering combustion may be initiated and will propagate at an inner face of a pair of closely spaced parallel plane slabs.

By studying the smouldering of dusts prepared from these rigid foams, Ohlemiller and Rogers (1978) found that the peak smouldering temperatures in rigid polyurethanes were similar to those observed in flexible polyurethanes (∼400°C), but those in rigid polyisocyanurate foams were significantly higher (∼500°C) (see also Ohlemiller and Lucca (1983)).

(g) *Phenolformaldehyde foams.* Certain phenolformaldehyde foams can undergo a smouldering process known as 'punking'. Ohlemiller and Rogers (1978) showed that in a configuration which was conducive to minimal heat loss, smouldering in a particular foam could be initiated at a much lower temperature than necessary for a rigid polyurethane foam, although the peak smoulder temperatures of these materials (in powder form) were very similar. Once initiated, the smouldering may continue until the sample has been completely consumed. No smoke is produced but some gaseous volatiles are released, which are reported to have a pleasant antiseptic smell. No systematic study of the phenomenon has been made, although it appears that a highly rigid, almost brittle char is required; moreover, there is evidence that the open cell foams are more likely to 'punk' than the closed cell ones. Concern over the possibility of smouldering in this type of material, which is used extensively for insulation, led to the investigation of a test suitable for assessing 'punking tendency' (British Standards Institution, 1980).

8.2.2 Transition from Smouldering to Flaming Combustion

Flaming combustion can only be established at the surface of a combustible solid if the flowrate of volatiles exceeds a certain critical flux (Section 6.2.3). Thus the application of a small flame to the visible products rising from a smouldering source in still air is unlikely to lead to flaming combustion. Yet transition from smouldering to flaming is known to occur spontaneously even under still air conditions. In these circumstances the flux of volatiles must have increased to at least the critical value and the volatiles must either be ignited by some external means (pilot ignition) or ignite spontaneously. Observation would appear to favour the latter mechanism (Palmer, 1957).

Increasing the temperature of Zone 2 will increase the rate of the production of the volatiles in Zone 1. This is observed in smouldering of coarse dusts (see Table 8.5) and loose cellulosic insulation (Ohlemiller, 1990) if the velocity of the ambient air is increased (Figure 8.15). Ohlemiller and Rogers (1978) have reported spontaneous flaming to develop from smouldering initiated on the inside faces of closely spaced parallel slabs of rigid polyurethane and polyisocyanurate foams. Under these conditions, the conservation of heat, combined with an increased convective flow of air, enhance the temperature and rate of combustion in Zone 2. Transition to flaming in an item of upholstered furniture occurs only after smouldering has been underway for a substantial period of time (e.g., Babrauskas, 1979). The limited studies that have been carried out seem to suggest that propagation of the smouldering wave down through the material is a prerequisite of the transition. As the smouldering wave breaks through the underside of the cushioning material, a "chimney" is created through which air can be drawn as a buoyancy-induced flow. If the cushion is resting on a solid base, the transition is inhibited as this flow is prevented (Salig, 1981).

Torero and Fernandez-Pello (1995) observed the transition to flaming following upward (forward) smoulder through a polyurethane foam under conditions of natural convection. The upward velocity was found to increase as the smoulder wave approached the upper surface. They attributed this to an increasing buoyancy-driven flow through the fuel bed as the resistance to flow decreased with progressive conversion of virgin foam to char, which has a much greater porosity, thus allowing an enhanced rate of burning of the char. The transition was observed only for the deepest fuel beds used (300 mm) (Figure 8.18). Under these circumstances of forward smoulder, the vapours driven off Zone 1 would be highly flammable pyrolysis products, and would ignite spontaneously when they achieved a sufficient temperature and concentration in air above the free surface. (In these carefully controlled experiments, air was free to enter through the lower face of the fuel bed. This is unlikely to be the case in practical situations, e.g., smouldering initiated at the foot of a deep layer of sawdust (Palmer, 1957), but more vigorous smouldering will occur as the smoulder wave approaches the surface, thus gaining freer access to fresh air. This has not been studied quantitatively.)

Fernandez-Pello and his co-workers have proposed that the transition to flaming during the forward smouldering of flexible polyurethane foam may be initiated by spontaneous flaming that has been observed to occur within the open structure of the char behind the active smoulder front (Tse et al., 1996; Bar-Ilan et al., 2005; Putzeys et al., 2007).[8] In his study of smouldering in cellulose insulation, Ohlemiller (1990) reported blue flames in the openings in the char at increased airflows; these are noted in Figure 8.15 ('More frequent flames in char') for both forward and reverse smouldering. Full transition to flaming ('Pyrolysis flames') was observed at velocities ≥ 2.5 m/s, but only for forward smouldering. For reverse smoulder, flames were observed in the open char structure above 4 m/s, but transition to flaming did not occur <5 m/s (this was the limit to the airflow velocity in the equipment used). In smouldering sawdust of large particle size

---

[8] These experiments involved relatively small samples of foam, with an exposed vertical face heated radiatively to support the smouldering process. There was an internal (upward) airflow as well as an airflow parallel to the exposed surface.

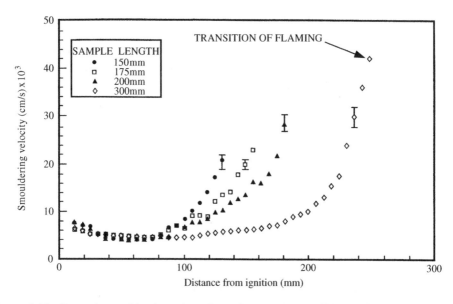

Figure 8.18 Rate of smoulder in polyurethane foam under conditions of natural convection. Upward spread following ignition at the lower face of a cylindrical sample (forward smoulder). The data represent the variation of smouldering velocity with distance from the ignition source for four samples of different height (150, 175, 200 and 300 mm). Transition to flaming was observed in the 300 mm sample only (Torero and Fernandez-Pello, 1995). By permission

(>O(1 mm)), Palmer (1957) noted glowing of the char at increased airflows, but did not report 'internal flames' within the structure of the char. Indeed, he was of the opinion that flame appeared spontaneously in the gas phase. Clearly, the mechanism of the transition from smouldering to flaming deserves further study.

8.2.3 Initiation of Smouldering Combustion

There have been very few experimental studies of the initiation of this mode of combustion. It is well known that smouldering can follow flaming combustion of many cellulosic materials if the flame is extinguished without cooling the fuel (Section 8.3). A burning cigarette can initiate smouldering in a susceptible fuel (e.g., British Standards Institution, 2006), but there is no readily available information on other possible modes of ignition. The basic requirement is the provision of a source of heat that will produce some char and initiate its oxidation, but it would appear that the temperature of the source and/or the rate of supply of heat (e.g., from a radiant heater) are important, more so with some materials than others (Anderson *et al.*, 2000). For example, a rubber latex foam pillow can be ignited within a few minutes by direct contact with a burning 40 W light bulb. Smouldering can commence within the latex foam with little more than a smoke stain visible on the inner and outer cotton covers (Drysdale, 1981). Fibreboard cannot be ignited in this way.

The various experimental studies of smouldering have involved a variety of means of ignition, from the application of a flame to the use of electrically heated coils and plates (e.g., Ohlemiller and Lucca, 1983; Torero and Fernandez-Pello, 1996). Walther *et al.* (2000) examined the relationship between the strength of the 'ignition source' (an electrically heated nichrome wire held between two 5 mm thick ceramic discs) expressed as a heat flux (kW/m^2) and the time to achieve ignition, defined as the onset of smouldering that was self-sustaining. They found a remarkable similarity to the relationship between the time to piloted ignition of a combustible solid and the external heat flux (Figure 6.21), allowing them to identify a heat flux below which smouldering would not propagate (cf. the critical radiant heat flux for piloted ignition, Table 6.5). They concluded from their results that a critical depth of char had to be formed during the period of heating to ensure propagation of smouldering combustion.

A similar conclusion was drawn by Anderson *et al.* (2000), who studied downward smoulder in samples of (standard) flexible PUF initiated by exposing the upper surface to a source of radiant heat. They found that self-sustained smouldering only occurred following exposure to a flux within a very narrow range of heat fluxes, *viz.* 6.1–6.7 kW/m^2, and was weakly dependent on the duration of exposure (c. 1000 s). From this, they deduced that a minimum temperature was required to initiate the process (approximately 300°C), although a self-sustained smoulder could not be initiated above 6.8 kW/m^2. They attributed this to the high rate of oxidation of the char at these fluxes, which prevented the formation of a layer that was thick enough to insulate the reaction zone and allow the smoulder wave to propagate. This need for a minimum thickness of char for smouldering to propagate has been discussed by others (see, for example, Ohlemiller (1985, 2008)).

Initiation of smouldering in this manner can be interpreted in terms of the mechanism of spontaneous combustion as applied to the ignition of a layer of sawdust lying on a hot surface, as discussed in Section 8.1.3 (Figure 8.6). Here, the boundary condition (Equation (8.3b)) is expressed in terms of the surface temperature of the plate, but both Anderson *et al.* (2000) and Walther *et al.* (2000) have shown that a critical heat flux can be determined: this may be suitable for defining the 'ease of smouldering ignition' for a given material, although as with flaming ignition the value would be very sensitive to configuration and heat losses (Section 6.3).

It should be noted that smouldering – or a smoulder-like process – can occur on the surface of non-porous combustible material exposed to an external heat flux, or under conditions where heat is conserved. Thus, Ohlemiller (1991) observed the propagation of a surface smoulder in wooden channels (white pine and red oak) 51 mm wide and 114 mm deep, but only under forced flow conditions. (Transition to flaming occurred at higher flows.) It was concluded that cross-radiation from the smouldering surfaces was an essential factor in this process. Beyler *et al.* (2006) have shown that under imposed radiant heat fluxes, the surface of plywood will begin to undergo a smoulder-like process, and that self-heating will occur in the sub-surface layers.

By way of summary, the following situations may give rise to smouldering combustion in porous materials:

(i) contact with a smouldering material;
(ii) uniform heating (spontaneous ignition within the material);
(iii) unsymmetrical heating (dust on a hot surface, material exposed to a heat flux on one face); and
(iv) the development of a hot spot within the mass of material, e.g., as the result of an electrical fault or a buried light fitting.

8.2.4 The Chemical Requirements for Smouldering

The principal requirement for smouldering is that the material must form a rigid char when heated. Materials yielding non-rigid chars or fluid tarry products will tend not to smoulder. This suggests that the molecular structure and mode of degradation of the parent material is of fundamental importance in determining its behaviour. For example, McCarter (1976) has noted that flexible foams made from polyols grafted with acrylonitrile/styrene produce very rigid chars in high yield and have a high propensity to smoulder; similar observations have been made by Rogers and Ohlemiller (1978). The implication of work of this kind is that flexible foams with low propensity to smoulder could be 'designed' chemically or simply selected from materials now available for applications in which a low smoulder tendency is desirable. The importance of the mechanism of degradation can best be illustrated by referring to the smouldering of cellulosic materials. Although smouldering is most often associated with cellulosic materials, pure α-cellulose has little or no tendency to burn in this way. This is compatible with the fact that pure cellulose yields very little char on heating (Section 5.2.2), compared with cotton and rayon, both of which are known to smoulder. The difference can be explained by the catalytic activity of the inorganic impurities which promote the char-forming reaction, some of which are listed in Table 8.6. This has been examined in detail by McCarter (1977, 1978). Dwyer et al. (1994) carried out tests on 500 commercially available upholstery fabrics and found that their propensity to ignite from a smouldering cigarette increased with the cellulose content of the fabric, and with the amount of sodium and potassium salts present.

The reduction of smoulder potential can be achieved by selection of materials either which do not form rigid chars or whose char yield can be depressed. An alternative is

Table 8.6 Inorganic compounds,[a] which enhance smouldering in cellulosics (McCarter, 1977)

Lithium[b]	Rubidium chloride
chloride	Caesium chloride
hydroxide	
Sodium[b]	
chloride	Silver nitrate
bromide	
iodide	
hydroxide	Ferrous chloride
nitride	Ferrous sulphate
nitrate	
carbonate	Ferric chloride
sulphate	
acetate	Lead nitrate

[a] Applied to the material by soaking in appropriate solutions.
[b] Potassium ions are present in commercial upholstery fabrics (Dwyer et al., 1994) and appear to be more effective than sodium ions in promoting smouldering in cotton fabrics (Kellogg et al., 1998).

Table 8.7 Powders which inhibit smouldering in cellulosics
(McCarter, 1977)

Sulphur	Copper (I) chloride
Aluminium chloride 6H$_2$O	Iron (II) chloride 4H$_2$O
Antimony chloride	Iron (III) chloride 6H$_2$O
	Magnesium chloride 6H$_2$O
Calcium	Nickel chloride 6H$_2$O
chloride	Stannous chloride 2H$_2$O
chloride 2H$_2$O	Zinc chloride
Chromium (III) chloride 6H$_2$O	

to use a smoulder suppressant. Boron- and phosphorus-containing compounds are known to inhibit smouldering in cellulosic materials: thus, Ohlemiller (1990) found that 15% boric acid reduced the smoulder velocity through a layer of cellulosic insulation, and increased the airflow velocity necessary to induce flaming combustion (see Section 8.2.1). However, the mechanism of action is not clear as many of these promote the formation of char (see Table 5.13). McCarter (1978) suggests that the reactive sites on the char are blocked, but more work is required not only to elucidate the mechanism but also to improve our understanding of the surface oxidation reaction. McCarter discovered a range of compounds which were capable of suppressing smoulder potential in cellulosic materials if they were dispersed as fine powders: one of the best of these is elemental sulphur (see Table 8.7), although the mechanism is unclear (see Gann *et al.*, 1981).

8.3 Glowing Combustion

Glowing combustion is associated with the surface oxidation of carbonaceous materials or char. The onset of glowing combustion at the surface of plywood exposed to relatively low levels of radiant heat flux has been studied by Beyler *et al.* (2006) and Swann *et al.* (2008). They identified a limiting heat flux for the onset of 'glowing' which was of the order of 8 kW/m^2, although glowing was not self-sustaining in the absence of the radiative flux. Almost all char-forming fuels – including coal, anthracites, etc. – will undergo flaming combustion, leaving behind the involatile char which will continue to burn slowly, even after cessation of flaming, provided that high temperatures are maintained. A substantial amount of the total heat of combustion of the original material is associated with the char (approximately 30% in the case of wood (Section 5.2.2)), but glowing combustion will only continue if heat is conserved at the reacting surface. If a heap of glowing charcoal is opened out to expose the burning surfaces and eliminate cross-radiation, the radiative and convective heat losses can overwhelm the rate of heat production if the individual pieces are large enough and have not become uniformly heated (cf. Section 6.3.2). Under these circumstances, the burning surfaces will cool and the residual fire will self-extinguish.

The term 'afterglow' describes the residual glowing combustion of the char that is left after burnout of a piece of fabric, splint of wood (e.g., a match) or similar item. It is normally of short duration and can easily be eliminated by means of a suitable fire retardant treatment (Lyons, 1970).

Problems

8.1 Given the following data on the ignition of cubes of activated charcoal, calculate by extrapolation the depth of a 'semi-infinite slab' (i.e., a layer) of this material that would be a spontaneous ignition risk at 40°C.

r_0 (half side of cube) (mm)	T_a (crit) (K)
25.40	408
18.60	418
16.00	426
12.50	432
9.53	441

If the thermal conductivity of the activated charcoal was 0.05 W/m K, is the Frank-Kamenetskii model strictly applicable for this range of sample sizes?

8.2 A serious fire broke out in the cable trays in the machine hall of a large factory producing tissue paper. The trays, 60 cm wide, were located below the ceiling at a height of 15 m, and had accumulated a fibrous fluffy deposit over a substantial period of time: the deposit was thought to be 25 cm deep. The ambient temperature below the ceiling was known to be about 40°C during operation of the plant below. The cables (covered by the dust) were working well below their design capacity and there was no evidence of an electrical fault having caused the fire. Spontaneous ignition was suggested. A series of tests was carried out which gave the following data on the critical temperatures of four sizes of sample in the form of cubes.

Cube size ($2r_0$, mm)	Critical temperature (°C)
25	188
50	166
100	142
250	119

Use these data to estimate whether or not spontaneous ignition of the dust layer (at 40°C) could have caused the fire. Assume as a first approximation that the layer can be treated as an infinite slab.

9

The Pre-flashover Compartment Fire

The term 'compartment fire' is used to describe a fire which is confined within a room or similar enclosure within a building. The overall dimensions are important, but for the most part we shall be directing our attention to room-like volumes of the order of $100\,\mathrm{m}^3$. Fire behaviour in elongated chambers, or in very large spaces ($>1000\,\mathrm{m}^3$), will depend very much upon the geometry of the enclosure: comments will be made as appropriate.

Following ignition, while still small, the fire will be burning freely: following the definition of Walton and Thomas (2008), the pyrolysis rate and the energy release rate are affected only by the burning of the fuel itself and not by the presence of the boundaries of the compartment. If it can grow in size, either as a result of flame spread over the item first ignited, or by spreading to nearby objects, a stage will be reached when confinement begins to influence its development. If there is sufficient ventilation to allow the growth to continue, then its course can be described in terms of the rate of heat release (power) of the fire as a function of time (Figure 9.1). The form is similar to a diagram that is often presented in which the *average* compartment temperature is plotted against time (e.g., Drysdale, 1985b; Walton and Thomas, 2008).

Although purely schematic, Figure 9.1 illustrates that the compartment fire can be divided into three stages:

(i) The growth or pre-flashover stage, in which the average compartment temperature is relatively low and the fire is localized in the vicinity of its origin.

(ii) The fully developed or post-flashover fire, during which all combustible items in the compartment are involved and flames appear to fill the entire volume.

(iii) The decay period, often identified as that stage of the fire after the average temperature has fallen to 80% of its peak value.

While the average temperature in stage (i) is low, high local temperatures exist in and around the burning zone and the temperature under the ceiling begins to rise at an early stage (Section 4.3.3.2). During the growth period, the fire increases in size, to and beyond the point at which interaction with the compartment boundaries becomes significant. The transition to the fully developed fire (stage (ii)) is referred to as 'flashover' and involves

---

An Introduction to Fire Dynamics, Third Edition. Dougal Drysdale.
© 2011 John Wiley & Sons, Ltd. Published 2011 by John Wiley & Sons, Ltd.

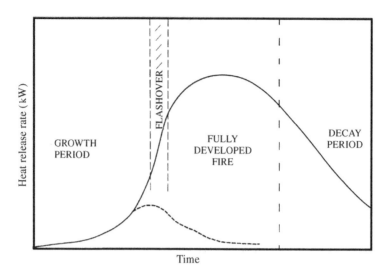

Figure 9.1 The course of a well-ventilated compartment fire expressed as the rate of heat release as a function of time. The broken line represents either depletion of fuel before flashover has occurred or, under very poorly ventilated conditions, reducing rates of burning as oxygen depletion occurs

a rapid spread from the area of localized burning to all combustible surfaces within the room. The transition is normally short in comparison with the duration of the main stages of the fire and is sometimes considered to be a well-defined physical event, in the same way that ignition is an event. This is not the case (Section 9.2), although in many circumstances it is tempting to treat it as such, particularly as it marks the beginning of the fully developed fire (Chapter 10). Anyone who has not escaped from a compartment before flashover is unlikely to survive. Rasbash (1991) has drawn attention to the role that flashover has played in a number of major fire disasters.

During the fully developed stage of a fire, the rate of heat release reaches a maximum and the threat to neighbouring compartments – and perhaps adjacent buildings (Section 2.4.1) – is greatest. Flames may emerge from any ventilation opening, spreading fire to the rest of the building, either internally (through open doorways) or externally (through windows). In addition to the obvious threat to occupants remaining within the building, it is during this stage that structural damage may occur, perhaps leading to partial or total collapse of the building. During the cooling period (stage (iii)), the rate of burning diminishes as the fuel is depleted of its volatiles. Flaming will eventually cease, leaving behind a mass of glowing embers that will continue to burn, albeit slowly, for some time, maintaining high local temperatures (Section 8.3).

An understanding of the pre-flashover stage of the fire is directly relevant to life safety within the building. Once flashover has occurred in one compartment, the occupants of the rest of the building may be threatened directly. This is best illustrated by comparing the 'required safe egress time' (RSET) with the 'available safe egress time' (ASET). For successful escape:

$$\text{RSET} \leq \text{ASET}$$

RSET can be broken down into its component parts, as may be seen by considering the inequality discussed by Marchant (1976) and Nelson and Mowrer (2002):

$$t_p + t_a + t_{rs} \leq t_u \tag{9.1}$$

where t_p is the time elapsed from ignition to the moment at which the fire is detected, t_a is the delay between detection and the beginning of the 'escape activity', t_{rs} is the time to move to a place of relative safety and t_u is the time (from ignition) for the fire to produce untenable conditions at the location under consideration (closely related to ASET). This provides the basis of 'timeline analysis', which has been adopted in several Fire Safety Engineering Codes of Practice (e.g., CIBSE, 2003; BSI, 2004; ISO, 2005) and is discussed by Proulx (2008). While automatic detection can reduce t_p, in some cases very significantly, successful escape will ultimately depend on how rapidly the fire grows, i.e., on t_u. Consequently, the time to flashover is a significant factor in determining the fire hazard associated with a particular scenario. Multiple-fatality fires are commonly associated with very rapid fire growth rates that create untenable conditions in a short period of time (e.g., Rasbash, 1991; Grosshandler et al., 2005). Clearly, delaying flashover (i.e., increasing t_u) will provide greater opportunity for detection and suppression (manual or automatic) and for safe evacuation of the occupants of the building.

Smoke detectors would be expected to operate during the earlier stages of the pre-flashover fire before it has grown much beyond ~10 kW, which is large enough to carry sufficient combustion products to ceiling level in a 3 m high enclosure (Section 4.3.1).[1] Operation of fixed temperature heat detectors and conventional sprinkler heads depends on how quickly the sensing element under the ceiling can be raised to a specific operating temperature (e.g., 68°C). This is determined by the response time index (RTI) (see Sections 2.2.2 and 4.4.2). A relatively large fire is required to produce a flow of sufficiently hot gas (Section 4.4.2) and, given the additional delay due to the relatively high thermal mass of the sensing elements used in conventional sprinklers, the fire may be fast approaching the end of the growth period before activation occurs (e.g., 0.5–1.0 MW in an average sized room). The advantage of fast response sprinklers, with sensing elements of low thermal mass, is clear.

This chapter is devoted to the early stages of the compartment fire, up to and including the onset of flashover. The criteria necessary for flashover to occur and the factors which determine the duration of the growth period will be identified and discussed.

9.1 The Growth Period and the Definition of Flashover

After localized burning has become established, one of three things may happen:

(i) The fire may burn itself out without involving other items of combustible material, particularly if the item first ignited is in an isolated position (see Figure 9.1).
(ii) If there is inadequate ventilation, the fire may self-extinguish or continue to burn at a very slow rate dictated by the availability of oxygen (the scenario which may lead to 'backdraught' (Section 9.2.1)).

---

[1] There can be a significant delay before a smouldering fire is detected as the smoke is 'cool' and insufficiently buoyant to be carried rapidly to ceiling level.

(iii) If there is sufficient fuel and ventilation, the fire may progress to full room involvement in which all exposed combustible surfaces are burning.

The following discussion will concentrate mainly on the third scenario. A number of issues need to be reviewed, but these may be grouped into two broad areas: the conditions associated with the onset of flashover, and the factors which determine the duration of the growth period. The latter is extremely important as it has a major impact on life safety: if the 'time to flashover' is short, the time available for escape (t_u in Equation (9.1)) may be inadequate.

The conditions that are necessary and sufficient for the fire to develop to the fully developed stage need to be understood and quantified. 'Flashover' is identified in Figure 9.1 and will be described in the following text. Its duration will depend on a number of factors, including the nature and distribution of the combustible contents, as well as the size and shape of the compartment. In a typical living room with upholstered furniture containing standard polyurethane foam (which is no longer available in the UK) the growth period may be short and the flashover process may last only 15–30 s (Building Research Establishment, 1989). However, in very large spaces, such as a warehouse, it is likely to take longer, and indeed it is conceivable that the area first involved may burn out before the most distant area becomes involved.

Flashover is generally associated with enclosed spaces. However, the Bradford Football Stadium fire, which occurred in May 1985 (Popplewell, 1986; Building Research Establishment, 1987), revealed that the term is also applicable to relatively open situations where confinement is provided by little more than the roof of the structure.[2] (The development of such a fire is illustrated in Figure 9.2, which relates to a laboratory-scale experiment carried out at FMRC (Friedman, 1975).)

The basic features of the growth period and the onset of flashover may be understood qualitatively in terms of the elements of fire behaviour which have been discussed in earlier chapters. In the open, an isolated fuel bed will burn at a rate determined by the heat flux from the flame to the surface, $\dot{Q}''_F$, i.e.

$$\dot{m}'' = \frac{\dot{Q}''_F - \dot{Q}''_L}{L_v} \tag{9.2}$$

where $\dot{Q}''_F$ will be dominated by radiation if the fuel bed is greater than about 0.3 m in diameter (Section 5.2.1). For most fuels, approximately 30% of the heat liberated in the flame is radiated to the environment (see Section 4.4.1) and the rest is dispersed convectively in the buoyant plume. If the item is burning in a compartment, this heat is not totally lost from the environs of the fuel as fire gases will be deflected and trapped below the ceiling (Section 4.3.4), which will become heated. If the size of the fire is such that the natural flame height is greater than the height of the room, then the flame will extend as a ceiling jet (Section 4.3.4) and thus contribute significantly to the heat transfer to the ceiling. The radiant heat flux back to the fuel will increase as the temperature of the ceiling rises, but once the horizontal flow of the ceiling jet is interrupted by the compartment boundaries a layer of hot smoky gases will form under the ceiling and radiate downwards with increasing intensity as the smoke concentration, the layer thickness and

---

[2] The fire at the Bradford Football Stadium was recorded during a live TV broadcast (BRE, 2008).

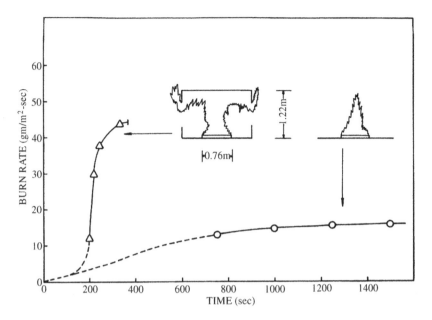

Figure 9.2 The effect of enclosure on the rate of burning of a slab of polymethylmethacrylate (0.76 m × 0.76 m) (Friedman, 1975)

the temperature all increase. The effect on the burning fuel will be to increase the rate of burning, according to the equation:

$$\dot{m}'' = \frac{\dot{Q}_F'' + \dot{Q}_E'' - \dot{Q}_L''}{L_v} \tag{9.3}$$

However, more significantly, it will promote flame spread over the item first ignited (Section 7.2.5) and to contiguous surfaces and adjacent items (see Section 4.3.4), thereby increasing the area of burning (A_F) and consequently the rate of generation of fuel vapours (Equation (9.4)):

$$\dot{m} = \left[\frac{\dot{Q}_F'' + \dot{Q}_E'' - \dot{Q}_L''}{L_v} \right] A_F \tag{9.4}$$

Thomas (1981) has argued that the rate of increase in the area of burning is more important than the rate of increase of $\dot{m}''$ (Equation (9.3)) in determining the rate of fire development to the flashover stage. Any scenario which leads to fast fire spread – and hence a rapidly increasing area of burning – will promote the onset of flashover. An extreme example of this was encountered in the fire involving a wooden escalator at the King's Cross Underground Station, London, in November 1987 (Fennel, 1988; Rasbash, 1991), but there are many other examples, such as the Station Nightclub fire in West Warwick, Rhode Island on 27 February 2003 (Grosshandler et al., 2005).

Some of these features are well illustrated in Figure 9.2, which shows the effect of an enclosure on the burning of a slab of polymethylmethacrylate (Friedman, 1975). A hood (acting as a ceiling and the upper parts of the walls) above the burning slab deflects the fire plume, and causes the radiative heat feedback to the fuel surface to increase: air had unrestricted access to all four sides in these experiments. The maximum rate of burning is three times greater than that in the open and appears to be achieved in about one-third of the time. This effect is quite general, although its magnitude depends on the nature and distribution of the fuel and the size of the compartment. For example, the rate of burning of an alcohol fire in a small enclosure can be up to eight times greater than in the open (Thomas and Bullen, 1979; Takeda and Akita, 1981). There is evidence that the increase will be self-limiting, as the fuel vapours rising from the surface attenuate the radiation reaching the fuel (Thomas and Bullen (1979); see also Section 5.1.1).

Given that flashover marks the beginning of the fully developed fire, this term must be defined more precisely in order that the factors which determine the duration of the growth period can be examined. Many definitions have appeared in the literature (Thomas, 1981, 1982; Peacock et al., 1999), the most common of which are:

(i) The transition from a localized fire to the general conflagration within the compartment when all fuel surfaces are burning.
(ii) The transition from a fuel-controlled fire to a ventilation-controlled fire.[3]
(iii) The sudden propagation of flame through the unburnt gases and vapours collected under the ceiling.

Martin and Wiersma (1979) point out that (ii) is the result of (i) and is not a fundamental definition. The third definition is based partly on an observation that flames often emerge from the window or other ventilation openings at around the time full room involvement commences, but as stated would seem to imply premixed burning. This is not correct, but it is likely that the burning of the smoke layer plays a significant role as one of several mechanisms that contribute to the transition to the fully developed fire. This is discussed in greater detail in the following sections.

The first definition (i) is the one that is being used in this text, although Thomas (1974b) has pointed out that it may not apply to very long or 'deep' compartments in which it may be physically impossible for all the fuel to become involved at the same time.

9.2 Growth to Flashover

9.2.1 Conditions Necessary for Flashover

The conditions necessary for flashover to occur have been reviewed in detail by Peacock et al. (1999). The earliest investigation of flashover is that of Waterman (1968), who carried out a series of experiments in a compartment 3.64 m × 3.64 m × 2.43 m high, in which items of furniture were burned. The onset of the fully developed fire was defined by the ignition of paper 'targets' located on the floor. He concluded that a heat flux of about 20 kW/m^2 at floor level was required for flashover to occur. While this may be sufficient to bring about spontaneous ignition of paper (Martin and Wiersma, 1979), it

---

[3] Babrauskas et al. (2003) suggest that a convenient visual marker of the 'flashover event' could be taken as the time at which the bottom of the smoke layer (the 'thermal discontinuity') drops below the mid-height of the compartment.

is well below that necessary for 'thick' pieces of wood and other combustible solids. However, it is more than enough to promote piloted ignition and rapid flame spread over the surface of most combustible materials (Sections 6.3 and 7.2.5). Waterman postulated that most of this heat flux came from the heated upper surfaces of the room, rather than directly from the flames above the burning fuel. He also observed that flashover did not occur in his experiments unless the rate of burning exceeded 40 g/s.

Experimental studies of flashover carried out since this work have tended to define the 'onset of flashover' in terms of either Waterman's criterion (20 kW/m^2 at low level – e.g., Babrauskas (1979)) or a ceiling temperature of approximately 600°C, which has been observed in several studies (Hägglund *et al.*, 1974; Fang, 1975b), although Hägglund *et al.* (1974) actually defined flashover as the moment at which flames emerged from the opening. The ceiling height of their compartment was 2.7 m: if the mechanism of flashover involves heat radiated to the lower levels, it might be anticipated that the ceiling temperature at flashover would be a function of height. There is limited evidence to support this conclusion: for example, Heselden and Melinek (1975) report 450°C for 'flashover' in a small-scale experimental compartment 1.0 m high, but of course relating the temperature to ceiling height neglects the smoke layer as a significant source of thermal radiation.

In addition to the vertical flames above the fire, there are three sources of radiative flux, namely:

(i) from the hot surfaces in the upper part of the enclosure;
(ii) from flames under the ceiling;
(iii) from hot combustion products trapped under the ceiling.

The relative importance of these will vary as a fire develops, and which one is predominant at flashover will depend on the nature of the fuel and the nature and degree of ventilation. If a clean-burning fuel such as methanol is involved, heat feedback will be dominated by radiation from the upper surfaces as the flames and combustion products are of low emissivity (Section 2.4.2). However, smoke is produced in almost all fires and ultimately the thickness and temperature of the hot layer of combustion products under the ceiling will determine the behaviour (Section 2.4.3). The 'effective height' of the source of radiation cannot be determined with any accuracy and it is perhaps unreasonable to expect there to be a well-defined upper layer temperature associated with the onset of flashover. A range of temperatures is reported in the literature (Peacock *et al.*, 1999): for example, in their study of fires in mobile homes, Budnick and Klein (1978) found that flashover (indicated by the ignition of crumpled newspaper at floor level) occurred at upper layer temperatures above 673°C.

For convenience, two situations will be considered, as illustrated in Figure 9.3. In the first (Figure 9.3(a)), the fire plume is deflected to form a one-dimensional ceiling jet, thus modelling a corridor ceiling with no barrier to limit horizontal spread in the direction of flow. The depth of layer will then depend only on the width of the 'corridor' and the rate of flow of hot gas into the ceiling layer (Section 4.3.3.2). If the fire is of sufficient size that the flames are deflected, then a range of flame behaviour can be observed, encompassing the two regimes defined by Hinkley *et al.* (1968) (Section 4.3.3.2) as 'fuel lean' and 'fuel rich'. In the 'fuel-lean' situation, the flame in the ceiling jet will be of limited extent and burn close to the surface of the ceiling. With the 'fuel-rich' situation, the flame under

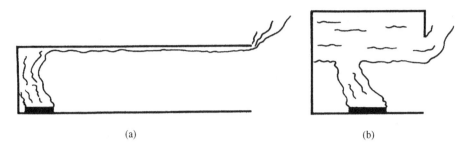

(a)   (b)

Figure 9.3 Interaction of a fire plume with the ceiling of a compartment: (a) corridor ceiling, with unrestricted horizontal flow in one direction (cf. Figure 4.29(a)); (b) compartment ceiling, showing the formation of a smoke layer under conditions of adequate ventilation

the ceiling needs more air to complete the combustion process and flaming occurs at the lower boundary of the layer where air entrainment is taking place. The difference between these two extremes is shown clearly by the vertical temperature distributions below the ceiling (Figure 4.34). The net result of the horizontal deflection of the fire plume is to increase the radiative heat flux at floor level at points remote from the fire. This is shown schematically in Figure 9.4, where the radiative heat flux at floor level is shown as a function of distance from the centreline of a crib fire of constant heat output (see Figure 4.33(a)) for three situations: a free burning fire with the ceiling removed ($h = \infty$) and for two ceiling heights, $h = 0.97$ m and $h = 0.66$ m. Hinkley *et al.* (1968) estimated from a detailed heat balance that long, fuel-rich flames lost as much as 55% of the heat of combustion released in the layer by radiation. If a combustible lining is involved, the flame extension is increased, and the heat flux to the lower levels is correspondingly greater (Hinkley and Wraight, 1969).

The second scenario, illustrated in Figure 9.3(b), shows the formation of a smoke layer under the ceiling of a compartment. Immediately after burning has become established, a fire plume will develop and grow in strength as the fire grows in size. It will form a ceiling jet, which will flow outwards until it reaches the walls (Section 4.3.4),[4] after which a smoke layer will quickly develop under the ceiling, deepening as the fire continues to burn. As the layer descends below the top of any opening, such as a door or window, hot combustion products will flow from the compartment. The height of the smoke layer will stabilize at a level below the soffit of the opening, reducing the height through which cool air can be entrained into the vertical plume (Quintiere, 1976; Beyler, 1984b, 1986a; Zukoski, 1986, 1995): consequently, the temperature of the layer will progressively increase. The plume will continue to penetrate to the ceiling, maintaining a hot ceiling jet. Although the temperature of the layer is not strictly uniform (Babrauskas and Williamson, 1978), it is commonly assumed to be so, particularly in zone models (Quintiere, 1989, 2008). As the walls heat up relatively slowly, the layer of gas close to the wall will cool and become negatively buoyant with respect to the surrounding gas (Cooper, 1984; Jaluria,

[4] If the ceiling is of unlimited extent, the outward flow will eventually cease as the gases cool and lose buoyancy, when they will tend to 'fall' and mix with the air below. This is a recognized problem regarding smoke control in shopping centres (Section 11.3.2).

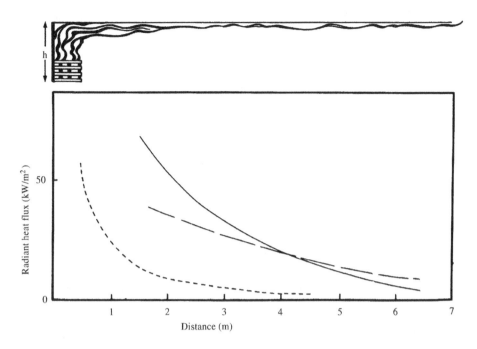

Figure 9.4 Radiative heat flux at floor level as a function of distance from a crib fire of constant heat output located at the closed end of a corridor. The dashed line (. . . . .) represents the distribution of heat flux from the same fire in the absence of the ceiling (vertical fire plume only). The other lines represent different values of *h* (the height of the ceiling above the crib base): _____, *h* = 970 mm; _ _ _, *h* = 660 mm (from Thomas, 1975b). Reproduced by permission of The Controller, HMSO. © Crown Copyright

1995). A downward flow will be generated which will carry smoke into the clear, cool air below – when it will then become positively buoyant and rise towards the layer again. This will contribute to some mixing of smoke in the lower layer. The temperature will remain highest near the ceiling, and the ceiling and upper walls will be heated, initially by convection, although radiation will come to dominate as the temperature of the layer rises and the concentration of the smoke particles increases.

As the fire grows in size, the flame will eventually penetrate the smoke layer. When this happens, the upper part of the flame will entrain increasingly vitiated air, already contaminated with fire products. This will lead to an increase in the rate of accumulation of partially burnt fuel vapours and particulate smoke (Beyler, 1985; Zukoski, 1986; Mowrer, 2008). As the temperature of the layer continues to rise, it will radiate downwards with increasing intensity, enhancing not only the rate of burning of the existing fire (Equation (9.3)), but also the rate of flame spread over contiguous surfaces (Section 7.2.4). At the same time, the smoke layer will continue to descend, albeit more slowly than in the early stages, enveloping an increasing proportion of the growing flame. This is the situation that is likely to exist at the onset of flashover, when the fire begins its rapid transition to its fully developed stage. It is interesting to consider the fact that an upper layer temperature of 600°C has been identified as one indicator of the onset of flashover, while the average temperature at the tip of a buoyant diffusion flame is c. 550°C. A logical conclusion is that once the tips of flame reach the ceiling, the onset of flashover will not be far behind (Rasbash, 1974).

Before considering the subsequent events, it is important to consider the nature and composition of the upper layer at this stage. These have been examined by carrying out experiments in which a stable, buoyant diffusion flame has been allowed to burn beneath a hood, designed to mimic the situation described above in which the upper part of the flame penetrates the smoke layer (Figure 9.5). It is also an experimental idealization of the 'zone model' in which the upper and lower layers are considered separately. By varying the supply rate of gaseous fuel and the distance of the burner below the lower edge of the hood, Beyler (1984b) was able to examine the effect of varying the equivalence ratio (ϕ) on the composition and behaviour of the layer. This quantity (ϕ) is the ratio of fuel to air in the mixture entering the layer, normalized to the stoichiometric ratio (see Section 1.2.4, Equation (1.26)): it is determined by the amount of 'clean' air that is entrained into the vertical fire plume below the base of the hot layer.[5] He found that the layer would begin to burn when the equivalence ratio exceeded approximately 1.8. The entrainment of hot smoke, containing a progressively decreasing concentration of oxygen, into the upper part of the flame inhibits the combustion process and causes the concentration of unburned and partially burnt fuel vapours in the layer to increase as ϕ increases. In particular, the level of carbon monoxide increases significantly as the equivalence ratio increases above 0.6 (Figure 9.6). Its yield increases asymptotically to a high level ($>10\%$

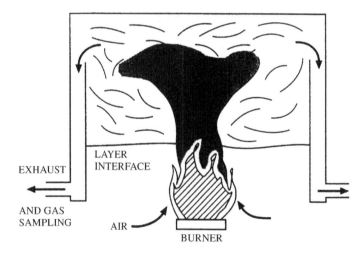

Figure 9.5   Diagram showing the two-layer system used by Beyler (1984b) to study the smoke layer that forms under the ceiling during a compartment fire. By permission of the Society of Fire Protection Engineers

[5] This is determined from the rate of extraction required to maintain the layer at a constant height above the burner (see Figure 9.5). The rate of air entrainment is calculated by subtracting the rate of fuel supply from the total mass flowrate (cf. Equation (9.12)).

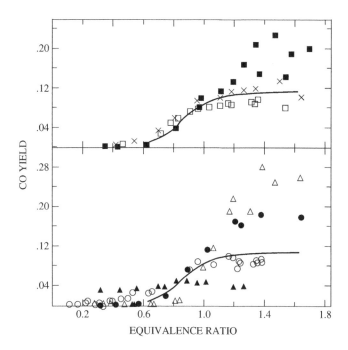

Figure 9.6 Normalized carbon monoxide yields as a function of equivalence ratio (ϕ) for propane (___); propene (○); hexanes (□); toluene (▲); methanol (△); ethanol (●); isopropanol (×); and acetone (■) (Beyler, 1986a). By permission of Elsevier

of its theoretical maximum value for $\phi > 1.2$). This corresponds to levels of CO of up to 6%, and is the source of the highly toxic smoke that can flow from the room of origin through an open door into the rest of the building during the period immediately preceding flashover (Beyler, 1986a; Zukoski *et al.*, 1991).

In a series of steady state experiments, Beyler (1984b) found that when the equivalence ratio exceeded 0.8 to 1.0, sufficient unburned and partially burned products had accumulated for independent flaming to occur at the lower face of the smoke layer, located close to the region where the main fire plume penetrated. If the rate of fuel supply was increased so that $\phi > 1.8$, then flame was observed to spread across the interface, independently of the original fire plume. Beyler was unable to pursue this further with his experimental facility, but he identified an equivalency ratio of this order as marking a significant stage in the development of a fire. The development of burning at the interface will cause the layer to become unstable and undergo mixing with air from below. It can be argued that this is part of the flashover process, consistent with the observation of Hagglund *et al.* (1974) that flames emerge from the openings 'when flashover occurs'. Subsequently, Gottuk *et al.* (1992a) found that in a series of compartment fire experiments the onset of sustained external flaming was associated with $\phi > 1.7$ (see Section 11.1.4).

There is now good evidence that a large proportion of the heat radiated to the lower part of a compartment during fire growth is from the layer of hot smoky combustion products (Croce, 1975; Quintiere, 1975a). This is discussed briefly by Tien *et al.* (2008), but earlier, Orloff *et al.* (1979) described a sophisticated model for calculating radiative output from the smoky gases under the ceiling of a poorly ventilated compartment, taking into account the stratification of temperature and combustion products within the smoke layer. The problem is complicated by the fact that although radiation is being emitted by H_2O, CO_2 and soot throughout the layer, the lower levels of the layer will be cooler and will tend to absorb radiation emitted from the hotter levels nearer the ceiling.[6] Orloff *et al.* (1979) compared their theoretical model with the results of a full-scale test fire in which a slab of polyurethane foam was burned under conditions of restricted ventilation. They found excellent agreement between measured and predicted radiant heat flux levels at the floor during the development stage of the fire and from their calculations showed that only a small proportion (less than 15% of the radiant heat reaching the floor) came from the ceiling and upper walls of the enclosure; during the pre-flashover stage, radiation from the layer of smoky gases dominated. Of course, this represents a particular case. The contribution from the hot walls and ceiling would be greater if the layer was relatively thin and/or the yield of particulate smoke was significantly less, although the net effect would not be distinguishable.

Waterman's original criterion for the onset of flashover (20 kW/m^2 at floor level) can only be achieved when the fire has grown to a size sufficient to create conditions under the ceiling which can provide this heat flux. Normally the latter will increase monotonically as the fire grows, but if the fuel burns out or oxygen depletion occurs, then flashover may not be reached (see Figure 9.1).

Oxygen depletion will be significant if the rate of ventilation to the compartment is insufficient to replenish oxygen as it is consumed by the developing fire. In an extreme case, a fast burning fire (e.g., one involving a hydrocarbon pool, or an upholstered chair bursting into flame after a prolonged period of smouldering) in a small, poorly venti-lated compartment, may self-extinguish or (more likely) die down and burn slowly at a rate dictated by the ingress of air. Provided the ambient temperature within the compart-ment remains high, weak flaming can continue under severely vitiated conditions (e.g., at oxygen concentrations as low as 8%,) but the combustion process will be incomplete and unburned and partially burned pyrolysis products will continue to accumulate in the space. Bullen (1978) considered the conditions under which this would be important in domestic fires and concluded (on the assumption that the rate of burning will be con-trolled by the rate of ventilation (see Section 10.1)) that, theoretically, burning could continue, supported by leakage of air around reasonably well-fitting doors and windows. Under these circumstances, the fire could create additional ventilation (e.g., by burning through a wooden door), which will permit a progressive increase in intensity. However, in modern homes there is much less 'leakage' due to concerns about energy conservation (e.g., double glazing units rather than traditional sash windows) (Foster, 1997) and there is more chance that a flaming fire will effectively self-extinguish, perhaps dying down to a smouldering fire, if doors and windows are closed. Under these circumstances, if addi-tional ventilation is provided suddenly (e.g., by the opening of a door or the breaking of a window), the ingress of fresh air may be followed by rapid burning of the accumulation of unburned smoke and fuel vapours, producing a backdraught.

This process was studied in detail by Fleischmann *et al.* (1994) (Figure 9.7) and at around the same time research began in Sweden, Japan and the UK. An early review was

---

[6] This effect is responsible for the reduction in the emissivity that is observed for flames of large diameter hydrocarbon pool fires (see Sections 2.4.3 and 4.4.1 (Figure 4.35)).

(a)

(b)

Figure 9.7 Development of backdraught (Fleischmann *et al.*, 1994). A 70 kW methane flame was burned in a sealed chamber, the flame eventually self-extinguishing due to oxygen starvation. The vent was opened (on the right): a continuous ignition source was present towards the rear of the enclosure. (a) 5.6 s after opening the vent; (b) 7.1 s after opening the vent; (c) 8.0 s after opening the vent. By permission

(c)

Figure 9.7   (*continued*)

prepared by Chitty (1994) on behalf of the UK Home Office (the government department with responsibility for the UK Fire Service). This surge of interest was driven by concerns for the safety of firefighters and the realization that the mechanism of backdraught was poorly understood. Fleischmann demonstrated that a key factor in the development of a backdraught is the 'gravity current' of cold air that flows into the compartment at low level as hot buoyant (fuel-rich) gases flow out through the opening at high level. The consequences of this can be seen in Figure 9.7(b). More recent studies (e.g., Bengtsson and Karlsson, 2001; Gojkovich and Karlsson, 2001) of the phenomenon have confirmed these conclusions and explored other variables that affect the outcome, such as the position, size and shape of the ventilation opening (Weng *et al.*, 2003) and the fuel mass fraction within the compartment when backdraught is initiated (Gottuk *et al.*, 1999). In general, relatively low overpressures are created, but forces capable of causing structural damage can be generated. It is likely that this will be associated with small ventilation openings, vigorous mixing during the development of the gravity current and ignition at a location remote from the ventilation opening (compare these remarks with the discussion of vented explosions in Section 3.6). It has been noted (Foster, 1997) that more severe backdraughts are obtained if the temperature of the compartment is increased.

The term 'smoke explosion' has been used by some to describe such events: these were reviewed by Croft (1980/81). An oft-quoted example of a smoke explosion is the incident (explosion followed by fire) which occurred at Chatham Dockyard in 1975, in a warehouse in which rubber latex mattresses were stored (Anon., 1975; Woolley and Ames, 1975; Croft, 1980/81) (see Section 8.2).

There is a very clear distinction between flashover and backdraught. Backdraught is preceded by burning under conditions of very poor ventilation and is initiated by the sudden influx of air when a door opens or a window breaks. The development of flashover occurs under conditions where a fire in a compartment already has adequate ventilation (e.g., see Figure 9.2) (Drysdale, 1996).

Thomas *et al.* (1980) have shown how flashover can be regarded as a case of thermal instability within the compartment, analogous in many ways to the Semenov model for spontaneous ignition (Section 6.1). The starting point is to assume that the rate of burning is a function of temperature but is limited by the rate of supply of air. (This limit is the ventilation-controlled fire in which the rate of burning is determined by the maximum rate at which air can flow into the compartment during fully developed burning (Section 10.1).) These authors have considered a quasi-steady state model of the fire by considering the energy balance of the hot layer under the ceiling (Figure 9.8). The rates of heat generation and heat loss are compared, recalling that both are functions of temperature of the upper layer (T), thus:

$$\dot{Q}_c(T, t) = \dot{L}(T, t) \tag{9.5}$$

where $\dot{Q}_c(T, t) = R$ is the rate of heat release – i.e., the rate of energy gain by the upper layer – and $\dot{L}(T, t) = L$ is the rate of heat loss as shown in Figure 9.9, where several simplifying assumptions are made, e.g.

$$L = \dot{L}(T, t) = h_k A_T (T - T_0) + (\dot{m}_a + \dot{m}_f) c_p (T - T_0) \tag{9.6}$$

where h_k is an effective heat transfer coefficient (referring to heat losses through the compartment boundaries) and A_T is the appropriate internal surface area through which heat is lost (McCaffrey *et al.*, 1981). For the simplest case (constant fuel bed area), the rate of heat release is determined by radiative heat transfer from the upper layer, and is thus a

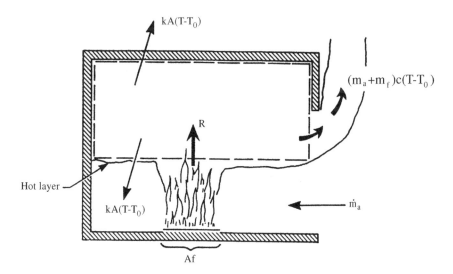

Figure 9.8 Diagram showing the energy balance for the hot layer in an enclosure fire (from Thomas *et al.*, 1980). Reproduced by permission of the Combustion Institute

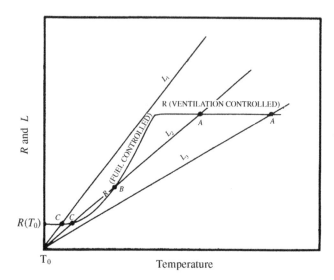

Figure 9.9 Flashover as a thermal instability. The line marked 'R' represents the rate of heat release within the compartment as a function of temperature, while the lines marked 'L' are representative of heat loss functions. Reproduced by permission of the Combustion Institute from Thomas *et al.* (1980)

strongly non-linear function of temperature. However, it is limited by the rate of supply of air, as shown schematically in Figure 9.9, where the rate of heat release curve is compared with three representative heat loss curves, L_1, L_2 and L_3, which correspond (for example) to decreasing compartment sizes. Three types of intersection for the heat release (R) and heat loss (L) curves are possible, namely A, B and C (cf. Figure 6.1). While B is unstable, A corresponds to a 'steady' ventilation-controlled fire (Chapter 10), corresponding to a high rate of burning, and C represents the small localized fire that is not influenced to any significant extent by the heat feedback from the upper parts of the enclosure. With a growing fire, both R and L will change and can exhibit a criticality in which any slight increase in rate of burning would cause a substantial jump in both temperature and rate of burning to point A. Thomas *et al.* (1980) drew attention to the fact that this behaviour is 'suggestive of the phenomenon of flashover' and take the model much further to show how various types of instability can be explained on the basis of such diagrams. Their prediction that 'oscillating' fires might occur has been confirmed experimentally (Takeda and Akita, 1981). Subsequently, Bishop *et al.* (1993), Graham *et al.* (1995) and others have explored how the techniques of non-linear dynamics can be used to analyse this model quantitatively and predict how flashover depends on parameters such as the ventilation conditions and the thermal characteristics of the compartment boundaries (Holborn *et al.*, 1993).

9.2.2 Fuel and Ventilation Conditions Necessary for Flashover

Although the fully developed fire will be discussed in detail in Chapter 10, one result is required here, namely that the rate of air inflow into a fully developed fire is given by the expression:

$$\dot{m}_a = 0.5 A_w H^{1/2} \text{ kg/s} \tag{9.7}$$

where A_w and H are the area and height of the ventilation opening, respectively (see Section 10.1). (The term 'ventilation factor' $A_w H^{1/2}$ was originally identified by Kawagoe (1958) in his classic study of compartment fires (Figure 10.1).) Hägglund and his co-workers (1974) carried out a series of experiments in which they burned wood cribs in a compartment measuring 2.9 m × 3.75 m × 2.7 m high, monitoring the rate of burning continuously by weighing the fuel. They presented their data graphically as a plot of $\dot{m}$ vs. $A_w H^{1/2}$ and found that those fires that 'flashed over' (i.e., flames out of the door and temperatures in excess of 600°C under the ceiling) occupied a narrowly defined region of the diagram. These are identified as solid symbols in Figure 9.10. Flashover was not observed for rates of burning less than about 80 g/s (twice that quoted by Waterman (1966)), although this limit increased as the ventilation was increased in accordance with the empirical expression:

$$\dot{m}_{\text{limit}} = 50.0 + 33.3 A_w H^{1/2} \text{ g/s} \tag{9.8}$$

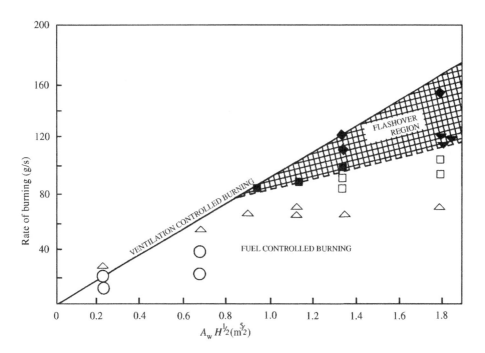

Figure 9.10 Compartment fire burning rate data of Hägglund *et al.* (1974) as a function of the ventilation parameter $A_w H^{1/2}$. Fuel: wood cribs, mass 18–36 kg (○), 45–55 kg (△), 60–74 kg (□), 80–91 kg (▽), 98–103 kg (◆). Solid symbols indicate flames out of the doorway and ceiling gas temperatures ≥ 600°C (from Hägglund *et al.*, 1974). Quintiere (1976), reproduced by permission of ASTM

While flashover was not observed for values of $A_w H^{1/2}$ less than $0.8\,\mathrm{m}^{5/2}$, only a limited range was studied and the upper limit for Equation (9.8) was not determined. As will be shown in Section 10.1, the rate of burning becomes independent of ventilation at large values of $A_w H^{1/2}$ when the transition to the fully developed fire (i.e., 'flashover') may be ill-defined, or not occur (e.g., Takeda and Akita, 1981) (see Figure 10.12).

Although this work was confined to a single compartment of height 2.7 m, it does suggest a more general principle, that a limiting burning rate must be exceeded – and presumably maintained for a period of time – before flashover can occur. Indeed, single items of furniture can lead to flashover if their rates of burning are high enough (Klitgaard and Williamson, 1975; Babrauskas, 1979): thus, Babrauskas (1979) observed that flashover (defined as 20 kW/m^2 at floor level) was achieved in 280 s by a single modern chair consisting of block polyurethane foam covered by acrylic fake-fur burning in a compartment 2.8 m high. The burning rate 'peaked' at 150 g/s. A leather chair also exhibited a high rate of burning, peaking at 112 g/s, but too briefly for the flashover criterion to be met.

While the concept of a limiting burning rate for flashover to occur is based on experimental observation, it would seem more logical to examine critical rates of heat release within the compartment, particularly in view of Thomas' contention that flashover represents a thermal instability (Thomas et al., 1980). Babrauskas (1980b) examined results from 33 room fire tests and found that the rate of heat release that corresponded to the onset of flashover lay between 23% and 86% (average 50%) of the rate of heat release predicted by Equation (9.7) if stoichiometric burning took place within the compartment (see Deal and Beyler (1990)). The stoichiometric rate of heat release is calculated as $\Delta H_{c,\mathrm{air}} \dot{m}_a$, where $\Delta H_{c,\mathrm{air}}$ can be assumed to be 3000 kJ/kg(air) (see Table 1.13), and is therefore $1500 A_w H^{1/2}$ kW. Taking the average of 50%, the rate of heat release at the onset of flashover becomes:[7]

$$\dot{Q}_{FO} = 750 A_w H^{1/2}\ \mathrm{kW} \tag{9.9}$$

Thomas (1981) carried out a heat balance on the upper layer, introducing a number of simplifying assumptions, and incorporating heat losses through the compartment boundaries (total area A_T) and convective losses through the ventilation opening. He derived the following expression:

$$\dot{Q}_{FO} = 7.8 A_T + 378 A \sqrt{H} \tag{9.10}$$

which is compared with the other correlations below (Figure 9.13).

McCaffrey et al. (1981) have taken this further, starting with a model developed previously by Quintiere et al. (1978). This also involved a simple heat balance applied to the layer of hot gases below the ceiling (Figure 9.8). Combining Equations (9.5) and (9.6), we have:

$$\dot{Q}_c = (\dot{m}_a + \dot{m}_f) c_p (T - T_0) + h_k A_T (T - T_0) \tag{9.11}$$

---

[7] Babrauskas (1980b) derived a very similar equation ($\dot{Q}_{FO} = 600 A_w H^{1/2}$ kW) from a heat balance for the upper layer as given in Equation (9.5) and a set of very simple assumptions, viz. at flashover, the upper layer temperature is 600°C, and that heat losses are by radiation to the floor and the walls below the hot layer.

where T and T_0 are the temperatures of the upper layer and the ambient atmosphere, respectively. It is assumed that the layer is well mixed and its temperature is uniform. Writing

$$\dot{m}_g = \dot{m}_a + \dot{m}_f \tag{9.12}$$

and rearranging Equation (9.11) gives:

$$\frac{\Delta T}{T_0} = \frac{\dot{Q}_c/(c_p T_0 \dot{m}_g)}{1 + h_k A_T/(c_p \dot{m}_g)} \tag{9.13}$$

The mass flowrate of gas leaving the compartment above the neutral plane (Figure 9.8: see also Section 10.1) can be approximated by the expression developed by Rockett (1976) for flow of air induced by a small fire:

$$\dot{m}_g = \frac{2}{3} C_d A_w H^{1/2} \rho_0 \left(2g \frac{T_0}{T} \left(1 - \frac{T_0}{T} \right) \right)^{1/2} \left(1 - \frac{h_0}{H} \right)^{3/2} \tag{9.14}$$

where h_0 is the height of the neutral plane and C_d is the discharge coefficient. For the present purposes (see below), Equation (9.14) may be reduced to the proportionality:

$$\dot{m}_g \propto g^{1/2} \rho_0 A_w H^{1/2} \tag{9.15}$$

where $g = 9.81$ m/s and ρ_0 is the density of ambient air. Using this relationship, Equation (9.13) can then be cast in general terms, giving $\Delta T/T_0$ as an unspecified function of two dimensionless groups, namely:

$$\frac{\Delta T}{T_0} = f \left(\frac{\dot{Q}_c}{g^{1/2}(c_p\rho_0)T_0 A_w H^{1/2}}, \frac{h_k A_T}{g^{1/2}(c_p\rho_0)A_w H^{1/2}} \right) \tag{9.16}$$

or

$$\frac{\Delta T}{T_0} = C \cdot X_1^N \cdot X_2^M \tag{9.17}$$

where X_1 and X_2 represent the two dimensionless groups in Equation (9.16), and the constant C and exponents N and M remain to be determined from experimental data.

McCaffrey et al. (1981) analysed data from more than 100 experimental fires (from eight series of tests involving several types of fuel) in which steady burning rates were achieved, but upper gas layer temperatures did not exceed 600°C. Above 600°C, flaming can occur intermittently within the layer and many of the assumptions inherent in the model break down. To enable these data to be cast in the form suggested by Equation (9.17), it is necessary to obtain appropriate values for h_k which depend on the duration of the fire and the thermal characteristics of the compartment boundary. For a fire which burns with a characteristic time (t_c) greater than the 'thermal penetration time' (t_p) of the boundary ($t_p = \delta^2/4\alpha$, where α is the thermal diffusivity ($k/\rho c$) and δ is the boundary thickness – see Equation (2.41) et seq.) – h_k can be approximated by

$$h_k = k/\delta \tag{9.18}$$

where k is the thermal conductivity of the material from which the compartment boundaries have been constructed (i.e., heat is conducted in a quasi-steady state manner to the exterior). If, on the other hand, t_c is less than t_p, the boundary will be storing heat during the fire and little will be lost through the outer surface. Normally, this would require detailed solution of the transient heat conduction equations (Section 2.2.2), but a simplification can be achieved by replacing δ by $(\alpha t_c)^{1/2}$, the effective depth of the lining material that is heated significantly during the course of the fire (the 'thermal thickness' – see Equation (2.24b) et seq.). In these circumstances:

$$h_k = \frac{k}{(\alpha t_c)^{1/2}} = \left(\frac{k\rho c}{t_c}\right)^{1/2} \tag{9.19}$$

For a compartment bounded by different lining materials, the overall value of h_k must be weighted according to the areas: thus, if the walls and ceiling (W,C) are of a different material to the floor (F), then, if $t_c > t_p$:

$$h_k = \frac{A_{W,C}}{A_T} \cdot \frac{k_{W,C}}{\delta_{W,C}} + \frac{A_F}{A_T} \cdot \frac{k_F}{\delta_F} \tag{9.20a}$$

but if $t_c < t_p$, then:

$$h_k = \frac{A_{W,C}}{A_T} \left(\frac{(k\rho c)_{W,C}}{t_c}\right)^{1/2} + \frac{A_F}{A_T} \left(\frac{(k\rho c)_F}{t_c}\right)^{1/2} \tag{9.20b}$$

where A_T is the total internal surface area.

The experimental data were found to fit satisfactorily to the form of Equation (9.17): multiple linear regression analysis on its logarithm allowed the constants N and M to be evaluated. Figure 9.11 shows the correlation between ΔT and the dimensionless variables X_1 and X_2 when the floor is included in the calculation of h_k and A_T. If the floor is excluded, the correlation is equally good, if slightly different. McCaffrey et al. (1981) suggest that the expression:

$$\Delta T = 480 \, X_1^{2/3} \cdot X_2^{-1/3} \tag{9.21}$$

(in which ΔT is calculated with $T_0 = 295$ K) describes both situations adequately, and that providing the thermal characteristics of the floor and of the wall and ceiling linings are not too dissimilar, it does not make a significant difference if floor losses are ignored.

Equation (9.21) can be used to estimate the size of fire necessary for flashover to occur. If a temperature rise of 500 K is taken as a *conservative* criterion for the upper layer gas temperature at the onset of flashover (cf. Hägglund et al., 1974) then substitution for X_1 and X_2 in Equation (9.21) (see Equation (9.16)) gives (after rearrangement):

$$\dot{Q}_c = \left[g^{1/2}(c_p\rho_0)T_0^2\left(\frac{\Delta T}{480}\right)^3\right]^{1/2} (h_k A_T A_w H^{1/2})^{1/2} \tag{9.22}$$

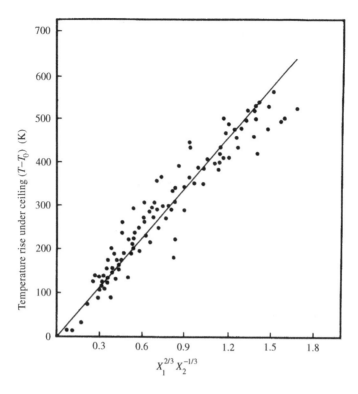

Figure 9.11 Correlation of the upper room gas temperatures with two dimensionless variables (no floor). The line represents the least squares fit of eight sets of data. From McCaffrey *et al.* (1981), by permission

With $\Delta T = 500$ K (which assumes a critical layer temperature of $525°C^8$) and taking appropriate values for g, etc.

$$\dot{Q}_{FO} = 610(h_k A_T A_w H^{1/2})^{1/2} \tag{9.23}$$

where h_k is in kW/m²·K, A_T and A_w are in m² and H is in m, where $\dot{Q}_{FO}$ (kW) is the rate of heat output necessary to produce a hot layer at approximately $525°C$ beneath the ceiling. The square root dependence indicates that if there is 100% increase in any of the parameters h_k, A_T or A_w, then the fire will have to increase in heat output by only 40% to achieve the flashover criterion as defined. The effect of changing the thermal properties of the lining material is shown in Figure 9.12. For highly insulating linings, such as fibre insulating board, or expanded polystyrene, the size of the fire sufficient to produce flashover is greatly reduced, even if the linings are assumed to be inert and any contribution to the rate of heat release is neglected.

These correlations are based on a limited set of data and would not be expected to apply to fires in large compartments, significantly different in size and shape from those

---

[8] Peacock *et al.* (1999) take $\Delta T = 575$ K and thus obtain the equation $\dot{Q}_{FO} = 740(h_k A_T A_w H^{1/2})^{1/2}$. Here, the upper layer temperature at flashover is assumed to be $600°C$. This is shown in Figure 9.13.

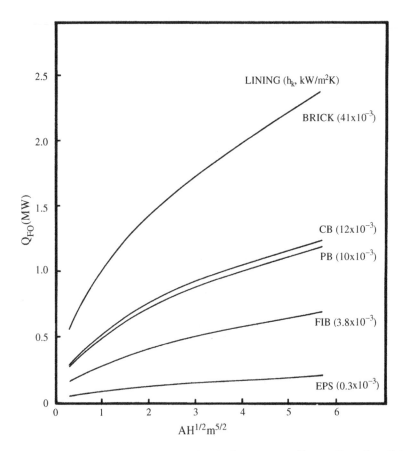

Figure 9.12 Rate of heat release necessary for flashover (according to Equation (9.23)) in a compartment 4 m × 6 m × 2.4 m high as a function of $A_w H^{1/2}$ for different boundary materials. CB, chipboard: PB, plasterboard; FIB, fibre insulating board; EPS, expanded polystyrene. The lining materials were assumed to be inert and thermally thick, and the characteristic time of the fire was taken to be $t_c = 1000$ s. The values for h_k are suspect (see original references). From McCaffrey *et al.* (1981), by permission

involved in the correlation.[9] Although two subsets of the data were obtained in small-scale experiments, others refer to near-cubical full-scale compartments with heights in the range 2.4 ± 0.3 m. As compartment height is not specified *per se* in these equations (it is incorporated into A_T), Equation (9.23) should not be applied to elongated geometries. Similarly, it is likely to break down for compartments with restricted ventilation in which the assumption of a uniform temperature within the layer is no longer tenable (Orloff *et al.*, 1979). Moreover, the data were obtained from fires set in the centre of the compartment. If

---

[9] Babrauskas *et al.* (2003) have observed that higher than expected values of $\dot{Q}_{FO}$ are recorded for rapidly developing fires in which the time to flashover is less than 100–150 s.

Table 9.1   Variation of $\dot{Q}_{FO}$ with location of the fire[a]

Location of fire	$\dot{Q}_{FO}/kW$
In the centre	475
Next to a wall	400
In a corner	340

[a]$3\,m \times 3\,m \times 2.3\,m$ high compartment (Lee, 1982).

the fire is next to a wall, or in a corner, the minimum burning rate for flashover is reduced for reasons developed in Section 4.3.3. This is illustrated by results obtained by Lee (1982), who studied experimental fires in full-scale and quarter-scale compartments. The relevant information is summarized in Table 9.1: this refers to a full-scale compartment, although most of the data were scaled up from the small-scale tests.

Mowrer and Williamson (1987) have shown how this effect can be quantified using plume theory (Morton *et al.*, 1956), although they predict much greater differences in the ratios $\dot{Q}_{FO}$ (wall)/$\dot{Q}_{FO}$ (centre) and $\dot{Q}_{FO}$ (corner)/$\dot{Q}_{FO}$ (centre) than were suggested by Lee (1982).

Equations (9.8), (9.9), (9.10) and (9.23) are compared in Figure 9.13. It should be noted that Hägglund's experiments (Hägglund *et al.*, 1974) provided one of McCaffrey's databases (McCaffrey *et al.*, 1981). The difference between these two correlations (lines C and E) may be explained by the fact that there is no 'safety factor' incorporated into Hagglund's equation (line E), whereas the correlation of McCaffrey *et al.* (line C) is very conservative. Instead of assuming a critical temperature of 600°C in the upper layer,

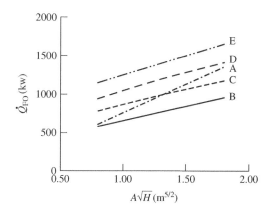

Figure 9.13   Comparison of the correlations for $\dot{Q}_{FO}$: **A**, Equation (9.9) (Babrauskas, 1980b); **B**, Equation (9.10) with $A_T = 35\,m^2$ (Thomas, 1981); **C**, Equation (9.23) with $h_k A_T = 2.03$ kW/K (including floor) (McCaffrey *et al.*, 1981); **D**, as **C**, but with $\Delta T = 575$ K in Equation (9.22) (Peacock *et al.*, 1999); **E**, Equation (9.8) with the heat of combustion of wood equal to 15 kJ/g (Hägglund *et al.*, 1974)

they chose 525°C. Peacock *et al.* (1999) modified McCaffrey's correlation by reverting to 600°C (see footnote 6), generating a correlation (line D) closer to that of Hagglund, but still (apparently) 'conservative'. The equations derived by Babrauskas (1980b) and Thomas (1981) led to even more cautious predictions of the rates of heat release that would lead to flashover.

Clearly, more data are required to test these correlations, but Equation (9.23) offers a very conservative approach to identifying the maximum rate of heat release that may be permitted in a room with a view to *avoiding* flashover. However, its application to the assessment of the flashover potential of 'real' compartments requires information on the rates of heat release during the burning of the contents, both as separate items and in combination, as $\dot{Q}_{FO}$ may only be achieved if the fire has spread to involve more than one item. A simple logic diagram to illustrate this point is shown in Figure 9.14.

Before the development of the technique of oxygen consumption calorimetry (Huggett, 1980), rates of heat release from individual items of furniture had to be calculated from data on their mass loss rates (e.g., Quintiere, 1976), assuming an 'effective' heat of combustion (Section 1.2.3). It is now possible to determine the rate of heat release of items commonly found in buildings, including furniture, office equipment, wall linings, etc. on a full scale using large heat release calorimeters of the type described by Babrauskas *et al.* (1982); Nordtest (1991); Babrauskas (1992b, 2008a). The method involves burning an item below a hood from which the fire products are removed through a duct (Figure 9.15(a)). The temperature and flowrate of the gases in the duct are measured continuously, as are

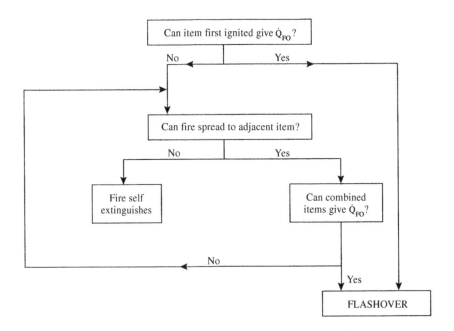

Figure 9.14 Logic diagram for flashover in a compartment

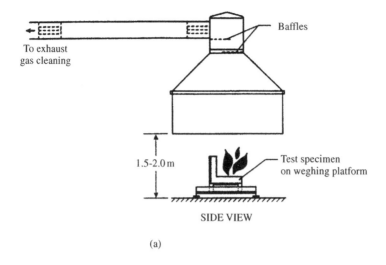

1.5-2.0 m

To exhaust gas cleaning

Baffles

Test specimen on weghing platform

SIDE VIEW

(a)

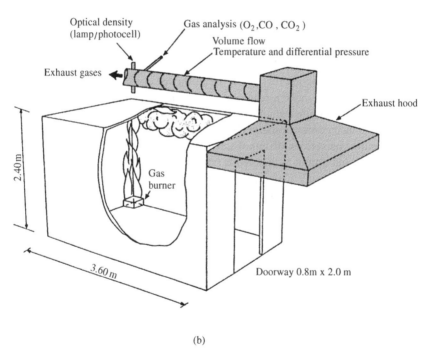

Optical density (lamp/photocell)

Gas analysis (O_2, CO , CO_2)

Volume flow
Temperature and differential pressure

Exhaust gases

Exhaust hood

2.40 m

Gas burner

3.60 m

Doorway 0.8m x 2.0 m

(b)

Figure 9.15 (a) The Nordtest furniture calorimeter (Nordtest, 1991). By permission of Elsevier. (b) The ISO room-corner test (ISO 9705:1993) was developed for testing wall lining materials. The gas burner is run for 10 min at 100 kW and then (if necessary) for another 10 min at 300 kW (ISO, 1993b). 'Flashover' is said to have occurred when flames emerge from the doorway. By permission of John Wiley & Sons

the concentrations of oxygen, carbon dioxide and carbon monoxide. The rate of heat release is calculated on the basis of the oxygen deficiency, taking into account incomplete combustion of CO to CO_2 (Section 1.2.3). As the combustion products are removed continuously, there is no thermal feedback from the enclosing boundary or an accumulating smoke layer. However, Babrauskas (1980b) has argued that the effect of thermal feedback will be small up to the end of the growth period. Thomas *et al.* (1980) suggest an increase in the rate of burning of c. 50%, while Quintiere (1982) quotes evidence that shortly before flashover the rate of burning is 'nearly equal to the free burn value'.

These statements are in apparent conflict with Alpert's conclusion based on the third full-scale bedroom fire test (Alpert, 1976) that the rate of burning of individual items just prior to flashover had increased by 150–200%. The difference may lie in the definitions for the 'onset' of flashover, as well as the complex nature of the items of 'fuel'. Alpert was reporting on the results of full-scale, furnished bedroom burn tests that were allowed to proceed to fully developed, ventilation-controlled burning (Section 10.1). In contrast, Babrauskas (1984a) made a careful comparison of the rates of heat release from upholstered items that were burning freely (as in Figure 9.15(a)) with identical items burning in an enclosure similar to that shown in Figure 9.15(b). These enclosure fires were designed with sufficient ventilation to prevent the fires becoming ventilation controlled after 'flashover' (see Section 10.1).[10] Under these circumstances, critical heat fluxes of 20 kW/m^2 were reached at floor level, but the rates of heat release were only slightly greater than those for the freely burning items. In these enclosure fires, Babrauskas (1984a) noted that when the flashover criterion was reached (20 kW/m^2 at floor level) the rate of heat release began to increase very rapidly, thus making it difficult to derive an accurate value for $\dot{Q}_{FO}$.

Supporting evidence comes from the CBUF project – the European research programme into the combustion behaviour of upholstered furniture (Sundström, 1995) – in which it was found that the maximum heat release rates for 27 items of upholstered furniture in the ISO Room Test (Figure 9.15(b)) (ISO, 1993b) were on average only 10% higher than values found in the furniture calorimeter[11] (Nordtest, 1991), although it should be noted that none of the CBUF items gave rise to flashover in the room test. Nevertheless, this is not inconsistent with the observations of Babrauskas (1980b, 1984a) and Quintiere (1982).

Examples from the CBUF Report of the heat release rate curves of two armchairs are given in Figure 9.16, partly to illustrate typical rate of heat release results, but also to demonstrate how performance depends on the materials used in the construction of upholstered furniture, in this case the change from a high resilience foam to a 'combustion modified' foam. The possibility of deducing full-scale fire behaviour of items of upholstered furniture from data obtained from the cone calorimeter was examined in some detail in the CBUF project (Sundström, 1995), but at the present time it must be concluded that it is not possible to predict reliably the rate of heat release curves from such bench-scale tests. Earlier, Babrauskas and Walton (1986) proposed a correlation method which was based on a rather limited, contemporary set of data, but in view of the large variety of upholstered items that are now in existence, there seems to be no reliable alternative to carrying out full-scale burns at the present time.

---

[10] This is a slightly 'grey area' that emphasizes the difficulty in developing a robust definition of flashover. One such definition links flashover to the switch-over to a ventilation-controlled fire. This is discussed in Section 10.1.

[11] The difference was negligible for low rates of heat release (<150 kW) but increased to >20% at 1000 kW.

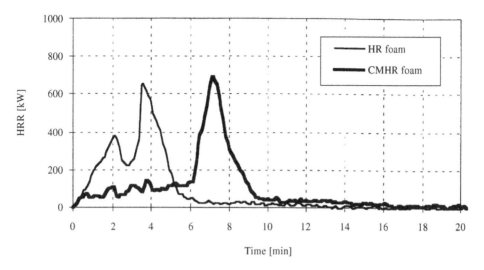

Figure 9.16 Heat release curves of two armchairs covered with cotton fabric, one filled with high resilience PU foam, the other with combustion modified HR foam (Sundström, 1995). © Interscience Communications

It will be noted that a significant period of steady state burning for a given item may not occur (e.g., see Figure 9.16), and the peak rate of heat release may not be sufficient to create the conditions for flashover: however, such information will be indicative of the potential of the item to contribute to the flashover process. The logic diagram of Figure 9.14 can be applied to assess the magnitude of the problem in a given space, recalling that flame spread to adjacent items may require close proximity ($< 1\,m$, according to Babrauskas (1982), see Section 9.2.4) or the involvement of a combustible floor covering. When two or more items are burning, there may be enhancement of the total rate of heat release due to cross-radiation (see Section 9.2.3). This cannot be generalized, but it has to be taken into account in the assessment of the flashover potential of a given space.

Data are now available on the rate of heat release of a variety of items of furniture and building contents, including upholstered furniture, mattresses, pillows, wardrobes, television sets, curtains, waste containers and Christmas trees (Babrauskas, 2008). To illustrate this point, some of these are give in Table 9.2, but the reader is referred to the source documents for full information.

The importance of wall linings in fire development has been recognized for several decades, but the first tests were designed before there was a proper understanding of fire behaviour. As a consequence, several different tests were developed independently, on a national basis, all purporting to be able to achieve the same objective – to identify wall lining materials which would present a threat to life in the event of fire. These included (*inter alia*) two British Standard Tests ('Surface Spread of Flame Test' (BSI, 1997) and 'Fire Propagation Test' (BSI, 1989)), the German 'Brandschacht Test' (DIN, 1998) and the US 'Steiner Tunnel Test' (ASTM, 2009). Unfortunately, they were quite unsuitable for engineering applications as they could only be used to 'rank' materials according to their behaviour in each test. Quite arbitrary limits had to be set on what was considered to be

Table 9.2 Maximum (peak) rates of heat release from some typical items (see also Babrauskas, 1995a, 2008)

Item	Maximum rate of heat release (kW)	Source
Chair, wooden frame, HR foam, cotton fabric	650	Sundström (1995)
Chair, wooden frame, CMHR foam, cotton fabric	700	Sundström (1995)
Latex foam pillow, 50/50 cotton/polyester fabric	117	Babrauskas (1984/85)
Wardrobe, 68 kg, 12.7 mm thick plywood	3500	Lawson et al. (1983)
Curtain (closed), 117 g/m^2 cotton/polyester	267	Moore (1978)
Curtain (open), 117 g/m^2 cotton/polyester	303	Moore (1978)
Christmas tree (dry), 7.0 kg	650	Ahonen et al. (1984)
Waste container (0.63 kg polyethylene) with empty milk cartons (0.41 kg)	13	Ahonen et al. (1984)

'acceptable behaviour'. The rankings were highly apparatus-dependent, as demonstrated by Emmons (1974) for six of the European standard fire tests. This led directly to the move towards a new generation of tests – the 'reaction to fire tests' – initiated by the International Organization for Standardization (ISO, 1993, 1997a, 2002). These tests are now well established and, for the first time, provide data which can be used in engineering calculations. The principal tests are the cone calorimeter (Babrauskas, 1984b, 1995b; British Standards Institution, 1993; ISO, 2002) and the LIFT apparatus developed by Quintiere (1981) (ISO, 2006; ASTM, 2008).

There has been considerable success in using data from the cone calorimeter (CC) to model upward flame spread on vertical surfaces (Section 7.3), which has prompted work to investigate if this test protocol could be used to assess the likely fire performance of wall lining materials at full scale. There have been two approaches: in one, the CC data are used as the basis for a correlation that can be used to rank materials, while in the second, the data are used in an appropriate model to predict the rate of fire development in a room-corner assembly. Both approaches have been studied with reference to standard room-corner tests such as Nordtest Method NT FIRE 025 (ISO, 1993b) (Figure 9.15(b)). Thus, Östman and Nussbaum (1988) found that the quantity

$$2.76 \times 10^6 \times \frac{\tau_{25}\sqrt{\rho}}{Q''_{pp,50}} - 46 \text{ seconds}$$

(where τ_{25} is the time to ignition (s) at 25 kW/m^2, $Q''_{pp,50}$ is the heat released during the peak period (J/m^2) at 50 kW/m^2 in the CC and ρ is the density of the material (kg/m^3)) produced a ranking order for 13 materials which correlated very well with that based on the time to flashover[12] when the same materials were tested according to NT FIRE 025 (Figure 9.17). If such a correlation proved to be robust and reliable, it would be possible to adopt the cone calorimeter as the means of ranking lining materials, avoiding the expense of full-scale testing. However, it still only produces a ranking order, with respect to one scenario, viz a single item burning in a corner.

---

[12] Taken as the time at which flames first emerge from the door opening. The test is terminated at this point, when flashover is deemed to have occurred.

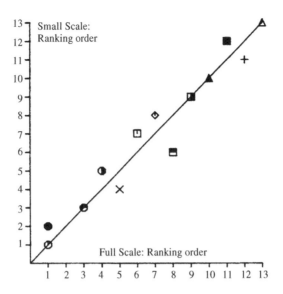

Figure 9.17 Comparison of the ranking orders obtained for wall covering materials in the cone calorimeter and in the ISO Room Fire Test (Östman and Nussbaum, 1988). By permission

Consequently, this method has the same inflexibility as the original tests. Ideally, we should be able to use mathematical models to predict the rate of fire growth in a real scenario from bench-scale test data. As a first step, a number of models have been developed to predict the rate of flame spread over lining materials in the room-corner tests (ASTM, 1982; ISO, 1993b) (e.g., Magnusson and Sundstrom, 1985; Cleary and Quintiere, 1991; Karlsson, 1993; Quintiere, 1993) (see Section 7.3). Karlsson (1993) has shown how a graphical representation developed by Baroudi and Kokkala (1992) can be used to identify lining materials which will give flashover in the room-corner test, essentially by examining whether or not vertical flame spread will accelerate (Figure 9.18). For those materials which gave flashover in the room-corner test, it was found that

$$\lambda_{50}\tau_{50} \le (1 - \sqrt{a})^2 \tag{9.24}$$

where λ_{50} (s^{-1}) is the decay coefficient as shown in Figure 9.19, τ_{50} is the time to ignition (s) and $a = K\dot{Q}''_{max,50}$ ($\dot{Q}''_{max,50}$ being the maximum rate of heat release and K a 'flame length coefficient' based on work by Parker (1982) and Saito *et al.* (1985) and taken to be 0.015 m^2/ kW), all at 50 kW/m^2 in the cone calorimeter. Although this is also limited to a single scenario, it suggests a useful way forward. However, the goal should be to develop models that are able to predict fire growth rates from 'flammability data' obtained in bench-scale tests. This would provide a more rational approach to the selection of materials in fire safety engineering design and analysis (Lautenberger *et al.*, 2006).[13]

---

[13] It is interesting to compare where we are now (2010/11) with the predictions published in 1984 by Emmons (1984).

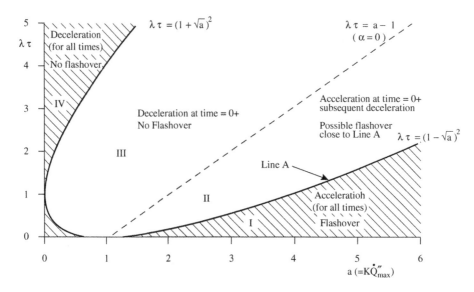

Figure 9.18   Regions of flame front acceleration and deceleration (see Equation (9.24)) (Karlsson, 1993). By permission of Elsevier

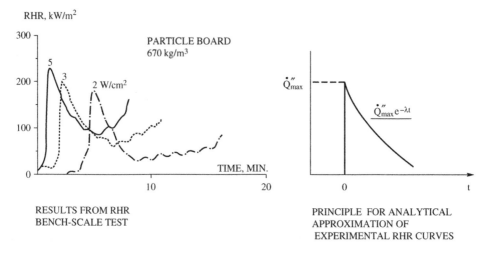

Figure 9.19   Representation of the rate of heat release from a material (Karlsson, 1993). By permission of Elsevier

9.2.3   Factors Affecting Time to Flashover

As the duration of the pre-flashover fire has direct relevance to the safety of life (Equation (9.1) *et seq.*), it is important to know how the fuel and ventilation parameters affect the initial rate of growth. In a major research programme (the 'Home Fire Project') (e.g., Croce, 1975) carried out jointly by Harvard University and Factory Mutual Research

Table 9.3 CIB study of the pre-flashover fire[a]

Variable	Level 1	Level 2
Shape of compartment[b]	$1 \times 2 \times 1$	$2 \times 1 \times 1$
Position of ignition source[c]	Rear corner	Centre
Fuel height	160 mm	320 mm
Ventilation opening	Full width	Quarter width
Bulk density of fuel[d]	20 mm	60 mm
Fuel continuity	One large crib	21 small cribs
Lining (walls and ceiling)[e]	None	Hardboard
Ignition source area[c]	16 cm^2	144 cm^2

[a]The fuel was in the form of wood cribs (beech). The sticks were 20 mm square and the moisture content was ~10%.
[b]Figures refer to width × depth × height (all in metres).
[c]The ignition source was a tray of alcohol.
[d]Figures refer to stick spacing.
[e]The inner surface of the compartment (unlined) was asbestos millboard.

Corporation,[14] a wealth of data was obtained on the growth of fire in a full-scale, fully furnished bedroom.[15] While this has encouraged further study of many of the factors recognized as contributing to fire growth (e.g., Quintiere *et al.*, 1981) and has enabled sophisticated mathematical models of fire behaviour to be developed (e.g., Mitler and Emmons, 1981; Mitler, 1985; Friedman, 1992; Quintiere, 2008; Walton *et al.*, 2008), there are so many potential variables it is difficult to determine which are the most significant parameters. A systematic study of all the variables requires that a very large number of experiments be carried out. The Fire Commission (W14) of the Conseil Internationale du Bâtiment (CIB) launched such a programme in the late 1960s in which nine laboratories around the world participated (Japan, The Netherlands, Australia, USA (FMRC[14] and NBS[16]), UK, Germany, Canada and Sweden). The secretariat, which was located at the Fire Research Station, UK, coordinated the experimental programme as well as the collection and analysis of the data (Heselden and Melinek, 1975). It was decided to carry out tests using small-scale compartments with wood cribs forming the fuel bed, looking at the effects of changing each of the eight variables listed in Table 9.3. Each was studied at two levels, as shown. The number of individual tests carried out was $2^8 = 256$, which were allocated randomly around the nine laboratories: each one was duplicated and some additional tests were performed at FMRC and NBS. The times at which each of the following events occurred were noted in each experiment, namely:

t_f = the time for flames to reach the ceiling;
t_2 = the time to the final transition from slow to fast rate of spread over the upper surface of the crib(s); and
t_3 = the time to flaming over the whole of the upper surfaces of the crib(s).

---

[14] Now FM Global.
[15] These tests pre-dated the development of oxygen consumption calorimetry.
[16] Now NIST.

The last (t_3) was taken as the time to flashover (Heselden and Melinek, 1975). A multiple regression analysis of the results led to the following conclusions:

(i) t_3 was not significantly affected by the shape of the compartment;
(ii) t_3 was only slightly dependent on the size of the ventilation opening and the continuity of the fuel, although with the former there may have been some interaction with the laboratory conditions;
(iii) t_3 was much more sensitive to the position and the area of the ignition source, the height of the fuel bed, the bulk density of the fuel and the nature of the lining material. Of these, the last was *not* the most significant.

The effects of the latter group can be understood in terms of a simple fire growth model. The following refer to first-order effects (interactions are considered subsequently):

(a) *Ignition source*: t_3 was shorter with a central ignition source because the area of the initial fire increases more rapidly. Similarly, the larger area of ignition gave a shorter t_3 as initially a larger area of fuel was involved.
(b) *Fuel height*: With a high fuel bed, flames reach the ceiling much more rapidly, thereby promoting the spread of fire over the combustible surfaces at an earlier stage.
(c) *Bulk density*: The cribs of low bulk density tend to spread fire more rapidly (Section 7.4), hence the diameter of the fire increases at a greater rate and flashover is achieved much earlier. In terms of a real fire, this might correspond to fire spread between adjacent items of low thermal capacity.
(d) *Lining material*: A combustible lining material reduces the time to flashover, but this variable was not the most significant. Bruce (1953) noted that in a full-scale compartment fire with central ignition, combustible wall linings did not become involved until the fire was well advanced (after the flames touched the ceiling).

Several important interactive effects were noted, some of which are intuitively obvious. The most significant was that between the position of the ignition source and the nature of the wall lining material. If the lining is combustible and becomes involved as the result of direct ignition from a source in the corner, then the time to flashover is greatly reduced. A similar, but less dramatic, interaction exists between height and bulk density of the fuel bed. While these results refer to small-scale tests, relevance to 'real' fires is readily deduced.

However, not all factors that might be deemed to be important were examined. The most serious omission was the height of the ceiling, but in addition the effects of moisture content of the fuel, the relative humidity and air movement (draughts) within the enclosure were not considered. During the period 1973/74, three tests were carried out at Factory Mutual Research Corporation[13] in association with Harvard University on the development of fire in a bedroom, each test furnished and ignited identically (Croce, 1975). These showed a variation in the times to flashover (Table 9.4), the greatest difference being between Test 1 and Tests 2 and 3. It was found that although the furnishings between Tests 1 and 2 were similar, they were not identical. The furnishings for Test 3 were carefully

Table 9.4 Times to flashover in the 'bedroom fire tests' carried out at FMRC (Croce, 1975)

Test number	Date	Time to flashover (s)[a]
1	11 July 1973	1055
2	24 July 1974	429
3	31 July 1975	391

[a]Flashover in Test 1 was well defined. Those given for Tests 2 and 3 are averaged values of time to 'equivalent involvement' (e.g., ignition of a book lying on the floor) as the transition was not as 'dramatic' (Croce, personal communication).

selected to be the same as in Test 2, yet there was still a difference in the time to flashover. This was thought to be due largely to a difference in relative humidity – by a process of elimination. These results were the first to indicate that the early stages of a developing fire are very sensitive to random variations in a number of factors (such as relative humidity) to the extent that it may not be possible to predict the earliest stages of development of a fire with any certainty. Indeed, in the CBUF project (Sundström, 1995), the investigators found that for fires involving upholstered furniture, the stage from ignition to 50 kW was not reproducible, although great care was taken to try to achieve identical conditions in each replicate test. The stage from 50 kW to 400 kW was much more repeatable. The implication here is that once a fire has grown beyond a certain size (taken as 50 kW in this case), it is powerful enough to overcome minor stochastic influences.

Another factor that can affect the time to flashover is the thermal inertia ($k\rho c$) of the compartment boundary (cf. Figure 9.12). This was first reported in the 1950s, following some full-scale fires carried out at the Fire Research Station, UK, in a compartment 4.5 m square by 2.7 m high, containing wooden furniture at a density of $22 \, kg/m^2$ ($5 \, lb/ft^2$). The time to flashover varied enormously with the density of the wall lining, as can be seen in Table 9.5 (Joint Fire Research Organisation, 1962). Similar results from small-scale compartment fires were obtained by Waterman (Pape and Waterman, 1979). Limited evidence of the effect was found in some extra experiments carried out at NBS as part of the CIB programme (Heselden and Melinek, 1975). Theoretical work by Thomas and Bullen (1979) suggests that the time to flashover is directly proportional to the square root of the thermal inertia, but with a lesser dependence for a fast growing fire.

The importance of this effect on fire development in buildings can be overstated. The need to conserve energy has led to increased thermal insulation in all types of building, but it is only if the insulating material (of low $k\rho c$) is exposed as the internal surface of the enclosure that a marked effect on the time to flashover would be observed. Normally the insulation is protected by at least a layer of plasterboard so that, providing the thermal penetration time (t_p) of the protective layer is greater than the characteristic time of the fire (t_c) as defined by McCaffrey and co-workers (1981), then the insulation will not affect the time to flashover significantly. On the other hand, insulation (e.g., cavity wall insulation) will tend to increase the 'severity' of a fully developed fire (Section 10.3).

Table 9.5 Variation of time to flashover with the density of the wall lining material (Joint Fire Research Organization, 1962)

Nature of wall finish	Density[a] (kg/m^3)	Time to flashover (min)
Brick	1600	23.5
Lightweight concrete A	1360	23.0
Lightweight concrete B	800	17.0
Sprayed asbestos	320	8.0[b]
Fibre insulating board[c]	($\sim$300)	6.75

[a]Details are not available on k and c to allow the thermal inertia to be calculated. However, density on its own is an adequate indicator.

[b]A repeat of this experiment with the same finish which had been completely dried during the first test gave a time to flashover of 4.5 min.

[c]This combustible lining is included for comparison. Its density is not known, but it is noted that the sprayed asbestos was chosen to have the same density and thermal conductivity as the fibre insulating board (Joint Fire Research Organization, 1961).

9.2.4 Factors Affecting Fire Growth

The CIB programme involved wood crib fires in small-scale compartments, and although fuel continuity was varied by comparing the behaviour of one single crib with that of a large number of small cribs, the results give little information about fire spread between isolated 'packets' of fuel in a real fire situation. Unless the item first ignited is capable of producing (and at least sustaining) the necessary heat output for flashover to occur, then development to flashover will require that other items of fuel become involved (Figure 9.14). Only in this way can the rate of burning increase.

Whether or not an adjacent item will ignite depends on its separation from the item already burning. It may be sufficiently close and in a suitable configuration for direct flame impingement to occur (Section 4.3.4), but if this is not possible, fire can only spread by a mechanism involving radiant heat transfer. This was originally considered by Theobald (1968), who showed that the radiation from a fire involving a traditional upholstered chair could ignite cotton cloth 0.15 m away, while a burning wardrobe could ignite the same material 1.2 m away. Pape and Waterman (1979) have commented, on the basis of a series of full-scale tests, that fire will not spread from a single upholstered chair to a neighbouring one if the separation is more than 30 cm.

However, this statement is too general as the properties of both the burning item and the 'target' must determine the outcome. Fang (1975b) and Babrauskas (1982) have attempted to 'map' the distribution of radiant flux around items of burning furniture ranging from stacking chairs to storage cabinets. Not unexpectedly, it was found that the radiant heat flux at a distance was a function of the rate of burning. Items which burned rapidly produced large flames that were capable of giving substantial radiant heat fluxes at up to 1 m from their leading edges. In Figure 9.20, the distribution of radiant heat flux near

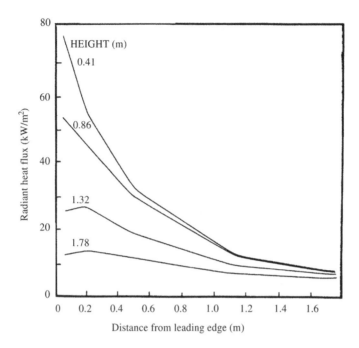

Figure 9.20   Distribution of radiant heat flux at various heights near a couch at maximum rate of burning (polyurethane foam with polypropylene fabric, wicker frame). Babrauskas (1982), by permission

a hospital couch (wicker frame, polyurethane foam, covered by polypropylene fabric) at peak burning, is shown as a function of distance from the front edge and height above the platform on which the couch was resting. In principle, this type of information could be obtained for all items of furniture, and compared with the 'ignitability' of target materials, perhaps expressed as the heat flux necessary for pilot ignition to occur within a certain period of time (say 40 s). While this will depend substantially on the thermal properties of the 'target' (Section 6.3), Babrauskas (1982) concluded from the results of a wide range of items of furniture that if the 'target item' was more than 1 m away from a burning upholstered chair, ignition would be unlikely. Even with the most rapidly burning item of furniture (in Babrauskas' paper, this was the couch described in Figure 9.20), a heat flux exceeding 20 kW/m² was not found beyond 0.88 m. Information of this type is required to allow an assessment to be made of the potential for fire to spread directly from one fuel 'package' to the next. While it may be possible to use the theoretical approach described in earlier chapters to calculate rates of burning and radiant heat flux from flames above burning items (Section 4.4.1), there is a continuing need at present to use empirical data. Additional information of this type is still required, but combined with ignitability data, it should be possible to examine the flashover potential of a compartment from a more rational point of view.

However, fire spread from the item first ignited may proceed by direct flame impingement between contiguous items or across a combustible floor covering, such as a synthetic carpet. The rate of development will be strongly dependent on the configuration of the combustible surfaces involved and will be enhanced by radiative heat transfer from the item(s) that are already burning, e.g., to the adjacent floor covering. There are particular scenarios which are associated with rapid flame spread, the most obvious of which is the combustible wall lining. A lining material may be deemed to be 'acceptable' according to a standard test, but under certain conditions it could be exposed to high heat fluxes capable of overcoming any inherent resistance to ignition.[17] This can occur if an item of furniture is burning close to, or against the wall: indeed, the heat flux is very sensitive to the 'stand-off' distance, or gap between the item and the wall, as demonstrated by Williamson *et al.* (1991). Indeed, any configuration in which heat is conserved in the vicinity of the burning surface(s) – e.g., by cross-radiation – is likely to be hazardous. Thus, Morris and Hopkinson (1976) observed very rapid development of a fire that was initiated in a waste paper basket between a bed and the side of a wooden wardrobe.

Another hazardous scenario that should be considered involves those synthetic materials that either melt or produce a molten product that will form a pool on the floor and continue to burn. Certain polyurethane foams behave in this way. If an upholstered item or a bed with one of these foams is ignited, a burning pool will form in the enclosed space under the item. Very vigorous burning will result and it is possible that molten polymer will flow beyond the 'footprint' of the item, thereby assisting the spread of the fire.

Despite these uncertainties, it is found that the rates of development of many fires approximate to a parabolic growth ('t^2 fire') after an initial incubation period (Heskestad, 1982), thus:

$$\dot{Q} = \alpha_f (t - t_0)^2 \tag{9.25}$$

where α_f is a fire growth coefficient (kW/s^2) and t_0 is the length of the incubation period (s). This is shown schematically in Figure 9.21. The coefficient α_f lies in the range 10^{-3} kW/s^2 for very slowly developing fires to 1 kW/s^2 for very fast fire growth (Table 9.6). The incubation period (t_0) is difficult to quantify (see Section 9.2.3) and will depend *inter alia* on the nature of the ignition source, its location and the properties of the item first ignited. Data are now available on fire growth rates on a number of commodities and fuel arrays which may be expressed in these terms (Babrauskas, 1979, 1983a, 2008a; National Fire Protection Association, 2010). This is useful in engineering calculations relating, for example, to the response of ceiling-mounted heat detectors to a growing fire when it is necessary to take the growth rate into account (Alpert, 2008; National Fire Protection Association, 2010).

Although the nature of the materials can sometimes determine the outcome, a common factor in many 'fast fire spread' scenarios is localized confinement of combustible materials, such as the bed/wardrobe arrangement referred to above (Morris and Hopkinson, 1976). Unusual or unexpected flow dynamics can be created which generate high rates of heat transfer and promote rapid flame spread: a good example is the Kings Cross escalator fire (see Section 7.2.1; Drysdale and Macmillan (1992)). Indeed, combustible materials ignited in confined spaces will burn vigorously due to cross-radiation between hot, burning surfaces (e.g., a fire developing beneath a bed or under a floor). This mechanism can

---

[17] This is less likely to be a problem for lining materials that have been tested for the full 20 minutes in the room-corner test (ISO 1993b) in which heat fluxes in excess of 100 kW/m^2 are generated.)

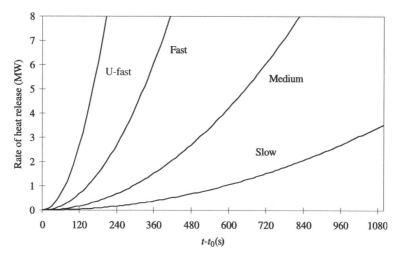

Figure 9.21 Parabolic fire growth from the end of the incubation period ($t = t_0$) according to Equation (9.25) for the fire growth coefficients given in Table 9.6

Table 9.6 Parameters used for 't-squared fires' (Alpert, 2008)

Description	Typical scenario	α_f (kW/s^2)
Slow	Densely packed paper products[a]	0.00293
Medium	Traditional mattress/boxspring[a]	0.01172
	Traditional armchair	
Fast	PU mattress (horizontal)[a]	0.0469
	PE pallets, stacked 1 m high	
Ultrafast	High-rack storage	0.1876
	PE rigid foam stacked 5 m high	

[a]National Fire Protection Association (2010).

account for rapid fire development in buildings where there are service ducts, vertical shafts and ceiling voids containing combustible materials. Many multiple-fatality fires in buildings have been shown to have started (often unseen) in hidden spaces and cavities: when the fire eventually breaks out, not only is it large but it may have spread far beyond its origin. Examples include the Summerland Leisure Complex, Isle of Man (Silcock and Hinkley, 1974), the MGM Grand Hotel fire of 1981 (Best and Demers, 1982), the Bradford Football Stadium fire (Popplewell, 1986) and the Dusseldorf Airport Terminal fire (NFPA, 1996). Further discussion may be found in Section 10.7.

Problems

9.1 Estimate how long it would take for the concentration of oxygen in a poorly ventilated room, 5 m $\times$ 8 m $\times$ 3 m high, to fall to 10% by volume (the concentration below which flaming combustion is unlikely to continue under these circumstances) if a settee is burning on its own with an *average* rate of heat release of 300 kW. Assume that the average temperature of the room reaches 500 K and that there is no air entering the room during this period. (It is also necessary to assume that $\Delta H_{c,ox}$ remains constant at 13 kJ/g of oxygen.)

9.2 How much oxygen will remain in this room at 10 minutes if the CMHR foam chair referred to in Figure 9.16 burns out in the closed room of Problem 9.1, without involving any other furniture or furnishings? Approximate the rate of heat release as 100 kW for six minutes and 300 kW for four minutes.

9.3 Using Equation 9.23, calculate the size of fire necessary to give flashover in the following rooms, assuming that the basic structure comprises brick walls and concrete slab floor and ceiling:
(a) 4 m × 6 m × 2.5 m high with a single ventilation opening (a door) 0.9 m × 2 m high. The walls and ceiling are lined with plaster.
(b) As (a) but with a ceiling height of 3.5 m.
(c) As (a) but with the walls and ceiling lined with 20 mm thick fire-retardant fibre insulation board (ignore possible involvement of the lining).
(d) As (a) but with one window open (area 2.0 m × 1.5 m high) (see Equation (10.37)).
Assume in all of these that the characteristic time of the fire is 10^3 s.

10

The Post-flashover
Compartment Fire

After flashover has occurred, the exposed surfaces of all combustible items in the room of origin will be burning and the rate of heat release will develop to a maximum (Figure 9.1). High temperatures will be achieved, typically 900–1100°C, but under certain conditions much higher temperatures can be reached.[1] This is the fully developed fire which will continue until the rate of generation of flammable volatiles begins to decrease as a result of fuel consumption. It is during the post-flashover stages of the fire that elements of a building structure can become heated to temperatures at which they are no longer able to fulfil their function, e.g., carry their design load. Depending on the type of structure, this may cause local or more general collapse (see, for example, Buchanan (2002)). Compartment walls (boundaries) which may or may not be load-bearing may also 'fail' by permitting fire spread into adjacent spaces by flame penetration or excessive transmission of heat (BSI, 1987).

Traditionally, fire resistance is measured empirically as the time to failure of the element of structure in a standard furnace test (ISO, 1975; British Standards Institution, 1987; ASTM, 2008b). This was originally designed to model the thermal environment to which an element of structure would be exposed in a 'real' fire. However, the validity of this method needs to be examined in the light of our knowledge of the behaviour of fully developed fires. Ventilation to the fire compartment and the nature, distribution and quantity of fuel all have a significant effect on duration and severity: the concepts of 'fire severity' and 'fire resistance' in relation to real fires require careful definition. These will be introduced and discussed in subsequent sections.

10.1 Regimes of Burning

The first systematic study of the behaviour of the fully developed compartment fire was carried out in Japan in the late 1940s. Kawagoe and his co-workers (Kawagoe, 1958)

---

[1] Temperatures in excess of 1300°C are occasionally reached, but the conditions must be right. For example, the Summit Rail Tunnel fire (Department of Transport, 1986) produced sufficiently high temperatures to cause the faces of brick-lined ventilation shafts to fuse.

An Introduction to Fire Dynamics, Third Edition. Dougal Drysdale.
© 2011 John Wiley & Sons, Ltd. Published 2011 by John Wiley & Sons, Ltd.

measured the rate of burning of wood cribs contained within compartments with different sizes of ventilation opening. Full-scale and reduced-scale tests were carried out, the characteristic dimension of the smallest compartment being less than 1 m. The burning rate ($\dot{m}$) was found to depend strongly on the size and shape of the ventilation opening, the results correlating very well with the relationship

$$\dot{m} = 5.5 A_{\mathrm{w}} H^{1/2} \text{ kg/min}$$

$$= 0.09 A_{\mathrm{w}} H^{1/2} \text{ kg/s} \tag{10.1}$$

where A_{w} and H are the area (m^2) and height (m) of the ventilation opening, respectively (Figure 10.1). However, the numerical constant is somewhat ill-defined (Thomas *et al.*, 1967a; Thomas and Heselden, 1972) and it is found that the correlation only holds over a limited range of values of $A_{\mathrm{w}} H^{1/2}$. The conventional interpretation is that within this range the rate of burning is controlled by the rate at which air can flow into the compartment: such a fire is said to be 'ventilation-controlled'. However, if the ventilation opening is enlarged, a condition will be reached beyond which the rate of burning becomes independent of the size of the opening and is determined instead by the surface area and burning characteristics of the fuel. Indeed, it is found that this change from ventilation to

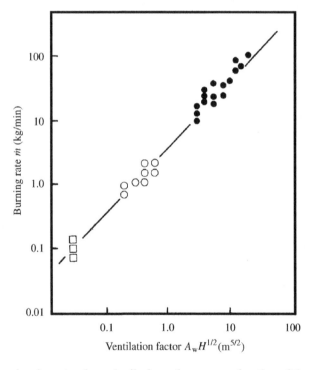

Figure 10.1 Mass burning rate of wood cribs in enclosures as a function of the ventilation factor, $A_{\mathrm{w}} H^{1/2}$ for ventilation-controlled fires (Equation (10.1)): ●, full-scale enclosures; ○, intermediate-scale models; ⊓, small-scale models (Kawagoe and Sekine, 1963). Reproduced by permission of Elsevier Applied Science Publishers Ltd

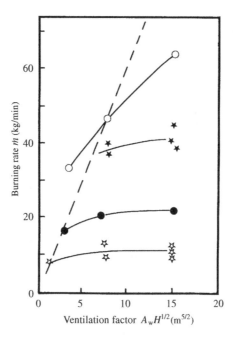

Figure 10.2 Variation of mass burning rate with $A_w H^{1/2}$ for large ventilation openings and different fire loads (wood cribs): ✩, 7.5 kg/m²; ●, 16 kg/m²; ✦, 30 kg/m²; ○, 60 kg/m². Dashed line $(- - -)$ represents Equation (10.1) for the ventilation-controlled fire (Thomas *et al.*, 1967a). Reproduced by permission of The Controller. HMSO. © Crown copyright

fuel control depends predominantly on the surface area of combustible material: this was noted by Gross and Robertson (1965) and by Thomas *et al.* (1967a) (see Figure 10.2), who made a formal distinction between the two regimes of burning by referring to them as 'Regime I' (ventilation-controlled) and 'Regime II' (fuel-controlled).

While the ventilation factor $A_w H^{1/2}$ was deduced semi-empirically by Kawagoe (1958), it can be derived by a theoretical analysis of the flow of gases in and out of a burning compartment. The following assumptions must be made:

(a) The gases in the compartment behave as if they are 'well stirred', i.e., their properties are uniform throughout the volume (some support for this is to be found in Figure 10.3, in which it is shown that, except near the floor, temperature gradients virtually disappear at flashover (Croce, 1975; Nakaya *et al.*, 1986)).
(b) There is no net flow created by buoyancy *within* the compartment.
(c) Hot gases leave the compartment above a neutral plane and cold air enters below it (Figure 10.4).
(d) The flow of gases *in* and *out* of the compartment is driven by buoyancy forces.
(e) There is no interaction between the inflowing and outflowing gases.

The compartment is thus modelled as a 'well-stirred reactor', as shown in Figure 10.4. The induced flows are caused by the differences in pressure between the compartment and the outside, which are a direct consequence of the high internal temperature: this has

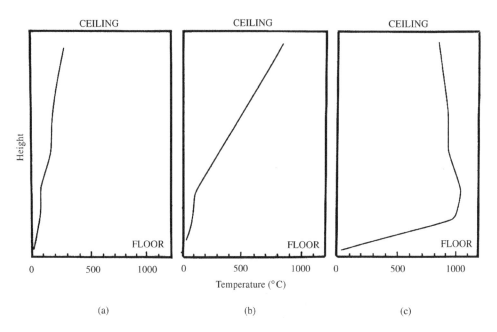

Figure 10.3 Vertical temperature distribution in a compartment fire (data from the Home Fire Project (Croce, 1975)): (a) pre-flashover (5.67 min after ignition); (b) incipient flashover (6.11 min); (c) 'time of observed flashover' (7.05 min) (Babrauskas and Williamson, 1978). Reproduced with permission

been studied experimentally by McCaffrey and Rockett (1977), whose results vindicate the theoretical treatment that follows.

The horizontal flow out of the compartment along any streamline above the neutral plane can be calculated from Bernoulli's equation if the pressures inside and outside are known (see Emmons and Tanaka, 2008). Following the treatment presented by Babrauskas and Williamson (1978), the pressure inside the compartment at a height y above the neutral plane (point 1 in Figure 10.4) will be:

$$P_1 = P_0 - \rho_1 g y \qquad (10.2a)$$

where P_0 is the atmospheric pressure on the neutral plane (i.e., at $y = 0$). Just outside the ventilation opening, at point 2, the pressure of the issuing jet will be equal to the atmospheric pressure at that level, i.e.

$$P_2 = P_0 - \rho_0 g y \qquad (10.2b)$$

(Note that on the neutral plane $P_1 = P_2 = P_0$ and there is no net flow.) Bernoulli's equation can be used to relate the conditions at points 1 and 2 thus:

$$\frac{P_1}{\rho_1} + \frac{v_1^2}{2} = \frac{P_2}{\rho_2} + \frac{v_2^2}{2} \qquad (10.3)$$

where v_1 and v_2 are the net horizontal flowrates at the two points. Well away from the incoming jet, there will be no directional flow as the mixture inside the compartment is highly turbulent. Thus, v_1 will be zero, and Equation (10.3) can be written:

$$\frac{P_0 - \rho_1 g y}{\rho_1} = \frac{P_0 - \rho_0 g y}{\rho_1} + \frac{v_2^2}{2} \qquad (10.4)$$

in which it is assumed that the gas issuing from the opening at position 2 will be at the same temperature (and density, ρ_1) as the gas at position 1. This rearranges to give:

$$v_2 = \left(\frac{2(\rho_0 - \rho_1) g y}{\rho_1} \right)^{1/2} \qquad (10.5)$$

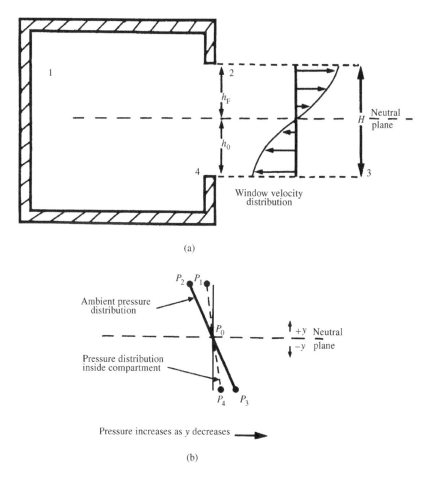

(a)

(b)

Figure 10.4 Buoyancy-driven flows through a ventilation opening during a fully developed fire. (a) Vertical section showing window flows; (b) pressure distributions outside (———) and inside (- - - - -) the compartment (after Babrauskas and Williamson, 1978). Reproduced with permission

A similar analysis can be carried out for the inflowing air, where

$$P_3 = P_0 - \rho_0 g y \tag{10.6a}$$

and

$$P_4 = P_0 - \rho_4 g y \tag{10.6b}$$

y being negative below the neutral plane. Using the subscripts 'F' (for the compartment gases) and '0' (for the ambient air), we then have two equations:

$$v_F = \left(\frac{2(\rho_0 - \rho_F)g y}{\rho_F} \right)^{1/2} \tag{10.7}$$

and

$$v_0 = \left(\frac{2(\rho_F - \rho_0)g y)}{\rho_0} \right)^{1/2} \tag{10.8}$$

which refer to the horizontal flows at any height y above or below the neutral plane, respectively. These velocities are normally quite low, on the order of 5–10 m/s. Equations (10.7) and (10.8) can then be used to calculate the mass flowrates:

$$\text{Inflow}: \quad \dot{m}_{air} = C_d B \rho_0 \int_{-h_0}^{0} v_0 \, dy \tag{10.9}$$

$$\text{Outflow}: \quad \dot{m}_F = C_d B \rho_F \int_{0}^{h_F} v_F \, dy \tag{10.10}$$

where C_d is a discharge coefficient, B is the width of the window (m), $\dot{m}$ is the mass flow (kg/s) and $h_F + h_0 = H$, as shown in Figure 10.4. These lead to:

$$\dot{m}_{air} = \frac{2}{3} C_d B (h_0)^{3/2} \rho_0 \left(2g \frac{\rho_0 - \rho_F}{\rho_0} \right)^{1/2} \tag{10.11}$$

and

$$\dot{m}_F = \frac{2}{3} C_d B (h_F)^{3/2} \rho_F \left(2g \frac{\rho_0 - \rho_F}{\rho_0} \right)^{1/2} \tag{10.12}$$

If the overall chemical reaction taking place within the compartment can be expressed as

$$1 \, \text{kg fuel} + r \, \text{kg air} \rightarrow (1 + r) \, \text{kg products}$$

or more generally, if the burning is non-stoichiometric

$$1 \, \text{kg fuel} + \frac{r}{\phi} \text{kg air} \rightarrow \left(1 + \frac{r}{\phi} \right) \text{kg products}$$

where ϕ is a correction factor, then

$$\frac{\dot{m}_F}{\dot{m}_{air}} = \frac{1 + r/\phi}{r/\phi} = 1 + \frac{\phi}{r} \tag{10.13}$$

The height of the neutral plane (h_0) may be expressed as a fraction of the total height of the ventilation opening (H) by substituting $\dot{m}_{air}$ and $\dot{m}_F$ from Equations (10.11) and (10.12) into Equation (10.13). Writing $h_F = H - h_0$ and rearranging:

$$\frac{h_0}{H} = \frac{1}{1 + [(1 + (\phi/r))^2 \cdot \rho_0/\rho_F]^{1/3}} \tag{10.14}$$

Using typical values of ϕ, r and ρ_F, the ratio h_0/H works out to be 0.3–0.5, consistent with general observations of the fire plume emerging from the ventilation opening of a compartment (Figure 10.7(b)). If it is assumed as an approximation that $\dot{m}_F = \dot{m}_{air}$ (i.e., $\phi/r = 0$), then substituting h_0 from Equation (10.14) into Equation (10.11) gives

$$\dot{m}_{air} \approx \frac{2}{3} A_w H^{1/2} C_d \rho_0 (2g)^{1/2} \left(\frac{(\rho_0 - \rho_F)/\rho_0}{[1 + (\rho_0/\rho_F)^{1/3}]^3} \right)^{1/2} \tag{10.15}$$

As the ratio ρ_0/ρ_F normally lies between 1.8 and 5 for post-flashover fires (Babrauskas and Williamson, 1978), the square root of the density term can be approximated by 0.21. Then with $\rho_0 = 1.2$ kg/m^3, $C_d = 0.7$ (Prahl and Emmons, 1975; Steckler *et al.*, 1984[2]) and $g = 9.81$ m/s^2, the rate of inflow of air can be approximated by:

$$\dot{m}_{air} \approx 0.52 A_w H^{1/2} \text{ kg/s} \tag{10.16}$$

If stoichiometric burning occurs within the compartment (i.e., $\phi = 1$ in Equation (10.13)), then for wood the rate of burning must be:

$$\dot{m}_b \approx \frac{0.52}{5.7} A_w H^{1/2} = 0.09 A_w H^{1/2} \text{ kg/s}$$
$$= 5.5 A_w H^{1/2} \text{ kg/min} \tag{10.17}$$

since the stoichiometric air requirement for the combustion of wood is approximately 5.7 kg air/kg wood (Section 4.4.3).[3]

The remarkable agreement with Kawagoe's original correlation (Equation (10.1)) must be regarded as fortuitous in view of the many simplifying assumptions that are made, but the emergence of the 'ventilation factor' $A_w H^{1/2}$ is significant.

However, by assuming that stoichiometric burning occurs within the compartment (Equation (10.17)), it is implied that the rate of burning is directly coupled to the rate of air inflow. On consideration, it would appear to be surprising that the burning rate correlates directly with the air inflow as it is known that in a confined situation the rate of release of fuel vapours is strongly dependent on radiative heat feedback from the surroundings (see Section 9.1, Figure 9.2 and Equation (9.4)). It is not clear how the ventilation conditions could influence the thermal feedback in such a manner. Indeed, Bullen and Thomas (1979) have demonstrated clearly that they are not coupled (see Section 10.2) and that it seems likely that the relationship described in Equation (10.1) may only apply to wood cribs in which the internal burning surfaces are effectively shielded from the influence of the compartment (see Section 5.2.2).

---

[2] Steckler *et al.* (1984) derived mean values of C_d of 0.68 and 0.73 for the inflow and the outflow, respectively.
[3] As char burns slowly in comparison, it would be more logical to take the stoichiometric air requirement for wood volatiles (4.6 kg/kg), which leads to $\dot{m}_b = 6.8 A_w H^{1/2}$ kg/min (see Table 5.14).

Harmathy (1972, 1978) developed this concept further, proposing that the energy responsible for producing 'the volatiles' during the steady burning period comes principally from surface oxidation of the char within the structure of the crib. He analysed data from a large number of wood crib fires in compartments and showed that the results plotted as $\dot{m}/A_f$ vs. $\rho g^{1/2} A_w H^{1/2}/A_f$ (where A_f is the surface area of the fuel) show a clear distinction between the 'ventilation-controlled regime' and a 'fuel-controlled regime' in which $\dot{m}$ is independent of the ventilation factor (Figure 10.5). Harmathy has recommended the following be used to distinguish between ventilation- and fuel-controlled fires involving cellulosic (i.e., wood or wood-based) fuels:

$$\text{Ventilation control} : \quad \frac{\rho g^{1/2} A_w H^{1/2}}{A_f} < 0.235 \tag{10.18a}$$

$$\text{Fuel control} : \quad \frac{\rho g^{1/2} A_w H^{1/2}}{A_f} > 0.290 \tag{10.18b}$$

the transition, or 'cross-over', being ill-defined (see Figure 10.5). However, as these apply to wood crib fires in compartments, their applicability to 'real' fires is not clear (see Figure 10.11). Harmathy argued most cogently about the unusual properties of char-forming fuels such as wood, but his analysis is too restrictive, particularly as it takes no account of radiative feedback from the environment within the compartment (Thomas, 1975).

It is important to be able to distinguish between these two regimes as the fuel-controlled fire is generally less severe than the ventilation-controlled fire, except in situations in which the ventilation is very restricted (e.g., $A_T/A_w H^{1/2} > 40$ in Figure 10.6). This is illustrated by data obtained in an international programme of small-scale tests coordinated by CIB and designed to investigate the factors which influence the behaviour of the fully developed fire (Thomas and Heselden, 1972) (see Section 10.2). Figure 10.6 shows a plot of 'average' steady state gas temperature inside the compartment[4] during a number of fully developed fires vs. $A_T/A_w H^{1/2}$, where A_T is the area of the walls and ceiling of the compartment, excluding the ventilation area, A_w. Values of $A_T/A_w H^{1/2}$ less than $8–10 \text{ m}^{-1/2}$ correspond to fuel-controlled fires.

In the fuel-controlled regime (Regime II), the excess air entering the compartment has the effect of moderating the temperature. The cross-over point between the two regimes will depend on the relationship between $\dot{m}$, the rate of gasification or the 'burning rate' of the fuel (kg/s), and the rate of inflow of air (kg/s). Ideally, if:

$$\frac{\dot{m}_{air}}{\dot{m}} < r \text{ kg/s} \tag{10.19}$$

where r is the stoichiometric air/fuel ratio (Section 1.2.3), and $\dot{m}_{air}$ is given by Equation (10.16), then the burning is ventilation-controlled (Regime I). Similarly, the fire is

---

[4] It should be noted that the values of temperature (θ_c) plotted in Figure 10.6 relate to point measurements taken at 75% of the height of the compartment, above the centre of the floor (Thomas and Heselden, 1972).

fuel-controlled (Regime II) if:

$$\frac{\dot{m}_{air}}{\dot{m}} > r \text{ kg/s} \tag{10.20}$$

although this argument tacitly assumes that mixing is instantaneous and the rate of the reaction between the fuel volatiles and oxygen from the air is infinitely fast. Neither of these assumptions is valid, as is shown by the fact that flames may be seen to issue from the ventilation opening even under conditions when a fire is still developing and the rate of entry of air is in excess of the stoichiometric air requirement (Thomas $et\ al.$, 1967a; Harmathy, 1978). Because mixing is not perfect and the volatiles burn at a finite rate, burning gases flow from the upper part of the ventilation opening and are visible as relatively 'shallow' flames. On the other hand, if $\dot{m} > \dot{m}_{air}/r$ (see Equation (10.19)), excess vapours flow from the opening, burning as they mix with air outside the compartment (Section 10.6). If the fire grows and develops to a Regime I post-flashover fire, these

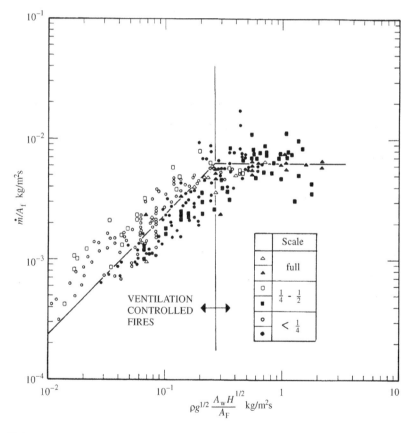

Figure 10.5 Identification of the transition between ventilation-controlled and fuel-controlled burning for wood cribs, according to Harmathy (1972)

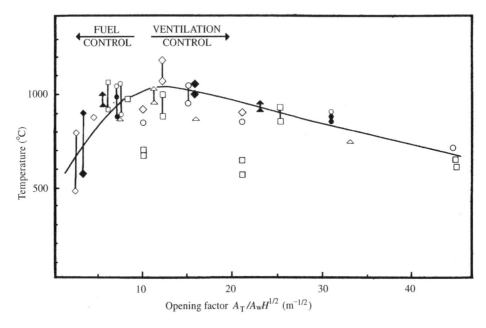

Figure 10.6 Average compartment temperatures during the steady burning period for wood crib fires in model enclosures as a function of the 'opening factor' $A_T/A_w H^{1/2}$. Symbols refer to different compartment shapes (see Table 9.3): $\circ$, $1 \times 2 \times 1$; $\triangle$, $2 \times 2 \times 1$; $\diamond$, $2 \times 1 \times 1$; $\square$, $4 \times 4 \times 1$. Solid points are means of 8–12 experiments (Thomas and Heselden, 1972). Reproduced by permission of The Controller, HMSO. © Crown copyright

flames become increasingly 'deep', eventually filling the upper half to two-thirds of the ventilation opening when the fire is fully developed (see Figure 10.7b).

10.2 Fully Developed Fire Behaviour

Most of our knowledge of the behaviour of compartment fires comes from experiments with near-cubical compartments, with characteristic dimensions ranging from c. 0.5 m to c. 3 m (e.g., Kawagoe, 1958; Thomas and Heselden, 1972; Steckler *et al.*, 1982), which of course are very different in shape and size compared with typical spaces in modern commercial buildings. Early studies involved the use of wood cribs as the fuel bed, simply because this was the only means available for producing reproducible fires. However, as noted earlier, their internal burning surfaces are shielded from the environment within the compartment and consequently the rate of burning is relatively insensitive to the thermal environment. Indeed, Thomas and Nilsson (1973) have pointed out that a third regime must be considered in which the rate of burning is controlled by the crib structure: this will occur only if the crib 'density' is high (Section 5.2.2(b)).

When 'real' fire loads are used, particularly if these involve non-cellulosic materials, there is no reason to expect coupling between the rate of burning and the rate of ventilation. This point is emphasized in Figure 10.7, which shows marked differences in behaviour between two fires in identical compartments, one involving twice as much fuel (wood) as

Figure 10.7 The effect of a large exposed fuel surface area on fire behaviour. (a) Fuel control regime, 15 kg/m². Fuel in the form of wood cribs, $A_f = 55$ m²: no external flaming. (b) Ventilation control regime, 7.5 kg/m². Fuel was fibre insulating board, lining the walls and ceiling, $A_f = 65$ m²; external flaming lasted for 5.5 minutes (Butcher *et al.*, 1968). Reproduced by permission of The Controller, HMSO. © Crown copyright

the other. In the former, the wood was present in the form of cribs (with a surface area of 55 m², *including* the internal surfaces – see Figure 5.20), while in the other it was present as the wall lining material (exposed surface area 65 m²) (Butcher *et al.*, 1968). The large area of fuel directly exposed to the fire in the latter case produced flashover followed by Regime I burning with flames emerging from the window, while the wood cribs burned as a fuel-controlled fire (Regime II). Harmathy's method (Equation (10.18)) does not distinguish between these two scenarios.

Any realistic theoretical treatment of the post-flashover fire must consider rate of burning and rate of ventilation separately (Bullen, 1977a; Babrauskas and Williamson, 1978, 1979). Bullen (1977a) carried out an analysis of steady state compartment fires involving liquid fuels in which detailed heat and mass balances were considered. While of limited application, this model is useful for demonstrating the significance of different parameters. Thus, the rate of burning is calculated from Equation (5.24), i.e.

$$\dot{m} = \frac{\dot{Q}_F'' + \dot{Q}_E'' - \dot{Q}_L''}{L_v} A_f \text{ kg/s} \tag{10.21}$$

and the rate of air inflow from

$$\dot{m}_{air} = 0.5 A_w H^{1/2} \text{ kg/s} \tag{10.16}$$

which are quite independent. If the burning was under fuel-rich conditions, Bullen assumed that all the air was 'burned' within the compartment, giving $\dot{m}_{air} \cdot \Delta H_{c,air}$ as the rate of heat release within the compartment, where $\Delta H_{c,air}$ is the heat of combustion per unit mass of air consumed (3000 kJ/kg) (Section 1.2.3). The net heat flux to the fuel surface is estimated by assuming that the gas within the compartment is 'grey' and that the walls are also radiating (both emissivities assumed to be 0.8). Two simultaneous equations in T_g (the gas temperature) and T_1 (the wall temperature) were solved, and $\dot{m}$ calculated from Equation (10.21). The effects of varying $\Delta H_{c,air}$ and the heat required to produce the volatiles (L_v) on $\dot{m}$ and T_g were calculated: the results are shown in Figures 10.8

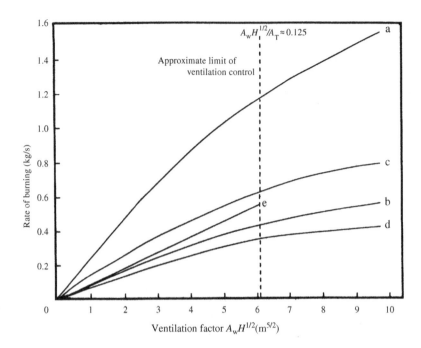

Figure 10.8 Variation of mass burning rate with the ventilation factor $A_w H^{1/2}$ for a range of fuels: (a) $\Delta H_{c,air} = 3$ kJ/g, $L_v = 0.5$ kJ/g; (b) $\Delta H_{c,air} = 3$ kJ/g, $L_v = 2.0$ kJ/g; (c) $\Delta H_{c,air} = 2.5$ kJ/g, $L_v = 0.85$ kJ/g (values for industrial methylated spirits); (d) $\Delta H_{c,air} = 1.5$ kJ/g, $L_v = 0.5$ kJ/g; (e) mass burning rate according to Equation (10.1). The limit of ventilation control $A_w H^{1/2}/A_T \approx 0.125$ is shown as the vertical dashed line (Bullen, 1977a). Reproduced by permission of The Controller, HMSO. © Crown copyright

and 10.9. In Figure 10.8, the variation of burning rate with $A_w H^{1/2}$ is shown for four liquid fuels (three of which are hypothetical). Curves (a) and (b) refer to two fuels with the same $\Delta H_{c,air}$ but with widely differing values of L_v. As would be expected, the fuel with the lower value of L_v (i.e., curve a) gives a much higher burning rate ($\dot{m}$). Similarly, reducing $\Delta H_{c,air}$ while keeping L_v constant (compare (a) and (d)) has a very dramatic effect on $\dot{m}$, although in fact the 'heat of combustion of air' does not vary much from 3000 kJ/kg (Table 1.13). However, this observation would have relevance to situations involving vitiated air (less than 21% oxygen by volume).

The effect on steady state gas temperatures within the compartment is shown in Figure 10.9. Bullen's calculations show that the fuel with the lower value of L_v gives a lower temperature (cf. curves a and b). The reason for this is that the excess volatiles released from the fuel leave the compartment unburnt, thereby increasing the 'convective' heat loss and also the volume of outflow, which restricts the inflow of fresh air (see Equation (10.14)).[5] Thus the rate of heat release within the compartment is reduced.

---

[5] Early press speculation following the attack on the World Trade Centre in September 2001 attributed the collapse of WTC1 and WTC2 to the high temperatures produced during the burning of the aviation fuel. Not only did a significant proportion of the fuel burn outside the building, but Bullen's results suggest that the fire would have been relatively 'cool' at this stage.

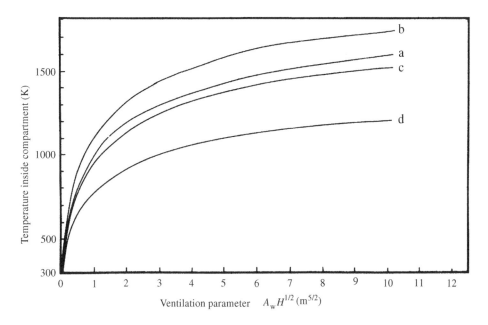

Figure 10.9 Compartment temperature as a function of the ventilation factor $A_w H^{1/2}$ for the fuels indicated in Figure 10.8 (Bullen, 1977a). The limit of ventilation control is as shown in Figure 10.8. Reproduced by permission of The Controller, HMSO. © Crown copyright

(On the basis of the experimental work quoted earlier (Figure 10.6), T_g should fall with increasing $A_w H^{1/2}$ once the boundary between ventilation control and fuel control has been crossed ($A_w H^{1/2} \approx 10$ m$^{5/2}$). However, this is not taken into account in the model.)

Bullen (1977a) also showed that, of a number of other variables, the thermal conductivity of the wall and the area of the fuel bed had the most significant effect on $\dot{m}$. The importance of radiation as the dominant mode of heat transfer in determining $\dot{m}$ was confirmed in a later publication by Bullen and Thomas (1979). They burned 'pool fires' of ethanol (IMS), polymethylmethacrylate and polyethylene in a compartment 2 m wide × 1 m × 1 m, with three different ventilation openings, corresponding to $A_w H^{1/2}/A_T$ of 0.032, 0.067 and 0.14 m$^{1/2}$, and measured $\dot{m}$ (g/s), temperatures inside and outside the compartment, and the radiant heat flux at the ventilation opening. In Figure 10.10, the ratio $\dot{m} / A_w H^{1/2}$ (kg/m$^{5/2}$ s) – which Kawagoe found to be constant for crib fires – is plotted against fuel bed area (A_f m^2) for the ethanol fires, showing clearly that $\dot{m}$ is not determined by $A_w H^{1/2}$ alone (cf. Equation (10.1)) and that the effects of $A_w H^{1/2}$ and A_f are not independent. Although not immediately obvious, Figure 10.10 also shows that $\dot{m}$ increases as A_f is increased, as anticipated (compare the values of $A_w H^{1/2}/A_f$ assigned to each curve).

In Figure 10.11, data from a number of sources are plotted on the basis of the correlation proposed by Harmathy (1972) (Figure 10.5). This shows that the rates of burning of non-cellulosic materials can be substantially higher than that predicted by Equation (10.1) and suggests that flashover could be achieved with some non-cellulosic fuels with surface

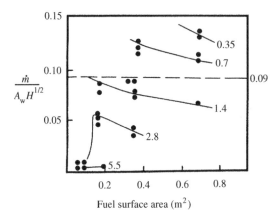

Figure 10.10   Variation of $\dot{m}/A_w H^{1/2}$ with fuel area (A_f) for ethanol pool fires in a small-scale compartment (2 m wide $\times 1$ m $\times 1$ m). The numbers assigned to the curves are values of $A_w H^{1/2}/A_f$ (Bullen and Thomas, 1979). Reproduced by permission of The Controller, HMSO. © Crown copyright

areas only one-tenth of that required for wood. The divergence of the rate of burning from 'Kawagoe behaviour' is illustrated schematically in Figure 10.12 (Bullen and Thomas, 1979). In the 'ventilation-controlled regime', $\dot{m}$ can be greater than predicted by Equation (10.1) (the stoichiometric line), while for large values of $A_w H^{1/2}$, the rate of burning will be similar to that in the open, although with enhancement due to radiation feedback.[6]

Bullen and Thomas (1979) examined the influence of radiant heat flux within the compartment by plotting the burning rate ($\dot{m}$, g/s) of a number of fuels vs. IA_f/L_v, where I (kW/m^2) is the intensity of radiant heat as measured at the ceiling of the compartment during steady burning (Figure 10.13). Although the data are scattered, there is a reasonable correlation about the line $\dot{m} = IA_f/L_v$, strongly suggesting that radiation dominates the heat transfer to the fuel surface. This relationship should be compared with Equation (10.21). Many of the data points corresponding to high burning rates of ethanol fall below the line, but this is consistent with the observation that there is a substantial layer of vapour above the fuel surface which would attenuate the radiant heat reaching the liquid (de Ris, 1979). The same comment might account for the one single datum for polyethylene.

The behaviour in the ventilation-controlled regime illustrated in Figure 10.12 is observed with many synthetic fuels and with cellulosic fuels which have extended surface areas, such as wall linings (Figure 10.7(b)). The rate at which air enters the compartment is insufficient to burn all the volatiles and the excess will be carried through the ventilation opening with the outflowing combustion products. This is normally accompanied by external flaming. Bullen and Thomas (1979) compared burning rates ($\dot{m}$) of their small-scale liquid and plastic fuel fires with rates of air inflow calculated

---

[6] Experimental results from a study of methanol pool fires in a small compartment by Takeda and Akita (1981) show this effect very dramatically.

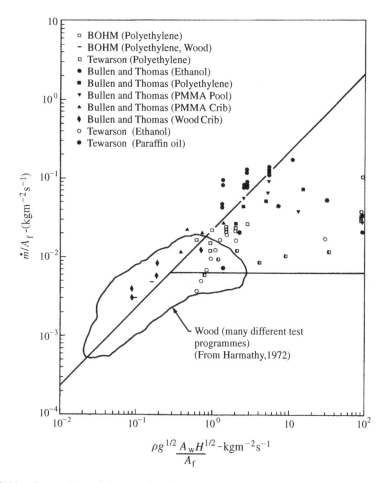

Figure 10.11 Comparison of the mass burning rates of different fuels in compartments, using Harmathy's (1972) correlation (Bullen and Thomas, 1979) (cf. Figure 10.6). Data from Bohm (1977), Tewarson (1972) and Bullen and Thomas (1979). Reproduced by permission of The Controller, HMSO. © Crown copyright

using Equation (10.16). They defined an excess fuel factor (f_{ex}) which is zero for stoichiometric burning ($\dot{m}_{air} = r \cdot \dot{m}$) and positive if there are unburnt volatiles leaving the compartment (assuming that all the oxygen in the entrained air has been consumed within the compartment):

$$f_{ex} = 1 - \frac{\dot{m}_{air}}{r} \cdot \frac{1}{\dot{m}} \tag{10.22}$$

Some of the results are quoted in Table 10.1: burning rates of the same fuel beds in the open are included for comparison. This illustrates clearly how the excess fuel factor depends on the area of the fuel surface, and the ventilation factor $A_w H^{1/2}$, and also

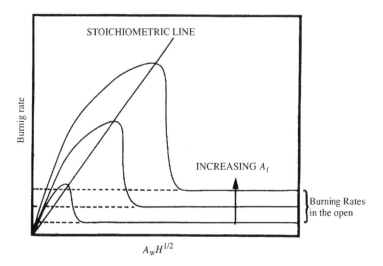

Figure 10.12 Schematic diagram showing the variation of mass burning rate with ventilation factor $A_w H^{1/2}$ and fuel bed area A_f (Bullen and Thomas, 1979). Reproduced by permission of The Controller, HMSO. © Crown copyright

Table 10.1 Pool fire burning rates in the open and in compartments (Bullen and Thomas, 1979)

Fuel[a] type	Fuel area (m²)	Open fire $\dot{m}_{ave}$ (g/s)	Compartment fire $A_w H^{1/2}$					
			0.25			0.5		
			$\dot{m}$ (g/s)	f_{ex}	$\theta (°C)$[b]	$\dot{m}^c$ (g/s)	f_{ex}	$\theta (°C)$[b]
IMS	0.186	3.5	19	0.41	880	26	0.09	1060
PE	0.186	3.7	10	0.25	980	NF	−0.9	–
PMMA	0.186	5.1	12	0.025	910	NF	−2.2	–
IMS	0.372	8.8	30	0.63	780	42	0.47	950
PE	0.372	8.4	14	0.48	890	26	0.45	1150
PMMA	0.372	7.4	21	0.45	820	31	0.30	1030

[a] IMS, industrial methylated spirits; PE, polyethylene; PMMA, polymethylmethacrylate.
[b] θ is the maximum temperature rise over ambient under the ceiling.
[c] NF, no external flames.

how the 'severity' of the fire (as judged by the maximum temperature under the ceiling) depends on both f_{ex} and $A_w H^{1/2}$.

Temperature profiles in the external flames from three of these small-scale compartment fires (Bullen and Thomas, 1979) are shown in Figure 10.14. The fuel was IMS and only the area of the fuel 'bed' was varied. The tip of the flame was assumed to be represented by the 550°C contour, although photographs taken during the experiments suggest that this might underestimate the height by ∼10%. External flaming is an important mechanism of fire spread and can cause damage to the external load-bearing structural

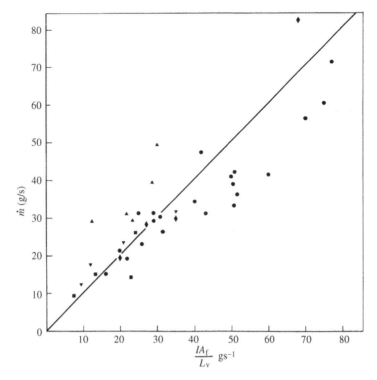

Figure 10.13 Correlation of mass burning rate ($\dot{m}$) with radiant intensity at ceiling level (I) (see text): •, ethanol ($L_v = 850$ J/g); ▾, PMMA pool (1600 J/g); ▪, polyethylene (2200 J/g); ♦, wood (1340 J/g); ▴, PMMA crib (Bullen and Thomas, 1979). Reproduced by permission of The Controller, HMSO. © Crown copyright

members (Section 10.7). There is a very rough correlation between the radiant heat flux to which the building facade is exposed and the excess fuel factor, f_{ex}. This is illustrated in Figure 10.15, in which the radiant heat flux is shown as a function of height above the soffit of the ventilation opening for the three small-scale compartment fires described in Figure 10.14. This should only be taken as indicative of large-scale behaviour, particularly as these results refer to IMS which is fuel that burns very cleanly. Fuels which generate a lot of smoke will give flames of lower temperature but higher emissivity (see Table 5.4).

Before continuing the topic of fully developed fire behaviour, it is worth revisiting the concept of 'flashover'. The preferred definition used in this text is 'the transition from a localized fire to a general conflagration within the compartment when all fuel surfaces are burning'. Figure 10.7(a) shows a compartment fire in which all surfaces are burning, yet there is no external flaming and it has all the attributes of a 'fuel-controlled fire'. This is a fully developed, Regime II fire and begs the question 'has flashover occurred?' The fire has certainly spread to involve all the fuel and on the basis of the first 'definition' of flashover given in Chapter 9 (see page 351 et seq.), flashover has occurred, although there was no transition from fuel control to ventilation control. Clearly, 'Regime II, post-flashover fires' can occur, but their significance is not yet known, although they may have relevance to the phenomenon of 'travelling fires' that is discussed below.

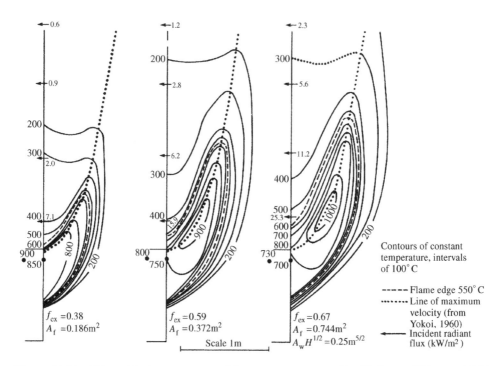

Figure 10.14   Effect of fuel area on external flame temperature profiles (Bullen and Thomas, 1979). Reproduced by permission of The Controller, HMSO. © Crown copyright

10.3   Temperatures Achieved in Fully Developed Fires

While there have been a number of efforts to develop a theoretical model for the fully developed fire, only one significant attempt has been made to resolve the important parameters experimentally. This will be discussed briefly before the theoretical work is reviewed.

10.3.1   *Experimental Study of Fully Developed Fires in Single Compartments*

A major experimental programme was undertaken in the 1960s to improve our understanding of the behaviour of the fully developed compartment fire (Thomas and Heselden, 1972; Thomas, 1972b). The objective was to provide information which would allow the development of a rational approach to the problems of fire severity and fire resistance. Fire research laboratories of eight countries collaborated in this work, which was carried out under the auspices of the Conseil Internationale du Bâtiment (CIB). It involved over 400 experiments with small-scale compartments, using wood cribs as the fuel, and varying the compartment size and shape, the ventilation factor $A_w H^{1/2}$ and the fire load density. The effect of wind was also considered.

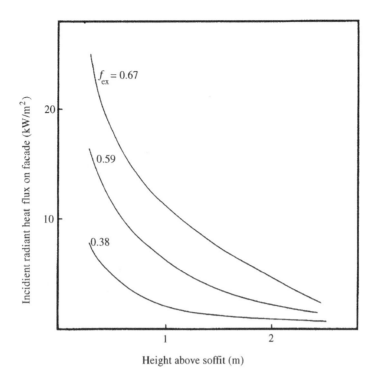

Figure 10.15 Variation of radiant flux in the plane of the opening of a small-scale compartment as a function of height above the soffit for the flames shown in Figure 10.14 (from which this diagram has been derived)

The conclusions of this study have had wide-reaching implications *vis-à-vis* compartment fire testing but need not be enumerated in detail here. However, three are of relevance to the present discussion:

(i) The ratio $(\dot{m}/A_wH^{1/2})$ is not a constant, as indicated in Equation (10.1), but depends on compartment shape (depth (D) and width (W)) and scale (particularly A_T, the internal surface area *excluding* the ventilation opening and the floor). A correlation between $(\dot{m}/A_wH^{1/2})$, $(D/W)^{1/2}$ and $A_T/A_wH^{1/2}$ was found, indicating the importance of the internal dimensions. The dependence of burning rate on $A_wH^{1/2}/A_T$ (the 'opening factor') was derived independently by Odeen (1963), and has been incorporated into theoretical models of compartment fires (e.g., Pettersson *et al.*, 1976).

(ii) The intensity of radiation emitted through the ventilation opening (as viewed from directly in front of the compartment) is sensitive to shape (particularly the ratio of compartment width to height) for all ventilation sizes. It correlates strongly with the rate of burning, except for small ventilation openings.

(iii) All other things being equal, the maximum temperature for a given compartment fire scenario is observed just inside the ventilation-controlled regime (Figure 10.6), and corresponds to $A_T/A_w H^{1/2} \approx 8-15$ (see (i) above).

Despite the fact that only small-scale compartments were involved, this CIB programme provided a substantial body of data which has been used to develop models relating to fire resistance requirements (Law, 1971) and behaviour of flames outside ventilation openings (Thomas and Law, 1974).

10.3.2 Mathematical Models for Compartment Fire Temperatures

Several research groups have turned their attention to developing ways of predicting the likely temperature–time history of a potential compartment fire (Kawagoe and Sekine, 1963; Pettersson *et al.*, 1976; Babrauskas and Williamson, 1978; Cadorin and Franssen, 2003). The ultimate objective of such an exercise is to be able to specify for design purposes the thermal stress to which elements of structure would be exposed in the event of fire in a particular space, thus providing an alternative to dependence on the fire resistance tests which are enshrined in the prescriptive building codes and regulations. In Sweden, Pettersson and co-workers (Pettersson *et al.*, 1976) developed a method of calculating the 'fire resistance requirements' which was adopted initially by the Swedish regulatory authorities. Although it was a conceptually simple 'one-zone model' in which the compartment is treated as a well-stirred reactor (as in Figure 10.4) and has been superseded in the Eurocodes by the so-called 'parametric fire curve' (Equation 10.44: see also Franssen and Zahria, (2005)), the treatment developed by Pettersson *et al.* for calculating the temperature–time history of a fire remains the clearest exposition of the method and merits discussion. (The second stage of Pettersson's procedure – calculating the thermal response of elements of structure – is considered in Section 10.5.)

As we are concerned only with that period of the fire during which structural damage can occur, the pre-flashover stage may be neglected as average temperatures are relatively low.[7] Thus, in subsequent calculations, $t = 0$ refers to the start of the fully developed fire. The compartment is regarded as a calorimeter and its temperature obtained by solving the following heat balance (Figure 10.16):

$$\dot{q}_C = \dot{q}_L + \dot{q}_W + \dot{q}_R + \dot{q}_B \tag{10.23}$$

where:

$\dot{q}_C$ = rate of heat release due to combustion;
$\dot{q}_L$ = rate of heat loss due to replacement of hot gases by cold;
$\dot{q}_W$ = rate of heat loss through the walls, ceiling and floor;
$\dot{q}_R$ = rate of heat loss by radiation through the openings;
$\dot{q}_B$ = rate of heat storage in the gas volume (neglect).

---

[7] The method developed by Cadorin and Franssen (2003) ('OZone') takes the pre-flashover stage into account.

The following assumptions are made to simplify the model:

(i) Combustion is complete and takes place entirely within the confines of the compartment.
(ii) The temperature is uniform within the compartment at all times (cf. Figure 10.3).[8]
(iii) A single surface heat transfer coefficient may be used for the entire inner surface of the compartment.
(iv) The heat flow to and through the compartment boundaries is unidimensional, i.e., corners and edges are ignored and the boundaries are assumed to be 'infinite slabs'.

The terms in Equation (10.23) are as follows:

$\dot{q}_C$ –rate of heat release

Pettersson *et al.* (1976) assume that the fire will be ventilation-controlled and that the Kawagoe relationship (Equation (10.1)) can be applied directly. Should the fire happen to be in the fuel-controlled regime, then this assumption will lead to an overestimate of the rate of burning (cf. Figure 10.2) and consequently an underestimate of the duration of burning. The following expression is taken for the rate of heat release:

$$\dot{q}_C = 0.09 A_w H^{1/2} \cdot \Delta H_c \qquad (10.24)$$

where ΔH_c is the heat of combustion of the fuel. ΔH_c is taken to be the heat of combustion of wood (18,800 kJ/kg), and in the subsequent calculations other materials are expressed in terms of 'wood equivalents'.[9] Moreover, $\dot{q}_C$ is assumed to remain constant from $t = 0$ (i.e., immediately after flashover) until all the fuel has been consumed. This ignores the fact that any char produced will burn much more slowly than implied by Equation (10.24) (see footnote 3): this will lead to conservative temperatures.

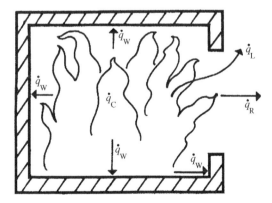

Figure 10.16 Heat losses during a fully developed compartment fire (Equation (10.23)). After Pettersson *et al.* (1976). Reproduced by permission of The Swedish Institute of Steel Construction

---

[8] While this may be approximately true for small compartments, it is very unlikely to apply to large compartments (see Section 10.3.3).
[9] The rate of heat release could be expressed as $\dot{m}_{air} \times \Delta H_{c,air}$, i.e. (from Equation (10.16)) $\dot{q}_C = 0.52 A_w H^{1/2} \times 3000 = 1560 A_w H^{1/2}$ kW. (Note that Equation 10.24 gives $1692 A_w H^{1/2}$.)

$\dot{q}_R$ – heat loss by radiation through the openings

From the Stefan–Boltzmann law (Section 2.4):

$$\dot{q}_R = A_w \varepsilon_F \sigma (T_g^4 - T_0^4) \, \text{kW} \tag{10.25}$$

where A_w is the total area of the openings (m^2), T_g is the gas temperature within the compartment (K) and T_0 is the outside (ambient) temperature (K).

As $T_g \gg T_0$, this can be written:

$$\dot{q}_R = A_w \varepsilon_F \sigma T_g^4 \tag{10.26}$$

where ε_F is the effective emissivity of the gases within the compartment. This can be calculated from

$$\varepsilon_F = 1 - \exp(-K x_F) \tag{10.27}$$

where x_F is the flame thickness (m) (normally taken as the depth of the room) and K is the emission coefficient (m^{-1}) (see Table 2.10). Pettersson *et al.* (1976) use $K = 1.1 \, \text{m}^{-1}$, quoting work by Hägglund and Persson (1976a) on wood crib fires.

$\dot{q}_L$ – heat loss due to convective flow

$$\dot{q}_L = \dot{m}_F c_p (T_g - T_0) \tag{10.28}$$

where $\dot{m}_F$ is the rate of outflow of fire gases (see Equation (10.10)). Assuming that $\dot{m}_F \approx \dot{m}_{air}$ (i.e., ignoring fuel volatilization), then letting $\chi = \dot{m}_{air} / A_w H^{1/2}$ (which is assumed to be approximately constant at $\sim$0.5 kg/m$^{5/2}$.s, see Equations (10.15) and (10.16)):

$$\dot{q}_L = \chi c_p (T_g - T_0) A_w H^{1/2} \tag{10.29}$$

$\dot{q}_W$ – heat loss through the compartment boundaries

The rate of heat loss through the boundaries will depend on both the gas temperature within the compartment (T_g) and the internal surface temperature (T_i). Heat transfer by conduction through the boundaries must be solved numerically, as described briefly in Section 2.2.3. The enclosing boundary of the compartment (see assumption (iv) above) is divided into n layers, each of thickness Δx (see Figure 10.17). A series of equations can then be written, one expression for each layer.

Exposed surface layer:

$$\Delta x c_1 \rho \frac{\Delta T_1}{\Delta t} = \frac{T_g - T_1}{\dfrac{1}{\gamma_i} + \dfrac{\Delta x}{2k_1}} - \frac{T_1 - T_2}{\dfrac{\Delta x}{2k_1} + \dfrac{\Delta x}{2k_2}} \tag{10.30}$$

Inner layer j (describing $n-2$ layers):

$$\Delta x c_j \rho \frac{\Delta T_j}{\Delta t} = \frac{T_{j-1} - T_j}{\dfrac{\Delta x}{2k_{j-1}} + \dfrac{\Delta x}{2k_j}} - \frac{T_j - T_{j+1}}{\dfrac{\Delta x}{2k_j} + \dfrac{\Delta x}{2k_{j+1}}} \tag{10.31}$$

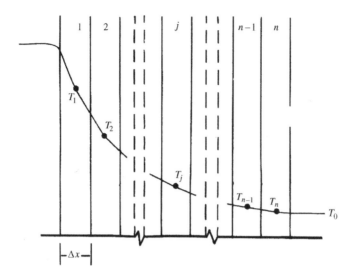

Figure 10.17 Transient heat conduction through the compartment boundaries. Boundary assumed to be an infinite slab, divided into elements $1-n$. Pettersson *et al.* (1976). Reproduced by permission of The Swedish Institute of Steel Construction

Outer layer:

$$\Delta x c_n \rho \frac{\Delta T_n}{\Delta t} = \frac{T_{n-1} - T_n}{\dfrac{\Delta x}{2k_{n-1}} + \dfrac{\Delta x}{2k_n}} - \frac{T_n - T_0}{\dfrac{\Delta x}{2k_n} + \dfrac{1}{\gamma_u}} \tag{10.32}$$

where T_i is replaced by T_1, the temperature of the innermost layer (to which Equation (10.30) refers). Both c and k are functions of temperature and have to be assigned values according to the local temperature. T_g and T_0 are the compartment gas temperature (K) and the ambient temperature (K), respectively. The two heat transfer coefficients γ_i and γ_u are given by Pettersson *et al.* (1976) as:[10]

$$\gamma_i = \frac{\varepsilon_r \sigma}{T_g - T_i}(T_g^4 - T_i^4) + 0.023 \, \text{kW/m}^2.\text{K} \tag{10.33}$$

where ε_r is the resultant emissivity

$$\varepsilon_r = \left(\frac{1}{\varepsilon_F} + \frac{1}{\varepsilon_i} - 1\right)^{-1} \tag{10.34}$$

with the subscript i referring to the inner surface (equivalent to $j = 1$ in Figure 10.17) and

$$\gamma_u = 3.3 \times 10^{-5} T_u - 3.09 \times 10^{-4} \, \text{kW/m}^2.\text{K} \tag{10.35}$$

where T_u is the temperature of the outside surface (K) (equivalent to $j = n$).

---

[10] Pettersson uses the symbols α_i and α_u instead of γ_i and γ_u in Equations (10.33) and (10.35). The change has been made to avoid confusion with α, which is used here for thermal diffusivity.

The first-order difference Equations (10.30)–(10.32) are solved numerically for each time step, and corresponding values of $\dot{q}_W$ calculated from

$$\dot{q}_W = (A_t - A_W)\left(\frac{1}{\gamma_i} + \frac{\Delta x}{2k_1}\right)^{-1}(T_g - T_i) \qquad (10.36)$$

where A_t is the total area of the boundary surfaces (walls, ceiling and floor), including the area of the ventilation openings, A_W (m^2). Enclosing boundaries of different materials can be dealt with by a simple adaptation of these equations.

Calculation of the ventilation factor $(A_W H^{1/2})$

When there is more than one opening in the walls of a compartment, the ventilation factor is given by:

$$A_\Sigma H_m^{1/2} = \sum_i A_i H_i^{1/2} \qquad (10.37)$$

where A_Σ is the sum of the areas of the openings and H_m is a mean height, as defined by Equation (10.37). This is unlikely to apply to unusual ventilation conditions, such as several small openings at different heights (Babrauskas and Williamson, 1978) or if an opening exists in the roof. The latter situation is discussed by Pettersson et al. (1976), who provide a nomogram that allows a corrected ventilation factor to be calculated.

Calculation of $T_g(t)$

Now that the terms in the heat balance equation have been identified, substitution of Equations (10.24), (10.25), (10.29) and (10.36) into Equation (10.23) gives, after rearrangement:

$$T_g = \frac{\dot{q}_c + 0.09c_p A_W H^{1/2} T_0 + (A_t - A_W)\left[\dfrac{1}{\gamma_i} + \dfrac{\Delta x}{2k}\right]^{-1}(T_g - T_1) - \dot{q}_R}{0.09c_p A_W H^{1/2} + (A_t - A_W)\left[\dfrac{1}{\gamma_i} + \dfrac{\Delta x}{2k}\right]^{-1}} \qquad (10.38)$$

T_g is calculated by numerical integration, e.g., using the Runge–Kutta procedure (e.g. Margenau and Murphy, 1956). T_1 depends on T_g and is obtained by solving the set of equations described above, iterating on T_g five times for each time step. For each solution, the value of T_g used is that calculated at the end of the previous time interval using Equation (10.38). Note that $\dot{q}_R$, γ_i and c_p are also functions of T_g. The time interval is chosen to be short enough for the surface heat transfer coefficients to be regarded as constant during each increment. The duration of burning is taken to be:

$$t_d = \frac{M_f}{0.09 A_W H^{1/2}} \text{ s} \qquad (10.39)$$

where M_f is the fire load in kg 'wood equivalent'. After this time, $\dot{q}_c$ is set equal to zero.

The validity of the assumptions relating to the heat losses in this model was tested by using data on the burning rates in experimental compartment fires to predict temperature–time histories using Equation (10.38). These were then compared with

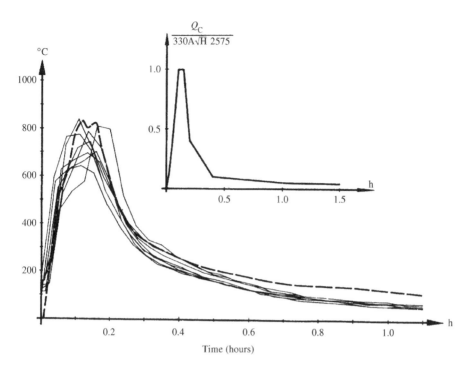

Figure 10.18 Gas temperature–time curves in full-scale fires. Solid lines represent experimental results from a number of full-scale tests using furniture as the fire load (96 MJ/m^2). Opening factor $A_w H^{1/2}/A_t = 0.068$ m$^{1/2}$. Dashed line (- - - - -) is the temperature–time curve calculated using the measured rate of burning ($\dot{m}$) to give $\dot{q}_C$ as a function of time (see inset), which in turn was incorporated into Equation (10.38) (Pettersson *et al.*, 1976). Reproduced by permission of The Swedish Institute of Steel Construction

the measured temperature–time curves, and were found to be in satisfactory agreement (Figure 10.18).

Typical temperature–time curves derived from the above equations are shown in Figure 10.19 for a 'standard compartment' constructed from materials of 'average thermal properties'. Each set of curves shown corresponds to a different 'opening factor' $A_w H^{1/2}/A_t$ (m$^{1/2}$) (Section 10.3.1), while individual curves correspond to different 'fire loads' expressed in terms of their net heat of combustion, $q_f = M.\Delta H$. If several fuels are involved, then

$$q_f = \sum_i M_i \Delta H_i \qquad (10.40)$$

The data are also presented in Table 10.2. The effect of changing the thermal properties of compartment boundaries is illustrated in Figure 10.20: insulating materials like lightweight concrete tend to give hotter fires. (This is confirmed experimentally by results of Latham *et al.* (1987).) In Pettersson's method, this can be incorporated by applying an empirical correction factor to q_f and $A_w H^{1/2}/A_t$ before selecting the appropriate temperature–time

Table 10.2 Compartment temperature–time curves according to Pettersson et al. (1976) (by permission) (Pettersson's 'Standard compartment', wall thickness 0.2 m, $k = 0.8$ W/m.K, $\varrho c = 1700$ kJ/m^3.K)

$A_w\sqrt{H}/A_t = 0.08$ m$^{1/2}$

t (h)	25	50	75	100	200	300	400	500
				Fire load (MJ/m^2)				
0.05	528	528	528	528	528	528	528	528
0.10	742	742	742	742	742	742	742	742
0.15	423	733	746	746	746	746	746	746
0.20	359	697	750	750	750	750	750	750
0.25	268	594	761	761	761	761	761	761
0.30	163	478	732	777	777	777	777	777
0.35	161	439	668	792	792	792	792	792
0.40	154	390	595	758	806	806	806	806
0.45	148	338	515	706	820	820	820	820
0.50	141	282	481	647	832	832	832	832
0.55	134	269	442	584	843	843	843	843
0.60	127	254	400	534	854	854	854	854
0.65	120	238	356	509	864	864	864	864
0.70	113	222	309	481	860	874	874	874
0.75	106	205	298	453	839	883	883	883
0.80	99	190	285	423	816	891	891	891
0.85	92	173	273	392	789	899	899	899
0.90	85	156	261	360	761	907	907	907
0.95	77	138	248	326	730	914	914	914
1.00	69	119	236	315	698	907	920	920
1.10	53	88	210	293	627	886	933	933
1.20	51	83	185	270	585	861	944	944
1.30	48	79	160	248	551	831	951	954
1.40	46	75	132	226	515	801	943	964
1.50	44	71	104	203	478	765	934	972
1.60	43	68	94	181	440	727	923	980
1.70	41	65	90	158	400	685	903	974
1.80	40	62	86	134	360	644	881	967
1.90	39	60	82	109	345	612	858	958

$A_w\sqrt{H}/A_t = 0.12$ m$^{1/2}$

t (h)	37.5	75	112.5	150	300	450	600	750
				Fire load (MJ/m^2)				
0.05	602	602	602	602	602	602	602	602
0.10	854	854	854	854	854	854	854	854
0.15	481	845	858	858	858	858	858	858
0.20	403	802	862	862	862	862	862	862
0.25	296	673	873	873	873	873	873	873
0.30	211	537	836	888	888	888	888	888
0.35	173	489	759	903	903	903	903	903
0.40	165	432	671	860	916	916	916	916
0.45	158	370	574	798	928	928	928	928
0.50	150	304	532	727	940	940	940	940
0.55	142	289	485	650	950	950	950	950
0.60	135	271	436	589	960	960	960	960
0.65	127	253	383	558	969	969	969	969
0.70	118	235	328	525	961	977	977	977
0.75	111	216	314	491	934	985	985	985
0.80	103	197	301	456	904	992	992	992
0.85	95	179	287	419	873	999	999	999
0.90	86	160	273	381	839	1005	1005	1005
0.95	78	139	259	342	803	1011	1011	1011
1.00	68	118	245	329	763	1001	1017	1017
1.10	50	82	216	304	678	973	1027	1027
1.20	47	77	188	279	628	941	1036	1036
1.30	45	73	159	254	587	905	1040	1045
1.40	43	69	128	229	545	866	1028	1052
1.50	41	65	96	204	502	824	1014	1059
1.60	40	62	86	180	458	779	999	1065
1.70	38	59	82	154	412	729	973	1055
1.80	37	57	78	127	366	681	946	1043
1.90	36	54	74	99	350	644	918	1031

1017	887	606	335	91	71	52	35	2.00	948	833	580	331	101	79	58	38	2.00
708	708	708	708	708	708	708	708	0.05	649	649	649	649	649	649	649	649	0.05
1020	1020	1020	1020	1020	1020	1020	1020	0.10	928	928	928	928	928	928	928	928	0.10
1021	1021	1021	1021	1021	1021	1008	571	0.15	931	931	931	931	931	931	917	520	0.15
1023	1023	1023	1023	1023	1023	950	465	0.20	934	934	934	934	934	934	868	432	0.20
1031	1031	1031	1031	1031	1031	797	332	0.25	944	944	944	944	944	944	732	313	0.25
1044	1044	1044	1044	1044	978	619	186	0.30	959	959	959	959	959	901	574	180	0.30
1055	1055	1055	1055	1055	880	554	183	0.35	972	972	972	972	972	815	520	179	0.35
1065	1065	1065	1065	994	772	482	175	0.40	984	984	984	984	922	718	456	171	0.40
1074	1074	1074	1074	914	648	405	167	0.45	995	995	995	995	852	609	387	163	0.45
1082	1082	1082	1082	827	593	324	158	0.50	1005	1005	1005	1005	775	562	315	155	0.50
1090	1090	1090	1090	730	534	305	149	0.55	1015	1015	1015	1015	688	509	298	146	0.55
1097	1097	1097	1097	653	472	285	140	0.60	1023	1023	1023	1023	620	454	279	138	0.60
1103	1103	1103	1103	612	407	265	131	0.65	1031	1031	1031	1031	585	396	260	129	0.65
1108	1108	1108	1089	571	341	243	122	0.70	1039	1039	1039	1021	548	336	240	121	0.70
1114	1114	1114	1051	529	326	222	113	0.75	1046	1046	1046	989	510	321	220	112	0.75
1118	1118	1118	1012	485	309	200	104	0.80	1052	1052	1052	955	471	306	200	104	0.80
1123	1123	1123	970	441	294	179	95	0.85	1058	1058	1058	919	431	292	180	95	0.85
1127	1127	1127	927	395	278	157	85	0.90	1063	1063	1063	881	389	277	159	86	0.90
1130	1130	1130	881	349	263	133	75	0.95	1068	1068	1068	841	347	262	138	77	0.95
1134	1134	1115	834	335	247	108	63	1.00	1073	1073	1055	799	334	247	114	67	1.00
1140	1140	1074	728	306	214	66	41	1.10	1081	1081	1022	704	307	216	76	47	1.10
1145	1145	1030	667	279	183	61	39	1.20	1089	1089	984	649	281	187	71	44	1.20
1150	1126	984	618	252	150	57	37	1.30	1095	1091	944	603	255	156	67	42	1.30
1154	1105	935	568	224	115	54	35	1.40	1101	1076	900	558	228	123	63	40	1.40
1158	1083	883	517	197	78	51	34	1.50	1107	1058	854	511	201	89	60	38	1.50
1161	1050	829	465	170	67	48	33	1.60	1111	1040	805	463	176	79	56	37	1.60
1146	1015	770	412	141	63	46	32	1.70	1099	1011	750	414	149	74	54	36	1.70
1129	979	712	359	110	59	44	31	1.80	1085	980	697	365	120	71	51	34	1.80
1112	942	668	344	79	56	42	30	1.90	1070	948	657	349	91	68	49	33	1.90
1094		624	330	69	54	41	29	2.00	1055	915	616	334	82	64	47	32	2.00

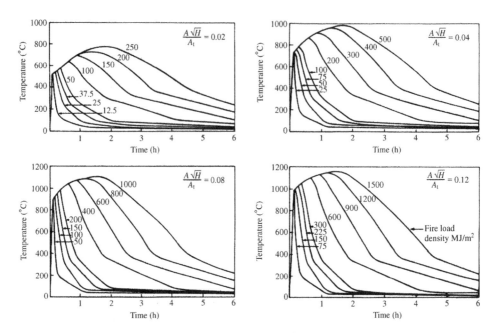

Figure 10.19 Theoretical temperature–time curves for compartment fires with different fire load densities (MJ/m^2) and opening factors, $A_w H^{1/2}/A_t$ (m$^{1/2}$) (Pettersson *et al.*, 1976). Reproduced by permission of The Swedish Institute of Steel Construction. (Pettersson's 'standard compartment', wall thickness 0.2 m, $k = 0.8$ W/m.k, $\rho c = 1700$ kJ/m^3.K)

curve.[11] Application of these data to the calculation of fire resistance requirements for elements of structure is discussed below (Section 10.5).

As noted above, Pettersson's model assumes ventilation-controlled burning throughout the fully developed stage, from $t = 0$ until the fuel is consumed (Equation (10.24) *et seq.*). Consequently, it will overestimate the burning rate of a fire for which fuel control conditions actually exist. This will generate much higher temperatures, albeit for a shorter period of time (Equation (10.39)) and may give a more severe fire with respect to its potential for damaging the structure of the building.[12] Babrauskas and Williamson (1978, 1979) developed a similar model which requires details of the nature and distribution of the fuel as input, incorporating the means of identifying when the fire has progressed from ventilation control to fuel control as the fuel is consumed. The rate of heat release ($\dot{q}_C$) can be calculated as either

$$\dot{q}_C = \dot{m}_{air} \cdot \frac{\Delta H_c}{r} \quad \text{(ventilation controlled)} \tag{10.41}$$

---

[11] The correction factors are not given here. Table 10.2 refers to temperature–time curves for compartment boundaries with thermal properties corresponding to 'average values for concrete, brick and lightweight concrete'. For example, if the compartment boundaries consisted of 100% lightweight concrete, the equivalent fire load and opening factor are obtained by multiplying the actual fire load and opening factor by a correction factor = 3.0. The result is a hotter fire (Figure 10.20), consistent with the observation of Latham *et al.* (1987).

[12] Lamont *et al.* (2004) have compared the effects of "short hot fires" and "long cool fires" on a composite steel frame structure.

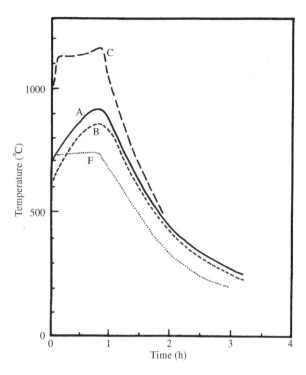

Figure 10.20 Theoretical temperature–time curves for fully developed fires in compartments with different boundaries: A, materials with thermal properties corresponding to the average values for concrete, brick and lightweight concrete; B, concrete ($\rho \approx 2400$ kg/m³); C, lightweight concrete ($\rho \approx 500$ kg/m³); F, 80% uninsulated steel sheeting, 20% concrete. (In all cases, the fire load was 250 MJ/m² and $A_w H^{1/2}/A_t = 0.04$ m$^{1/2}$.) Pettersson *et al.* (1976). Reproduced by permission of The Swedish Institute of Steel Construction

or

$$\dot{q}_C = \dot{m} \cdot \Delta H_c \quad \text{(fuel controlled)} \tag{10.42}$$

(implying that the rate of pyrolysis of the fuel (i.e., $\dot{m}$) must be known at all times). It is assumed that the heat of combustion is released entirely within the compartment and that $\dot{m}_{air}$ can still be calculated from Equations (10.15) and (10.16), even under fuel-controlled conditions when it may no longer be applicable. However, the model has great flexibility and was found to correlate well with some contemporary full-scale compartment fire tests.

Cadorin and Franssen (2003) and Cadorin *et al.* (2003) developed a model ('OZone') which has many similarities to that of Pettersson *et al.* (1976), designed to calculate 'natural' temperature–time curves, including the pre-flashover stage. The work was carried out under the European 'Natural Fire Safety Concept' project (see Schleich and Cajot, 2001). The pre-flashover fire is assumed to follow a t^2 relationship (cf. Equation (9.25)) of the form:

$$\dot{Q}(t) = 10^6 (t/t_\alpha)^2 \text{ W} \tag{10.43}$$

where t_α is the time taken for the fire to reach 1 MW. Values of t_α are given for slow ($t_a = 600$ s), medium ($t_a = 300$ s), fast ($t_a = 150$ s) and ultrafast ($t_a = 75$ s) fires. (Equation 10.43 gives a slightly lower rate of increase of $\dot{Q}$ than that given by Equation (9.25).) It is a zone model, initially running with two zones (the upper and lower layers, as described in Section 4.4.5) but converting to a single layer (one zone) when one of four criteria is met, each one taken as an indicator of 'flashover'. The pre-flashover fire is defined by Equation (10.43) and the maximum rate of heat release is limited either by the ventilation control (stoichiometric burning, cf. Equation (10.41)) or by a maximum rate of heat release which is specified by the user (e.g., in the Eurocodes, according to occupancy type). The computer code, 'OZone' is used to generate temperature–time curves equivalent to those presented by Pettersson *et al.* (1976) in tabular form (Table 10.2).

An analytical formula (Equation (10.44)) is given in Eurocode 1 to allow parametric temperature–time curves to be calculated for compartments with floor areas up to 100 m^2 and compartment heights up to 4.5 m. It is not explained how this equation was derived and its pedigree is not divulged, although it is likely that it is based on the Pettersson model.[13] In view of its widespread use in structural fire safety design, it is included here for completeness. However, it cannot be used effectively without continual reference to Eurocode 1, Part 1–2, as may be deduced from Table 10.3 in which the parameters involved in Equation (10.44) are defined (but not justified or referenced).

$$T_g = 1325 \left(1 - 0.324e^{-0.2t^*} - 0.204e^{-1.7t^*} - 0.472e^{-19t^*}\right) \tag{10.44}$$

where T_g is the gas temperature in the compartment (°C) and t^* (in hours) is the time, weighted to take into account the ventilation parameter and the thermal properties of the compartment boundaries. Typical T_g–t curves are given in Figure 10.21, the parameters of which are taken from Lennon and Moore (2003). As part of the ECSC programme, a series of eight full-scale tests were carried out to ascertain whether or not the equation was able to predict the T_g–t curves satisfactorily. The results are summarized by Lennon and Moore (2003), who concluded that there were 'a number of areas in which the method of calculating temperatures in DD ENV1991-2-2 could be improved'. For example, the duration of burning was underpredicted for all eight tests (see, for example, Figure 10.22).

The methods described above (Kawagoe and Sekine, 1963; Pettersson *et al.*, 1976; Babrauskas and Williamson, 1978; Cadorin and Franssen, 2003) all require lengthy or awkward computation to generate temperature–time curves that can be used to calculate heat transfer rates to elements of structure (see Section 10.5). As only Pettersson *et al.* (1976) have converted their results into tables of temperature–time relationships that can be used directly (e.g., in Table 10.11), these are used in Section 10.5, despite the fact that many of the assumptions associated with the basic model are at best uncertain, as will become clear in the next section. Indeed, our understanding of the post-flashover fire is incomplete and the above models must be used with circumspection. Further research is required to develop more robust tools for estimating 'fire severity'.

---

[13] As may be seen in Table 10.3, the opening factor is defined in exactly the same way as Pettersson *et al.* (1976). Also in common with Pettersson, the unit of time is the hour, which is unusual.

Table 10.3 Definition of the terms required in Equation (10.44) (Eurocode 1)

T_g	= gas temperature in the compartment (°C)
$t*$	= $t.\Gamma$ (hours)

with

t	= time (hours)
Γ	= $[O / b]^2 / (0.04 / 1160)^2$ (−)[a]
b	= $\sqrt{(\rho c k)}$ (J/m^2.s$^{1/2}$.K) (where $100 \leq b \leq 2200$)
ρ	= density of the boundary of the enclosure (kg/m^3)
c	= specific heat of the boundary of the enclosure (J/kg.K)
k	= thermal conductivity of the boundary of the enclosure (W/m.K)
O	= opening factor: $A_w H^{1/2} / A_t$ (m$^{1/2}$) (where $0.02 \leq O \leq 0.20$)
A_w	= total area of vertical openings on all walls (m^2)
H	= weighted average of window heights on all walls (m)
A_t	= total area of enclosure (walls, ceiling and floor, including openings)

The maximum temperature T_{max} occurs when $t^* = t^*_{max}$, where

$$t^*_{max} = t_{max}\Gamma \text{ (hours)}$$

$$t_{max} = \max\ [(0.2 \times 10^{-3} \times q_{t,d}/O); t_{lim}] \text{ (hours)}$$

and

$q_{t,d}$ = the design value of the fire load density related to the total surface area A_t of the enclosure whereby $q_{t,d} = q_{f,d} \times A_f / A_t$ (MJ/m^2). The following limits should be observed: $50 \leq q_{t,d} \leq 1000$ (MJ/m^2)

$q_{f,d}$ = the design value of the fire load density related to the surface area A_f of the floor (MJ/m^2) – this is to be found in Annex E of BS EN 1991-1-2 (2002)

t_{lim} = See definition of $t_{max} \cdot t_{lim}$ applies to the fuel-controlled fire – in the case of slow fire growth rate, $t_{lim} = 25$ min. For medium and fast fire growth rates, $t_{lim} = 20$ and 15 minutes, respectively (these must be converted to hours)

The temperature–time curve during the cooling phase depends on the value of t^*; three expressions have to be considered:

for $t^*_{max} \leq 0.5$	$T_g = T_{max} - 625(t^* - t^*_{max}x)$
for $0.5 < t^*_{max} < 2$	$T_g = T_{max} - 250(3 - t^*_{max})(t^* - t^*_{max}x)$
for $t^*_{max} \geq 2$	$T_g = T_{max} - 250(t^* - t^*_{max}x)$

where $t^*_{max} = (0.2 \times 10^{-3}q_{t,d}/O).\Gamma$ and $x = 1.0$ if $t_{max} > t_{lim}$, or $x = t_{lim}.\Gamma / t^*_{max}$ if $t_{max} = t_{lim}$

[a]If $\Gamma = 1$, Equation (10.44) approximates to the standard temperature–time curve (Equation (10.45)).

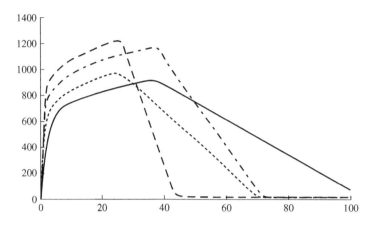

Figure 10.21 Parametric temperature–time curves for the four compartment conditions described by Lennon and Moore (2003)

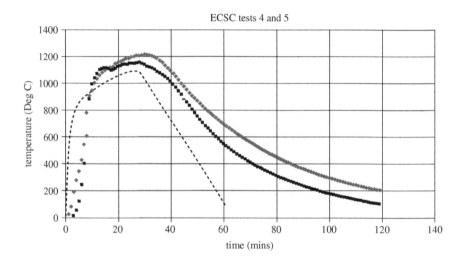

Figure 10.22 Comparison between the results of ESCS tests 4 (◆) and 5 (■) and the T_g–t curves predicted by Equation 10.44 (------). The key parameters are $O = 0.07$ m$^{1/2}$ and $b = 720$ J/m^2.s$^{1/2}$.K (i.e., $\Gamma = 7.95$) (Lennon and Moore, 2003). Tests 4 and 5 differed only in that the fire loads were 100% wood and 80% wood/20% plastic, respectively: each corresponded to 40 kg wood (equivalent)/unit floor area. By permission of Elsevier

10.3.3 Fires in Large Compartments

The assumption that during the post-flashover stage the temperature is uniform throughout the entire compartment rests on limited experimental evidence (e.g., Figure 10.3). The earliest experiments were carried out mostly on a small scale and very few temperature measurements were taken within the compartments (e.g., Kawagoe and Sekine, 1963;

Thomas and Heselden, 1972). Even when full-scale room-size experiments were performed (e.g., Steckler *et al.*, 1982), the instrumentation was very limited and the spatial resolution was poor: the original 'Steckler fire' has been widely used as a bench-mark against which computer models have been tested,[14] but the vertical temperature distribution was measured in only two locations, insufficient to test the 'well-stirred reactor' assumption associated with zone models. Much higher resolution is required for a reliable 'benchmark experiment' – particularly if it is to be used to test CFD or 'field models', of which FDS (Fire Dynamics Simulator) is perhaps the best-known example.

The full-scale tests that were carried out to test the validity of the parametric fire curve (Equation (10.44)) captured temperatures at 64 locations during every one of eight fires (40 kg/m^2 of wood and wood/plastic cribs) in a compartment measuring 12 m × 12 m by 3 m high (Lennon and Moore, 2003). Average temperatures were used to compare with the parametric fire curve (Figure 10.22), but there was a considerable scatter in the temperature measurements. Stern-Gottfried *et al.* (2010) found that the data exhibited a normal distribution, with the mean standard deviations ranging from 50 to 100°C.[15] The spatial resolution of these data has not been fully analysed, although clear variations have been observed. As might be expected, the most vigorous burning occurs near the ventilation opening, where fresh air first mixes with fuel vapours. This is not a new observation: Harmathy (1986) noted that when wood cribs were used as the fuel in a small compartment, the cribs were consumed progressively from the front (nearest the ventilation opening) to the rear, indicating the importance of the inflow of air.

It seems very unlikely that the uniform temperature assumption can be applied to large compartments. This effect was revealed much more dramatically in some large experi-mental fires carried out in 'long' compartments measuring 23 m × 6 m by 2.7 m high (Kirby *et al.*, 1994). These involved an extensive array of wood cribs and showed that the fire could 'travel', burning vigorously where there was a supply of air and spreading to unburnt fuel as each area progressively burned out. In one experiment with restricted ventilation, when only the row of cribs at the back wall (furthest from the ventilation open-ing) was ignited, the fire developed locally, then rapidly propagated forward and became established on the row of cribs closest to the opening. The rate of burning towards the rear diminished substantially due to low oxygen levels and over a period of time the fire spread back into the compartment as the rows of cribs burned out one by one. This phenomenon has been studied systematically on a small scale by Thomas and Bennetts (1999) and on a larger scale by Thomas *et al.* (2005). They confirmed that the temperature within the compartment was not uniform and that the fire was much more severe in the vicin-ity of the ventilation opening than at the rear. Concern is now being expressed about the consequences of non-uniform temperature fields on structural fire safety design (e.g., Stern-Gottfried *et al.*, 2010). This may be exacerbated by the phenomenon of the 'travel-ling fire' when during the fully developed stage the fire appears to move from one 'fuel package' to the next. This behaviour has been noted recently in a number of fires that have

---

[14] This procedure is frequently and incorrectly described as 'validation'. This is highly misleading in this con-text as it implies much more than is delivered. 'Validation' implies that a model has been rigorously – and satisfactorily – tested against a very large number of scenarios. No compartment fire model has been subjected to such scrutiny.

[15] A similar analysis of the temperature records from Dalmarnock Test 1, which involved a fully furnished room (3.5 m ×4.75 m ×2.45 m high) with 270 thermocouples (Rein *et al.*, 2009) revealed a standard deviation of 132°C.

caused major structural collapse, e.g., the World Trade Centre Tower 7 (Stern-Gottfried *et al.*, 2009).

10.4 Fire Resistance and Fire Severity

As stated at the start of this chapter, the term 'fire resistance' is associated with the ability of an element of building construction to continue to perform its function as a barrier[16] or structural component during the course of a fire. Conventionally, it is determined by testing a full-scale sample (under load, if appropriate) to failure as it is subjected to a 'standard fire', defined by the temperature–time variation of the fire gases within a large furnace. The concept was first introduced in 1916, based on observations of the temperatures of wood fires used in early *ad hoc* testing (Babrauskas and Williamson, 1980a,b). The standard curve has changed only slightly over the years. In the USA it is specified by a set of data points (ASTM, 2008b) (Table 10.4), although it is more common to define it mathematically, thus:

$$T = T_0 + 345 \log(0.133t + 1) \tag{10.45}$$

where T_0 and T are the temperatures (°C or K) at time $t = 0$ and $t = t$ (s), respectively (British Standards Institution, 1987): this is plotted in Figure 10.23. A very similar curve has been adopted by the International Organization for Standardization (ISO, 1999a).

The required temperature–time curve in the standard test is obtained by controlling the rate of fuel supply to the furnace as the test proceeds to achieve the standard curve, monitoring the gas temperature by means of an array of thermocouples. Unfortunately, the actual 'fire exposure' created is sensitive to the nature of the fuel used[17] and the physical

Table 10.4 Definition of the standard temperature–time curve according to ASTM E119 (ASTM, 2008b)

Time (min)	Temperature (°C) (ASTM E119)	Temperature (°C) (BS 476 Part 20)[a]
5	538	583
10	704	683
30	843	846
60	927	950
120	1010	1054
240	1093	1157
≥480	1260	1261[b]

[a]Calculated from Equation 10.45 (British Standards Institution, 1987).
[b]At 480 minutes.

---

[16] The strategy of 'fire compartmentation' was developed to limit the maximum fire loss in a building by ensuring that the boundaries of defined 'fire compartments' acted as barriers to fire spread, surviving the duration of the fire by virtue of having the appropriate 'fire resistance'.

[17] Oil-fired furnaces have much more emissive gases (i.e., flames) than those fuelled by natural gas.

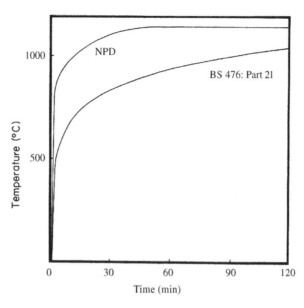

Figure 10.23 Comparison between the 'standard' temperature–time curve BS 476 Part 21 and the hydrocarbon fire curve adopted by the Norwegian Petroleum Directorate (Shipp, 1983)

properties (and emissivity) of the furnace walls. It is believed that the dominant mode of heat transfer to the specimen is by radiation from the walls (Paulsen and Hadvig, 1977). If two furnaces have difference lining materials, the surface temperature of the walls of the furnace with the lining with the lower thermal inertia will increase more rapidly than the walls of the other furnace (see Figure 2.10), resulting in a more severe fire exposure, even if the furnaces are otherwise identical (Malhotra, 1982). It has been suggested that the thermal properties of the furnace linings should be specified (Bohm, 1982), but this would be much too expensive to implement. It would be more logical to define a heat flux–time curve for the test, but there are no heat flux sensors that are robust enough to be used in a routine test such as this. A compromise solution has been proposed, based on the 'plate thermometer' developed by Wickstrom (1994, 1997). This consists of a thin metal plate (100 mm × 100 mm × 0.7 mm thick) with a layer of insulation covering the rear (unexposed) face. Its temperature is monitored by means of a thermocouple welded to the centre of the rear face. This is more responsive to radiant heat than a shielded thermocouple and is therefore a more suitable device to monitor the furnace temperature. It is now routinely used in ISO 834 (ISO, 1999) and the European Standard (European Committee for Standardization, 1999), but has not been adopted for ASTM E119 (Sultan, 2004).

It is normal to equate the 'fire resistance' of an element of structure with the time to a specified failure in the standard test (e.g., buckling of a column, deflection of a beam, etc.), although the exposure in a real fire can be very different. A qualitative measure of this can be seen in Figures 10.24 and 10.25. In the former, the standard temperature–time curve (Equation 10.45) is compared with the temperatures measured in a series of full-scale compartment fires with different fire loads and ventilation openings; the fuel was in the form of wood cribs. In Figure 10.25, the effect of introducing a thermoplastic (polypropylene) is shown.

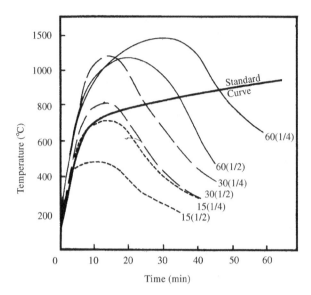

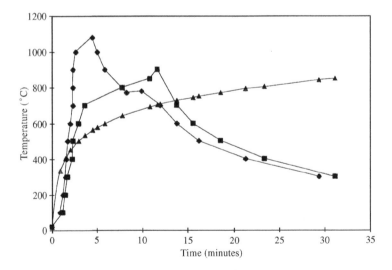

Figure 10.24 Comparison of the standard temperature–time curve with the temperatures measured during compartment fires. The fire load density is given in kg/m^2 and the ventilation as a fraction of one wall. 60(1/2) implies 60 kg/m^2 and 50% of one wall open (Butcher *et al.*, 1966). Reproduced by permission of The Controller, HMSO. © Crown copyright

Figure 10.25 Average combustion gas temperature inside a compartment for a fire load density of 15 kg/m^2 and 1/4 ventilation, for wood (■) and wood + polypropylene (◆). The standard temperature–time curve is shown for comparison (▲) (Latham *et al.*, 1987)

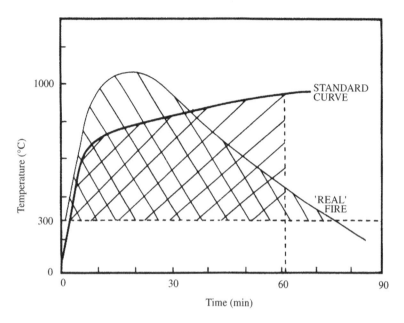

Figure 10.26 Ingberg's equal area hypothesis. If the shaded areas are equal then the fires are said to have equal 'severity'. Thus the fire resistance requirement for the 'real' fire shown corresponds to ~60 min exposure to the standard test

When the disparity between the standard test and 'real fires' was recognized in the 1920s, there was considerable discussion about the validity of continuing to use the standard curve. However, Ingberg (1928) carried out a series of tests which identified 'fire load' (combustible content per unit floor area) as an important factor in determining the potential fire severity. He proposed that the 'severity' of a fire could be related to the fire resistance requirement, using what came to be known as the 'equal area hypothesis', in which it is assumed that if the areas under the temperature–time curves (above a baseline of 300°C) of two fires are equal, then the severities are equal. If one of these 'fires' is the standard temperature–time curve, then 'severity' and 'fire resistance' can be equated. This is illustrated in Figure 10.26.

On this basis, Ingberg developed a table relating fire load and 'fire severity' (Table 10.5), from which the fire resistance requirement of a particular compartment can be obtained directly from the measured, or anticipated, fire load. Little information on fire load densities was available to Ingberg, but surveys have been carried out subsequently and are available in the literature (e.g., Pettersson *et al*., 1976; Conseil Internationale du Bâtiment, 1986; National Fire Protection Association, 2008a). Following this hypothesis, fire resistance requirements can be specified for an occupancy type if the potential (or actual) fire load is known. However, in addition to the practical question of whether or not the fire load will change significantly during the lifetime of the building, there is no theoretical justification for the hypothesis. The dependence of radiative heat flux on T^4 makes simple scaling impossible when heat transfer is dominated by radiation: for example, assuming Ingberg's baseline of 300°C, 10 minutes at 900°C will *not* have the same effect as 20

Table 10.5 Ingberg's fuel load/fire severity relationship

Combustible content[a] (wood equivalent)		Equivalent[a,b] (MJ/m^2)	Standard fire duration (h)
1b/ft^2	kg/m^2		
10	49	900	1
15	73	1340	1.5
20	98	1800	2
30	146	2690	3
40	195	3590	4.5
50	244	4490	6
60	293	5390	7.5

[a]Calculated on the basis of *floor* area.
[b]Heat of combustion of wood taken as 18.4 kJ/g.

minutes at 600°C (cf. Figure 10.26). Furthermore, in the last few decades, changes in building design and construction have resulted in buildings of much lower thermal mass, which are consequently much more responsive to heat transfer from a fire.

These difficulties were tacitly ignored, and the apparent success of fire load as a surrogate for 'fire severity' was accepted by default: apparently, no significant building failures have occurred which have shown the method to be unsafe.[18] Although it cannot be defended scientifically or technically, the philosophy is embedded into the common approach to structural fire safety, through the standard fire resistance test. A number of attempts have been made to relate 'real fires' to the standard tests, using the concept of 'equivalent fire severity'. This has to be based on a comparison of the response of an element of structure in the real fire and in the test furnace. Law (1971) sought such a relationship by analysing the thermal responses of insulated steel columns exposed to the standard temperature–time curve and to real fires. A steel temperature of 550°C was taken as marking a critical point in the fire exposure at which the strength of steel is reduced to about 50% of its ambient value.[19] The real fire temperature was modelled as a constant temperature, equal to the peak fire temperature, sustained for a period $\tau = M_f / \dot{m}$, where M_f is the total fire load (kg wood equivalent) and $\dot{m}$ is the mean burning rate (kg/min). Drawing on data gathered in the CIB study of fully developed compartment fires (Thomas and Heselden, 1972), Law found the following correlation to hold:

$$t_f = K' \frac{M_f}{(A_w A_T)^{1/2}} \tag{10.46}$$

where t_f is the fire resistance period (minutes) associated with this set of parameters (Figure 10.27). With the areas in m^2, K' is a constant whose value is close to unity. (It should be noted that this expression is based on an analysis of insulated columns and will not hold for exposed steelwork.)

---

[18] This statement was probably true before 11 September 2001, but the collapse of the WTC Towers has emphasized the need to consider the response of the whole structure to fire, and not just single elements.
[19] The "critical temperature" depends on the grade of steel.

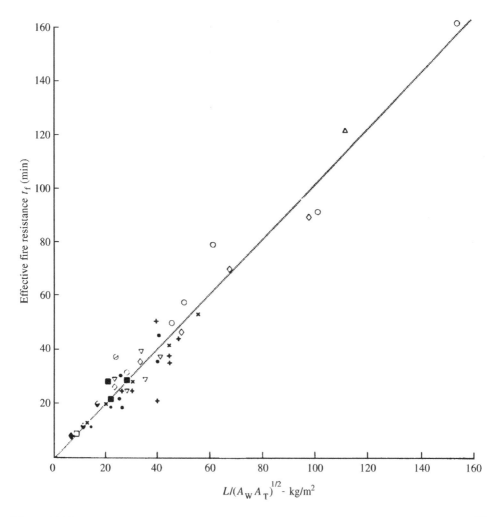

Figure 10.27 Law's (1971) correlation between fire resistance requirement (t_f) and $L/(A_w A_T)^{1/2}$, where L is the fire load (kg), A_w is the area of the ventilation and A_T is the total internal surface area of the compartment (m²) (Equation (10.46)). Reproduced by permission of The Controller, HMSO. © Crown copyright

Law's analysis – unlike that of Ingberg – showed clearly that ventilation to the fire is an important factor in determining 'severity', and could not be ignored. However, as comparative studies of the behaviour of elements of structure in different fire resistance furnaces have shown, the thermal properties of the compartment boundaries can have a significant effect. This is demonstrated clearly in the results of Butcher *et al*. (1966) and Latham *et al*. (1987). Pettersson developed the following relationship:

$$t_f = 0.31 C \frac{M_f}{(A_t A_w \sqrt{H})^{1/2}} \tag{10.47}$$

Table 10.6 Values of C relevant to Equation 10.47

$\sqrt{(k\rho c_p)}$ $(\mathrm{J/m^2 \cdot s^{1/2} \cdot K})$	C $(\mathrm{h \cdot m^{3/4}/\, kg})$
≤ 720	0.09
720–2520	0.07
≥ 2520	0.05

where C is a factor which depends on the thermal inertia of the compartment boundaries (Table 10.6) and is based on the analysis developed by Pettersson and co-workers (Pettersson, 1976).[20] A similar formula is given in DIN 18230:

$$t_f = 0.087(M_f /\, A_F)m.w.c \tag{10.48}$$

where m is a dimensionless burning factor (taken as unity for wood), w is dependent on the ratio A_w/A_F and c depends on the thermal inertia of the compartment boundaries (see Table 10.7).

These formulae are compared by Harmathy (1987), who has introduced the concept of 'normalized heat load' (H) (Harmathy and Mehaffey, 1982), which is defined as:

$$H = \frac{1}{\sqrt{k\rho c_p}} \int_0^\tau \dot{q}'' \cdot dt \tag{10.49}$$

where $\dot{q}''$ is the heat flux through the building element $(\mathrm{kW/m^2})$, t is time (s) and τ is the fire duration (s). The 'normalization' is with respect to the thermal inertia of the wall lining materials.

These five formulae give widely differing results (Table 10.8). While Ingberg's method may be rejected out of hand for the reasons given above, it is not clear how the others can best be interpreted. Law's formula does not bring the nature of the compartment boundaries into account, but on the other hand the inclusion of relevant correction terms into the DIN method and that of Pettersson is based on very limited data. The methods of Law and Harmathy give more conservative figures than the others and may provide useful methods for estimating fire resistance requirements (based on the standard test),

Table 10.7 Values for Equation 10.48

A_w/A_F	w	$\sqrt{(k\rho c_p)}$ $(\mathrm{J/m^2 \cdot s^{1/2} \cdot K})$	c
0–0.05	3.2	0–720	0.25
0.05–0.10	2.0	720–2520	0.20
0.10–0.15	1.5	>2520	0.15
0.15–0.20	1.2		

---

[20] The 'equivalent time of fire exposure' included in Eurocode 1 is also based on Pettersson's formula (e.g., British Standards Institution, 2002).

Table 10.8 Comparison of the formulae for equivalent fire exposure[a] (Harmathy, 1987)

Method	Prediction (hours)
Ingberg (1928)	0.62
Law (1971)	1.22
Pettersson (CIB, 1986)	1.02
DIN (1978)	1.04
Harmathy (1987)	1.29

[a]The values are as follows: $A_F = 20$ m^2, $A_t = 85$ m^2, $H = 1.2$ m, $A_w = 1.8$ m^2, thermal inertia of the boundaries $(k\rho c) = 10^6$ J/m^2.s$^{1/2}$.K and fire load density = 30 kg/m^2.

but may be too conservative. The alternative is to abandon the furnace test completely and rely on calculating the amount of fire protection necessary on the basis of a predicted temperature–time curve (such as Pettersson *et al.*, 1976). This is discussed in the following section.

It must be emphasized that fire resistance as defined above refers only to compartment fires involving combustible solids. Flammable liquid fires, particularly those encountered in the open in the refining and petrochemical industries, exhibit totally different characteristics. A hydrocarbon spill fire (Section 5.1.2) or a running liquid fire can engulf an item of process plant in flame within seconds, thus exposing the structure to dangerous levels of heating very quickly (Figure 10.28). The standard fire resistance test is not relevant in this situation, and consequently more appropriate tests have been developed in which the temperature–time curve rises very much more steeply than indicated by Equation 10.45. The 'hydrocarbon temperature–time curve' originally adopted by the Norwegian Petroleum Directorate is shown in Figure 10.23 (Shipp, 1983), but others exist.[21] For example, in the Dutch RWS (Rijkswaterstaat) test which was devised for severe fires involving fuel tankers in tunnels (van de Leur, 1991; Carvel, 2005), the temperature rises rapidly to over 1100°C within 5 minutes, peaking at 1350°C at 60 minutes then decreasing to 1200°C at 120 minutes. Apparently, the test is normally terminated at this point.

10.5   Methods of Calculating Fire Resistance

It has been shown in previous sections how 'fire severity' has come to be identified with fire resistance requirements, and thus with the results of the standard tests. While this seems to have proved satisfactory in the past, there are numerous problems associated with fire resistance testing, including the obvious ones of cost, lack of interlaboratory reproducibility and the questionable relevance of the standard temperature–time curve to real fire exposure, pointing to the need for a more rational approach to structural fire protection (Milke, 2008a). This requires failure criteria to be identified and incorporated into a suitable mathematical model. Failure of most structural components in a fire can

---

[21] A standard test for determining the resistance of passive fire protection systems to jet fires has recently been developed (British Standards Institution, 2007b).

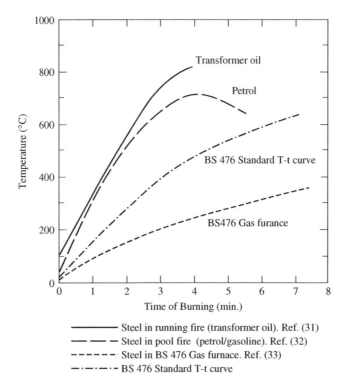

- ——————— Steel in running fire (transformer oil). Ref. (31)
- — — — Steel in pool fire (petrol/gasoline). Ref. (32)
- — — — — Steel in BS 476 Gas furnace. Ref. (33)
- —·—·—·· BS 476 Standard T-t curve

Figure 10.28 Comparison of the temperatures attained by identical steel members exposed to a standard fire (—————·), a liquid pool fire (— — —) and a running liquid fire (———————). The standard temperature–time curve is shown as (—·—·—··) (Stark, 1972). Reproduced by permission of the Institution of Chemical Engineers

be related to the loss of strength at high temperature. Thus, it is known that steel begins to lose strength above ~550°C, but how quickly structural steel will fail depends on many factors: indeed, if designed appropriately, a steel column or beam can continue to perform its function at significantly higher temperatures. This will depend, *inter alia*, on the loading, the support conditions, the degree of restraint, the dependence of the material properties on temperature and the temperature gradient within the section (British Standards Institution, 2003). (It should be noted that the test does not take into account interaction between elements in the final structure – in particular, the effect of restraint arising from the expansion of individual elements as the temperature rises.[22])

In general, if a bare steel member is exposed to a high heat flux, it will tend to fail early, although the time to failure will depend on its thermal mass (see below). Structural members may be protected from fire by encasing them with a suitable insulating material (Elliot, 1974; National Fire Protection Association, 2008a), the thickness of which will determine the 'fire resistance' of the assembly, as defined in the standard test (British Standards Institution, 1987; ASTM, 2008b). In reinforced concrete, the reinforcing bars

---

[22] Further discussion of this issue is beyond the scope of this text, but the interested reader may wish to consult Gillie *et al.* (2002) and Wang (2002).

are protected by the concrete, although this will itself lose strength as dehydration and loss of water of crystallization occur. Spalling may also occur under fire conditions and expose the reinforcing bars. The effect of temperature on building materials in general is discussed at some length by Lie (1972), Abrams (1979) and Kodur and Harmathy (2008).

A discussion of the behavioural response of structures to fire conditions is beyond the scope of this text (Buchanan, 2002; Wang, 2002), but to illustrate a number of important points, it will be assumed that there is a critical temperature above which a specific steel member in a structure can be regarded as significantly weakened. The critical temperature is not a constant, but for the sake of this discussion, 550°C will be taken as the criterion for the onset of failure. The 'fire resistance' of a structural element may then be obtained from basic heat transfer calculations, if the temperature–time curve for the fire exposure is known. The standard test may be taken as a simple example (Equation 10.45). Consider a hollow, square cross-section column, of outside dimensions 0.2 m × 0.2 m, constructed from 0.015 m thick plate, as shown in Figure 10.29, exposed in the column furnace. To enable the 'fire resistance' of this column to be calculated, the following must be known:

(a) the thermal capacity of the column;
(b) the mode of heat transfer within the furnace; and
(c) the emissivity of the furnace and the column.

This is a problem of transient heat transfer, requiring numerical solution (Section 2.2.3). Manual, iterative calculation is possible, but the problem may best be approached using a spreadsheet, writing a short computer program or making use of a commercial package such as MATLAB. It may be reduced to a single dimension by assuming that the column is of infinite length and heated uniformly. If the steel behaves as a 'lumped thermal capacity' (Section 2.2.2), then the heat transferred to the column (for each unit of length) during a short time interval Δt is:

$$Q_{\text{in}} = \varepsilon_r \sigma A_s [(T_F^{j+1})^4 - (T_s^j)^4] \Delta t + h A_s (T_F^{j+1} - T_s^j) \Delta t \qquad (10.50a)$$

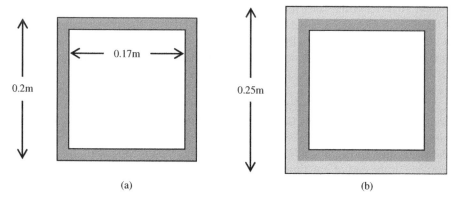

(a)   (b)

Figure 10.29   Cross-section of the hollow column to which Equations (10.50a), (10.50b) and (10.51) and Figure 10.30 apply

This quantity of heat (Q_{in}) causes the temperature of the steel to rise by an amount determined by the thermal capacity of the column (per unit length):

$$Q_{stored} = V_s \rho_s c_s (T_s^{j+1} - T_s^j) \tag{10.50b}$$

where A_s and V_s are the surface area (m^2) and volume (m^3) of the column per unit length, T_s^{j+1} and T_s^j are the temperatures (K) of the steel at the end of the $(j+1)$th and jth time intervals, respectively, and T_F^{j+1} is the temperature of the furnace (or the fire) at the end of the $(j+1)$th time interval. Equations (10.53a) and (10.53b) can be equated (i.e., $Q_{in} = Q_{stored}$ for each time interval Δt) to give a value for the steel temperature at the end of the $(j+1)$th timestep, as follows:

$$T_s^{j+1} = T_s^j + \frac{A_s}{V_s} \left\{ \frac{\varepsilon_r \sigma}{\rho_s c_s} \left[\left(T_F^{j+1}\right)^4 - \left(T_s^j\right)^4 \right] + \frac{h}{\rho_s c_s} \left[T_F^{j+1} - T_s^j \right] \right\} \Delta t \tag{10.51}$$

The importance of the ratio A_s/V_s should be noted: it is equal to the ratio of the exposed perimeter to the cross-sectional area of the element of structure (H_p/A), which is used in current design guidance in the UK to compare the heating rates of structural elements of different section (e.g., Latham *et al.*, 1987; British Standards Institution, 2003). In the USA, the ratio W/D is used in a similar manner (Law, 1972; Stanzak and Lie, 1973; Milke, 2008a), where W is the mass of steel per unit length (i.e., $\rho_s V_s$) and D is the heated perimeter (i.e., H_p), which is numerically equal to A_s. As the density of steel is constant, this ratio is equivalent to $(A_s/V_s)^{-1}$. The greater the ratio W/D (the smaller A_s/V_s), the greater will be the fire resistance: this can be deduced from Equation 10.51.

Numerical values may be substituted into Equation 10.51. Thus, the resultant emissivity ε_r is calculated from the emissivities of the furnace and the steel (ε_F and ε_s, respectively) using Equation (10.34): typically, $\varepsilon_F = 0.7$ and $\varepsilon_s = 0.9$, so that $\varepsilon_r = 0.65$, and the convective heat transfer coefficient, h, can be taken as 0.023 kW/m^2.K (Pettersson *et al.*, 1976).[23] The density (ρ_s) and thermal capacity (c_s) of steel are 7850 kg/m^3 and 0.46 kJ/kg.K, respectively, and are assumed to be independent of temperature. By choosing a suitably small time interval, Δt, and iterating the calculation from $t = 0$ (i.e., $j = 0$), values of T_s^{j+1} can be worked out from the furnace temperature T_F^{j+1} (Equation 10.45) and the steel temperature at the end of the previous time interval, T_s^j. The iteration must be continued until $T_s^{j+1} > 550°C$ (823 K). If Δt is taken as 120 s, only a few iterations are required to find the solution: this could be carried out manually, but the calculations are best done using a spreadsheet. The calculations are displayed in detail in Table 10.9 with $\Delta t = 120$ s and the results are plotted in Figure 10.30. This gives the time to achieve 550°C as just under 14 minutes (840 s). With a spreadsheet, it is possible to reduce the timestep and obtain a more accurate answer. Thus, if $\Delta t = 30$ seconds, the time to reach 550°C is slightly less than 15.5 minutes, but 32 iterations are required.

The calculation shown in Table 10.9 underestimates the time to failure because the steel temperature at the end of one timestep (j) is held constant during the duration of the next timestep ($j + 1$). Consequently, the heat transfer to the steel is overestimated because the value of T_F is taken as the value at the *end* of timestep ($j + 1$). Thus the steel temperature

---

[23] A value of $h = 0.025$ kW/m^e.K is recommended in EN-1991-1-2 (British Standards Institute, 2002).

Table 10.9 Calculation of fire resistance of a hollow square column (Figure 10.29) using Equation (10.51)

Column width = 0.2 m	Density of steel = 7850 kg/m³
Column thickness = 0.015 m	Thermal capacity of steel = 0.46 kJ/kg
Volume = 0.111 m³/m	Resultant emissivity = 0.65
Area = 0.8 m²/m	Convective HT coefficient = 0.023 kW/m².K
Initial temperature = 20°C	Timestep Δt = 120 s

Standard fire resistance test temperature–time curve $(T_F) = 20 + 345\log(0.133t + 1)$

Time step j	Time t (s)	T_F (K)	ΔT_s (K)	T_s (K)	T_s (°C)
0	0	293	0	293	20
1	120	717	46.1	339	66
2	240	816	67.4	406	133
3	360	876	80.3	487	214
4	480	918	88.4	575	302
5	600	951	92.8	668	395
6	720	978	93.3	761	488
7	840	1002	89.3	851	578
8	960	1021	80.5	931	658
9	1080	1038	66.9	998	725

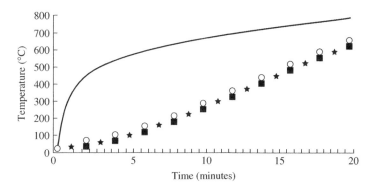

Figure 10.30 Calculated temperature–time curves for the hollow column shown in Figure 10.29(a), exposed to the standard curve. Equation (10.51) and Table 10.9 apply. ○, Data from Table 10.9 ($\Delta t = 120$ s); ★, $\Delta t = 30$ s; ■, $\Delta t = 120$ s, but at each iteration, $T_F = (T_F^j + T_F^{j-1})/2$

is overestimated, as may be deduced by inspecting Equation (10.50a). With spreadsheets, and an analytical formula for the temperature–time relationship, a timestep can be chosen to give whatever accuracy is required. However, Pettersson's temperature–time curves are presented in digital form (Table 10.2), with temperatures given at 3 minute intervals. As will be shown later, these data can be used in a spreadsheet, improving the accuracy by taking the average value of T_F (i.e., $(T_F^j + T_F^{j+1})/2$) over each timestep. The effectiveness of this is shown in Figure 10.30, when $\Delta t = 120$ s and T_F is taken to be $(T_F^j + T_F^{j+1})/2$ for each iteration: the result compares favourably with that obtained with $\Delta t = 30$ s.[24]

A simple modification can be made to the above model to allow for the temperature dependence of c_s, etc., but it is much more difficult to improve the heat transfer model. As discussed earlier (Section 10.4), the column receives heat radiated from the furnace linings as well as convected and radiated heat from the hot gas within the furnace. As the former will inevitably lag behind the gas temperature, it would be necessary to incorporate the thermal response of the furnace walls into the model. This would require much more sophisticated numerical techniques.

The time to failure of an insulated column can also be deduced by an iterative procedure, but simplifying assumptions make the calculation more straightforward. If the Biot number (Section 2.2.2) is large, it may be assumed that the temperature of the exposed surface of the insulation is equal to that of the furnace at all times during the test. Moreover, it is also assumed that in the iteration, quasi-steady state is achieved over each time increment Δt, so that the rate of heat transfer through the insulation is given by:

$$\dot{q}'' = \frac{k}{L}(T_F^{j+1} - T_s^j) \tag{10.52}$$

where k is the thermal conductivity of the insulation material (kW/m.K) and L is its thickness. The energy entering the steel per unit length of column in time interval Δt is then

$$Q_{in} = \frac{k}{L}(T_F^{j+1} - T_s^j) \cdot A_i \cdot \Delta t \tag{10.53}$$

where A_i is the internal surface area of the insulation per unit length of column through which heat is being transferred, assuming perfect thermal contact. The temperature rise of the steel during the time interval Δt is then:

$$T_s^{j+1} - T_s^j = \frac{Q_{in}}{V_s \rho_s c_s} \tag{10.54}$$

The calculation must be iterated until $T_s^{j+1} \geq 550°C$ (or until T_s^{j+1} begins to fall), using the procedure laid down in Table 10.9.

Using the same failure criterion, the 'fire resistance' of structural steel members exposed to real fires can be calculated if the temperature–time curve is known. The models developed by Pettersson *et al.* (1976) and Babrauskas and Williamson (1978, 1979), as well as the parametric temperature–time curve (Equation 10.44), can all be used to provide input to iterative calculations as described above. In fact, Pettersson has developed the method beyond this point to analyse the stability of the structural element (under load) at its maximum temperature, thus requiring a knowledge of the temperature dependence of those factors which determine the strength of the steel (modulus of elasticity, yield stress, etc.). This avoids defining a specific 'failure temperature' such as 550°C which, as we have already noted, is only indicative of the temperature at which the element will begin to lose strength significantly. This aspect cannot be pursued further here.

---

[24] If the column had been solid steel, rather than hollow, the same method may be used to show that it would take 34.5 minutes to reach 550°C.

The temperature of a steel element as a function of time during a fire can be calculated using Equation 10.51, in which T_F can be taken from Pettersson's model (interpolation of the data presented in Table 10.2) or indeed from the parametric temperature–time curve (Equation (10.44)). Pettersson *et al*. (1976) recommend certain values for ε_r which they refer to as 'resultant emissivities' – although they apply to different configurations of structural element, they are in fact composite terms incorporating both an emissivity and a geometric factor. These are quoted in Table 10.10.

Consider the following example in which the hollow steel column shown in Figure 10.29(a) is exposed on all sides to a fully developed compartment fire described in Pettersson's tables (see Table 10.2) for a fire load of 200 MJ/m² and an opening factor of $A_w H^{1/2} / A_t = 0.08$ (Pettersson *et al*., 1976). The temperature rise in the steel in a time interval Δt is given in Pettersson's formulation[25] by:

$$\Delta T_s = \frac{\gamma_i}{\rho_s c_s} \cdot \frac{A_s}{V_s} (T_F^{j+1} - T_s^j) \Delta t \qquad (10.55)$$

Table 10.10 Resultant emissivities ε_r (Pettersson *et al*., 1976)

Type of construction	ε_r
1. Column exposed to fire on all sides	0.7
2. Column outside façade	0.3
3. Floor girder with floor slab of concrete, only the underside of the bottom flange being directly exposed to fire (see Figure 10.31)	0.5
4. Floor girder with floor slab on the top flange	
I section girder, width/depth ratio < 0.5	0.7
I section girder, width/depth ratio > 0.5	0.5
Box girder and lattice girder	0.7

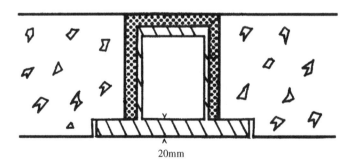

20mm

Figure 10.31 Steel floor girder carrying precast concrete floor units on the bottom flange (see Table 10.9). Pettersson *et al*. (1976). Reproduced by permission of The Swedish Institute of Steel Construction

---

[25] This can be expanded to give Equation 10.51.

Table 10.11 Calculation of the temperature–time curve for the hollow steel column shown in Figure 10.29 exposed to a 'real fire' temperature–time curve (Pettersson *et al.*, 1976) (see Figure 10.32)

Line	Time (s)	T_F) (°C)	T_F(ave) (°C)	γ_i (kW/m².K)	$T_F - T_s$ (°C)	ΔT_s (°C)	T_s (°C)
1	0	20					20
2			334.5	0.0393	314.5	9.4	
3	180	649					29
4			788.5	0.0889	759.1	51.3	
5	360	928					81
6			929.5	0.1200	848.8	77.5	
7	540	931					158
8			932.5	0.1295	774.3	76.2	
9	720	934					234
10			939.0	0.1408	704.6	75.4	
11	900	944					302
12			951.5	0.1549	641.7	75.6	
13	1080	959					385
14			965.5	0.1711	580.1	75.5	
15	1260	972					461
16			947.0	0.1802	486.2	66.6	
17	1440	922					527
18			887.0	0.1776	359.6	48.5	
19	1620	852					576
20			813.5	0.1691	237.5	0	
21	1800	775					607

where γ_i is given by Equation (10.33), with $T_g = T_F^{j+1}$ and $T_i = T_s^j$. Numerically, A_s/V_s is the ratio of the exposed perimeter to the cross-sectional area of the steel, so that $A_s/V_s = 0.8/0.0111 = 72.07$ m^{-1}, and with ρ_s and c_s as given above, Equation 10.55 can be written:

$$T_s^{j+1} = T_s^j + 1.97 \times 10^{-2}\gamma_i(T_F^{j+1} - T_s^j)\Delta t \tag{10.56}$$

The temperature–time curve for the steel column is calculated with Equation 10.56 by an iterative process using a spreadsheet, as shown in Table 10.11. Because the fire temperature data presented in Table 10.2 are at 3 minute (0.05 hour) intervals, it is necessary to use the method illustrated in Figure 10.30, i.e., taking the average fire temperature $(T_F(\text{ave}) = (T_F^{j+1} + T_F^j)/2)$ at each timestep. The result is shown in Figure 10.32, where T_F and T_s are plotted as functions of time. The steel reaches a maximum temperature of 625°C at 35 minutes, long after the fire has peaked at 972°C (at 21 minutes). The maximum steel temperature is achieved during the decay period of the fire, exceeding the nominal failure temperature of 550°C. Whether or not failure will occur depends on conditions of loading and the details of the design of the beam.

For completeness, the temperature of the same column insulated with 25 mm thick insulation board (Figure 10.29(b)) exposed to exactly the same fire has been calculated

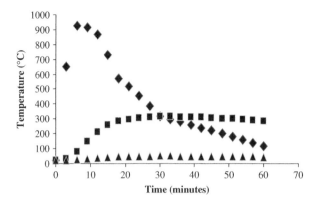

Figure 10.32 Temperature response of a hollow steel column (Figure 10.29) $(\blacksquare)$ exposed to a fire $(\blacklozenge)$ (200 MJ/m^2, $A_w H^{1/2} / A_t = 0.08$ m$^{1/2}$), showing the effect of 25 mm ceramic fibreboard insulation $(\blacktriangle)$

using Equations (10.56) and (10.54). The result is shown for comparison in Figure 10.32. The effectiveness of the insulation is clear.

10.6   Projection of Flames from Burning Compartments

External flaming which accompanies most fully developed compartment fires not only provides a mechanism for fire spread to upper floors, perhaps involving combustible cladding (Oleszkiewicz, 1989; Lougheed and Yung, 1993) but also poses a threat to the stability of external structural members. To enable an assessment of this problem to be made, the general behaviour of such flames needs to be understood. Bullen and Thomas (1979) related flame size (and radiative heat flux to the building façade) to an excess fuel factor f_{ex}, given by Equation (10.22) (Figures 10.14 and 10.15), but the correlation has not yet been explored sufficiently to be applied generally to estimating flame sizes. Indeed, even with an 'excess fuel factor' of zero (stoichiometric) or less, some external flaming can occur due to the time taken for the flame gases to burn out due to inefficient mixing and finite chemical kinetics (cf. the 'fuel-lean' deflected flames described by Hinkley *et al*. (1968) (Section 9.2.1)).

Consequently, it is necessary to rely on an empirical correlation derived by Thomas and Law (1972) from data on flame projection from burning buildings by Yokoi (1960), Webster *et al*. (1959), Seigel (1969) and others (Thomas and Heselden, 1972). The data were correlated using dimensional analysis of the buoyant plume, and led to the following expressions which were found to apply approximately to crib fires in enclosures in the absence of wind. The scatter of data is large.

$$z + H = 12.8(\dot{m}/B)^{2/3} \tag{10.57}$$

$$x/H = 0.454/n^{0.53} \tag{10.58}$$

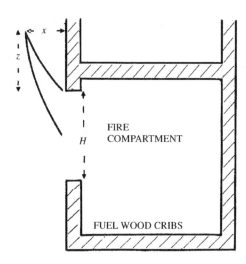

Figure 10.33 Projection of flame from a compartment during fully developed burning. Symbols refer to Equations (10.60) and (10.61)

The symbols are identified in Figure 10.33 and are defined (with the relevant units) as follows:

H, B height and width of the ventilation opening, respectively (m);
  z height of flame tip above the window soffit (m);
  $\dot{m}$ rate of burning (kg/s);
  x horizontal reach from face of building (m); and
  n 2B/H (the 'shape factor').

The width of the flame may be assumed to be equal to B, although some spreading is likely to occur for wide openings when the exit velocity is relatively low. The horizontal projection (x in Figure 10.33) is strongly influenced by the velocity of this emerging jet, which in turn is dependent on the 'shape factor' n (Equation 10.58). This behaviour has also been reported by Lougheed and Yung (1993) in a study of the radiation exposure experienced by adjacent structures.

The tip of the flame is identified as the point on the flame axis where $T_f = 540°C$ (Figure 10.14). Law and O'Brien (1981) have incorporated these into a design guide for protection of exterior steel structures, and recommend that the burning rate ($\dot{m}$) is taken from either

$$\dot{m} = \frac{M}{1200} \text{ kg/s} \qquad (10.59)$$

where M is the fire load (on the assumption that most fires burn out in 20 minutes) or (Thomas and Heselden, 1972):

$$\frac{\dot{m}}{A_w H^{1/2}} \cdot \left(\frac{D}{W}\right)^{1/2} = 0.18\left[1 - \exp\left(-\frac{0.036 A_T}{A_w H^{1/2}}\right)\right] \qquad (10.60)$$

where D and W are the depth and width of the compartment, whichever gives the lower value. Note that in Equation 10.60, A_T is the internal area (floor, ceiling and walls) less the area of the ventilation opening. The method was extended for conditions of 'through draught' and 'no through draught', and for a lateral wind. Under the latter conditions, the maximum flame deflection may be assumed to be 45° (Law and O'Brien, 1981).

However, the uncertainties in some of the definitions of the parameters used in the given correlations mean that the use of this method is not always straightforward. This work deserves revisiting, taking into account the few relevant experimental results that have been reported in the intervening years (e.g., Sugawa *et al.*, 1997; Klopovic and Turan, 2001a,b; Rein *et al.*, 2008).

It should be noted that the correlations (Equations 10.57 and 10.58) break down under the following circumstances (Thomas and Law, 1972):

(a) If there are substantial heat losses from the projecting flame to the façade of the building.
(b) If there is a wind (the flame will be deflected and reduced in length).[26]
(c) If there is a fire on a lower floor (flames will lengthen due to oxygen depletion by the rising combustion products. Merging of flames from different floors can occur – cf. the São Paulo fires of 1972 and 1974).
(d) If the fuel bed has a very large surface area ($\dot{m}$ will be higher than anticipated by Equations 10.57 and 10.60 and the flames will be longer (Figure 10.7(b))).
(e) If the fuel is non-cellulosic and has a low value of L_v (the heat required to produce the volatiles) – as (d).

Situations (d) and (e) correspond to large excess fuel factors (Equation (10.22)) and indicate the need to incorporate this parameter into the flame height correlations (Figure 10.14).

In general, the correlations described above are considered to be conservative, although the presence of non-cellulosic fuels on the size of external flames has not been addressed. As hydrocarbon polymers have a much higher air requirement than cellulosic fuels, it is almost inevitable that external flaming will be more significant, all other things being equal. It should also be remembered that thermoplastics will tend to burn as pools in the fully developed fire, with the potential to produce large areas of burning surface.

10.7 Spread of Fire from a Compartment

The spread of fire from one compartment into neighbouring compartments, ultimately to involve an entire building, has received little direct experimental study. Information available at the time was reviewed by Quintiere (1979), who concluded that the growth was exponential. This was based on observations of the spread of fires set in buildings earmarked for demolition and led to the concept of a 'volumetric fire growth rate', i.e.

$$\frac{dV_F}{dt} = k'V_F \qquad (10.61)$$

---

[26] The effect of wind on the behaviour of the external plume is addressed by Sugawa *et al.* (1997).

where V_F, the volume of the fire, is a measure of its extent at any moment of time. This indicates that the fire grows at an exponential rate ($V_F \propto \exp(k't)$), provided that the 'fuel bed' is uniform: a similar assumption based on fire area is often made regarding the growth of fire within a compartment (Section 9.2.3).

If there is an open door leading into an adjacent corridor, smoke and hot gases will flow out of the fire compartment from a relatively early stage. Flames will escape into the corridor as the fire grows to the fully developed stage, and begin to flow under the corridor ceiling. If they are combustible, the wall and ceiling linings will ignite and the combustible surfaces in the lower part of the corridor will be heated to the firepoint condition. Quintiere and his co-workers (Quintiere, 1974) have studied this scenario in some detail, particularly in relation to the behaviour of carpets in corridors and their contribution to fire spread (see Table 7.5). Flashover in the corridor leads to a further restriction of air flowing to the room of origin, resulting in fuel-rich burning which will result in an increase in the quantity and toxicity of the smoke produced (Section 11.1.4). There will be very substantial flame extension below the ceiling under these circumstances (Section 4.3.4; Hinkley and Wraight (1969)) and it is possible for the unburnt volatiles to burn vigorously when the horizontal ceiling flow turns to enter a stairwell, thus forming a vertical fire plume in which enhanced burning will take place. This can be at a point that is relatively remote from the room of origin.

Fire can spread from the room of origin through gaps or holes in the compartment boundaries above the neutral plane where the positive pressure differential (Figure 10.4) will force flame and hot gases into the adjacent space(s) (Quintiere, 1979) (see also Section 11.2.1(a)). These may be in existence before the fire,[27] for example gaps around service penetrations (carrying water pipes, electrical cables, etc.) which have not been 'fire stopped' (e.g., Buchanan, 2002). Breaches to horizontal compartment boundaries (especially the ceiling) by service ducts are particularly vulnerable to fire attack. In addition to the need for fire-stopping around the penetration, fire may enter the duct unless it has the necessary level of fire resistance. If this happens, fire will spread rapidly upwards, perhaps the entire height of the building if the duct serves all floors. The situation will be exacerbated if the service duct contains any combustible material, such as electrical cables. (Practical issues of preventing fire spread between fire compartments are discussed by Buchanan (2002).) In general, concealed spaces, such as ducts and false ceilings, are potentially hazardous as they can carry the fire far beyond the location of the original locus of origin. Fires can start in such spaces and result in serious threat to life, simply because they can go undetected for a considerable period of time and when they break out of their initial confinement they are large enough to present an immediate threat to the occupants of the building (Rasbash, 1991). Examples include the Summerland Leisure Complex fire of 1974 (Silcock and Hinkley, 1974), the Stardust Club fire of 1981 (Rasbash, 1984), the Bradford Football Stadium fire of 1985 (Popplewell, 1986) and the Dusseldorf Airport Terminal fire of 1996 (NFPA, 1996).

Compartment boundaries (which include closed doors and shutters and fire-stopped service penetrations) can fail after prolonged exposure to fire, sufficient to overcome the 'resistance' of the barrier. The most obvious type of failure ('loss of integrity') is the

---

[27] It goes without saying that the compartment boundary must reach the structural ceiling (or roof) and not stop at a false ceiling. A continuous void extending over a large area of a building (such as the roof space) will allow fire to spread unchecked once it enters the void (e.g., Beard, 1981; NFPA, 1996).

development of full-thickness cracks or openings through which hot gases and flames may pass; these are usually the consequence of poor workmanship (Buchanan, 2002). Another, less common failure mode involves conductive heat transfer through the boundary, generating high temperatures on the other side which may be capable of causing ignition of items or materials in the adjacent compartment[28] (see British Standards Institution, 1987).

External fire spread by flames projecting from windows has been referred to in the previous section. Convective and radiative heating by the fire plume may break windows and ignite combustible materials on the floor above. The use of horizontal projections between floors to deflect the flames away from the façade of the building has been investigated. Early small-scale work in Japan (Yokoi, 1960) showed that a relatively small projection was sufficient if the shape factor $(2B/H)$ was small, but quite substantial projections were necessary where wide windows are involved and the thrust of the flame is less. In general, this principle has been confirmed in large-scale tests (Ashton and Malhotra, 1960; Moulen, 1971; Harmathy, 1979).

Problems

10.1 Calculate the fire resistance of a load-bearing unprotected hollow steel column of square cross-section (250 mm × 250 mm) and thickness 25 mm assuming the standard temperature–time curve to be given by Equation 10.45. Use the same data as for Equation (10.51).

10.2 If the column described in Problem 10.1 is protected by 25 mm thick insulation board, what temperature will the steel achieve in 30 min? Assume that the thermal conductivity of the board is $k = 0.1$ W/m.K.

10.3 The column described in Problem 10.1 is located within a compartment measuring 8 m × 6 m × 3 m high with a ventilation opening 2 m high and 4 m wide on one long wall. The fire load corresponds to 150 MJ/m². Using Table 10.2, ascertain whether or not the column would continue to perform its load-bearing function for the duration of the fire. What is the maximum temperature achieved by the steel?

10.4 What is the temperature that would be achieved by the steel after 30 minutes if the column of Problem 10.2 is exposed to the fire of Problem 10.3?

10.5 Use Law's formula (Equation (10.46)) to estimate the fire resistance required for an insulated column exposed to the fire described in Problem 10.3. Assume that the heat of combustion of wood is 19.5 MJ/kg.

10.6 Calculate how long the column described in Problem 10.1 would survive if it became engulfed in flames arising from a large spillage of hydrocarbon liquid (e.g., crude oil). For simplicity, assume a constant flame temperature of 900°C and take the emissivity to be 1.0. Compare your answer with that of Problem 10.1.

---

[28] This is more commonly encountered in ships, where the 'fire compartments' are formed by steel bulkheads (e.g., Rushbrook, 1985; White *et al.*, 1996).

11

Smoke: Its Formation, Composition and Movement

Gross *et al.* (1967) define smoke as 'the gaseous products of burning organic materials in which small solid and liquid particles are also dispersed'. This is wider than most common definitions, one of which refers to smoke as 'the visible volatile products from burning materials' (*Shorter Oxford English Dictionary*), but it has its limitations. In particular, it does not address the fact that what the observer sees as 'smoke' will contain a substantial quantity of air that has been entrained into the fire plume. The definition given in NFPA 92B (National Fire Protection Association, 2009) and quoted by Milke (2008) is much more comprehensive in that it includes entrained air.[1] However, it is the combination of obscuration and toxicity that presents the greatest threat to the occupants of a building involved in fire. Indeed, statistics collected in the UK and the USA suggest that more than 50% of all fatalities can be attributed to the inhalation of '(particulate) smoke and toxic gas'[2] (National Fire Protection Association, 2008a; Department for Communities and Local Government, 2009).

People die in fires in buildings because they are unable to reach a place of safety before being overcome by untenable conditions created by the fire. It is possible to identify an 'available safe egress time' (ASET, see Pages 350–351), based on the rate of fire growth and how quickly the smoke layer descends to head height (see Section 4.4.4). Successful escape requires that the escape time is significantly less than the available safe egress time. This can be illustrated using time line analysis (e.g., British Standards Institution, 2001; Proulx, 2008) but the concept can be explained in terms of the simple inequality introduced in Chapter 9 (see Equation (11.1)), which has been used by many to demonstrate the important parameters in escape route design (e.g., Marchant, 1976; Nelson and Mowrer, 2008):

$$t_p + t_a + t_{rs} < t_u \qquad (11.1)$$

[1] Smoke: 'The airborne solid and liquid particulates and gases evolved when a material undergoes pyrolysis or combustion, together with the quantity of air that is entrained or otherwise mixed into the mass' (NFPA, 2009).
[2] The UK statistics attribute 44% of fire fatalities to 'overcome by gas or smoke' and 20% to 'burns and overcome by gas or smoke'.

An Introduction to Fire Dynamics, Third Edition. Dougal Drysdale.
© 2011 John Wiley & Sons, Ltd. Published 2011 by John Wiley & Sons, Ltd.

The left-hand side of Equation (11.1) is the 'required safe egress time' (RSET) and distinguishes the components of the escape time, namely t_p (the time from ignition to perception that a fire exists[3]), t_a (the time from perception to the start of the escape action) and t_{rs} (the time taken to move to a place of relative safety). The term t_u is the time from ignition to the production of untenable conditions (i.e., the available safe egress time, or ASET). Provided that this is greater than RSET, successful escape will be possible.

While t_p and t_a are largely dependent on state of awareness, t_{rs} is influenced by many factors, including the agility of the individual, the building geometry (including travel distance to a place of (relative) safety) and the degree to which the person is affected by the products of the fire. It was argued earlier (Chapter 9) that the chance of successful escape can be enhanced by providing means of early detection (which will reduce t_p) and by avoiding materials and configurations that would lead to rapid growth (which will increase t_u). Additionally, measures can be taken to reduce t_a and t_{rs}. These would include provision of well-marked and well-designed escape routes, which could be kept free of smoke for as long as it takes to evacuate the building (see Section 11.2). However, should smoke invade the escape paths, its effects on the occupants may delay escape due to reduced visibility and the presence of toxic and irritant gases. The increased duration of exposure to these harmful products may lead to incapacitation and death.

For clarity, particulate 'smoke' will be dealt with separately from the gaseous combustion products – although they are inhaled together.[4] Thus, existing standard methods of test for 'smoke' measure only the potential for materials to produce particulate matter under the specific test conditions (American Society for Testing and Materials, 1993b, 1994a, 2009d) (see Sections 11.1.1–11.1.3). Different types of test have been considered for toxicity (see Section 11.1.4), although it has been argued that the problem of toxicity of fire products cannot be resolved in this manner (e.g., Punderson, 1981). The yields of particulate smoke and harmful gaseous components are extremely sensitive to changes in the 'fire scenario', and it is unrealistic to expect small-scale fire tests to provide a simple classification of combustible materials according to their potential to generate a smoke hazard (obscuration and/or toxicity) in a real fire. This has now been recognized in the relevant standards. For example, in BS ISO 19706 (British Standards Institution, 2007a) three categories of fire are specified: (1) non-flaming (smouldering); (2) well-ventilated flaming; and (3) underventilated flaming. Categories (2) and (3) are compatible with pre-flashover and post-flashover compartment fires, respectively, and may be identified in terms of the ratio $[CO]/[CO_2]$ as indicated in Table 11.1. This ratio relates to the 'equivalence ratio (ϕ)' which is discussed below.

The first three sections of this chapter are devoted to a summary of the formation and measurement of particulate smoke (Sections 11.1.1–11.1.3), followed by a brief discussion of toxicity (Section 11.1.4). Sections 11.2 and 11.3 deal with smoke movement and control. These topics are of considerable importance in relation to fire safety design of large buildings and are included here for completeness.

[3] This can be broken down further as the time from ignition to detection plus the time to notify the occupants that there is a fire (Nelson and Mowrer, 2008).
[4] If smoke particles are found in the respiratory tract of a fire victim, it is unambiguous evidence that he/she was still alive when the fire started (Anderson et al., 1981).

Table 11.1   The three main fire types defined in BS ISO 19706 (British Standards Institution, 2007)

	Fire type[a]	Fuel/air equivalence ratio	[CO]/[CO$_2$] (v/v)
1	Non-flaming (smouldering)	–	0.1–1
2	Well-ventilated flaming	<1	<0.05
3	Underventilated flaming	>1	0.2–0.4

[a]Types 1 and 3 are subdivided further in the Standard.

11.1   Formation and Measurement of Smoke

Particulate smoke is a product of incomplete combustion. It is generated in both smouldering and flaming combustion, although the nature of the particles and their modes of formation are very different. Smoke from smouldering is similar to that obtained when any carbon-based material is heated to temperatures at which there is chemical degradation and evolution of volatiles. The high molecular weight fractions condense as they mix with cool air to give an aerosol consisting of minute droplets of tar and high-boiling liquids. These will tend to coalesce under still air conditions to give a distribution of particle size with a mass median diameter of the order of 1 μm (Bankston *et al*., 1978; Mulholland, 2008) and will deposit on surfaces to give an oily residue.

Smoke from flaming combustion is different in nature and consists almost entirely of solid particles. While a small proportion of these may be produced by ablation of a solid under conditions of high heat flux, most are formed in the gas phase as a result of incomplete combustion and high temperature pyrolysis reactions at low oxygen concentrations (see reviews by Rasbash and Drysdale (1982), Glassman (1988), Kennedy (1997) and Mulholland (2008)). Particulate matter can be generated even if the original fuel is a gas or a liquid.

Both types of smoke are combustible, due at least in part to the unburned and partially burned fuel vapours which are inevitably present along with the particles: under certain circumstances, explosions or explosion-like events can occur (Croft, 1980/81). One of the best documented cases is the 'Chatham mattress fire', which involved prolonged smouldering of latex rubber mattresses in a large warehouse (Woolley and Ames, 1975) (Section 8.2). A related phenomenon is backdraught (Section 9.2.1), which involves rapid burning when ventilation is provided to a compartment in which fuel-rich smoke has accumulated during conditions of poorly ventilated flaming combustion (Fleischmann *et al*., 1994).

11.1.1   Production of Smoke Particles

Under conditions of complete combustion, fuel will be converted into stable gaseous products, carbon dioxide and water (Section 1.2.3) but this is rarely, if ever, achieved in diffusion flames. In a typical fire, mixing of the fuel vapours and air within the flame involves buoyancy-induced turbulent flows in which substantial temperature and concentration gradients exist (Section 4.3). In regions of low oxygen concentration such as the fuel-rich core of a large hydrocarbon pool fire (see Figure 4.8), a proportion of the volatiles may undergo a series of pyrolysis reactions which lead to the formation of unsaturated molecular species. These include acetylene (ethyne) which, under these conditions, can undergo quasi-polymerization reactions (Section 1.1.1) to produce aromatic species, the simplest of which is benzene (C_6H_6). Benzene is the precursor of polycyclic hydrocarbons, which in turn grow to produce minute particles of soot within the flame (Glassman, 1989; Griffiths and Barnard, 1995; Richter and Howard, 2000).

Although a discussion of the chemical mechanism of smoke formation is beyond the scope of this text (see Richter and Howard (2000)), some of the possible steps in this complex process are shown in simplified format in Figure 11.1 (from Glassman (1989)).[5] All these steps are reversible, and any of the intermediate species may undergo oxidation although, once formed, the aromatic 'nucleus' (C_6H_6) is remarkably stable at high temperatures and has a degree of resistance to oxidation. It is the presence of 'soot particles' within the flame that gives the diffusion flame its characteristic yellow luminosity (Section 2.4.3) and provides the principal mechanism for radiative heat loss. The concentration of soot particles determines the effective emission coefficient (K in Equation (2.83), Section 2.4.3). It is difficult to measure, but can be related to the soot volume fraction (f_v), which is the fraction of the flame volume occupied by soot particles[6] (Pagni and Bard, 1979; Tien *et al.*, 2008). These minute particles (10–100 nm diameter) may oxidize within the flame and be destroyed (Wagner, 1979) but if the temperature and oxygen concentration are not high enough they will tend to grow in size and agglomerate to give substantially larger particles which will escape from the flame as 'smoke'. A critical height of a laminar diffusion flame can be measured at which smoke is first observed – this is known as the smoke point (see below). Smoke particles are complex agglomerations of minute soot particles which have come together to form chains and clusters, growing to an overall size in excess of 1 μm (Figure 11.2). When their size becomes of the same order as the wavelength of light (0.3–0.7 μm), they cause obscuration and reduced visibility by a combination of absorption and light scatter.

The tendency of a fuel to produce smoke may be assessed by measuring its 'smoke point'. It is defined as the minimum laminar flame height (or fuel mass flowrate) at which smoke first escapes from the flame tip. It can be determined relatively easily for gases and vapours simply by adjusting the flowrate of fuel from a simple burner, but its determination for combustible liquids and solids requires a much more sophisticated device that allows the flowrate of fuel vapour to be controlled by irradiation with a CO_2 laser (de Ris and Cheng, 1994). It has been shown to correlate very well with combustion efficiency, the radiative fraction (χ_r in Table 5.12: see also Markstein (1984)) and the yields of CO and 'particulate smoke' (Tewarson, 2008), all of which are of course inter-related.

The chemical composition and structure of the parent fuel are of considerable importance (Rasbash and Drysdale, 1982; Glassman, 1989). A small number of pure fuels (namely carbon monoxide, formaldehyde, metaldehyde, formic acid and methyl alcohol) burn with non-luminous (soot-free) flames and do not produce smoke. Other fuels, burning under identical conditions, give substantial yields, depending on their chemical nature. Oxygenated fuels such as ethyl alcohol (ethanol) and acetone give much less smoke than the hydrocarbons from which they are derived (Table 11.2), while for the hydrocarbons,

---

[5] The mechanism of soot formation is likely to be much more complex than that shown in Figure 11.1: it has been argued that ionic species may be involved (e.g., see Calcote and Keil (1990)).

[6] The magnitude of f_v is on the order of 10^{-6}.

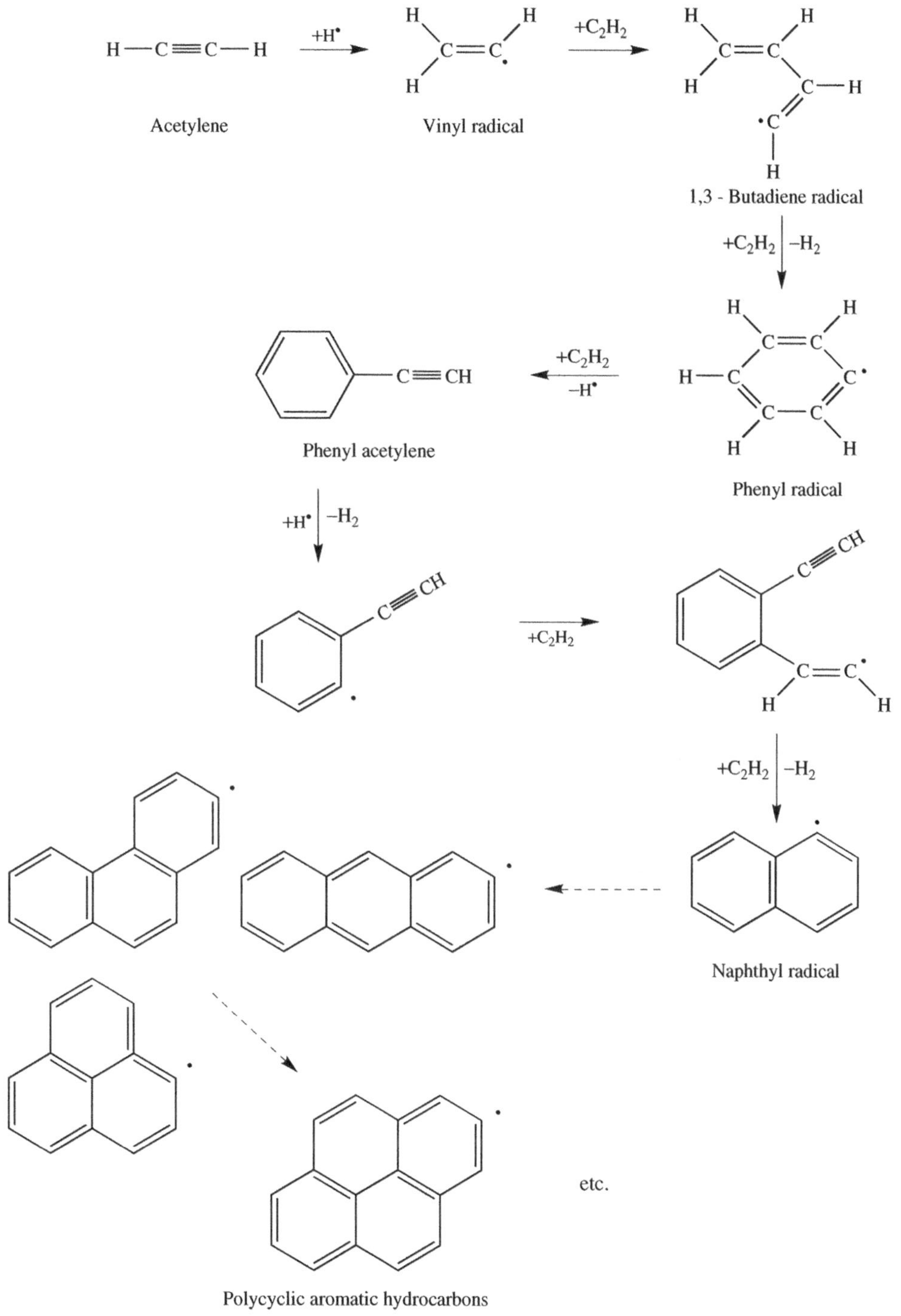

Figure 11.1 The chemical pathway to 'soot' (after Glassman, 1989). See footnote 5

AGGLOMERATE VIEWED AT TWO ANGLES

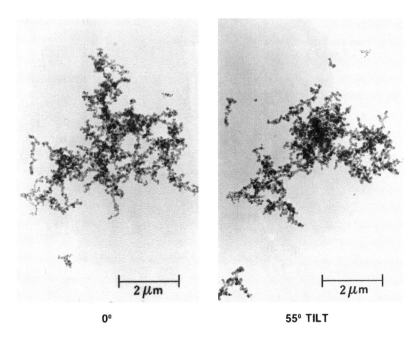

Figure 11.2 Transmission electron micrographs of a smoke particle. The overall size of the agglomerate is about 6 μm, and the diameter of the individual spherules is about 0.03 μm (from Mulholland (2008)). Photographs by Eric B. Steel, Chemical Science and Technology Laboratory, National Institute of Standards and Technology

Table 11.2 Smoke-forming tendency of simple fuels

Fuel type	Typical example		Smoke point (m)
Alcohols	Ethanol[a]	C_2H_5OH	0.227
Ketones	Acetone[a]	$CH_3.CO.CH_3$	0.205
n-Alkanes	n-Hexane[a]	$CH_3(CH_2)_4.CH_3$	0.118
iso-Alkanes	2,3-Dimethyl butane[a]	$(CH_3)_2.CH.CH.(CH_3)_2$	0.089
Alkenes	Propene[a]	$CH_3.CH=CH_2$	0.029
Alkynes	Propyne[a]	$CH_3.C\equiv CH$	–
Aromatics	Styrene[a]	⟨CH=CH₂ structure⟩	0.006
Polynuclear aromatics	Naphthalene[a]	⟨naphthalene structure⟩	0.005
Polymers:	Polymethylmethacrylate[b]	$-(CH_2-CH(COOCH_3))-$	0.105
	Polyethylene[b]	$-(CH_2-CH_2)-$	0.045
	Polypropylene[b]	$-(CH_2-CH(CH_3))-$	0.050
	Polystyrene[b]	$-(CH_2-CH(C_6H_5))-$	0.015

[a] A. Tewarson, personal communication (1986).
[b] Taken from Tewarson (2008).

there is a progression in the tendency to produce smoke with the introduction of branching (cf. *n*-alkanes and iso-alkanes), unsaturation (cf. alkanes, alkenes and alkynes) and aromatic character (cf. alkanes, aromatics and polynuclear aromatics). These observations on pure fuels are consistent with observations of the smoke production potential of solids (see Table 11.5). Under free burning conditions, oxygenated fuels such as wood and polymethylmethacrylate give substantially less smoke than hydrocarbon polymers such as polyethylene and polystyrene. Of the latter pair, polystyrene produces by far the greater yield as the volatiles from this polymer consist largely of styrene and some of its lower oligomers,[7] all of which contain the aromatic nucleus (Table 11.2).

In the next section, we shall consider how smoke yields can be measured and consider how results from standard methods of test for smoke production compare with smoke generation in 'real fires'.

11.1.2   Measurement of Particulate Smoke

The yield of particulate smoke from a burning material may be assessed by one of the following methods:

(a) Filtering the smoke and determining the weight of particulate matter (suitable for small-scale tests only) (ASTM, 1989 (now withdrawn)).
(b) Collecting the smoke in a known volume and determining its optical density (small and medium scale only) (Rasbash and Phillips, 1978; ISO, 1989; ASTM, 2009b).
(c) Allowing the smoke to flow along a duct, measuring its optical density where plug flow has been established, and integrating over time to obtain a measure of the total (particulate) smoke yield (Atkinson and Drysdale, 1989; Babrauskas, 2008b; ASTM, 2009c).

The quantity of particulate smoke produced in a test must be expressed as a yield, i.e., the amount generated per unit mass of fuel burned. For the gravimetric method (a), the mass of smoke particles from a known quantity of material is measured, and the yield may be quoted as mg particles/g fuel. However, methods (b) and (c) rely on measuring the optical density of the 'smoke' under the conditions of the test, involving some dilution by air. The smoke yield has to be related to the *concentration* of particles and the volume in which they have been dispersed. Fortunately, as will be shown below, optical density is directly proportional to the particle concentration (to within satisfactory limits of accuracy) and may be used as a surrogate measurement.

The definition of smoke given earlier specifically includes air that has been entrained. Thus, in the calculation of the flowrate of smoke 15 m above a 6.4 MW fire (see Section 4.4.4), the 'smoke' was almost entirely entrained air. The 'mass' (hence volume) of smoke increases with height according to Equation (4.48), in inverse proportion to the concentration of the smoke particles. Clearly, if optical density is to be used to quantify 'smoke yield', the volume in which the smoke particles are dispersed must be known.

Optical density may be determined by measuring the attenuation of a beam of light passing through the smoke (see Figure 11.3). If, in the absence of smoke, the intensity of

---

[7] Oligomer: a very short-chain 'polymer' containing only a few monomeric units (typically 2, 3 or 4 in this case) (see Figure 1.1(a)).

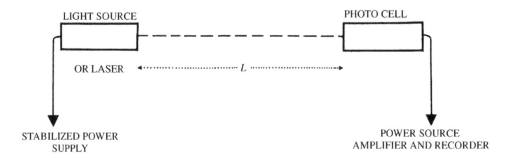

Figure 11.3   Representation of the apparatus required to measure optical density

light falling on the photocell is I_0, then in the presence of smoke, the reduced intensity (I) will be given by the Lambert–Beer law[8] (Equation (2.80)):

$$I = I_0 \exp(-\kappa CL) \tag{11.2}$$

where κ is the extinction coefficient, C is the mass concentration of smoke particles and L is the pathlength of the optical beam passing through the smoke. Optical density is then defined in terms of either natural logarithms (Babrauskas, 1984b):

$$D_e = -\log_e \left(\frac{I}{I_0} \right) = \kappa CL \tag{11.3}$$

or common logarithms (Rasbash and Phillips, 1978):

$$D_{10} = -10 \log_{10} \left(\frac{I}{I_0} \right) = \frac{10}{2.303} \cdot \kappa CL \tag{11.4}$$

where the factor of 10 is introduced to be consistent with the measurement of the attenuation of sound and electrical signals (decibels, or db) (e.g., see Halliday *et al.* (2008)).

Comparing these two expressions with Equation (11.2) shows that optical density is directly proportional to the concentration of smoke and is a much more satisfactory measurement than 'percentage obscuration', given by:

$$\frac{I_0 - I}{I_0} \times 100$$

(see Table 11.3).

Strictly speaking, Equations (11.3) and (11.4) apply to monochromatic light and laser beams are now commonly used in laboratory equipment such as the cone calorimeter,[9] although they have been used satisfactorily with white light[10] (Rasbash, 1967; Ostman and Tsantaridis, 1991). The absorption coefficient is dependent on wavelength, and on the particle size distribution which is likely to change as the smoke 'ages' with time (Seader and Chien, 1975). (The ageing process involves coalescence of particles, so that

---

[8] Also called Bouguer's law.

[9] A helium/neon laser is used in most equipment, with a wavelength of 632.8 nm (0.6238 μm).

[10] The laser is to be preferred as it gives a much more stable and reliable output.

Table 11.3 Relationship between percentage obscuration and optical density

Percentage obscuration	Optical density (Equation (11.3))	Optical density (db) (Equation (11.4))
10	0.11	0.46
50	0.69	3.01
90	2.30	10.00
95	3.00	13.01
99	4.61	20.00

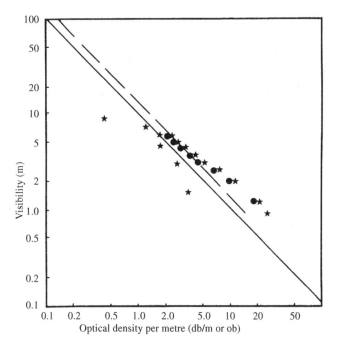

Figure 11.4 Relationship between visibility and optical density (per metre pathlength) for objects illuminated by scattered light: •, Rasbash (1951); ★, Malhotra (1967); _ _ _, Jin (1970, 1971); ——, simple correlation recommended by Butcher and Parnell (1979). By permission of E&F Spon Ltd from Butcher and Parnell, *Smoke Control in Fire Safety Design*

the number density will decrease. Ionization smoke detectors are particularly sensitive to the number of particles per unit volume and respond best to 'young smoke'. They can respond to particles as small as 10 nm in diameter, while scatter and obscuration devices do not respond until the particle size is of the same order as the wavelength of light (O(100 nm))).

Rasbash (1967) showed that the optical density, expressed in terms of unit pathlength (i.e., D_{10}/L, db/m), correlates reasonably well with visibility, 1 db/m corresponding approximately to a visibility of 10 m (Figure 11.4). This is for general visibility through smoke. For visibility of 'exit' signs in buildings, there is a substantial difference between those signs that are 'front illuminated' and those that are 'back illuminated'. On the basis of results of Jin (1978), Butcher and Parnell (1979) suggest that a sign which is self-illuminated can be seen at 2.5 times the distance of one that relies on surface illumination.

11.1.3 Methods of Test for Smoke Production Potential

In view of the serious nature of the problems presented by smoke generation in fires, it is hardly surprising that there has been considerable pressure from those sectors of local and national government concerned with public safety for a method of test whereby materials can be rated according to their potential to produce smoke. However, in searching for such a test, it has been tacitly assumed that smoke yield is a property of the material and that the 'fire environment' has only a second-order effect. This assumption is incorrect: relatively little attention has been paid to the fact that the particulate smoke yield is strongly dependent on the 'fire scenario' – in particular, the ventilation conditions that exist at the time, and whether the fire is pre- or post-flashover. This is true also for the yields of toxic and corrosive products (Section 11.1.4).

There are many smoke tests in existence (see Table 11.4). Of these, the 'NBS Smoke Density Chamber' (ASTM, 2009b) has probably been the most widely used. It involves exposing a 75 mm square sample of material in a vertical configuration to a nominally constant radiant heat flux of 25 kW/m². Flaming combustion is initiated and sustained by an array of six small pilot flames impinging on the face of the sample. The smoke is confined within the volume of the apparatus (0.51 m³) and the optical density measured

Table 11.4 Summary of existing smoke tests

Name	Type	Reference
Rohm-Haas XP-2	F,O,S	ASTM (2004)
NBS test	R,O,S	ASTM (2009b)
Arapahoe test	F,G	ASTM (1989)
Steiner tunnel	F,O,D	ASTM (2009a)
Radiant panel test	R,O,D/G[a]	ASTM (2009b)
OSU calorimeter	R,O,D	ASTM (2008d)
ISO dual chamber	R,O,S	ISO (1990)
Cone calorimeter	R,O,D	ASTM (2009c)
FTA	R,O,D	ASTM (2009d)
SBI test	F,O,D	BSI (2002b)

Key: F = sample exposed to flame only; R = sample exposed to radiant heat flux (flame may or may not be present); O = smoke determined by obscuration of a light beam; G = smoke determined gravimetrically; S = smoke allowed to accumulate in a known volume; D = smoke measured as it flows from the test apparatus.
[a] This involves a gravimetric method in which smoke is sampled from the duct through a filter.

by means of a vertical light beam, $L = 0.914\,\text{m}$. The standard specifies that the results are expressed as the maximum 'specific optical density', defined as:

$$D_\text{m} = \frac{1}{10}\frac{D_{10}V}{LA_\text{s}} \tag{11.5}$$

where D_{10} is defined as in Equation (11.4), V is the volume of the test chamber (m^3) and A_s is the exposed area of the sample (m^2). As written, D_m is non-dimensional.

Typical examples for flaming and non-flaming degradation are given in Table 11.5. However, as only the area of the sample is specified, in principle D_m needs to be determined as a function of thickness. The reproducibility and repeatability of this test are better than many others, although a variability of $\pm25\%$ is still to be anticipated (ASTM, 2009b). Rasbash and Phillips (1978) carried out a series of measurements of the smoke produced from a range of materials, generating the smoke using a furnace modelled on that used in the original NBS test (ASTM, 2009b), but collecting it in a much larger volume $(13\,\text{m}^3)$. The results were expressed in terms of a 'smoke potential' D_0:

$$D_0 = \frac{D_{10}}{L}\cdot\frac{V}{W_l}(\text{db/m})\cdot\text{m}^3/\text{g} \tag{11.6}$$

where W_l is the mass of material (g) vaporized (burned) during the test. There are a number of advantages in expressing the results in this way. D_0 is the volume of smoke of unit optical density per metre (i.e., visibility $10\,\text{m}$) that would be produced from one gram of material (in a standard test). Rasbash suggested that the term 'obscura' (ob) be used for db/m (Rasbash and Phillips, 1978) not only to emphasize the significance of db/m, but also to aid comprehension of the units of D_0: values of D_0 for some typical materials under conditions of flaming and non-flaming degradation are given in Table 11.6.[11] Most

Table 11.5 Comparison of smoke yields from flaming and non-flaming exposure in the NBS Smoke Chamber Test (ASTM, 1994a)

	Specific optical density[a]	
	Flaming	Non-flaming
Polyethylene	62	414
Polypropylene	96	555
Polystyrene	717	418
Polymethylmethacrylate	98	122
Polyurethane	684	426
Polyvinylchloride	445	306
Polycarbonate	370	41

[a]The maximum specific optical density is defined in Equation (11.5). Although apparently dimensionless, it has units $(\text{bels/m})\text{m}^3/\text{m}^2$, where 1 bel/m corresponds to a 10-fold reduction in light intensity over a pathlength of 1 metre.

[11] Despite its logical derivation, the 'obscura' has not been widely adopted (but see Ostman and Tsantaridis (1991)). It is included here because it focuses attention on the actual significance of the 'specific extinction area' developed for use in the cone calorimeter (Equation (11.10)).

Table 11.6 Smoke potentials of different materials (Rasbash and Phillips, 1978)

	Smoke potential (ob.m^3/g)^a	
	Flaming	Non-flaming
Fibre insulation board	0.6	1.8
Chipboard	0.37	1.9
Hardboard	0.35	1.7
Birch plywood	0.17	1.7
External plywood	0.18	1.5
α-Cellulose	0.22	2.4
Rigid PVC	1.7	1.8
Extruded ABS	3.3	4.2
Rigid polyurethane foam	4.2	1.7
Flexible polyurethane foam	0.96	5.1
Plasterboard	0.042	0.39

^{a}Smoke potential (D_0) is defined in Equation (11.6). The unit 1 ob (i.e., 1 db/m) corresponds to a 10-fold reduction in light intensity over a pathlength of 10 m.

common materials give less smoke in flaming combustion mode, although there are some notable exceptions (Tables 11.4 and 11.5).

This method of expressing 'smoke production potential' has not been widely adopted, despite its relative ease of application. For example, consider what will happen if a polyurethane cushion weighing 0.5 kg burns in a room of volume 50 m^3. If 15% by weight of residual char remains after combustion (i.e., $W_l = 0.85 \times 0.5 = 0.425$ kg), the optical density of the smoke which accumulates in the room will be:

$$\frac{D_{10}}{L} = \frac{D_0 \cdot W_l}{V} = \frac{0.96 \times 425}{50} = 8.16 \text{ ob} \tag{11.7}$$

if mixing is complete. (According to Figure 11.4, this corresponds to a visibility of slightly less than 2 m.) In this calculation, the product $D_0 \times W_l$ (408 ob·m^3) is the total volume of smoke released from the polyurethane during the fire, which has an optical density D_{10}/L of unity (1 ob). This may be generalized to define the total 'smoke load' of a room or compartment as:

$$V = \sum D_0 \cdot W_l \tag{11.8}$$

Provided that D_0[12] and W_l are known for each combustible item present, this could be used to compare the smoke production potential of different compartments within a building, although it gives no information on rates of production (see below). A similar method based on specific optical density (D_m) was originally proposed by Robertson (1975).

---

[12] It is worth emphasizing here that the value of D_0 must be appropriate to the fire scenario involved (well-ventilated, post-flashover, etc.).

It is now generally accepted that the yield of smoke must be determined on a mass *loss* basis, rather than total mass or exposed surface area of the specimen tested. In the ISO test (ISO, 1990), the smoke yield is expressed as the 'mass optical density':

$$\text{MOD} = \frac{V}{LW_l}\left(\log_{10}\frac{100}{T}\right) \tag{11.9}$$

where T is the percentage of light transmission ($100 \times I/I_0$). If this is compared with Equations (11.4) and (11.6), it is seen to differ from the 'smoke potential' by a factor of 10: in Equation (11.9), optical density has the unit 'bel', compared with 'decibel' in Equation (11.4). Otherwise, the two approaches are identical.

However, since the development of the cone calorimeter and large-scale experimental rigs which use oxygen consumption calorimetry to determine the rate of heat release, there has been a move to standardize smoke measurement using natural logarithms in which optical density is defined as D_e (Equation (11.3)). This has been incorporated into the software produced for the dynamic measurement of smoke associated with these devices. The optical density of the smoke is measured continuously as it passes through the extract duct at a linear flowrate that is also monitored continuously. Using the cross-sectional area of the duct, the volumetric flow rate ($\dot{V}$), corrected to ambient temperature, is calculated. By combining this with the rate of mass loss ($\dot{m}$) from the specimen under test, the 'specific extinction area' (SEA) is obtained:

$$\text{SEA} = \frac{D_e}{L}\cdot\frac{\dot{V}}{\dot{W}_l} \tag{11.10}$$

As the mass loss is expressed in kg, SEA has units of m^2/kg, but only because the unit of optical density per metre is lumped together with the units of volume. It can be converted into MOD (Equation (11.9)) or 'smoke potential' D_0 (Equation (11.6)) by multiplying by 4.34×10^{-4} (i.e., $10^{-3}/\log_e 10$) or 4.34×10^{-3}, respectively. Measurement of smoke is discussed in detail by Östman (1992) and Whiteley (1994).

The relationship between the smoke yields determined in standard tests and those obtained under real fire conditions has received comparatively little attention (see, however, Heskestad and Hovde, 1994). Most of the tests are carried out to allow the sample to burn under well-ventilated conditions (e.g., the cone calorimeter), but with tests such as the NBS chamber (American Society for Testing and Materials, 1994a), it is necessary to limit the amount of material subjected to test because of the limited air available in the chamber. In fact, smoke yield is very sensitive to the conditions of burning. This has been demonstrated in the small-scale tests by varying the imposed heat flux and the orientation of the fuel, in addition to comparing the smoke from flaming combustion and non-flaming thermal decomposition (pyrolysis). In 'real fires', the full range of orientations are possible, and the imposed heat flux will exceed $30\,kW/m^2$ in the post-flashover fire. However, of greater significance is the fact that during the fully developed fire the fuel volatiles are reacting in a hot atmosphere in which the oxygen concentration may be low. A few early studies showed that under these conditions the smoke yields will be much greater than those predicted by well-ventilated small-scale tests by as much as a factor of 4–6 (Drysdale and Abdul-Rahim, 1985) and perhaps by as much as a factor of 10 (Fleischmann *et al.*, 1990). On the other hand, vigorous free burning under well-ventilated conditions

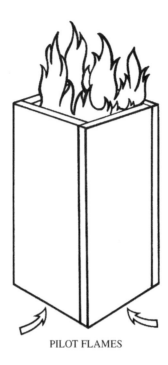

PILOT FLAMES

Figure 11.5 Chimney configuration used to generate smoke from freely burning combustible solids by Rasbash and Pratt (1980) (less smoke is produced than predicted by small-scale tests)

(i.e., pre-flashover) can produce less smoke than predicted in a small-scale test (Rasbash and Pratt, 1980) (see Figure 11.5). These issues need to be borne in mind when smoke test data are being considered as a means of assessing performance in real life. Heskestad and Hovde (1994) compared results on smoke yield from the cone calorimeter with data gathered from the ISO and CSTB rooms, and found no overall, simple correlation. It was only when smoke yield[13] results were correlated with the 'global equivalence ratio' (see Section 11.1.4) that some degree of order was created out of apparent chaos (Gottuk *et al.*, 1992a; Stec *et al.*, 2009).

While these effects have still to be quantified, it is the *rate* of smoke production that is a major factor in determining how rapidly smoke will spread from the fire area to affect other spaces and escape routes. For this reason, the emphasis on 'smoke yield' waned as interest switched to determining the rate of heat release (Babrauskas and Peacock, 1992). The rate of burning of a material (perhaps measured as the rate of heat release) must be considered with its smoke production potential (e.g., the results of Heskestad and Hovde (1994) referred to above). Take as an example the flexible polyurethane foam examined by Rasbash and Phillips (1978) (Table 11.6). Its smoke potential (0.96 ob m^3/g) is only 2–3 times greater than that for the wood products fibre insulating board, chipboard and hardboard, yet subjective impressions of fires involving these materials would suggest that polyurethane is in a different league as regards smoke production. This is due to the

---

[13] Defined as the mass of smoke produced per unit mass of fuel consumed.

rapid development of fire on the foamed plastic (Section 6.3.2) and the subsequent high burning rate (see Table 5.8), which will produce unacceptable levels of smoke much more rapidly than the cellulosic materials. Indeed, any material or configuration of materials which can develop fire quickly and burn vigorously presents a threat to life safety.

It has been suggested that a proper assessment of the hazard due to smoke should include a weighting factor to penalize those materials which release smoke quickly (e.g., Routley and Skipper, 1980). Babrauskas (1984) proposed the adoption of a smoke parameter based on data obtained in the cone calorimeter. This is the product of the average SEA for the first 5 minutes of flaming combustion and the maximum rate of release over the same period (measured at several imposed heat fluxes between 20 and $50\,kW/m^2$). A similar approach is adopted for the 'single burning item' test for wall lining materials (British Standards Institute, 2002) – the rate of smoke production is used to calculate the 'smoke growth rate index' (SMOGRA),[14] providing a ranking that is specific to the SBI test but is available for regulatory purposes within the member states of the European Union (see, for example, Smith and Shaw (1999)).

11.1.4   The Toxicity of Smoke

The above sections have dealt specifically with the particulate matter that is present in 'smoke', but the gaseous components are at least of equal importance. People can die during a fire in a building if they are unable to reach a place of safety before being overcome by conditions created by the fire itself (Equation (11.1)). Poor visibility will make it difficult for the occupants to find and navigate through the escape routes, thus increasing the length of time that they are exposed to the combustion products, many of which are harmful to humans. These may be subdivided into two groups, asphyxiants (which deprive the body of oxygen) and irritants (which inhibit breathing by stimulating pain receptors in the respiratory tract and can also cause severe eye irritation). The most important asphyxiants are carbon monoxide (CO) and hydrogen cyanide (HCN).[15] Unlike other asphyxiants known to be present, these have been measured in smoke at concentrations sufficient to cause significant acute toxic effects (Hartzell *et al.*, 1985). Their net effect is the same – reducing the rate at which oxygen is metabolized in the body – but the mechanisms are very different. Carbon monoxide competes with oxygen for haemoglobin in the bloodstream to form the relatively stable carboxyhaemoglobin and effectively reduces the oxygen-carrying capacity of the blood. Hydrogen cyanide, on the other hand, inhibits the utilization of oxygen in cellular metabolism. Both processes lead to a reduced supply of oxygen to the brain, perhaps with fatal consequences depending on the amount of each gas that has been inhaled. Typical irritants that may be encountered are acrolein (CH_2=CH–CHO), which is a constituent of wood smoke whose yield is scenario dependent, and hydrogen chloride (HCl), which will only be produced from chlorine-containing materials such as polyvinyl chloride (Section 1.1.2). The yield

---

[14] The optical density of the smoke is monitored as it flows through the duct and combined with the associated volumetric flowrate to give an average rate of smoke growth over a defined period (BSI, 2002). The derivation is too lengthy to include here and the interested reader is directed to the Standard.

[15] The toxicity of HCN is over 20 times that of CO, but it is produced in much lower concentrations and only if the 'fuel' contains nitrogen (e.g., nylon (Equation (1.R2)), wool and polyurethane). It is not formed from atmospheric N_2.

of HCl from PVC is not dependent on the fire scenario as it is released quantitatively in the chain-stripping described by Equation (1.R3).

What is important is the *dose* of the gas inhaled, which is expressed in terms of the product of concentration (C) and time (t). Work on the toxicity of gases has traditionally involved experiments in which laboratory animals (such as rats) are exposed to various concentrations for different periods of time and the effective dose (Ct) that causes incapacitation or death obtained.[16] According to Haber's rule (see Purser (2010)), the effective dose is constant: this is demonstrated in results for both incapacitation and death by Hartzell *et al.* (1985). Once values have been determined, they can be used to calculate the length of time an individual can be exposed to a particular toxicant before succumbing to its effects. If the 'fractional effective dose', expressed as:

$$\text{FED} = \frac{\text{Dose received at time } t\ (Ct)}{\text{Effective dose } Ct \text{ to cause incapacitation or death}} \tag{11.11}$$

is greater that 1.0, then the individual's chances of escaping unaided will be greatly reduced (Purser, 2008).

It is clear that two issues have to be considered – the length of time an individual is exposed to the smoke, and the concentration of the asphyxiant gases present. The length of exposure will be increased if the visibility is poor, or if the combustion products contain eye and/or respiratory irritants which make wayfinding through smoke extremely difficult and cause dramatic reductions in walking speed (Jin, 2008; Purser, 2010). In principle, the oxygen concentration should also be considered as concentrations significantly less than 21% (say, $< 15\%$) are known to be problematical. This would also enhance the effect of CO (see Purser (2008)), but as low oxygen levels are almost inevitably associated with lethal concentrations of carbon monoxide, it is considered unnecessary to take O_2 depletion into account in fire safety engineering calculations (ISO, 2007a). The other factor that must be considered is the effect of the increased carbon dioxide concentration in the smoke. At elevated levels (i.e., $>0.03\%$, see Table 1.11), it causes an increase in the rate of respiration,[17] which will in turn increase the rate of inhalation of smoke, producing a toxic response more rapidly.

If there are several toxic species present, then the sum of the individual fractional effective doses should be examined. If:

$$\sum_i \text{FED}_i > 1 \tag{11.12}$$

then escape will be compromised. This assumes that the effects of the individual components i are additive and not synergistic. There is no evidence for synergism between CO and HCN which are, in effect, the only *asphyxiants* that need to be considered in the fire products. Post-mortems of fire victims can determine the presence of CO in the body relatively easily (as carboxyhaemoglobin in the blood) and establish whether or not

---

[16] The time (t) is conventionally taken as the time at which 50% of the sample of animals has succumbed.

[17] This is known as hyperventilation. At 3% CO_2, the respiration rate is double that in a normal atmosphere, and triple at 5% CO_2 (Purser, 2008).

it has been a contributory factor in causing the fatality (e.g., see Nelson (1998)), but it is much more difficult to measure the amount of cyanide in the body as it tends to be converted to other species *post mortem* (Anderson *et al.*, 1979). In any case, it has been observed that high levels of cyanide tend to be found with levels of carboxyhaemoglobin that are known to be fatal, thus making it difficult to assess the role HCN may have had in causing the death.

Compared to the tests for particulate smoke discussed in the previous sections, completely different methods are required for the assessment of toxicity, many involving the exposure of animals to fire gases (Kimmerle, 1974; Potts and Lederer, 1977; Deutsches Institut für Normung, 1979; Purser, 2008; Stec, 2010). The quantities and distribution of the gaseous products of fire are sensitive to relatively minor changes in the fire scenario, as has been pointed out by Punderson (1981). Indeed, even in proposed standard tests (e.g., Potts and Lederer, 1977; Deutsches Institut für Normung, 1979), the distribution of products shows wide variation (Woolley and Fardell, 1982).

In Section 11.1.3, it was pointed out that under conditions of poor ventilation the yield of particulate smoke is increased significantly due to inefficient combustion. The yields of CO and HCN (as well as other products of incomplete combustion, many of which will be irritants) are known to increase markedly as a fire grows from well-ventilated to an underventilated regime (Table 11.1), thereby leading to an increased toxicity while at the same time the increased yield of smoke will cause a (further) reduction in visibility. The conditions under which materials burn – whether 'fuel lean' or 'fuel rich' – have a significant effect on the toxic hazard of the smoke.

The first systematic study that revealed the correlation between CO yield and ϕ was carried out by Beyler (1984) (see Section 9.2.1). He used the two-layer burner/hood assembly illustrated in Figure 9.5 to study the properties of the smoke layer under a ceiling and measured the yield of CO at various fuel/air ratios, converting the latter to equivalence ratios by normalizing them to the stoichiometric fuel/air ratio (Equation (1.26)) (The term 'global equivalence ratio' (GER) has come into common use in the fire community (see Pitts (1995)), but it has the same definition.[18]) The results are presented in Figure 9.6 and show that although there is a fuel dependency, when $\phi < 0.5$ the yield of CO is very low, but increases 10- to 20-fold when ϕ becomes >1.0. This result has been confirmed by Zukoski and co-workers (Zukoski *et al.*, 1991). CO appears to be the only product of incomplete combustion produced in significant quantities when ϕ is significantly less than 1.0 (Figure 9.6; Pitts (1995)).

Gottuk *et al.* (1992) extended this work to compartment fires and found similar results. The amount of air entrained into the fire was determined by separating the inflow of air to the compartment from the outflow of fire gases and measuring the inflow directly. They demonstrated that as a fire developed, the oxygen level in the compartment fell progressively and the CO concentration began to rise when the equivalence ratio ϕ (or GER) exceeded c. 0.8: this is illustrated in Figure 11.6. Subsequent work by Blomqvist and Lonnermark (2001) and Purser and Purser (2008) revealed the same behaviour for HCN. Moreover, the latter authors also showed that when HCN yields in large-scale fires were

---

[18] This statement is true, but estimating the entrainment of air into the upper layer (or the fully developed fire) is not always straightforward (see Gottuk and Lattimer (2009)).

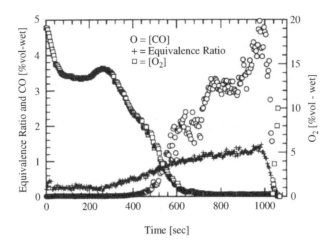

Figure 11.6 Experimental results from a PMMA compartment fire showing the concentrations of CO (○) and O_2 (□) as a function of time. The time variation of the equivalence ratio (+) is also shown (Gottuk *et al.*, 1992b). This may be compared with Figure 9.6 in which the CO yield is shown as a function of the equivalence ratio. Reproduced by permission of the Society of Fire Protection Engineers

compared with HCN measurements in the ISO 19700 tube furnace (ISO, 2007b) there was good agreement when both sets of data were plotted against the equivalence ratio.[19]

These experiments provide an understanding of the processes that take place in the upper layer of a pre-flashover, developing compartment fire. As the fire grows in size, the equivalence ratio (with respect to the upper layer) increases from a very low value after ignition to $\phi > 1.0$ once the flames are large enough to enter the deepening smoke layer and impinge on the ceiling. The oxygen concentration in the layer will be less than 2% (Figure 11.6; Beyler (1984)) so that there will be very little further oxidation within the layer, allowing high levels of smoke and CO to develop. This is the precursor to flashover in a compartment fire, and any smoke flowing from the room of origin into the rest of the building through an open door or other opening will be highly toxic. When a fire is so poorly ventilated that it dies right down to near extinction (the pre-backdraught situation), very high levels of CO would be expected to develop, with [CO]/[CO_2] greater than 0.4.

The equivalence ratio ϕ provides a means of classifying the conditions under which flaming combustion is taking place during a fire and has provided a convenient parameter for correlating yields of individual species in fire gases. Indeed, it has allowed comparisons to be made between analytical results from small-scale laboratory tests and full-scale room fires (e.g., Stec *et al.*, 2009). A simplified method of classifying the burning regime is used in ISO 19706 (ISO, 2004) which is based on the ratio [CO]/[CO_2] (Table 11.1).

As CO and HCN are the dominant asphyxiants, an assessment of toxic potential can be carried out using chemical analysis of the fire gases in standard tests rather than having to rely on animal experiments. Chemical analysis can also be used to examine the potential yields of irritants that may be present, such as HCl and acrolein.

---

[19] The equivalence ratio in full-scale fires (e.g., ISO room) may be determined by using a device to analyse the exhaust gases known as the 'phi meter', developed by Babrauskas *et al.* (1994).

11.2 Smoke Movement

The substantial majority of fire fatalities can be attributed to the inhalation of smoke and toxic gases (National Fire Protection Association, 2008; Department for Communities and Local Government, 2009). While many of these individuals may have been overcome by combustion products while asleep or otherwise incapacitated in the room of origin, a large number are found at points remote from the fire, perhaps attempting to effect their escape. Clearly, any life safety system designed into a building must incorporate means of protecting the occupants from smoke as they make their way to safety. For this to be achieved, the behaviour and movement of smoke within a building must be understood. In the following discussion, 'smoke' is as defined in NFPA 92B (see page 441), i.e., a suspension of particles in air which is also contaminated by gaseous combustion products, some of which are toxic. As the smoke flows away from the seat of the fire entraining more air, the concentration of these products (particles and vapours) falls proportionately, as does the temperature.

During the early stages of a fire in a single compartment, a smoke layer will form below the ceiling and descend at a rate determined by the rate of burning (rate of heat release) and (initially) by the height of the ceiling. As the smoke layer descends, the mass flow into the layer will decrease as the height of the free plume decreases. This logic was applied in Section 4.4.4 in the example relating to smoke extraction, and formed the basis of the computer programme developed by Cooper and Stroup (1985) to calculate the 'available safe egress time' from a single compartment (see also Mowrer (2008)). This is the simplest of the 'zone models' (see Section 4.4.5). Since the 1980s, a plethora of this type of model has appeared (Friedman, 1992; Walton *et al.*, 2008), some dealing with multiple interconnecting spaces in which the behaviour of smoke may be determined by factors that are unconnected with the fire itself. These are forces which are responsible for the movement of smoke in a building, and will be discussed before a brief review of smoke control systems is presented.

11.2.1 Forces Responsible for Smoke Movement

Smoke, and indeed any fluid, will move under the influence of forces which will be manifest as pressure gradients within the bulk of the fluid. For the movement of smoke within a building, such forces are created by:

(a) buoyancy created directly by the fire;
(b) buoyancy arising from differences between internal and external ambient temperature;
(c) effects of external wind and air movement; and
(d) the air handling system within the building.

Each of these will be considered in turn as all must be taken into account if the direction and rate of movement of the smoke are to be predicted.

(a) *Pressure generated directly by the fire.* Burning in a compartment generates high temperatures which produce the buoyancy forces responsible for hot fire gases being expelled through the upper portion of any ventilation opening (Figures 10.4 and 10.7) or through any other suitable leakage paths. If the temperature inside the fire compartment is known, then from Section 10.2 (Equations (10.2a) and (10.2b)), the value of Δp at any height (y) above the neutral plane can be calculated as ($P_1 - P_2$). Thus, if the fire temperature is 850°C (1123 K) and the ambient (external) temperature is 20°C (293 K) then, taking densities from Table 11.6, the pressure difference $y = 1.5$ m above the neutral plane will be c. 13 N/m^2. This will be proportionately greater for taller compartments, and is the source of the positive pressure that will force smoke, and ultimately flames, through unstopped service penetrations at high level (Section 10.7).

These calculations are in general agreement with measurements made by Fang (1980) in a study carried out to ascertain the magnitude of fire pressures that would have to be overcome to protect escape routes from smoke by pressurization (Section 11.3.3).

(b) *Pressure differences due to natural buoyancy forces.* In addition to the buoyancy forces generated by the fire itself, the stack effect must be taken into account in tall buildings. The concept of buoyancy was introduced and developed in Section 4.3.1. As long as the smoke is at a higher temperature than the surrounding air, it will rise, the buoyancy force per unit volume being given by the product $g(\rho_0 - \rho)$, where ρ_0 is the density of the ambient air, ρ is the density of the smoke (or smoke-laden air) and g is the gravitational acceleration constant. The fire provides the energy necessary to drive these buoyant flows which will dominate movement in the vicinity of the fire (Section 4.3). However, as one moves away from the locus of the fire, temperatures fall and the associated buoyancy force will diminish to a point where other forces can begin to dominate behaviour.

In tall buildings containing vertical spaces (stairwells, elevator shafts, etc.), if the internal and external temperatures are different, buoyancy-induced pressure differences will arise: this is the origin of the stack effect. If the temperature inside the building is uniform and greater than the external (ambient) temperature, then there will be a natural tendency for air to be drawn in at the lowest levels and expelled at the highest (see Figure 11.7(a)) (this should be compared with the arguments presented with Figure 10.4). Consider first a vertical column of air contained within a shaft of height H which is open only at the bottom (Figure 11.8) and take the internal and external temperatures to be T_i and T_0, respectively. If the external pressure at ground level is p_0 then the pressures at height H inside and outside the building will be:

$$p_i(H) = p_0 - \rho_i g H \qquad (11.13)$$

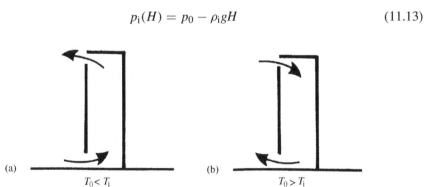

(a)   (b)

$T_0 < T_i$ $T_0 > T_i$

Figure 11.7 The stack effect in tall buildings: (a) external temperature (T_0) < internal temperature (T_i); (b) $T_0 > T_i$, showing accompanying flows (the 'reverse stack effect')

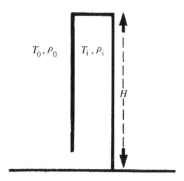

Figure 11.8 The origin of the stack effect (no flow)

and

$$p_0(H) = p_0 - \rho_0 g H \tag{11.14}$$

where ρ_i and ρ_0 are the densities of air at T_i and T_0, respectively. Thus the pressure difference between the inside and outside at the top of the building will be:

$$\Delta p = (\rho_0 - \rho_i)gH \tag{11.15}$$

i.e., the pressure will be greater within the building than outside it.

If the shaft has openings at the top and the bottom then, provided that $T_i > T_0$, there will be a net upward flow of air and a neutral plane will establish itself where $p_0 = p_i$. Carrying out the same analysis as before, then at any height h *above* the neutral plane, the pressure difference between the inside and outside will be:

$$\Delta p = (\rho_0 - \rho_i)gh \tag{11.16}$$

In a real building, of course, the openings will be distributed as a large number of small leakage areas around windows and doors, even if these are closed. Above the neutral plane, air (smoke) will tend to flow outwards from the shaft while below there will be a net inward flow, creating an upward movement within the shaft (Figure 11.7(a)). However, if the external temperature is greater than that inside the building, as would be the case in air-conditioned buildings in hot climates, then the situation will be reversed (Figure 11.7(b)). Air will tend to flow towards the bottom of such shafts, and should fire occur, the initial movement of smoke when the fire is small could be contrary to expectations.[20]

The pressure difference that will exist between inside and outside at height H in Figure 11.8 can be calculated by assuming ideal gas behaviour, i.e., the relationship $pV = nRT$ is obeyed: if the molecular weight of 'smoke' approximates to that of air (0.0289 kg/mol), then from Equation (11.15) (see Equation (1.10)):

$$\Delta p = 3.46 \times 10^3 \left(\frac{1}{T_0} - \frac{1}{T_i} \right) H \tag{11.17}$$

---

[20] This statement ignores the fact that there may be air movement associated with the air conditioning system.

For a 30 m high building for which $T_0 = 273$ K and $T_i = 293$ K, $\Delta p = 26$ N/m^2. However, the distribution of leakage paths to the outside will determine where the neutral plane lies. This is required for calculations relating to smoke movement, and can be calculated if the sizes of the openings are known. If h_1 and h_2 are the distances from the neutral plane to the lower and upper openings, respectively, Figure 11.7(a):

$$\frac{h_1}{h_2} = \frac{A_2^2}{A_1^2} \cdot \frac{T_1}{T_0} \tag{11.18}$$

where A_1 and A_2 are the cross-sectional areas of the lower and upper openings, respectively. This expression was originally derived for leakage via the top and bottom of the door to a fire compartment. Note that the greater the value of A_2, the higher the neutral plane will be above the lower opening.

The significance of the stack effect is that it can move relatively cool smoke around a high-rise building very efficiently, moving it rapidly to some areas, while protecting others (e.g., Figure 11.9). A fire in the lower part of a high-rise building can cause very rapid smoke logging of the upper levels of the building. There are many cases in which this has happened (Butcher and Parnell, 1979), one of the most tragic examples being the fire in the 26-storey MGM Grand Hotel, Las Vegas, in November 1980 in which 85 lives were lost. The fire was confined to the casino on the ground floor, but smoke spread to the upper floors via seismic joint shafts, elevator hoistways and the HVAC system. Of the 85 fatalities, 68 died on the upper floors from 'smoke and carbon monoxide inhalation' (Best and Demers, 1982). The stack effect had an equivalent effect during the fire in the

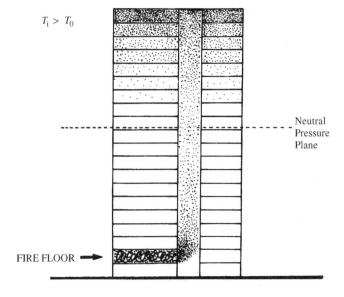

Figure 11.9 The effect of the stack effect on the movement of smoke in a high-rise building ($T_i > T_0$). Below the neutral plane there is a tendency for air to be drawn into the central core, while above it, there is a net outward flow (cf. Figure 11.7(a))

Garley Building, Kowloon in November 1996, but in this case it carried the fire to the upper levels. The set of four lifts was being refurbished and two of the liftshafts were completely open from the ground floor[21] to the 15th floor. The fire started on the second floor lift lobby and spread up both liftshafts. There was very severe damage to the lowest four floors (ground to 3rd level) and to the top three floors, but the 5th to 11th levels suffered only some smoke damage. The stack effect prevented the fire spreading out of the liftshafts below the neutral plane which lay between the 11th and 12th floors (Drysdale and Hamilton, 1999). There were 40 fatalities during the fire, 39 of which were on the top three floors.

(c) *Pressure differences generated by wind.* Natural wind can generate pressure distributions over the envelope of a building which are capable of influencing smoke movement within. The external pressure distribution depends on numerous factors including the speed and direction of the wind and the height and geometry of the building. The magnitude of the effects may be sufficient to override the other forces (natural and artificial) which influence smoke movement. Generally speaking, wind blowing against a building will produce higher pressures on the windward side and will tend to create air movement within the building towards the leeward side where the pressure is lower. The magnitude of the pressure difference is proportional to the square of the wind velocity (Benjamin *et al.*, 1977). The pressure at the surface of a building is given by:

$$p_{\mathrm{w}} = \frac{1}{2} C_{\mathrm{w}} \rho_0 u^2 \, \mathrm{N/m^2} \qquad (11.19)$$

where C_{w} is a dimensionless pressure coefficient (ranging from $+0.8$ to -0.8, for windward and leeward walls respectively), ρ_0 is the outside air density (kg/m^3) and u is the wind velocity (m/s). High windspeeds can generate pressure differentials which are sufficient to overcome other forces, both natural and those imposed artificially to control smoke movement (Section 11.3) (Klote, 2008).

The pressure distribution over the surface of a building is influenced strongly by the juxtaposition of the neighbouring buildings and by the geometry of the building itself. A common situation is a single-storey structure, such as a shopping centre, associated with a multi-storey building, e.g., an office block (Figure 11.10). The wind patterns around this particular geometry of building can be extremely complex and the pressure distribution over the surface of the roof of the shopping centre will vary considerably with changes in wind speed and direction (Marchant, 1984). Thus, while it is possible to protect a single-storey shopping centre from smoke logging during a fire by relying entirely on natural ventilation through smoke vents in the roof, the location and types of vent must be selected on the basis of the pressure distributions that are to be expected on the mall roof. If there is likely to be an area of relatively high pressure in the vicinity of the smoke vent under certain wind conditions, then natural venting will not be a reliable method of removing smoke from the shopping centre (see Section 11.2.3). However, it should be noted that for most other situations, the wind induces low pressures on roofs which will enhance the effectiveness of vents.

---

[21] The floor numbering is that used in the UK and Hong Kong. The stack effect would have resisted fire spreading onto the third floor from the liftshaft, but combustible panelling near the lift opening was ignited directly by thermal radiation.

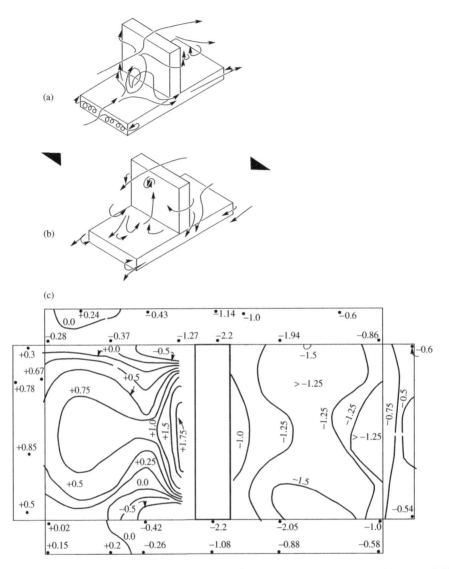

Figure 11.10 (a) and (b) show the airflow patterns around a single-storey shopping centre/office tower complex for one wind direction. (c) The wind-induced pressure patterns over the surface of the building for the airflow shown in (a) and (b). Overpressures in N/m^2. Marchant (1984), by permission

(d) *Pressure differences caused by air handling systems.* Many modern buildings contain air handling systems for the purpose of heating, ventilation and air-conditioning (HVAC). While the fans are idle, the ductwork can act as a network of channels through which smoke will move under the influence of the forces discussed above, including particularly the stack effect in multi-storey buildings. This will promote spread throughout the building, an effect which can be even greater if the system is running when fire breaks out. This situation can be avoided by automatic shutdown, activated by smoke detectors, should fire occur anywhere in that part of the building served by the HVAC system. Alternatively, on a different level of sophistication, the HVAC system can be designed to 'handle' the smoke, removing it from the building, while protecting other spaces by remote operation of isolation valves. This requires the facility for the system to be able to reverse the flow of air within the ductwork and would demand expert supervision and management. This type of dual-purpose system has been installed in some large department stores and other similar spaces. The role of HVAC systems in smoke movement is discussed briefly by Benjamin et al. (1977) and Klote (2008).

11.2.2 Rate of Smoke Production in Fires

As will be discussed in Section 11.3, there are various ways in which smoke can be 'managed' in large buildings to allow the occupants to escape, and the firefighters to find and extinguish the fire quickly and safely. The smoke may be extracted at a rate sufficient to keep the bottom of the smoke layer above head height (Sections 11.3.1 and 11.3.2), but to be able to design the smoke extraction system, it is necessary to be able to calculate the volume of smoke that has to be handled. In this context, the optical density of the smoke – while it may be useful to know – is not required for the calculations.

During the early stages of a compartment fire, when burning is localized, the combustion products will be progressively diluted as they rise vertically in the buoyant plume until deflected by the ceiling (Section 4.3.3). The hot smoke will then flow horizontally as a ceiling jet until it finds a discontinuity and can continue its upward movement, or as is more likely, until it encounters a vertical barrier, such as a wall, which will prevent further travel and cause the smoke layer to back up and deepen, confined by the ceiling and the compartment walls (cf. Section 9.2.1). The rate at which this layer thickens will depend partly on the rate of burning, but predominantly on the amount of air that is entrained into the fire plume before it enters the layer.

In Section 4.4.4, it was shown how Equation (4.49) could be used to calculate the mass flowrate of 'smoke' at any height above an axisymmetric fire whose rate of heat release is $\dot{Q}_c$ (kW):

$$\dot{M} = 0.071 \dot{Q}_c^{1/3} z^{5/3} [1 + 0.026 \dot{Q}_c^{2/3} z^{-5/3}] \tag{11.20}$$

where z is the height (in m) above the virtual source. If z is the height of the smoke layer, then this equation gives the rate of addition of smoke to the layer. If it is required to extract smoke at roof level to maintain the level of the layer at z m above the fire, then Equation (11.20) gives the 'rate of production of smoke' which must be balanced by the rate of extraction (see Figure 11.11). As discussed earlier, it is necessary to specify an appropriate 'design fire' which is based on the nature and distribution of combustible items present in the space. The temperature of the layer can be calculated from the equation:

$$\Delta T = \frac{\dot{Q}_c}{\dot{M} C_p} \tag{11.21}$$

where c_p is the heat capacity of air (c. $1.0 \, \text{kJ/kg·K}$) and ΔT is the temperature excess (i.e., above ambient). The latter information is required to specify the working temperature of the extract fan.

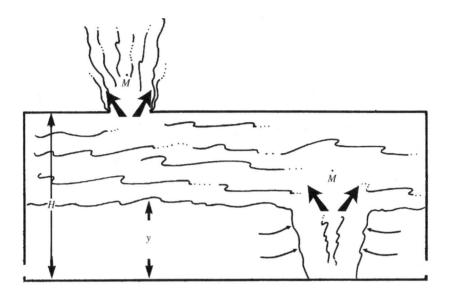

Figure 11.11 Venting of the hot gases from a fire. When the mass flowrate ($\dot{M}$) through the vent equals the mass flowrate from the fire plume into the smoke layer, the height of the smoke layer will stabilize at y

Equation (11.20) applies to the axisymmetric plume (see Section 4.3.2.4), which is the worst case scenario regarding the quantity of smoke that has to be removed by a smoke extraction system for a given design fire. As less air is entrained into 'attached plumes' (see Figure 4.25(b) and 4.25(c)), Milke (2008b) argues that the concept of the 'imaginary fire source' (Hasemi and Tokunaga, 1984) can be used to calculate the rate of smoke production ($\dot{M}$) from a fire (rate of heat release $\dot{Q}_c$) against a wall (see Figure 4.26). Because air is entrained through only half of the perimeter, the rate of smoke production is calculated as one-half of the rate of smoke production from a fire twice the size ($2\dot{Q}_c$). On this basis, the mass flowrate from a fire $\dot{Q}_c$ against a wall will be only 65% of the flowrate from an axisymmetric fire of the same heat release rate. However, a different scenario has to be considered when designing a smoke management system for open multi-level shopping centres and large buildings containing atria. A fire in an adjoining shop or office which communicates directly to a shopping centre or atrium produces the so-called 'spill plume' (or 'balcony spill plume'), releasing fire products along the width of the opening, similar to a line fire, but with additional complicating features. Air will be entrained at the opening, in the 'turning region', and thereafter the amount of entrainment will depend on whether the plume attaches itself to the wall above the opening or is projected clear of the wall (cf. Figure 10.33), as would be the case if the plume was spilling out from under a balcony. Entrainment into the spill plume is a problem that has still to be fully resolved (see Section 4.4.4) (see, for example, Thomas *et al*. (1998), Harrison and Spearpoint (2006), Kumar *et al*. (2010)).

The concentration of smoke in the layer is not required for smoke venting calculations, but it can be estimated if the rate of burning and the 'smoke yield' are known. To take the example from Section 4.4.4, if it is assumed that the principal fuel is polyurethane foam, then taking $\Delta H_c = 25$ kJ/g (an estimate) and $\dot{Q}_c = 6.4 \times 10^3$ kW, the rate of burning becomes $\dot{Q}_c/\Delta H_c = 256$ g/s. With $D_0 = $ c. 1 ob·m^3/g from Table 11.6, the rate of production of smoke is equivalent to 256 ob·m^3/s. It was shown in Section 4.4.4 that the rate of addition of smoke to the layer was equivalent to 131 m^3/s, indicating that the optical density per metre in the layer is $256/131 = 1.95$ ob, corresponding to a visibility of c. 5 m (Figure 11.4).

The earliest work on the removal of fire gases from buildings was to aid the design of 'fire vents' that would be capable of removing hot combustion gases with the purpose of preventing flashover in single-storey industrial buildings, storage facilities, etc. Thomas *et al.* (1963) developed the following expression which has been widely used in the UK for venting calculations, in which it was assumed that the total amount of air entering the rising plume was proportional to its 'surface area':

$$\dot{M} = 0.096 P_f \rho_0 y^{3/2} (g T_0 / T_f)^{1/2} \tag{11.22}$$

where P_f is the perimeter of the fire (m), y is the distance between the floor and the underside of the smoke layer below the ceiling (i.e., the height of clear air, in metres) (Figure 11.11), ρ_0 is the density of ambient air (kg/m^3), and T_0 and T_f are the temperatures of the ambient air and the fire plume, respectively (K). Taking $\rho_0 = 1.22$ kg/m^3, $T_0 = 290$ K and $T_f = 1100$ K, this becomes

$$\dot{M} = 0.188 P_f y^{3/2} \text{ kg/s} \tag{11.23}$$

Equation (11.23) is valid only for fires in which the ratio of flame height to fire diameter is of the order of one or less, and in principle applies only up to the flame tip, i.e., it can only be applied when the flames reach into the smoke layer. Nevertheless, it is claimed to give adequate predictions above the flame tip and the expression has been used in estimating the rate of smoke extraction that would be required to avoid smoke logging of large public areas with relatively low ceilings, such as single-storey shopping centres. As before, the objective would be to prevent the smoke layer descending below a critical height (say $y = 2$ m) (Figure 11.11). From Equation (11.23), the rate of extraction would have to be:

$$\dot{M} = 0.188 P_f \cdot (2)^{3/2} = 0.53 P_f \text{ kg/s} \tag{11.24}$$

or

$$V_s = \frac{0.53 P_f}{\rho_s} \tag{11.25}$$

where V_s is the volumetric flowrate and ρ_s is the density of the smoke at the point of extraction (the smoke vent). As density varies inversely with temperature, substantial volumes may have to be extracted. (Table 11.7 gives the density of air at temperatures up to 1100 K: this is a good approximation to 'smoke'.)

Clearly, a number of parameters need to be known to enable V_s to be calculated (P_f, y and T_s, the temperature of the smoke at the vent). The volumetric flow through the vent (which must not be less than V_s) will depend on the vent area, T_s, and the buoyancy head existing under the steady state conditions assumed for design purposes. This will be a function of T_s and the depth of the layer of hot smoke (($H - y$) in Figure 11.11).

Table 11.7 Density of air as a function of temperature

Temperature (K)	Density (kg/m^3)
280	1.26
290	1.22
300	1.18
500	0.70
700	0.50
1100	0.32

Thomas, Hinkley and co-workers (1963, 1964) (Butcher and Parnell, 1979) carried out a detailed analysis of the problem using scaled experimental models (Section 4.4.5) and developed a series of equations which may be used to calculate the required vent areas. These have been presented in the form of nomograms for use by the designer (Thomas *et al.*, 1963; Butcher and Parnell, 1979). It is required that there are sufficient air inlets near floor level to replenish the hot gases as they are vented.

Dividing Equation (11.22) by ρ_s, the density of the smoke layer, to give the volumetric rate of production (as in Equation (11.25)) and expressing $V_s = A(H - y)$, where A is the floor area of the enclosure, the resulting equation may be integrated to give:

$$t = 20.8 \frac{A}{P_f} \frac{T_0}{T_s} \left(\frac{T_f}{g T_0} \right)^{1/2} \left(\frac{1}{y^{1/2}} - \frac{1}{H^{1/2}} \right) \tag{11.26}$$

which Butcher and Parnell (1979) quote as

$$t = \frac{20A}{P_f g^{1/2}} \left(\frac{1}{y^{1/2}} - \frac{1}{H^{1/2}} \right) \tag{11.27}$$

assuming $T_s \approx 300°C$. This gives the time it would take for the smoke layer to descend to a height y m above the ground, assuming a fire of perimeter P_f burning in an enclosure of floor area A and height H. Clearly, the vents must open within this period of time if the smoke layer is to be arrested at a height of y metres. Figure 11.12 shows the variation of y with t for a fire with a 6 m perimeter in a 5 m high enclosure of area 100 m^2. This could be taken to represent a small hospital ward with a bed fully alight (ignoring the initial growth period).

If, by means of suitable venting, the smoke layer descends and is maintained at a height y above the floor, then the optical density within the layer will be given approximately by:

$$D = \frac{10^3 D_0 \cdot \dot{m} \cdot \rho_s}{0.188 P_f \cdot y^{3/2}} \tag{11.28}$$

if $\dot{m}$ is in kg/s and D_0 is an effective standard optical density relating to the materials and the conditions of burning. Apart from the effect of density (ρ_s), any effect of temperature on the obscuration potential of the smoke is tacitly ignored. This equation could be used to estimate the opacity of the smoke reaching a point remote from a fire, provided that the extent of further entrainment and dilution could be quantified.

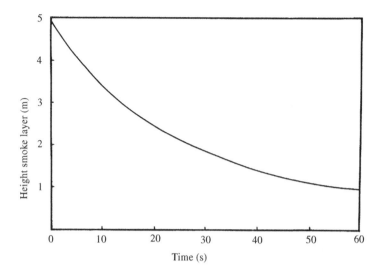

Figure 11.12 Decrease in the height of the lower boundary of the smoke layer with time for a fire of perimeter 6 m in a compartment 5 m high and of 100 m² floor area. It is assumed that the fire is burning steadily from $t = 0$. (Calculated from Equation (11.27), from Hinkley (1971))

Regarding the prediction of smoke movement in complex spaces, several modelling techniques are now available. The most successful are based on computational fluid dynamics (e.g., Fire Dynamics Simulator (FDS)), a technique which is now commonly used for this type of work. However, as was pointed out in Section 4.4.5, there is still a need to verify that the predictions are correct. More experimental data are required before any given model – zone or field – can be used outside the relatively narrow range of conditions for which it was developed. Nevertheless, this is the way forward, provided that the problem of reliable verification has been resolved and the tools are in qualified hands.

11.3 Smoke Control Systems

If it is considered necessary to prevent lethal concentrations of smoke accumulating in certain areas of a building to protect its occupants, there are two basic approaches that may be adopted at the design stage: either the smoke must be contained, or it must be extracted. Smoke containment may be achieved by the use of physical barriers such as walls, windows and doors, as well as smoke curtains or other purpose-designed smoke barriers. Service penetrations and other vertical shafts between compartments must be smoke stopped. Containment may also be achieved by the use of pressurization to provide adequate pressure to resist the flow of smoke, or by opposed flow which overcomes the flow of smoke by creating a flow of air in the opposite direction. Both of these principles are used in the design of pressurized stairs (Section 11.3.3), and the latter is a technique that is used to control smoke movement in tunnels (Jagger and Grant, 2005).

Smoke extraction is also referred to as smoke dilution, smoke clearance or smoke removal. The extraction system may be specified in terms of the number of air changes per hour, sufficient perhaps to ensure that tenable conditions in an area will be regained within a reasonable period. Alternatively, extracting smoke may be designed to retain the smoke above head height as discussed above, or to reduce the concentration of lethal products on escape routes to acceptable levels. In this instance, the extraction rate would be based on a specified design fire as previously discussed, although in reality this would be expected to vary with time. Smoke might be removed by natural venting or by mechanical means, but in either event a sufficient supply of make-up air has to be arranged in order for the extraction system to work as designed.

Some particular issues in smoke control design in buildings are discussed below.

11.3.1 Smoke Control in Large Spaces

Large, undivided single-storey buildings are normally designed around the operations which they are to house – e.g., assembly lines, etc. – where subdivision would hinder efficient operation. Such buildings may be fully sprinklered, so that any fire would be of limited extent, but smoke can still spread throughout the building. On arrival at the fireground, the first action of the fire brigade may be to vent the smoke by creating an opening in the roof, thereby improving visibility above floor level. If automatic vents of appropriate size open either by means of a fusible link or through a control system activated by smoke detectors, smoke logging of the building will not occur and fire brigade access to the seat of the fire will be much easier. In addition, there is no accumulation of hot smoke below the ceiling which could enhance the growth to flashover in the vicinity of the original fire. However, to be set against these points are the following:

(a) If the smoke vents are to operate effectively, there must be adequate ventilation at ground level to permit the hot gases to escape freely. It can be argued that this will enhance the fire growth and allow more rapid spread, which has been taken as an argument to reject fire venting as a general policy.
(b) If combined with sprinklers, early operation of the vents may divert the flow of hot gases away from the sprinkler head(s), thereby preventing their activation.

As with all these matters, there cannot be a definitive answer regarding the value of venting versus sprinklers: each case is different and must be judged on its merits. If there are good reasons for wanting to limit the spread of smoke within a large undivided space, then a combination of automatic vents and smoke 'curtains' to create smoke reservoirs underneath the ceiling might be an appropriate solution (Figure 11.13(a)). The number, size and location of vents necessary for efficient venting can be calculated using the procedures described briefly in the previous section (e.g., Butcher and Parnell, 1979). The theoretical background for natural venting is to be found in the original Fire Research Technical Papers (Thomas *et al.*, 1963; Thomas and Hinkley, 1964). Some of the factors that must be considered are:

 (i) the size of the fire;
 (ii) the height of the building;
 (iii) the type of roof; and
 (iv) the pressure distribution over the roof.

Of these, (i) and (ii) have already been discussed. The type of roof (flat, pitched, north-light, etc.) will determine the need for smoke 'curtains' or screens which will not only

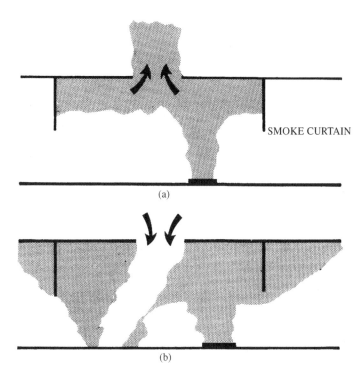

Figure 11.13 The operation of vents in association with smoke curtains located to form a 'reservoir': (a) neutral or negative pressure at roof level; (b) positive pressure at roof level (see Figure 11.10). In practice, it is better to have several smaller vents rather than one large one. (See Figure 11.14 (Butcher and Parnell, 1979))

limit the spread but also allow the smoke to build up a buoyancy head below the ceiling, which will enhance the flow through the vent (Figure 11.13(a)). If there is a positive pressure on the roof, created by the wind, then the effectiveness of the vent will at best be reduced: if this pressure is too great, the vent may operate in reverse (Figure 11.13(b)). Of course, this should not be a problem with mechanical venting.

11.3.2 Smoke Control in Shopping Centres

While the general rule is to keep smoke out of escape routes, there are certain situations in which this is not practicable. The most common example is the modern single-storey shopping centre, which is designed to allow free movement of people within a large enclosed space from which they can enter shops. The centre itself, if suitably designed and its contents controlled, will not present a fire hazard, but should a fire develop in any of the shops, the centre could become smoke-logged very quickly, making rapid egress impossible. There are two methods of controlling this situation (Hinkley, 1971):

(i) vent the smoke directly from the shop to the outside;
(ii) provide smoke 'reservoirs' fitted with automatic ventilators in the ceiling of the shopping centre.

The first of these is desirable if the shop is large, as by the time the smoke enters the mall it will have cooled considerably and will have lost much of the buoyancy necessary to allow natural venting to occur. Smoke reservoirs similar to those shown in Figure 11.13 should be formed within the shop to encourage direct venting (Butcher and Parnell, 1979). However, these must be large enough to prevent smoke entering the mall. Provision of an emergency smoke door or shutter dividing the shop from the centre is not practicable as the escape routes must remain clear. Consequently, it is physically possible to apply this method only to the largest shops. If there are sprinklers installed, their interaction with the smoke layer may cause some cooling and entrainment of smoke in the downward spray pattern, bringing cool smoke to ground level. However, this is not as great a problem as it might seem at first sight. If the smoke layer is greater than 1 m thick and is hot enough to activate the sprinkler, then the smoke will retain sufficient buoyancy to flow through the vents, at least until the sprinklers begin to reduce the intensity of the fire below (Bullen, 1977b). Also – initially at least – some of the smoke entrained in the sprinkler spray will be drawn back into the fire near floor level. The entrainment (and cooling) can be reduced by using a lower water pressure (Bullen, 1977b).

With all but the largest shops, it is inevitable that smoke will enter the centre, which must therefore contain its own smoke control system. If the smoke flows unhindered under the ceiling of the centre, it will cool and begin to descend towards the floor, particularly near the exits where it can mix with inflowing fresh air. Under these conditions, smoke-logging could occur very quickly and vents would not operate efficiently as the layer of smoke would be too thin (Figure 11.14). Consequently, the shopping centre must also be provided with smoke reservoirs which are deep enough (>1 m) to provide the buoyancy head required for natural venting. However, if the centre is associated with a high-rise building – or indeed with any multi-storey structure – then care must be taken that adverse pressure distributions over the roof structure will not negate the venting action (Figure 11.10). If this is a problem then power-operated fans may have to be used to guarantee the venting process.

An important general principle regarding all smoke control problems is that the total volume of smoke should be kept to a minimum. The provision of deep smoke reservoirs

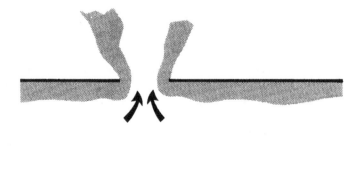

Figure 11.14 Operation of a vent in the absence of a substantial buoyant head (thin smoke layer). The same effect will occur with a deep layer if the area of the vent is too large

helps to achieve this. High temperatures (and buoyancy) are maintained and less 'smoke' has to be 'handled' and removed from the building. The problem is exacerbated in multi-storey atrium-like shopping centres where a fire in a shop at ground level will produce a plume which has a long way to rise before entering the smoke layer under the roof. The extra air entrainment that will occur during the vertical movement will greatly increase the effective smoke volume (Morgan *et al.*, 1976). The magnitude of this effect can be limited by restricting the initial width of the vertical plume, but the volume of the smoke reservoir and the area of the smoke vents have to be considerably greater than for a single-storey shopping centre.

11.3.3 Smoke Control on Protected Escape Routes

By definition, smoke must not enter protected escape routes. Smoke doors between the fire and the protected escape route will help to keep smoke back but this relies on (a) the door being closed at the time of the fire and (b) persons using the door to reach the escape route not keeping the doors open for a prolonged period. However, smoke will migrate with the natural movement of air within the building, and pressure differentials may encourage movement into the escapeways. One way to overcome this is to pressurize the escape route sufficiently so that even under the most unfavourable conditions smoke will not enter as there is a net flow of air from the escape route into the adjacent spaces (Klote, 2008). This technique has received a great deal of attention in the UK, Canada, USA and Australia, and has been adopted in several modern high-rise buildings around the world. The principles are illustrated in Figure 11.15.

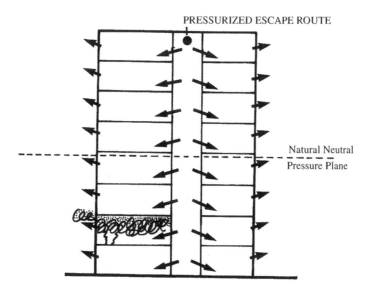

Figure 11.15   Pressurization as a means of keeping escape routes clear of smoke. Compare this with Figure 11.9

Early studies involved measuring the pressure differentials that can exist across doorways, etc., within a building as a result of the stack effect and the wind. It was shown that differentials of $25-50\,\text{N/m}^2$ would be sufficient to override the worst conditions that might arise naturally, although with a very tall building the stack effect might overcome even this. If the leakage of air around all doors, windows, etc., in the escape route is calculated, then the capacity of the fan necessary to maintain the desired overpressure can be found. In the British Standard Code of Practice (Butcher and Parnell, 1979; British Standards Institution, 1998) a pressure differential of $50\,\text{N/m}^2$ is called for under emergency conditions. This may be provided by a continuously operating fan which is integrated into the ventilation system, although normally providing a much lower overpressure. Leakage around doors into the remainder of the building will produce slight overpressures in the corridors leading to the protected escape route. This will help to reduce (but not prevent) the entry of smoke into these spaces.

The pressurization system must be designed to cope with one door being open continuously, providing an airflow through the open door of 0.75 m/s, although lower values are permitted if there is a lobby between the escape route (e.g., stairwell) and the corridor. This technique is described in detail by Butcher and Parnell (1979) and Klote (2008).

The subject of smoke movement and control has been studied extensively in the UK, USA, Canada and Australia. The various codes of practice that have been developed are based largely on the results of fundamental, small-scale studies which have been scaled up using established modelling procedures. Reviews of current practice may be found in the *SFPE Handbook* (Klote, 2008; Milke, 2008b) (USA) and in *CIBSE Guide E* (Chartered Institute of Building Services Engineers, 2003) (UK).

References

Abbasi, T. and Abbasi, S.A. (2007). 'The boiling liquid expanding vapour explosion (BLEVE): Mechanism, consequence, assessment and management'. *Journal of Hazardous Materials*, **141**, 489–519.

Abrams, M.S. (1979). 'Behaviour of inorganic materials in fire', in *Design of Buildings for Fire Safety*, ASTM STP 685 (eds E.E. Smith and T.Z. Harmathy), pp. 14–75. American Society for Testing and Materials, Philadelphia, PA.

Ahonen, A., Kokkala, M. and Weckman, H. (1984). 'Burning characteristics of potential ignition sources of room fires'. *VTT Research Report 285*, Valtion Teknillinen Turkimuskeskus, Espoo, Finland.

Akita, K. (1972). 'Some problems of flame spread along a liquid surface'. *Proceedings of the Combustion Institute*, **14**, 1075–1083.

Akita, K. and Yumoto, Y. (1965). 'Heat transfer in small pools and rates of burning of liquid methanol'. *Proceedings of the Combustion Institute*, **10**, 943–948.

Albini, F. (1983). 'Transport of firebrands by line thermals'. *Combustion Science and Technology*, **32**, 277–288.

Albini, F. (1997). 'An overview of research on wildland fire'. *Fire Safety Science*, **5**, 59–74.

Alderson, S. and Breden, L. (1976). 'Evaluation of the fire performance of carpet underlayments'. National Bureau of Standards, NBSIR 76–1018.

Alpert, R.L. (1972). 'Calculation of response time of ceiling-mounted fire detectors'. *Fire Technology*, **8**, 181–195.

Alpert, R.L. (1975a). 'Pressure modeling of fires controlled by radiation'. *Proceedings of the Combustion Institute*, **15**, 193–202.

Alpert, R.L. (1975b). 'Turbulent ceiling jet induced by large scale fires'. *Combustion Science and Technology*, **11**, 197–213.

Alpert, R.L. (1976). In 'The Third Full-Scale Bedroom Fire Test of the Home Fire Project' (July 30, 1975). Vol. II – Analysis of Test Results (ed. A.T. Modak). *FMRC Technical Report Serial* No. 21011.7, RC–B–48.

Alpert, R.L. (2008). 'Ceiling jet flows', in *SFPE Handbook of Fire Protection Engineering*, 4th edn (eds P.J. Di Nenno *et al.*), pp. 2.21–2.36. National Fire Protection Association, Quincy, MA.

An Introduction to Fire Dynamics, Third Edition. Dougal Drysdale.
© 2011 John Wiley & Sons, Ltd. Published 2011 by John Wiley & Sons, Ltd.

Alpert, R.L. and Ward, E.J. (1984). 'Evaluation of unsprinklered fire hazards'. *Fire Safety Journal*, **7**, 127–143.

Alvares, N.J. (1975). 'Some experiments to delineate the conditions for flashover in enclosure fires'. *International Symposium on Fire Safety of Combustible Materials*, pp. 375–383. University of Edinburgh.

Alvarez, N.J. (1985). Personal communication to E.E. Zukoski (1995).

American Gas Association (1974). LNG Safety Research Programme, Report IS 3-1.

American Society for Testing and Materials (1982). 'Proposed standard method for room fire test of wall and ceiling materials, and assemblies'. *1982 Annual Book of ASTM Standards*, Part 18. ASTM, Philadelphia, PA.

American Society for Testing and Materials (1989). 'Method for gravimetric determination of smoke particulates from combustion of plastic materials'. ASTM D4100–89. (Withdrawn in 1997.)

American Society for Testing and Materials (2002). 'Standard test method for evaluation of flame resistant clothing for protection against flash fire simulations using an instrumented manikin'. ASTM F1930–02.

American Society for Testing and Materials (2004). 'Method of test for density of smoke from the burning or decomposition of plastics'. ANSI/ASTM D2843–04.

American Society for Testing and Materials (2005a). 'Test method for flashpoint and firepoint by the Cleveland Open Cup'. ASTM D92–05.

American Society for Testing and Materials (2005b). 'Standard test method for autoignition temperature of liquid chemicals'. ASTM E659–78

American Society for Testing and Materials (2006). 'Test method for ignition characteristics of finished textile floor covering materials'. ASTM D2859–06.

American Society for Testing and Materials (2007). 'Guide for room fire experiments'. E603–07.

American Society for Testing and Materials (2008a). 'Test methods for flashpoint by the Pensky–Martens Closed Cup Tester'. ASTM D93–08.

American Society for Testing and Materials (2008b). 'Test methods for fire tests of building construction and materials'. ASTM E119–08.

American Society for Testing and Materials (2008c). 'Test method for surface flammability of materials using a radiant heat energy source'. ASTM E162–08.

American Society for Testing and Materials (2008d). 'Test method for heat and visible smoke release rates for materials and products using a thermopile method'. ASTM E906–08.

American Society for Testing and Materials (2008e). 'Test method for determining material ignition and flame spread properties'. ASTM E1321–08.

American Society for Testing and Materials (2009a). 'Test method for surface burning characteristics of building materials'. ASTM E84–09.

American Society for Testing and Materials (2009b). 'Test method for specific optical density of smoke generated by solid materials'. ANSI/ASTM E662–09.

American Society for Testing and Materials (2009c). 'Test method for heat and visible smoke release rates for materials and products using an oxygen consumption calorimeter'. ASTM E1354–09.

American Society for Testing and Materials (2009d). 'Standard test methods for measurement of synthetic polymer material flammability using a fire propagation apparatus (FPA)'. ASTM E2058–09.

American Society for Testing and Materials (2010). 'Standard test method for critical radiant flux of floor-covering systems using a radiant heat energy source'. ASTM E648–10.

Anderson, M.K., Sleight, R.T. and Torero, J.L. (2000). 'Downward smoulder of polyurethane foam: ignition signatures'. *Fire Safety Journal*, **35**, 131–147.

Anderson, R.A., Thompson, I. and Harland, W.A. (1979). 'The importance of cyanide and organic nitriles in fire fatalities'. *Fire and Materials*, **3**, 91–99.

Anderson, R.A., Watson, A.A. and Harland, W.A. (1981). 'Fire deaths in the Glasgow area: I. General considerations and pathology'. *Medicine, Science and the Law*, **21**, 175–183.

Andrews, C.L. (1987). 'Rebuilding a warehouse complex'. *Fire Prevention*, **207**, 28–32.

Andrews, G.E. and Bradley, D. (1972). 'Determination of burning velocities – a critical review'. *Combustion and Flame*, **18**, 133–153.

Anon. (1975). 'Fatal mattress store fire at Chatham Dockyard'. *Fire*, **67**, 388.

Anthony, E.J. and Greaney, D. (1979). 'The safety of hot, self-heating materials'. *Combustion Science and Technology*, **21**, 79–85.

Apte, V.B., Green, A.R. and Kent, J.H. (1991). 'Pool fire plume flow in a large scale wind tunnel'. *Fire Safety Science*, **3**, 425–434.

Ashton, L.A. and Malhotra, H.L. (1960). 'External walls of buildings. I – The protection of openings against spread of fire from storey to storey'. Fire Research Note No. 836.

Atkins, P. and de Paula, J. (2006). *Atkins Physical Chemistry*, 8th edn. Oxford University Press, Oxford.

Atkinson, G.T. and Drysdale, D.D. (1989). 'A note on the measurement of smoke yields'. *Fire Safety Journal*, **15**, 331–335.

Atkinson, G.T., Drysdale, D.D. and Wu, Y. (1995). 'Fire driven flow in an inclined trench'. *Fire Safety Journal*, **25**, 141–158.

Atreya, A. (2008). 'Convection heat transfer', in *SFPE Handbook of Fire Protection Engineering*, 4th edn (eds P.J. Di Nenno *et al.*), pp. 1.44–1.73. National Fire Protection Association, Quincy, MA.

Atreya, A. and Abu-Zaid, M. (1991). 'Effect of environmental variables on piloted ignition'. *Fire Safety Science*, **3**, 177–186.

Atreya, A., Carpentier, C. and Harkleroad, M. (1986). 'Effect of sample orientation on piloted ignition and flame spread'. *Fire Safety Science*, **1**, 97–109.

Babrauskas, V. (1979). 'Full scale burning behaviour of upholstered chairs'. National Bureau of Standards, NBS Technical Note No. 1103.

Babrauskas, V. (1980a). 'Flame lengths under ceilings'. *Fire and Materials*, **4**, 119–126.

Babrauskas, V. (1980b). 'Estimating room flashover potential'. *Fire Technology*, **16**, 94–104.

Babrauskas, V. (1982). 'Will the second item ignite?' *Fire Safety Journal*, **4**, 281–292.

Babrauskas, V. (1983a). 'Upholstered furniture heat release rates: measurements and estimation'. *Journal of Fire Sciences*, **1**, 9–32.

Babrauskas, V. (1983b). 'Estimating large pool fire burning rates'. *Fire Technology*, **19**, 251–261.

Babrauskas, V. (1984a). 'Upholstered furniture room fires – measurements, comparison with furniture calorimeter data and flashover predictions'. *Journal of Fire Sciences*, **2**, 5–19.

Babrauskas, V. (1984b). 'Development of the cone calorimeter – a bench scale heat release rate apparatus based on oxygen consumption'. *Fire and Materials*, **8**, 81–95.

Babrauskas, V. (1984/85). 'Pillow burning rates'. *Fire Safety Journal*, **8**, 199–200.

Babrauskas, V. (1992a). 'From Bunsen burner to heat release rate calorimeter', in *Heat Release in Fires* (eds V. Babrauskas and S.J. Grayson), pp. 7–29. Elsevier Applied Science, Barking.

Babrauskas, V. (1992b). 'Full-scale heat release rate measurements', in *Heat Release in Fires* (eds V. Babrauskas and S.J. Grayson), pp. 93–111. Elsevier Applied Science, Barking.

Babrauskas, V. (1983a). 'Upholstered furniture heat release rates: measurements and estimation'. *Journal of Fire Sciences*, **1**, 9–32.

Babrauskas, V. (2001). 'Ignition of wood: a review of the state of the art'. Interflam'01: Proceedings of the 9th International Fire Conference, Interscience Communications, London, pp. 71–88.

Babrauskas, V. (2003). *Ignition Handbook*, Fire Science Publishers, Washington, DC.

Babrauskas, V. (2008a). 'Heat release rates', in *SFPE Handbook of Fire Protection Engineering*, 4th edn (eds P.J. Di Nenno *et al.*), pp. 3.1–3.59. National Fire Protection Association, Quincy, MA.

Babrauskas, V. (2008b). 'The cone calorimeter', in *SFPE Handbook of Fire Protection Engineering*, 4th edn (eds P.J. Di Nenno *et al.*), pp. 3.90–3.108. National Fire Protection Association, Quincy, MA.

Babrauskas, V., Lawson, J.R., Walton, W.D. and Twilley, W.H. (1982). 'Upholstered furniture heat release rates measured with a furniture calorimeter', NBSIR 82–2604, National Bureau of Standards, Gaithersburg, MD.

Babrauskas, V. and Parker, W. (1987). 'Ignitability measurements with the cone calorimeter'. *Fire and Materials*, **11**, 31–43.

Babrauskas, V., Parker, W.J., Mulholland, G. and Twilley, W.H. (1994). 'The phi meter: a simple, fuel-independent instrument for monitoring combustion equivalence ratio' *Reviews of Scientific instrumentation*, **65**, 2367–2375.

Babrauskas, V. and Peacock, R. (1992). 'Heat release rate: the single most important variable in fire hazard'. *Fire Safety Journal*, **18**, 255–272.

Babrauskas, V., Peacock, R.D. and Reneke, P.A. (2003). 'Defining flashover for fire hazard calculations: Part II'. *Fire Safety Journal*, **38**, 613–622.

Babrauskas, V. and Walton, W.D. (1986). 'A simplified characterisation for upholstered furniture heat release rates'. *Fire Safety Journal*, **11**, 181–192.

Babrauskas, V. and Williamson, R.B. (1978). 'Post flashover compartment fires: basis of a theoretical model'. *Fire and Materials*, **2**, 39–53.

Babrauskas, V. and Williamson, R.B. (1979). 'Post flashover compartment fires: application of a theoretical model'. *Fire and Materials*, **3**, 1–7.

Babrauskas, V. and Williamson, R.B. (1980a). 'The historical basis of fire resistance testing – Part I'. *Fire Technology*, **14**, 184–194.

Babrauskas, V. and Williamson, R.B. (1980b). 'The historical basis of fire resistance testing – Part II'. *Fire Technology*, **14**, 304–316.

Back, E.L. (1981/82). 'Auto-ignition in hygroscopic organic materials'. *Fire Safety Journal*, **4**, 185–196.

Back, G., Beyler, C.L., Di Nenno, P. and Tatum, P. (1994). 'Wall incident heat flux distributions resulting from an adjacent fire'. *Fire Safety Science*, **4**, 241–252.

Bajpai, A.C., Calus, I.M. and Fairley, J.A. (1990). *Mathematics for Engineers and Scientists*, Volume 1. John Wiley & Sons, Ltd, Chichester.

Baker, R.R. (1977). 'Combustion and thermal decomposition regions inside a burning cigarette'. *Combustion and Flame*, **30**, 21–32.

Bamford, C.H., Crank, J. and Malan, D.H. (1946). 'The burning of wood'. *Proceedings of the Cambridge Philosophical Society*, **42**, 166–182.

Bankston, C.P., Cassanova, R.A., Powell, E.A. and Zinn, B.T. (1978). 'Review of smoke particulate properties data for burning natural and synthetic materials'. National Bureau of Standards, NBS-GCR-78–147.

Bar-Ilan, A., Putzeys, O.M., Rein, G., Fernandez-Pello, A.C. and Urban, D.L. (2005). *Proceedings of the Combustion Institute*, **30**, 2295–2302.

Baroudi, D. and Kokkala, M. (1992). 'Analysis of upward flame spread'. VTT Publication No. 89, Technical Research Centre of Finland, Espoo.

Bartknecht, W. (1981). *Explosions: Course, Prevention, Protection*. Springer-Verlag, Berlin.

Batchelor, G.K. (1967). *An Introduction to Fluid Dynamics*. Cambridge University Press, Cambridge.

Baulch, D.L. and Drysdale, D.D. (1974). 'An evaluation of the rate data for the reaction $CO + OH \rightarrow CO_2 + H$'. *Combustion and Flame*, **23**, 215–225.

Beard, A.N. (1981). 'A logic tree approach to the Fairfield Home Fire'. *Fire Technology*, **17**, 25–38.

Beard, A.N. (2000). 'On a priori, blind and open comparisons between theory and experiment'. *Fire Safety Journal*, **35**, 63–66.

Beaulieu, P.A. and Dembsey, N.A. (2001). 'Flammability characteristics at applied heat fluxes up to 200 kW/m^2'. *Fire and Materials*, **32**, 61–86.

Beever, P.F. (1990). 'Estimating the response of thermal detectors'. *Journal of Fire Protection Engineering*, **2**, 11–24.

Beever, P.F. (1995). 'Self heating and spontaneous combustion', in *SFPE Handbook of Fire Protection Engineering*, 2nd edn (eds P.J. Di Nenno *et al.*), pp. 2.180–2.189. National Fire Protection Association, Quincy, MA.

Bengtsson, L. and Karlsson, B. (2001), in 3rd International Seminar of Fire and Explosion Hazards (eds D. Bradley *et al.*), University of Central Lancashire, UK, pp. 143–154.

Benjamin, I.A., Fung, F. and Roth, L. (1977). 'Control of smoke movement in buildings: a review'. National Bureau of Standards, NBSIR 77–1209.

Best, R.L. (1978). 'Tragedy in Kentucky'. *Fire Journal*, **72**, 18–35.

Best, R.L. and Demers, D.P. (1982). 'Fire at the MGM Grand'. *Fire Journal*, **76**, 19–37.

Beyler, C.L. (1984a). 'A design method for flaming fire detection'. *Fire Technology*, **20**(4), 5–16.

Beyler, C.L. (1984b). 'Ignition and burning of a layer of incomplete combustion products'. *Combustion Science and Technology*, **39**, 287–303.

Beyler, C.L. (1986a). 'Major species production by diffusion flames in a two-layer compartment fire environment'. *Fire Safety Journal*, **10**, 47–56.

Beyler, C.L. (1986b). 'Fire plumes and ceiling jets'. *Fire Safety Journal*, **11**, 53–75.

Beyler, C.L. (1992). 'A unified model of fire suppression'. *Journal of Fire Protection Engineering*, **4**, 5–15.

Beyler, C.L. (2005). 'A brief history of the prediction of flame extinction based on flame temperature'. *Fire and Materials*, **29**, 425–427.

Beyler, C.L. (2008). 'Fire hazard calculations for large open hydrocarbon fires', in *SFPE Handbook of Fire Protection Engineering*, 4th edn (eds P.J. Di Nenno *et al.*), pp. 3.271–3.319. National Fire Protection Association, Quincy, MA.

Beyler, C.L., Gratkowski, M.T. and Sikorski, J. (2006). 'Radiant smouldering ignition of virgin plywood and plywood subjected to prolonged smouldering', International Symposium on Fire Investigation and Technology, Cincinnati.

Billmeyer, F.W. (1971). *Textbook of Polymer Science*, 2nd edn. John Wiley & Sons, Inc., New York.

Bishop, S.R., Holborn, P.G., Beard, A.N. and Drysdale, D.D. (1993). 'Nonlinear dynamics of flashover in compartment fires'. *Fire Safety Journal*, **21**, 11–45.

Blinov, V.I. and Khudiakov, G.N. (1957). 'The burning of liquid pools'. *Doklady Akademi Nauk SSSR*, **113**, 1094.

Blinov, V.I. and Khudiakov, G.N. (1961). Diffusive Burning of Liquids (English translation by US Army Engineering Research and Development Laboratories, T-1490a–c. ASTIA, AD 296 762)

Block, J.A. (1970). 'A theoretical and experimental study of non-propagating free-burning fires'. PhD Thesis, Harvard University.

Block, J.A. (1971). 'A theoretical and experimental study of non-propagating free-burning fires'. *Proceedings of the Combustion Institute*, **13**, 971–978.

Blomqvist, P. and Lonnermark, A. (2001). 'Characterisation of combustion products in large scale fire tests: comparison of three experimental configurations'. *Fire and Materials*, **25**, 71–81.

Bluhme, D.A. (1987). 'ISO ignitability test and proposed ignition criteria'. *Fire and Materials*, **11**, 195–199.

Board of Trade (1966). 'Carriage of dangerous goods in ships'. Report of the Standing Advisory Committee, Board of Trade, HMSO, London.

Boddington, T., Gray, P. and Harvey, D.I. (1971). 'Thermal theory of spontaneous ignition: criticality in bodies of arbitrary shape'. *Philosophical Transactions of the Royal Society*, **A270**, 467–506.

Bohm, B. (1977). PhD Thesis, Technical University of Denmark.

Bohm, B. (1982). 'Calculated thermal exposure of steel structures in fire test furnaces – a non-ε_{res} approach'. Laboratory of Heating and Air Conditioning, Technical University of Denmark, CIB W14/82/73 (DK).

Boonmee, N. and Quintiere, J.G. (2002). 'Glowing and flaming autoignition of wood.' *Proceedings of the Combustion Institute*, **29**, 289–296.

Botha, J.P. and Spalding, D.B. (1954). 'The laminar flame speed of propane/air mixtures with heat extraction from the flame'. *Proceedings of the Royal Society*, **A225**, 71–96.

Boucher, B. (2008). 'Aviation', in *NFPA Handbook*, Section 21, Chapter 9. National Fire Protection Association.

Bouhafid, A., Vantelon, J.P., Joulain, P. and Fernandez-Pello, A.C. (1988). 'On the structure at the base of a pool fire'. *Proceedings of the Combustion Institute*, **22**, 1291–1298.

Bowen, P.J. and Cameron, L.R.J. (1999). 'Hydrocarbon aerosol explosion hazards: a review'. *Transactions of the Institute of Chemical Engineers*, **77B**, 22–30.

Bowes, P.C. (1971). 'Application of the theory of thermal explosion to the self-heating and ignition of organic materials'. *Fire Research Note No. 867*.

Bowes, P.C. (1974). 'Fires in oil soaked lagging'. Building Research Establishment, *BRE Current Paper CP 35/74*.

Bowes, P.C. (1984). *Self-heating: Evaluating and Controlling the Hazards*. HMSO, London.

Bowes, P.C. and Cameron, A.J. (1971). 'Self-heating and ignition of chemically activated carbon'. *Journal of Applied Chemistry and Biotechnology*, **21**, 244–250.

Bowes, P.C. and Townsend, S.E. (1962). 'Ignition of combustible dusts on hot surfaces'. *British Journal of Applied Physics*, **13**, 105–114.

Bowman, C.T. (1975). 'Non-equilibrium radical concentrations in shock-initiated methane oxidation'. *Proceedings of the Combustion Institute*, **15**, 869–882.

Bradley, D. (1993). 'Is turbulent burning velocity a meaningful parameter?'. *Fizika Goreniya Vzryva*, **29**, 5–7.

Brandrup, J. and Immergut, E.H. (eds) (1975). *Polymer Handbook*, 2nd edn. John Wiley & Sons, Inc., New York.

Brannigan, F.L. (ed.) (1980). *Fire Investigation Handbook*. National Bureau of Standards Handbook No. 134, US Department of Commerce.

Brenden, J.J. (1967). 'Effect of fire retardant and other salts on pyrolysis products of Ponderosa pine at 280°C and 350°C'. *US Forest Service Research Paper FPL 80*. US Department of Agriculture.

Brenton, J.R., Thomas, G.O. and Al-Hassam, T. (1994). 'Small scale studies of water spray dynamics during explosion mitigation tests'. *Proceedings of the Institution of Chemical Engineers Symposium Series, Hazards XII: European Advances in Process Safety Conference No. 134*, I. Chem E., Rugby, pp. 393–403.

Brindley, J., Griffiths, J.F., McIntosh, A.C. and Zhang, J. (1998). 'The role of liquid-fuel vaporisation and oxygen diffusion in lagging fires'. *Proceedings of the Combustion Institute*, **27**, 2775–2782.

British Standards Institution (1959). 'Method of test for flameproof materials'. BS 3119: 1959.

British Standards Institution (1976). 'Method of test for flammability of vertically oriented textile fabrics and fabric assemblies subjected to a small igniting flame'. BS 5438: 1976.

British Standards Institution (1980). 'Method of test for determination of the punking behaviour of phenol-formaldehyde foam'. BS 5946: 1980.

British Standards Institution (1982). 'Flashpoint by the Abel apparatus (Statutory Method: Petroleum (Consolidation) Act 1928)'. BS 2000-33: 1982.

British Standards Institution (1987). 'Fire tests on building materials and structures. Part 20. Method determination of the fire resistance of elements of construction (general principles); Part 21. Method for determination of the fire resistance of loadbearing elements of construction; Part 22. Method for determination of the fire resistance of non-loadbearing elements of construction'. BS 476 Parts 20–22: 1987.

British Standards Institution (1989). 'Fire tests on building materials and structures. Part 6. Method of test for fire propagation for products'. BS 476-6: 1989.

British Standards Institution (1991). 'Fire tests on building materials and structures. Part 12. Method of test for ignitability of products by direct flame impingement'. BS 476-12: 1991.

British Standards Institution (1993). 'Fire tests on building materials and structures. Part 15. Methods of measuring the rate-of-heat release of products'. BS 476-15: 1993.

British Standards Institution (1997). 'Fire tests on building materials and structures. Part 7. Method of test to determine the classification of the surface spread of flame of products'. BS 476-7: 1997.

British Standards Institution (1998). 'Fire precautions in the design, construction and use of buildings. Code of practice for smoke control in protected escape routes using pressurisation'. BS 5588-4: 1998.

British Standards Institution (2001). 'Application of fire safety engineering principles to the design of buildings – Code of Practice'. BS 7974: 2001.

British Standards Institution (2002). 'Eurocode 1: Actions on structures – Part 1-2. General actions – actions on structures exposed to fire'. BS EN 1991-1-2.

British Standards Institution (2003). 'Structural use of steelwork in building. Code of practice for fire resistant design'. 78-1530BS 5950-8: 2003.

British Standards Institution (2004). 'Application of fire safety engineering principles to the design of buildings – Part 6. Human factors: fire safety strategies – occupant evacuation, behaviour and condition'. Published Document PD 7074-6: 2004.

British Standards Institution (2005). 'Electrical apparatus for gas explosive atmospheres'. BS EN 60079-15: 2005.

British Standards Institution (2006). 'Methods of test for assessment of the ignitability of upholstered seating by smouldering and flaming ignition sources'. BS 5852: 2006.

British Standards Institution (2007a). 'Guidelines for assessing the fire threat to people'. BS ISO 19706: 2007.

British Standards Institution (2007b). 'Determination of the resistance to jet fires of passive fire protection materials. Part 1: General requirements'. BS ISO 22899-1: 2007.

British Standards Institution (2010). 'Reaction to fire tests for building products – building products excluding flooring exposed to thermal attack by a single burning item'. BS EN 13823: 2010.

Britton, L.G. (1990). 'Spontaneous insulation fires', Paper 92c, 23rd Annual Symposium on Loss Prevention, AIChE, San Diego, CA, 19–22 August.

Brosmer, M.A. and Tien, C.L. (1987). 'Radiative energy blockage in large pool fires'. *Combustion Science and Technology*, **51**, 21–37.

Browne, F.L. and Brenden, J.J. (1964). 'Heats of combustion of the volatile pyrolysis products of Ponderosa pine'. *US Forest Service Research Paper FPL 19*, US Department of Agriculture.

Bruce, H.D. (1953). 'Experimental dwelling room fires'. Report No. D 1941, Forest Products Laboratory, US Department of Agriculture, Madison, WI.

Bryan, J.L. (1974). *Fire Suppression and Detection Systems*. Glencoe Press, Beverley Hills.

Brzustowski, T.A. and Twardus, E.M. (1982). 'A study of the burning of a slick of crude oil on water'. *Proceedings of the Combustion Institute*, **19**, 847–854.

Buchanan, A.H. (2002). *Structural Design for Fire Safety*. John Wiley & Sons, Ltd, Chichester.

Budavari, S. (ed.) (1996). *The Merck Index: an Encyclopedia of Chemicals, Drugs and Biologicals*, 12th edn. Merck Co. Inc., Whitehouse Station, NJ.

Budnick, E.K. and Klein, D.P. (1978). 'Mobile home fire studies: summary and conclusions'. NBSIR 78–1530.

Building Research Establishment, Garston, Watford (1982). Video: 'An anatomy of a fire'.

Building Research Establishment, Garston, Watford (2008). Video: 'Fire at Valley Parade, Bradford'.

Building Research Establishment, Garston, Watford (1989). Video: 'The front room fire'.

Bullen, M.L. (1977a). 'A combined overall and surface energy balance for fully developed ventilation controlled liquid fuel fires in compartments'. *Fire Research*, **1**, 171–185.

Bullen, M.L. (1977b). 'The effect of a sprinkler on the stability of a smoke layer beneath a ceiling'. *Fire Technology*, **13**, 21–34.

Bullen, M.L. (1978). 'The ventilation required to permit growth of a room fire'. Building Research Establishment, *BRE CP 41/78*.

Bullen, M.L. and Thomas, P.H. (1979). 'Compartment fires with non-cellulosic fuels'. *Proceedings of the Combustion Institute*, **17**, 1139–1148.

Burgess, D.S. and Zabetakis, M.G. (1973). 'Detonation of a flammable cloud following a propane pipeline break'. US Bureau of Mines, RI 7752.

Burgess, D.S., Strasser, A. and Grumer, J. (1961). 'Diffusive burning of liquids in open trays'. *Fire Research Abstracts and Reviews*, **3**, 177–192.

Burgoyne, J.H. and Cohen, L. (1954). 'The effect of drop size on flame propagation in liquid aerosols'. *Proceedings of the Royal Society (London)*, **A225**, 375–392.

Burgoyne, J.H. and Katan, L.L. (1947). 'Fires in open tanks of petroleum products: some fundamental aspects'. *Journal Institute of Petroleum*, **33**, 158–185.

Burgoyne, J.H. and Roberts, A.F. (1968). 'The spread of flame across a liquid surface, Part 2'. *Proceedings of the Royal Society (London)*, **A308**, 55–68.

Burgoyne, J.H. and Williams-Leir, G. (1949). 'Inflammability of liquids'. *Fuel*, **28**, 145–149.

Burgoyne, J.H., Newitt, D.M. and Thomas, A. (1954). 'Explosion characteristics of lubricating oil mist in crank cases'. *The Engineer, London*, **198**, 165.

Burgoyne, J.H., Roberts, A.F. and Alexander, J.L. (1967). 'The significance of open flash points'. *Journal of the Institute of Petroleum*, **53**, 338–341.

Burgoyne, J.F., Roberts, A.F. and Quinton, P.G. (1968). 'The spread of flame across a liquid surface, Part 1'. *Proceedings of the Royal Society (London)*, **A308**, 39–54.

Burke, S.P. and Schumann, T.E.W. (1928). 'Diffusion flames'. *Industrial and Engineering Chemistry*, **20**, 998–1004.

Bustamente Valencia, L., Rogaume, T., Guillaume, E., Rein, G. and Torero, J.L. (2009). 'Analysis of principal gas products during combustion of polyether polyurethane foam at different irradiation levels'. *Fire Safety Journal*, **44**, 933–940.

Butcher, E.G. and Parnell, A.C. (1979). *Smoke Control in Fire Safety Design*. E. & F.N. Spon Ltd., London.

Butcher, E.G., Bedford, G.K. and Fardell, P.J. (1968). 'Further experiments on temperatures reached by steel in building fires'. *Joint Fire Research Organization Symposium* No. 2. HMSO, London, pp. 2–17.

Butcher, E.G., Chitty, T.B. and Ashton, L.A. (1966). 'The temperature attained by steel in building fires'. *Fire Research Station Technical Paper* No. 15. HMSO, London.

Butler, B.W., Bartlette, R.A., Bradshaw, L.S., Cohen, J.D., Andrews, P.L., Putnam, T. and Mangan, R.J. (1998). 'Fire behaviour associated with the 1994 South Canyon Fire on Storm King Mountain, Colorado', *Research Paper RMRS-RP-9*, US Department of Agriculture, Forest Service, Rocky Mountain Research Station, Ogden, UT.

Butler, C.P. (1971). 'Notes on charring rates in wood'. *Fire Research Note No. 896*.

Cadorin, J.-F. and Franssen, J.-M. (2003). 'A tool to design steel elements submitted to compartment fires – OZone v.2. Part 1: pre-and post-flashover compartment fire model'. *Fire Safety Journal*, **38**, 395–427.

Cadorin, J.-F., Pintea, D., Dotreppe, J.-C. and Franssen, J.-M. (2003). 'A tool to design steel elements submitted to compartment fires – OZone v.2. Part 2: Methodology and applications'. *Fire Safety Journal*, **38**, 429–451.

Calcote, H.F. and Keil, D.G. (1990). 'The role of ions in soot formation'. *Pure and Applied Chemistry*, **62**, 815–824.

Carras, J.N. and Young, B.C. (1994). 'Self heating in coal and related materials: models, applications and test methods'. *Progress in Energy and Combustion Science*, **20**, 1–15.

Carslaw, H.S. and Jaeger, J.C. (1959). *Conduction of Heat in Solids*, 2nd edn. Oxford Science Publications, Oxford.

Carvel, R. (2005). 'Fire protection in concrete tunnels', in *The Handbook of Tunnel Fire Safety* (eds A. Beard and R. Carvel). Thomas Telford Ltd., London, pp. 110–126.

Catchpole, W.R., Catchpole, E.A., Butler, B.W., Rothermel, R.C., Morris, G.A. and Latham, D.J. (1998). 'Rate of spread of free-burning fires in woody fuels in a wind tunnel'. *Combustion Science and Technology*, **131**, 1–37.

Cetegen, B.M. and Ahmed, T.A. (1993). 'Experiments on the periodic instability of buoyant plumes and pool fires'. *Combustion and Flame*, **93**, 157–184.

Cetegen, B.M., Zukoski, E.E. and Kubota, T. (1984). 'Entrainment into the near and far field of fire plumes'. *Combustion Science and Technology*, **39**, 305–331.

Chartered Institute of Building Services Engineers (2003). *Fire Engineering: CIBSE Guide E*, 2nd edn. CIBSE, London.

Chase Jr., M.W. (ed.) (1998). 'NIST-JANAF Thermochemical Tables, 4th edn'. Journal of Physical and Chemical Reference Data, Monograph 9

Chatris, J.M., Quintela, J., Folch, J., Planas, E., Arnaldos, J. and Casal, J. (2001). 'Experimental study of burning rate in hydrocarbon pool fires'. *Combustion and Flame*, **126**, 1373–1383.

Chitty, R. (1994). 'A survey of backdraught'. *Fire Research and Development Group Publication* No. 5/94. Home Office, London.

Chitty, R. and Cox, G. (1979). 'A method of measuring combustion intermittency in fires'. *Fire and Materials*, **3**, 238–242.

Chuah, K.H., Kuwana, K. and Saito, K (2009). 'Modelling a fire whirl generated over a 5-cm diameter methanol pool fire'. *Combustion and Flame*, **156**, 1828–1833.

Clancey, V.J. (1974). 'The evaporation and dispersal of flammable liquid spillages'. *5th Symposium on Chemical Process Hazards*, Institution of Chemical Engineers, Rugby, pp. 80–98.

Cleary, T.G. and Quintiere, J.G. (1991). 'A framework for utilizing fire property tests'. *Fire Safety Science*, **3**, 647–656.

Colwell, J.D. and Reza, A. (2005). 'Hot surface ignition of automotive and aviation fluids'. *Fire Technology*, **41**, 105–123.

Conseil Internationale du Bâtiment W14 (1983). 'A conceptual approach towards a probability-based design guide on structural fire safety'. *Fire Safety Journal*, **6**, 24–79.

Conseil Internationale du Bâtiment W14 (1986). 'Design guide: structural fire safety'. *Fire Safety Journal*, **10**, 79–137.

Cook, S.J., Cullis, C.F. and Good, A.J. (1977). 'The measurement of the flammability of mists'. *Combustion and Flame*, **30**, 309–317.

Cooke, R.A. and Ide, R.H. (1985). *Principles of Fire Investigation*. Institution of Fire Engineers, Leicester.

Cooper, L.Y. (1984). 'On the significance of a wall effect in enclosures with growing fires'. *Combustion Science and Technology*, **40**, 19–34.

Cooper, L.Y. and Stroup, D.W. (1985). 'ASET – a computer program for calculating available safe egress time'. *Fire Safety Journal*, **9**, 29–45.

Cordova, J.L. and Fernandez-Pello, A.C. (2000). 'Convective effects on the endothermic gasification and piloted ignition of a radiatively heated combustible solid'. *Combustion Science and Technology*, **156**, 271–289.

Cordova, J.L., Walther, D.C., Torero, J.L. and Fernandez-Pello, A.C. (2001). 'Oxidiser flow effects on the flammability of solid combustibles'. *Combustion Science and Technology*, **164**, 253–278.

Corlett, R.C. (1968). 'Gas fires with pool-like boundary conditions'. *Combustion and Flame*, **12**, 19–32.

Corlett, R.C. (1970). 'Gas fires with pool-like boundary conditions: further results and interpretation'. *Combustion and Flame*, **14**, 351–360.

Corlett, R.C. (1974). 'Velocity distributions in fires', in *Heat Transfer in Fires* (ed. P.L. Blackshear), pp. 239–255. John Wiley & Sons, Inc., New York.

Coward, H.F. and Jones, G.W. (1952). 'Limits of flammability of gases and vapors'. *US Bureau of Mines Bulletin 503*.

Cox, G. (1983). 'A field model of fire and its application to nuclear containment problems'. *Proceedings of the CSNI Specialist Meeting on Interaction of Fire and Explosion with Ventilation Systems in Nuclear Facilities*, Los Alamos Report LA-9911C.

Cox, G. (1995). 'Basic considerations', in *Combustion Fundamentals of Fire* (ed. G. Cox), pp. 1–30. Academic Press, London.

Cox, G. and Chitty, R. (1980). 'A study of the deterministic properties of unbounded fire plumes'. *Combustion and Flame*, **39**, 191–209.

Cox, G. and Chitty, R. (1985). 'Some source-dependent effects of unbounded fires'. *Combustion and Flame*, **60**, 219–232.

Cox, G., Chitty, R. and Kumar, S. (1989). 'Fire modelling and the King's Cross Fire Investigation'. *Fire Safety Journal*, **15**, 103–106.

Cox, G., Kumar, S., Cumber, P., Thomson, V. and Porter, A. (1990). 'Fire simulation in the design evaluation process: an exemplification of the use of a computer field model'. *Interflam'90: Proceedings of the 5th International Fire Conference*, Interscience Communications, London, pp. 55–66.

Crauford, N.L., Liew, S.L. and Moss, J.B. (1985). 'Experimental and numerical simulation of a buoyant fire'. *Combustion and Flame*, **61**, 63–77.

Croce, P.A. (ed.) (1975). 'A study of room fire development: the second full scale bedroom fire test of the Home Fire Project'. Factory Mutual Research Corporation Serial 21011.4, RC 75–T–31.

Croce, P.A. (2003). 'The new FMGlobal Technology Centre'. *Fire Safety Science*, **7**, 105–106.

Croft, W.M. (1980/81). 'Fire involving explosions – a literature review'. *Fire Safety Journal*, **3**, 3–24.

Cullen, The Rt Hon. Lord. (1989). *The Public Inquiry into the Piper Alpha Disaster*. HMSO, London.

Cullis, C.F. and Hirschler, M.M. (1981). *The Combustion of Organic Polymers*. Clarendon Press, Oxford.

Custer, R.L.P., Meacham, B.J. and Schifiliti, R.P. (2008). 'Design of detection systems', in *SFPE Handbook of Fire Protection Engineering*, 4th edn (eds P.J. Di Nenno *et al.*), pp. 4.1–4.44. National Fire Protection Association, Quincy, MA.

Dakka, S.M., Jackson, G.S. and Torero, J.L. (2002). 'Mechanisms controlling the degradation of poly(methylmethacrylate) prior to piloted ignition'. *Proceedings of the Combustion Institute*, **29**, 281–287.

Damkohler, G. (1940). 'Influence of turbulence on the velocity of flames in gas mixtures'. *Z. Electrochem.*, **46**, 601–626.

Deal, S. and Beyler, C.L. (2003). 'Correlating preflashover room fire temperatures'. *Journal of Fire Protection Engineering*, **2**, 33–48.

Deepak, D. and Drysdale, D.D. (1983). 'Flammability of solids: an apparatus to measure the critical mass flux at the firepoint'. *Fire Safety Journal*, **5**, 167–169.

De Haan, J.D. (2007). *Kirk's Fire Investigation*, 6th edn. Brady/Prentice Hall, Englewood Cliffs, NJ.

Delichatsios, M.A. (1984). *Combustion Science and Technology*, **39**, 195–214.

Delichatsios, M.A. and Orloff, L. (1988). 'Effects of turbulence on flame radiation from diffusion flames'. *Proceedings of the Combustion Institute*, **22**, 1271–1279.

Delichatsios, M.A., Panagiotou, Th. and Kiley, F. (1991). 'The use of time to ignition data for characterising the thermal inertia and the minimum (critical) heat flux for ignition or pyrolysis'. *Combustion and Flame*, **84**, 323–332.

Department for Communities and Local Government (2009). *Fire Statistics: United Kingdom 2007*. DCLG, London.

Department of Transport (1986). *Report on the Derailment and Fire that occurred on 20th December 1984 at Summit Tunnel in the London Midland Region of British Railways*. HMSO, London.

de Ris, J.N. (1969). 'Spread of a laminar diffusion flame'. *Proceedings of the Combustion Institute*, **12**, 241–252.

de Ris, J.N. (1973). 'Modeling techniques for prediction of fires'. *Applied Polymer Symposium* No. 22, pp. 185–193. John Wiley & Sons, Inc., New York.

de Ris, J.N. (1979). 'Fire radiation – a review'. *Proceedings of the Combustion Institute*, **17**, 1003–1016.

de Ris, J. and Cheng, X-F. (1994). 'The rôle of smoke point in material flammability testing'. *Fire Safety Science*, **4**, 301–312.

de Ris, J. and Orloff, L. (2005). 'Flame heat transfer between parallel panels'. *Fire Safety Science*, **8**, 999–1010.

di Blasi, C. (2009). 'Combustion and gasification rates of lignocellulosic fuels'. *Progress in Energy and Combustion Science*, **35**, 121–140.

di Blasi, C., Crescitelli, S., Russo, G. and Fernandez-Pello, A.C. (1988). 'Model of the flow assisted spread of flames over a thin charring combustible'. *Proceedings of the Combustion Institute*, **22**, 1205–1212.

Di Nenno, P.J. *et al.* (eds) (2008). *SFPE Handbook of Fire Protection Engineering*, 4th edn. National Fire Protection Association, Quincy, MA.

Di Nenno, P.J. and Taylor, G.M. (2008). 'Halon and halon replacement agents and systems', in *NFPA Handbook*, 20th edn (ed. A.E. Cote), Section 17, Chapter 6, pp. 17.93–17.121. National Fire Protection Association, Quincy, MA.

DIN 18230 (1978). 'Structural fire protection in industrial building construction: Part 1: Required fire resistance period; Part 2: Determination of the burning factor *m*'. Deutsches Institut für Normung, Berlin.

DIN 53436 (1979). 'Erzeugung thermischer Zersetzungsprodukte von Werkstoffen unter Luftzufuhr und ihre toxikologische Prufung. Verfahren zur thermischen Zersetzung'. DIN 53436. Deutsches Institut für Normung, Berlin.

DIN 4102 (1990). 'Fire performance of building materials and components: Part 1 Building materials'. DIN 4102. Deutsches Institut für Normung, Berlin.

Dixon-Lewis, G. (1967). 'Flame structure and flame reaction kinetics. I. Solution of conservation equations and application to rich hydrogen–oxygen flames'. *Proceedings of the Royal Society (London)*, **A289**, 495–513.

Dixon-Lewis, G. and Williams, A. (1967). 'Some observations on the combustion of methane in premixed flames'. *Proceedings of the Combustion Institute*, **11**, 951–958.

Dixon-Lewis, G.L. and Williams, D.J. (1977). 'Oxidation of hydrogen and carbon monoxide', in *Comprehensive Chemical Kinetics*, Volume 17 (eds C.H. Bamford and C.F.H. Tipper). Elsevier, Amsterdam.

Dong, Y., Vagelopoulos, C.M., Spedding, G.R. and Egolfopoulos, F.N. (2002). 'Measurement of laminar flame speeds through digital particle image velocimetry: mixtures of methane and ethane with hydrogen, oxygen, nitrogen and helium'. *Proceedings of the Combustion Institute*, **29**, 1419–1426.

Dorofeev, S.B., Sidorov, V.P., Efimenko, A.A., Kochurko, A.S., Kuznetsov, M.S., Chaivanov, B.B., Matsukov, D.I., Pereverzev, A.K. and Avenyan, V.A (1995). 'Fireballs from deflagration and detonation of heterogeneous fuel-rich clouds'. *Fire Safety Journal*, **25**, 323–336.

Dosanjh, S.S., Pagni, P.J. and Fernandez-Pello, A.C. (1987). 'Forced co-current smouldering combustion'. *Combustion and Flame*, **68**, 131–142.

Drysdale, D.D. (1980). 'Aspects of smouldering combustion'. *Fire Prevention Science and Technology*, No. 23, pp. 18–28. Fire Protection Association, London.

Drysdale, D.D. (1981). Unpublished results.

Drysdale, D.D. (1983). 'Ignition: the material, the source and subsequent fire growth'. Society of Fire Protection Engineers, Technology Report 83–5. (Presented at the 1983 SFPE Seminar on Fire Protection Engineering, Kansas City, MI, May.)

Drysdale, D.D. (1985a). 'Fire behaviour of cellular polymers'. *Cellular Polymers*, **4**, 405–419.

Drysdale, D.D. (1985b). *Introduction to Fire Dynamics*, 1st edn. John Wiley & Sons, Inc., New York.

Drysdale, D.D. (ed.) (1992). 'Special Issue: The King's Cross Underground Fire'. *Fire Safety Journal*, **18**, 1–121.

Drysdale, D.D. (1996). 'The flashover phenomenon'. *Fire Engineers Journal*, **56**, 18–23.

Drysdale, D.D. (2008). 'Ignition of liquids', in *SFPE Handbook of Fire Protection Engineering*, 4th edn (eds P.J. Di Nenno *et al.*), pp. 2.211–2.228. National Fire Protection Association, Quincy, MA.

Drysdale, D.D. and Abdul-Rahim, F.F. (1985). 'Smoke production in fires. I. Small scale experiments', in *Fire Safety: Science and Engineering, ASTM STP 882* (ed. T.Z. Harmathy). American Society for Testing and Materials, Philadelphia, PA, pp. 285–300.

Drysdale, D.D. and Hamilton, S.E. (1999). 'Interpretation of evidence at the fire scene: the importance of fire dynamics'. *Interflam'99: Proceedings of the 8th International Fire Conference*, pp. 233–244. Interscience Communications, London.

Drysdale, D.D. and Kemp, N. (1982). 'Prevention and suppression of flammable and explosive atmospheres', in *Factories: Planning, Design and Modernisation* (ed. J. Drury). Architectural Press, London, pp. 266–286.

Drysdale, D.D. and Macmillan, A.J.R. (1992). 'Flame spread on inclined surfaces'. *Fire Safety Journal*, **18**, 245–254.

Drysdale, D.D. and Thomson, H.E. (1989). 'Flammability of plastics II. Critical mass flux at the firepoint'. *Fire Safety Journal*, **14**, 179–188.

Drysdale, D.D. and Thomson, H.E. (1990). 'Ignition of polyurethane foams: a comparison between modified and unmodified foams', in *Flame Retardants '90* (The British Plastics Federation), pp. 191–205. Elsevier Applied Science, Barking.

D'Souza, M.V. and McGuire, J.H. (1977). 'ASTM E-84 and the flammability of foamed thermosetting plastics'. *Fire Technology*, **13**, 85–94.

Dugger, G.L., Weast, R.C. and Heimel, S. (1955). 'Flame velocity and preflame reaction in heated propane–air mixtures'. *Industrial and Engineering Chemistry*, **47**, 114–116.

Dungan, K.W. (2008). 'Storage and handling of solid fuels'. *NFPA Handbook*, 20th edn. NFPA, Quincy, MA, pp. 7.3–7.13.

Dusinberre, G.M. (1961). *Heat Transfer Calculation by Finite Difference*. International Textbook Company, Scranton.

Dwyer, R.W., Fournier, L.G. *et. al.* (1994). 'The effects of upholstery fabric properties on fabric ignitabilities by smouldering cigarettes'. *Journal of Fire Sciences*, **12**, 268–283.

Eckhoff, R.K. (1997). *Dust Explosions in the Process Industries*, 2nd edn. Butterworth-Heinemann, Oxford.

Edwards, D.K. (1985). *Handbook of Heat Transfer Fundamentals*. McGraw Hill, New York, pp. 14–15.

Egerton, A.C., Gugan, K. and Weinberg, F.J. (1963). 'The mechanism of smouldering in cigarettes'. *Combustion and Flame*, **7**, 63–78.

Elliot, D.A. (1974). 'Fire and steel construction: protection of structural steelwork'. *CONSTRADO Publication 4/74*, Constructional Steel Research and Development Organization, London.

Emmons, H.W. (1965). 'Fundamental problems of the free-burning fire'. *Proceedings of the Combustion Institute*, **10**, 951–964.

Emmons, H.W. (1974). *Scientific American*, **231**, 21–27.

Emmons, H.W. (1984). 'The further history of fire science'. *Combustion Science and Technology*, **40**, 167–174.

Emmons, H.W. and Tanaka, T. (2008). 'Vent flows', in *SFPE Handbook of Fire Protection Engineering*, 4th edn (eds P.J. Di Nenno *et al.*), pp. 2.37–2.53. National Fire Protection Association, Quincy, MA.

Emmons, H.W. and Ying, S.-J. (1966). 'The fire whirl.' *Proceedings of the Combustion Institute*, **11**, 475–488.

Enright, P.A. and Fleischmann, C. (1999). Uncertainty of heat release rate calculation of the ISO 5660-1 cone calorimeter standard test method'. *Fire Technology*, **35**.

European Committee for Standardization (1999). 'Fire Resistance Tests – Part 1. General Requirements'. European Standard EN1363-1: 1999.

Evans, D.D. and Stroup, D.W. (1986). 'Methods to calculate the response time of heat and smoke detectors installed below large unobstructed ceilings'. *Fire Technology*, **22**(1), 54–66.

Fan, W.C., Hua, J.S. and Liao, G.X. (1995). 'Experimental study of the premonitory phenomena of boilover in liquid pool fires supported on water'. *Journal of Loss Prevention in the Process Industries*, **8**, 221–227.

Fang, J.B. (1975a). 'Fire build-up in a room and the role of interior finish materials'. National Bureau of Standards, *NBS Technical Note No. 879*.

Fang, J.B. (1975b). 'Measurement of the behaviour of incidental fires in a compartment'. National Bureau of Standards, NBSIR 75–679.

Fang, J.B. (1980). 'Static pressures produced by room fires', National Bureau of Standards NBSIR 80–1984.

Fay, J.A. (2003). 'Model of spills and fires from LNG and oil tankers'. *Journal of Hazardous Materials*, **96**, 171–188.

Fennell, D. (1988). *Investigation into the King's Cross Underground Fire*. HMSO, London.

Fernandez-Pello, A.C. (1977a). 'Downward flame spread under the influence of externally applied thermal radiation'. *Combustion Science and Technology*, **17**, 1–9.

Fernandez-Pello, A.C. (1977b). 'Upward laminar flame spread under the influence of externally applied thermal radiation'. *Combustion Science and Technology*, **17**, 87–98.

Fernandez-Pello, A.C. (1984). 'Flame spread modelling'. *Combustion Science and Technology*, **39**, 119–134.

Fernandez-Pello, A.C. (1995). 'The solid phase', in *Combustion Fundamentals of Fire* (ed. G. Cox), pp. 31–100. Academic Press, London.

Fernandez-Pello, A.C. and Hirano, T. (1983). 'Controlling mechanisms of flame spread'. *Combustion Science and Technology*, **32**, 1–31.

Fernandez-Pello, A.C. and Santoro, R.J. (1980). 'On the dominant mode of heat transfer in downward flame spread'. *Proceedings of the Combustion Institute*, **17**, 1201–1209.

Fernandez-Pello, A.C., Ray, S.R. and Glassman, I. (1981). 'Flame spread in an opposed forced flow: the effect of ambient oxygen concentration'. *Proceedings of the Combustion Institute*, **18**, 579–589.

Fernandez-Pello, A.C. and Williams, F.A. (1974). 'Laminar flame spread over PMMA surfaces'. *Proceedings of the Combustion Institute*, **15**, 217–231.

Field, P. (1982). *Dust Explosions*. Volume 4 of *Handbook of Powder Technology* (eds J.C. Williams and T. Allen). Elsevier Science Publishers, Amsterdam.

Fine, D.H., Gray, P. and Mackinven, R. (1969). 'Experimental measurements of self-heating in the explosive decomposition of diethylperoxide'. *Proceedings of the Combustion Institute*, **12**, 545–555.

Fishenden, M. and Saunders, O.A (1950). *Introduction to Heat Transfer*. Clarendon Press, Oxford.

Fire Protection Association (1972). *Fire and Related Properties of Industrial Chemicals*. Fire Protection Association, London.

Fleischmann, C.M., Dod, R.L., Brown, N.J., Novakov, T., Mowrer, F.W. and Williamson, R.B. (1990). 'The use of medium scale experiments to determine smoke characteristics', in *Characterisation and Toxicity of Smoke*, ASTM STP 1082 (ed. H.K. Hasegawa), pp. 147–164. American Society for Testing and Materials, Philadelphia, PA.

Fleischmann, C.M., Pagni, P.J. and Williamson, R.B. (1994). 'Quantitative backdraft experiments'. *Fire Safety Science*, **4**, 337–348.

Foley, M. and Drysdale, D.D. (1995). 'Heat transfer from flames between vertical parallel walls'. *Fire Safety Journal*, **24**, 53–73.

Fons, W.L. (1946). 'Analysis of fire spread in light forest fuels'. *Journal of Agricultural Research*, **72**, 93–121.

Foster, C.D. (1998). 'Investigation of gas phase explosions in buildings', in *Forensic Investigation of Explosions* (ed. A. Beveridge). Taylor and Francis, London, pp. 183–229.

Foster, J. (1997). 'The Fire Experimental Unit's backdraught simulator'. *Fire Research News 21*. Department of Communities and Local Government, London.

Frank-Kamenetskii, D.A. (1939). 'Temperature distribution in reaction vessel and stationary theory of thermal explosion'. *Journal of Physical Chemistry (USSR)*, **13**, 738–755.

Franssen, J.-F. and Zahria, R. (2005). 'Design of Steel Structures subjected to Fire: Background and Design Guide to Eurocode 3' (Les Editions de l'Universite de Liege, Belgique).

Friedman, R. (1968). 'A survey of knowledge about idealized fire spread over surfaces'. *Fire Research Abstracts and Reviews*, **10**, 1–8.

Friedman, R. (1971). 'Aerothermodynamics and modeling techniques for prediction of plastic burning rates'. *Journal of Fire and Flammability*, **2**, 240–256.

Friedman, R. (1975). 'Behaviour of fires in compartments'. *International Symposium on Fire Safety of Combustible Materials*, pp. 100–113, Edinburgh University.

Friedman, R. (1977). 'Ignition and burning of solids', in *Fire Standards and Safety. ASTM STP 614* (ed. A.F. Robertson), pp. 91–111. American Society for Testing and Materials, Philadelphia, PA.

Friedman, R. (1989). *Principles of Fire Protection Chemistry*, 2nd edn. National Fire Protection Association, Quincy, MA.

Friedman, R. (1992). 'An international survey of computer models for fire and smoke'. *Journal of Fire Protection Engineering*, **4**, 83–92.

Friedman, R. (2008). 'Chemical equilibrium', in *SFPE Handbook of Fire Protection Engineering*, 4th edn (eds P.J. Di Nenno *et al.*), pp. 1.101–1.111. National Fire Protection Association, Quincy, MA.

Fristrom, R.M. and Westenberg, A.A. (1965). *Flame Structure*. McGraw-Hill, New York.

Fung, F. (1973). 'Evaluation of a pressurised stairwell smoke control system for a twelve storey apartment building'. National Bureau of Standards, NBSIR 73–277.

Gann, R.G., Earl, W.L., Manka, M.J. and Miles, L.B. (1981). 'Mechanism of cellulose smouldering retardance by sulphur'. *Proceedings of the Combustion Institute*, **18**, 571–578.

Garo, J.P., Koseki, H., Vantelon, J.-P. and Fernandez-Pello, A.C. (2007). 'Combustion of liquid fuels floating on water'. *Thermal Science*, **11**, 119–140.

Garo, J.P., Vantelon, J.-P. and Fernandez-Pello, A.C. (1994). 'Boilover burning of oil spilled on water'. *Proceedings of the Combustion Institute*, **25**, 1481–1488.

Garo, J.P., Vantelon, J.-P. and Fernandez-Pello, A.C. (1996). 'Effect of fuel boiling point on the boilover burning of liquid fuels spilled on water'. *Proceedings of the Combustion Institute*, **26**, 1461–1467.

Garo, J.P., Vantelon, J.-P. and Koseki, H. (2006). 'Thin layer boilover: prediction of its onset and intensity'. *Combustion Science and Technology*, **178**, 1217–1235.

Gaydon, A.G. and Wolfhard, H.G. (1979). *Flames: Their Structure, Radiation and Temperature*, 4th edn. Chapman and Hall, London.

Gillie, M., Usmani, A.S. and Rotter, J.M. (2002). 'A structural analysis of the Cardington British Steel corner test'. *Journal of Constructional Steel Research*, **58**, 427–442.

Glassman, I. (1989). 'Soot formation in the combustion process'. *Proceedings of the Combustion Institute*, **22**, 295–311.

Glassman, I. and Dryer, F. (1980/81). 'Flame spreading across liquid fuels'. *Fire Safety Journal*, **3**, 123–138.

Glassman, I. and Hansel, J.G. (1968). 'Some thoughts and experiments on liquid fuel flames spreading, steady burning and ignitability in quiescent atmospheres'. *Fire Research Abstracts and Reviews*, **10**, 217–234.

GHS (2005). 'Globally Harmonised System of Classification and Labelling of Chemicals (GHS)' (First revised edition), UN Publication ST/SG/AC.10/30/Rev.1. United Nations, Geneva (ISBN 978-92-1-139121-3 (Vol. I).

Gojkovich, D. and Karlsson, B. (2001). 'Describing the importance of the mixing process in a backdraught situation using experimental work and CFD simulations', 3rd International Seminar on Fire and Explosion Hazards, University of Central Lancashire, Preston, pp. 167–178.

Gottuk, D.T. amd Lattimer, B.Y. (2008). 'Effect of combustion conditions on species production', in *SFPE Handbook of Fire Protection Engineering*, 4th edn (eds P.J. Di Nenno *et al.*), pp. 2.67–2.95. National Fire Protection Association, Quincy, MA.

Gottuk, D.T., Peatross, M.J., Farley, J.P and Williams, F.W. (1999). 'The development and mitigation of backdraught: a real-scale shipboard study'. *Fire Safety Journal*, **33**, 261–282.

Gottuk, D.T., Roby, R.J. and Beyler, C.L. (1992a). 'A study of carbon monoxide and smoke yields from compartment fires with external burning'. *Proceedings of the Combustion Institute*, **24**, 1729–1735.

Gottuk, D.T., Roby, R.J., Peatross, M.J. and Beyler, C.L. (1992b). 'Carbon monoxide production in compartment fires'. *Journal of Fire Protection Engineering*, **4**, 133–150.

Gottuk, D.T. and White, D.A. (2008). 'Liquid fuel fires', in *SFPE Handbook of Fire Protection Engineering*, 4th edn (eds P.J. Di Nenno *et al.*), pp. 2.337–2.357. National Fire Protection Association, Quincy, MA.

Graham, T.L., Makhviladze, G.M. and Roberts, J.P. (1995). 'On the theory of flashover development', *Fire Safety Journal*, **25**, 229–260.

Grant, G.B., Brenton, J.R. and Drysdale, D.D. (1999). 'Fire suppression by water sprays', *Progress in Energy and Combustion Science*, **26**, 79–130.

Grant, G.B. and Drysdale, D.D. (1995). 'Numerical modelling of early fire spread in warehouse fires'. *Fire Safety Journal*, **24**, 247–278.

Grant, G.B. and Drysdale, D.D. (1997). 'Estimating heat release rates from large scale tunnel fires'. *Fire Safety Science*, **5**, 1213–1224.

Gray, B.F. (2008). 'Spontaneous combustion and self-heating', in *SFPE Handbook of Fire Protection Engineering*, 4th edn (eds P.J. Di Nenno *et al.*), pp. 2.241–2.259. National Fire Protection Association, Quincy, MA.

Gray, B.F. and Halliburton, B. (2000). 'The thermal decomposition of hydrated calcium hypochlorite (UN 2880)'. *Fire Safety Journal*, **35**, 223–239.

Gray, P. and Lee, P.R. (1967). 'Thermal explosion theory', in *Oxidation and Combustion Reviews*, Volume 2 (ed. C.F.H. Tipper), pp. 1–183. Elsevier, Amsterdam.

Gray, W.A. and Muller, R. (1974). *Engineering Calculations in Radiative Heat Transfer*. Pergamon Press, Oxford.

Greenwood, C.T. and Milne, E.A. (1968). *Natural High Polymers*. Oliver and Boyd, Edinburgh.

Griffiths, J.F. (2004). 'Why cool flames are a hot prospect'. *New Scientist*, 5 June 2004.

Griffiths, J.F. (2008). 'Combustion kinetics', in *SFPE Handbook of Fire Protection Engineering*, 4th edn (eds P.J. Di Nenno *et al.*), pp. 1.220–1.230. National Fire Protection Association, Quincy, MA.

Griffiths, J.F. and Barnard, J.A. (1995). *Flame and Combustion*, 3rd edn. Blackie Academic and Professional, London.

Gross, D. (1962). 'Experiments on the burning of cross piles of wood'. *Journal of Research, National Bureau of Standards*, **66C**, 99–105.

Gross, D. (1989). 'Measurement of flame lengths under ceilings'. *Fire Safety Journal*, **15**, 31–44.

Gross, D. and Robertson, A.F. (1965). 'Experimental fires in enclosures'. *Proceedings of the Combustion Institute*, **10**, 931–942.

Gross, D., Loftus, J.J. and Robertson, A.F. (1967). 'Method for measuring smoke from burning materials'. *Symposium on Fire Test Methods – Restraint and Smoke, 1966*, ASTM STP 422 (ed. A.F. Robertson), pp. 166–204. American Society for Testing and Materials, Philadelphia, PA.

Grosshandler, W.L. and Modak, A.T. (1981). 'Radiation from non-homogeneous combustion products'. *Proceedings of the Combustion Institute*, **18**, 601–609.

Grosshandler, W., Bryner, N., Madrzykowski, D. and Kuntz, K. (2005). 'Report on the technical investigation of the Station Nightclub Fire'. NIST NCSTAR2: Volume 1.

Gugan, K. (1976). 'Technical lessons of Flixborough'. *The Chemical Engineer*, May.

Gugan, K. (1979). *Unconfined Vapour Cloud Explosions*. Institution of Chemical Engineers, Rugby.

Hadvig, S. and Paulsen, O.R. (1976). 'One dimensional charring rates in wood'. *Journal of Fire and Flammability*, **7**, 433–449.

Hägglund, B. and Persson, L.E. (1976a). 'An experimental study of the radiation from wood flames'. *FoU-Brand*, **1**, 2–6.

Hägglund, B. and Persson, L.E. (1976b). 'The heat radiation from petroleum fires'. *FOA Report C20126-D6 (A3)*. Forsvarets Forskningsanstalt, Stockholm.

Hägglund, B., Jansson, R. and Onnermark, B. (1974). 'Fire development in residential rooms after ignition from nuclear explosions'. *FOA Report C 20016–D6 (A3)*. Forsvarets Forskningsanstalt, Stockholm.

Hall, A.R. (1973). 'Pool burning: a review', in *Oxidation and Combustion Reviews*, Volume 6 (ed. C.F.H. Tipper), pp. 169–225. Elsevier, Amsterdam.

Hall, C. (1981). *Polymeric Materials: an Introduction for Technologists and Scientists*. Macmillan Press Ltd., London.

Hall, H. (1925). 'Oil tank fire boilover'. *Mechanical Engineering*, **47**, 540.

Halliday, D., Resnick, R. and Walker, J. (2008). *Fundamentals of Physics*, 8th edn. John Wiley & Sons, Inc., New Jersey.

Hamilton, D.C. and Morgan, W.R. (1952). *NACA Technical Note TN-2836*.

Hamins, A., Yang, T.C. and Kashiwagi, T. (1992). 'An experimental investigation of the pulsation frequency of flames'. *Proceedings of the Combustion Institute*, **24**, 1695–1702.

Harmathy, T.Z. (1972). 'A new look at compartment fires'. Parts I and II. *Fire Technology*, **8**, 196–219; 326–351.

Harmathy, T.Z. (1978). 'Mechanism of burning of fully-developed compartment fires'. *Combustion and Flame*, **31**, 265–273.

Harmathy, T.Z. (1979). 'Design to cope with fully developed compartment fires', in *Design of Buildings for Fire Safety* (eds E.E. Smith and T.Z. Harmathy). American Society for Testing and Materials, STP 685, pp. 198–276.

Harmathy, T.Z. (1987). 'On the equivalent fire exposure'. *Fire and Materials*, **11**, 95–104.

Harmathy, T.Z. and Mehaffey, J.F. (1982). 'Normalised heat load: a key parameter in fire safety design'. *Fire and Materials*, **6**, 27–31.

Harris, R.J. (1983). *The Investigation and Control of Gas Explosions in Buildings and Heating Plant*. E. & F.N. Spon Ltd., London.

Harrison, A.J. and Eyre, J.A. (1987). 'The effect of obstacle arrays on the combustion of large premixed gas/air clouds'. *Combustion Science and Technology*, **52**, 121–137.

Harrison, R. and Spearpoint, M. (2006). 'Entrainment of air into a balcony spill plume'. *Journal of Fire Protection Engineering*, **16**, 211–245.

Harrison, R. and Spearpoint, M. (2007). 'The balcony spill plume: entrainment of air into a flow from a compartment opening to a higher projecting balcony'. *Fire Technology*, **43**, 301–317.

Harrison, R. and Spearpoint, M. (2010). 'Physical scale modelling of adhered spill plume entrainment'. *Fire Safety Journal*, **45**, 149–158.

Hartzell, G.E., Priest, D.N. and Switzer, W.G. (1985). 'Modelling of toxicological effects of intoxication of rats by carbon monoxide and hydrogen cyanide'. *Journal of Fire Sciences*, **3**, 115–128.

Hasegawa, K. (1989). 'Experimental study on the mechanism of hot zone formation in open tank fires'. *Fire Safety Science*, **2**, 221–230.

Hasegawa, K. and Kashuki, K. (1991). 'A method for measuring upper flashpoint – practical method using Setaflash Closed Cup Apparatus'. *Report of the Fire Research Institute No. 71*, pp. 69–77.

Hasemi, Y. (1984). 'Experimental wall heat transfer correlations for the analysis of upward wall flame spread'. *Fire Science and Technology*, **4**, 75.

Hasemi, Y. (1988). 'Deterministic properties of turbulent flames and implications for fire growth'. *Interflam'88: Proceedings of the 4th International Fire Conference*, pp. 45–52. Interscience Communications, London.

Hasemi, Y. (2008). 'Surface flame spread', in *SFPE Handbook of Fire Protection Engineering*, 4th edn (eds P.J. Di Nenno *et al.*), pp. 2.278–2.290. National Fire Protection Association, Quincy, MA.

Hasemi, Y., Nam, D. and Yoshida, M. (2001). 'Experimental flame correlations and dimensional relations in turbulent ceiling fires'. *Proceedings of the 5th Asia–Oceania Symposium on Fire Science and Fire Technology*, pp. 379–390.

Hasemi, Y. and Nishihata, M. (1989). 'Fuel shape effect on the deterministic properties of turbulent diffusion flames'. *Fire Safety Science*, **2**, 275–286.

Hasemi, Y. and Tokunaga, T. (1984). 'Some experimental aspects of turbulent diffusion flames and buoyant plumes from fire sources against a wall and in a corner of walls'. *Combustion Science and Technology*, **40**, 1–17.

Hasemi, Y., Yokobayashi, S., Wakamatsu, T. and Ptchelintsev, A. (1995). 'Fire safety of building components exposed to a localized fire – scope and experiments on ceiling/beam system exposed to a localized fire'. *Proceedings of ASIAFLAM*, Kowloon, Hong Kong, pp. 351–361. Interscience Communications, London.

Hasemi, Y., Yoshida, M., Nohara, A. and Nakabayashi, T. (1991). 'Unsteady state upward flame spreading velocity along vertical combustible solid and influence of external radiation on the flame spread'. *Fire Safety Science*, **3**, 197–206.

Hasemi, Y., Yoshida, M., Takashima, S., Kikuchi, R. and Yokobayashi, Y. (1996). 'Flame length and flame heat transfer correlations in corner-wall and corner-wall-ceiling configurations'. *Interflam'96: Proceedings of the Seventh International Fire Conference*, pp. 179–188. Interscience Communications, London.

Hassan, M.I., Kuwana, K., Saito, K. and Wang, F.-J. (2005). 'Flow structure of a fixed-frame type fire whirl', *Fire Safety Science*, **8**, 951–962.

Hawthorne, W.R., Weddell, D.S. and Hottel, H.C. (1949). *3rd Symposium (International) on Combustion*, pp. 266–288. Williams and Wilkins, Baltimore, MD.

Health and Safety Executive (1980). 'Flame arresters and explosion reliefs'. *Health and Safety Series Booklet HS(G)*11. HMSO, London.

Health and Safety Executive (1996). 'Flame arresters: preventing the spread of fires and explosions in equipment that contains flammable gases and vapours'. *HSG158*. HMSO, London.

Health and Safety Executive (2002). 'Chemicals (Hazard Information and Packaging for Supply) Regulations 2002, Guidance on Regulations'. *HSE Booklet L131*. HMSO, London.

Health and Safety Executive (2006). 'A guide to the control of Major Hazard Regulations 1999 (as amended)'. HSE Books, London.

Hertzberg, M. (1982). 'The flammability limits of gases, vapours and dusts: theory and experiment', in *Symposium on Fuel-Air Explosions*, pp. 3–47. University of Waterloo Press, Waterloo, Ont.

Hertzberg, M., Johnson, A.L., Kuchta, J.M. and Furno, A.L. (1981). 'The spectral radiance, growth, flame temperatures and flammability behaviour of large scale, spherical combustion waves'. *Proceedings of the Combustion Institute*, **16**, 767–776.

Heselden, A.J.M. and Baldwin, R. (1976). 'The movement and control of smoke on escape routes in buildings', *Building Research Establishment, Current Paper CP13/76*.

Heselden, A.J.M. and Melinek, S.J. (1975). 'The early stages of fire growth in a compartment. A co-operative research programme of the CIB (Commission W14). First Phase'. *Fire Research Note No. 1029*.

Heskestad, A.W. and Hovde, P.J. (1994). 'Assessment of smoke production from building products'. *Fire Safety Science*, **4**, 527–538.

Heskestad, G. (1972). 'Similarity relations for the initial convective flow generated by fire'. American Society of Mechanical Engineers, Winter Annual Meeting, New York, November 26–30.

Heskestad, G. (1975). 'Physical modelling of fire'. *Journal of Fire and Flammability*, **6**, 253–273.

Heskestad, G. (1976). Unpublished results.

Heskestad, G. (1981). 'Peak gas velocities and flame heights of buoyancy-controlled turbulent diffusion flames'. *Proceedings of the Combustion Institute*, **18**, 951–960.

Heskestad, G. (1982). 'Engineering relations for fire plumes'. *Society of Fire Protection Engineers, Technology Report 82–8*.

Heskestad, G. (1983a). 'Luminous heights of turbulent diffusion flames'. *Fire Safety Journal*, **5**, 103–108.

Heskestad, G. (1983b). 'Virtual origins of fire plumes', *Fire Safety Journal*, **5**, 109–114.

Heskestad, G. (1986). 'Fire plume entrainment according to two competing assumptions'. *Proceedings of the Combustion Institute*, **21**, 111–120.

Heskestad, G. (1989). 'Note on maximum rise of fire plumes in temperature-stratified ambients'. *Fire Safety Journal*, **15**, 271–276.

Heskestad, G. (1991). 'A reduced scale mass fire experiment'. *Combustion and Flame*, **83**, 293–301.

Heskestad, G. (2008). 'Fire plumes, flame height and air entrainment', in *SFPE Handbook of Fire Protection Engineering*, 4th edn (eds P.J. Di Nenno *et al.*), pp. 2.1–2.20. National Fire Protection Association, Quincy, MA.

Heskestad, G. and Bill, R.G. (1988). 'Quantification of thermal responsiveness of automatic sprinklers, including conduction effects'. *Fire Safety Journal*, **14**, 113–125.

Heskestad, G. and Delichatsios, M.A. (1978). 'The initial convective flow in fire'. *Proceedings of the Combustion Institute*, **17**, 1113–1123.

Heskestad, G. and Hamada, T. (1993). 'Ceiling jets of strong fire plumes'. *Fire Safety Journal*, **21**, 69–82.

Heskestad, G. and Smith, H. (1976). 'Investigation of a new sprinkler sensitivity approval test: the plunge test'. *FMRC Serial* No. 22485. Factory Mutual Research Corporation, Norwood, MA.

Hillstrom, W.W. (1975). 'Temperature effects on flame spreading over fuels', *Paper to the Eastern States Section of the Combustion Institute Fall Meeting*.

Hinkley, P.L. (1971). 'Some notes on the control of smoke in enclosed shopping centres'. *Fire Research Note No. 875*.

Hinkley, P.L. and Wraight, H.G.H. (1969). 'The contribution of flames under ceilings to fire spread in compartments. Part II. Combustible ceiling linings'. *Fire Research Note No. 743*.

Hinkley, P.L., Wraight, H.G.H. and Theobald, C.R. (1968). 'The contribution of flames under ceilings to fire spread in compartments. Part 1: Incombustible ceilings'. *Fire Research Note 712*.

Hirano, T. and Tazawa, K. (1978). 'A further study of the effects of external thermal radiation on flame spread over paper'. *Combustion and Flame*, **32**, 95–105.

Hirano, T., Noreikis, S.E. and Waterman, T.E. (1974). 'Postulations of flame spread mechanisms'. *Combustion and Flame*, **22**, 353–363.

Hirschler, M.M. and Morgan, A.B. (2008). 'Thermal decomposition of polymers', in *SFPE Handbook of Fire Protection Engineering*, 4th edn (eds P.J. Di Nenno *et al.*), pp. 1.112–1.143. National Fire Protection Association, Quincy, MA.

Hirschler, M.M., Zicherman, J.B. and Umino, P.Y. (2009). 'Forensic evaluation of clothing flammability'. *Fire and Materials*, **33**, 354–364.

Hirst, R., Savage, N. and Booth, K. (1981/82). 'Measurement of inerting concentrations'. *Fire Safety Journal*, **4**, 147–158.

Hjertager, B.H. (1993). 'Computer modelling of turbulent gas explosions in complex 2D and 3D geometries'. *Journal of Hazardous Materials*, **34**, 173–197.

Hoffmann, N., Galea, E.R. and Markatos, N.C. (1989). *Applied Mathematical Modelling*, **13**, 298.

Holborn, P.G., Bishop, S.R., Drysdale, D.D. and Beard, A.N. (1993). 'Experimental and theoretical models of flashover'. *Fire Safety Journal*, **21**, 257–266.

Hollyhead, R. (1996). 'Ignition of flammable gases and liquids by cigarettes'. *Science and Justice*, **36**, 257–266.

Holman, J.R. (1976). *Heat Transfer*, 4th edn. McGraw-Hill, New York.

Holmes, F.H. (1975). 'Flammability testing of apparel fabrics'. *International Symposium on Fire Safety of Combustible Materials*, pp. 317–324. Edinburgh University.

Holmstedt, G. and Persson, H. (1985). 'Spray fire tests with hydraulic fluids'. *Fire Safety Science*, **1**, 869–879.

Holve, D.J. and Sawyer, R.F. (1974). 'Measurement of burning polymer flame structure and mass transfer numbers'. *Technical Report No. ME–74–2*, Department of Mechanical Engineering, University of California, Berkeley, CA.

Hopkins, D. and Quintiere, J.G. (1996). 'Material fire properties and predictions for thermoplastics'. *Fire Safety Journal*, **26**, 241–268.

Hottel, H.C. (1930). 'Radiant heat transmission'. *Mechanical Engineering*, **52**.

Hottel, H.C. (1959). 'Review: Certain laws governing the diffusive burning of liquids', by Blinov and Khudiakov (1957) (*Dokl. Akad. Nauk SSSR*, **113**, 1096). *Fire Research Abstracts and Reviews*, **1**, 41–43.

Hottel, H.C. (1961). 'Fire modelling', in *International Symposium on the use of models in fire research* (ed. W.G. Berl). National Academy of Sciences, Publication 786 (Washington), p. 32.

Hottel, H.C. and Egbert, R.B. (1942). *American Institution of Chemical Engineers Transaction*, **38**.

Hottel, H.C. and Hawthorne, W.R. (1949). *3rd Symposium (International) on Combustion*, pp. 255–266. Williams and Wilkins, Baltimore, MD.

Hottel, H.C. and Sarofim, A.F. (1967). *Radiative Transfer*. McGraw-Hill, New York.

Huggett, C. (1973). 'Habitable atmospheres which do not support combustion'. *Combustion and Flame*, **20**, 140–142.

Huggett, C. (1980). 'Estimation of rate of heat release by means of oxygen consumption measurements'. *Fire and Materials*, **4**, 61–65.

Huggett, C., von Elbe, G. and Haggerty, W. (1966). 'The combustibility of materials in O_2/He and O_2/N_2 atmospheres'. *Report SAM–TR–66–85*.

Inamura, T., Saito, K. and Tagivi, K.A. (1992). 'A study of boilover in liquid pool fires supported on water. Part 2. Effects of in-depth radiation absorption'. *Combustion Science and Technology*, **86**, 105.

Incropera, F.P., DeWitt, D.P., Bergman, T.L. and Lavine, A.S. (2007) *Fundamentals of Heat and Mass Transfer*, 6th edn. John Wiley & Sons, Inc., New York.

Ingason, H. (2005) 'Fire dynamics in tunnels', in *The Handbook of Tunnel Fire Safety* (eds A. Beard and R. Carvel). Thomas Telford, London, pp. 231–266.

Ingberg, S.H. (1928). 'Fire loads'. *Quarterly Journal of the National Fire Protection Association*, **22**, 43–61.

Iqbal, N. and Quintiere, J.G. (1994). 'Flame heat fluxes in PMMA pool fires'. *Journal of Fire Protection Engineering*, **6**, 153–162.

ISO (1975). *Fire Resistance Tests. Elements of Building Construction*. ISO 834, International Organization for Standardization, Geneva.

ISO (1981). *Fire Tests – Building Materials: Corner wall/room type test*. ISO/TC92N581, International Organization for Standardization, Geneva.

ISO (1984). *Fire Tests – Reaction to Fire – Ignitability of building products*. ISO/TC92/SC1/WG2-N54, International Organization for Standardization, Geneva.

ISO (1989). *Reaction to Fire Tests – Smoke generated by building products (dual chamber test)*. ISO TR5924:1989, International Organization for Standardization, Geneva.

ISO (1997). *Reaction to Fire Tests. Ignitability of building products using a radiant heat source*. ISO 5657:1997, International Organization for Standardization, Geneva.

ISO (1999a). *Fire Resistance Tests. Elements of Building Construction Part 1: General Requirements*. ISO 834-1, International Organization for Standardization, Geneva.

ISO (1999b). *Reaction to Fire Test – Ignitability of building products subjected to direct impingement of flame. Part 1: Guidance on ignitability*. ISO/TR 11925-1 (prepared by ISO TC92/SC1/WG2), International Organization for Standardization, Geneva.

ISO (2002a). *Reaction to Fire Tests – Heat release, smoke production and mass loss – Part 1: Heat release rate (cone calorimeter method)*. ISO 5660-1:2002, International Organization for Standardization, Geneva.

ISO (2002b). *Fire Tests – Full-scale room test for surface products*. ISO 9705:2002, International Organization for Standardization, Geneva.

ISO (2002c). *Reaction to fire tests for floorings – Part 1: Determination of the burning behaviour using a radiant heat source*. ISO 9239-1:2002, International Organization for Standardization, Geneva.

ISO (2005). *Fire Safety Engineering: Evaluation and behaviour of people*. Document 16738N19 Rev. 9, Draft Standard developed by Committee TC92 SC4 WG11, International Organization for Standardization, Geneva.

ISO (2006). *Reaction to Fire Tests. Spread of Flame – Part 2. Lateral spread on building and transport products*. ISO 5658-2:2006, International Organization for Standardization, Geneva.

ISO (2007a). *Life-threatening components of fire – Guidelines for the estimation of time available for escape using fire data*. ISO 13571:2007.

ISO (2007b). *Controlled equivalence ratio method for the determination of hazardous components of fire effluents*. ISO/TS 19700:2007.

Ishida, H. (1986). 'Flame spread over fuel-soaked ground'. *Fire Safety Journal*, **10**, 163–171.

Ishida, H. (1992). 'Initiation of fire growth on fuel-soaked ground'. *Fire Safety Journal*, **18**, 213–230.

Ito, A. and Kashiwagi, T. (1988). 'Characterisation of flame spread using holographic interferometry: sample orientation effects'. *Combustion and Flame*, **71**, 189–204.

Ito, A., Kudo, Y. and Oyama, H. (2005). 'Propagation and extinction mechanisms of opposed-flow flame spread over PMMA for different sample orientations'. *Combustion and Flame*, **142**, 428–437.

Ito, A., Masuda, D. and Saito, K. (1991). 'A study of flame spread over alcohols using holographic interferometry'. *Combustion and Flame*, **83**, 375–389.

Jackman, L.A., Nolan, P.F., Gardiner, A.J. and Morgan, H.P. (1992a). 'Mathematical model of the interaction of sprinkler spray drops with fire gases'. *Proceedings of the First International Conference on Fire Suppression Research*, 5–8 May 1992, Stockholm, pp. 209–227.

Jackman, L.A., Nolan, P.F. and Morgan, H.P. (1992b). 'Characterisation of water drops from a sprinkler spray'. *Proceedings of the First International Conference on Fire Suppression Research*, 5–8 May 1992, Stockholm, pp. 159–184.

Jagger, S. and Grant, G. (2005). 'Use of tunnel ventilation for fire safety', in *The Handbook of Tunnel Fire Safety* (eds A. Beard and R. Carvel). Thomas Telford Ltd., London, pp. 144–183.

Jaluria, Y. (1995). 'Natural convection wall flows', in *SFPE Handbook of Fire Protection Engineering*, 2nd edn (eds P.J. Di Nenno *et al.*), pp. 2.50–2.70. National Fire Protection Association, Quincy, MA.

James, J.J. (1991). 'A method for flammability testing of low flammability liquids'. MSc Thesis, University of Manchester, UK.

Janssens, M.L. (1991a). 'Piloted ignition of wood: a review'. *Fire and Materials*, **15**, 151–167.

Janssens, M.L. (1991b). 'Measuring rate of heat release by oxygen consumption'. *Fire Technology*, **27**, 234–249.

Janssens, M.L. (1991c). 'A thermal model for piloted ignition of wood including thermophysical properties'. *Fire Safety Science*, **3**, 167–176.

Janssens, M.L. (1993). 'Cone calorimeter measurements of the heat of gasification of wood'. *Interflam'93: Proceedings of the Sixth International Fire Conference*, pp. 549–558. Interscience Communications, London.

Janssens, M.L. (2008). 'Calorimetry', in *SFPE Handbook of Fire Protection Engineering*, 4th edn (eds P.J. Di Nenno *et al.*), pp. 3.60–3.89. National Fire Protection Association, Quincy, MA.

Janssens, M.L. and Parker, W.J. (1992). 'Oxygen consumption calorimetry', in *Heat Release in Fires* (eds V. Babrauskas and S.J. Grayson), pp. 31–59. Elsevier Applied Science, Barking.

Japan Institute for Safety Engineering (1982). Report on burning petroleum fires (in Japanese) (see Mudan (1984)).

Jin, T. (1970). 'Visibility through fire smoke'. *Building Research Institute, Tokyo, Report No. 30*.

Jin, T. (1971). 'Visibility through fire smoke'. *Building Research Institute, Tokyo, Report No. 33*.

Jin, T. (1978). 'Visibility through fire smoke'. *Journal of Fire and Flammability*, **9**, 135–155.

Jin, T. (2008). 'Visibility and human behaviour in fire smoke', in *SFPE Handbook of Fire Protection Engineering*, 4th edn (eds P.J. Di Nenno *et al.*), pp. 2.54–2.66. National Fire Protection Association, Quincy, MA.

Joint Fire Prevention Committee (1978). 'Report of the Technical Sub-Committee on the Fire Risks of New Materials, Central Fire Brigades' Advisory Council for England, Wales and Scotland', Home Office Fire Department, London.

Joint Fire Research Organization (1961). *Fire Research 1960: Report of the Director*. HMSO, London.

Joint Fire Research Organization (1962). *Fire Research 1961: Report of the Director*. HMSO, London.

Jost, W. (1939). *Explosions – und Verbrennungsvorgägne in Gasen*. Springer-Verlag, Berlin.

Joulain, P. (1998). 'Behaviour of pool fires: state of the art and new insights'. *Proceedings of the Combustion Institute*, **27**, 2691–2706.

Kandola, B.S. (2008). 'Introduction to the mechanics of fluids', in *SFPE Handbook of Fire Protection Engineering*, 4th edn (eds P.J. Di Nenno *et al.*), pp. 1.1–1.26. National Fire Protection Association, Quincy, MA.

Kanury, A.M. (1972). 'Ignition of cellulosic materials: a review'. *Fire Research Abstracts and Reviews*, **14**, 24–52.

Kanury, A.M. (1975). *Introduction to Combustion Phenomena*. Gordon and Breach, London.

Karlekar, B.V. and Desmond, R.M. (1979). *Heat Transfer*, 2nd edn. West Publishing Co., St Paul, MN.

Karlsson, B. (1993). 'A mathematical model for calculating heat release rate in the room corner test'. *Fire Safety Journal*, **20**, 93–113.

Kashiwagi, T. (1976). 'A study of flame spread over a porous material under external radiation fluxes'. *Proceedings of the Combustion Institute*, **15**, 255–265.

Kashiwagi, T. (1979). 'Effects of attenuation of radiation on surface temperature for radiative ignition'. *Combustion Science and Technology*, **20**, 225–234.

Kashiwagi, T. (1994). 'Polymer combustion and flammability – role of the condensed phase'. *Proceedings of the Combustion Institute*, **25**, 1423–1437.

Kashiwagi, T. and Newman, D.L. (1976). 'Flame spread over an inclined thin fuel surface'. *Combustion and Flame*, **26**, 163–177.

Kashiwagi, T. and Ohlemiller, T.J. (1982). 'A study of oxygen effects on non-flaming gasification of PMMA and PE during thermal irradiation'. *Proceedings of the Combustion Institute*, **19**, 815–823.

Kashiwagi, T., Ohlemiller, T.J. and Werner, K. (1987). 'Effects of external radiant heat flux and ambient oxygen concentration on non-flaming gasification rates and evolved products of white pine'. *Combustion and Flame*, **69**, 331–345.

Kashiwagi, T. and Omori, A. (1988). 'Effects of thermal stability and melt viscosity of thermoplastics on piloted ignition'. *Proceedings of the Combustion Institute*, **22**, 1329–1338.

Kawagoe, K. (1958). 'Fire behaviour in rooms'. *Report No. 27*, Building Research Institute, Tokyo.

Kawagoe, K. and Sekine, T. (1963). 'Estimation of temperature–time curves in rooms'. *Occasional Report No. 11*, Building Research Institute, Tokyo.

Kaye, G.W.C. and Laby, T.H. (1986). *Tables of Physical and Chemical Constants and Some Mathematical Functions*, 15th edn. Longman, London.

Kellogg, D.S., Waymack, B.E., McRae, D.D., Chen, P. and Dwyer, R.W. (1998). 'The initiation of smouldering combustion in cellulosic fabrics'. *Journal of Fire Sciences*, **16**, 90–104.

Kennedy, I.M. (1997). 'Models of soot formation and oxidation'. *Progress in Energy and Combustion Science*, **23**, 95–132.

Kent, J.H., Prado, G. and Wagner, H.Gg. (1981). 'Soot formation in a laminar diffusion flame'. *Proceedings of the Combustion Institute*, **18**, 1117–1126.

Khan, M.M., de Ris, J.L. and Ogden, S.D. (2008). 'Effect of moisture on ignition time of cellulosic materials'. *Proceedings of the 9th IAFSS Symposium*, pp. 167–178.

Kimmerle, G. (1974). 'Aspects and methodology for the evaluation of toxicological parameters during fire exposure'. *Journal of Fire and Flammability/Combustion Toxicology*, **1**, 4–51.

Kinbara, T., Endo, H. and Saga, S. (1968). 'Downward propagation of smouldering combustion through solid materials'. *Proceedings of the Combustion Institute*, **11**, 525–531.

Kirby, B.R., Wainman, D.E., Tomlinson, L.N., Kay, T.R. and Peacock, B.N. (1994). 'Natural Fires in Large Scale Compartments: a British Steel Technical – Fire Research Station Collaborative Project'. (British Steel Corporation).

Klitgaard, P.F. and Williamson, R.B. (1975). 'Impact of contents on building fires'. *Journal of Fire and Flammability/Consumer Product Flammability Supplement*, **2**, 84–113.

Klopovic, S. and Turan, O.F. (2001a). 'A comprehensive study of externally venting flames. Part I: Experimental plume characteristics for through-draught and no-through-draught ventilations, conditions and repeatability'. *Fire Safety Journal*, **36**, 99–133.

Klopovic, S. and Turan, O.F. (2001b). 'A comprehensive study of externally venting flames. Part II: Plume envelope and centre-line temperature comparisons, secondary fires, wind effects and smoke management systems'. *Fire Safety Journal*, **36**, 135–172.

Klote, J.H. (2008). 'Smoke control', in *SFPE Handbook of Fire Protection Engineering*, 4th edn (eds P.J. Di Nenno *et al*.), pp. 4.367–4.386. National Fire Protection Association, Quincy, MA.

Kodur, V.K.R. and Harmathy, T.Z. (2008). 'Properties of building materials', in *SFPE Handbook of Fire Protection Engineering*, 4th edn (eds P.J. Di Nenno *et al*.), pp. 1.167–1.195. National Fire Protection Association, Quincy, MA.

Kokkala, M.A. (1993). 'Characteristics of a flame in an open corner of walls'. *Interflam'93: Proceedings of the 6th International Fire Conference*, pp. 13–29. Interscience Communications, London.

Kokkala, M. and Rinkenen, W.J. (1987). 'Some observations on the shape of impinging diffusion flames'. *VTT Research Report 461*, Valtion Teknillinen Turkimuskeskus, Espoo, Finland.

Koseki, H. (1989). 'Combustion properties of large liquid pool fires'. *Fire Technology*, **25**, 241–255.

Koseki, H. (1993/94). 'Boilover and crude oil fire'. *Journal of Applied Fire Science*, **3**, 243–272.

Koseki, H., Kokkala, M. and Mulholland, G.W. (1991). 'Experimental study of boilover in crude oil fires'. *Fire Safety Science*, **3**, 865–874.

Krause Jr., R.F. and Gann, R.G. (1980). 'Rate of heat release measurements using oxygen consumption'. *Journal of Fire and Flammability*, **12**, 117–130.

Krause, U. and Schmidt, M. (2001). 'Smouldering fires in bulk materials'. *Proceedings of the 3rd International Seminar on Fire and Explosion Hazards*, University of Central Lancashire, pp. 389–400.

Kreith, F. (1976). *Principles of Heat Transfer*, 3rd edn. Intext Educational Publishers, New York.

Kumar, S., Cox, G. and Thomas, P.H. (2010). 'Air entrainment into balcony spill plumes'. *Fire Safety Journal*, **45**, 159–167.

Kumar, S., Heywood, G.M. and Liew, S.K. (1997). 'Superdrop modelling of a sprinkler spray in a two-phase CFD-particle tracking model'. *Fire Safety Science*, **5**, 889–900.

Kung, H.C. and Stravrianides, P. (1982). 'Buoyant plumes of large scale pool fires'. *Proceedings of the Combustion Institute*, **19**, 905–912.

Kuo, K.K. (2005). *The Principles of Combustion*, 2nd edn. John Wiley & Sons, Inc., New York.

Lam, C.S., Randsalu, E.J., Weckman, E.J., Brown, A.L., Gill, W. and Gritzo, L.A. (2004). 'Fuel regression rates of hydrocarbon pool fires in cross winds'. *Interflam'04: Proceedings of the Tenth International Fire Conference*, pp. 117–128. Interscience Communications, London.

Lamont, S., Usmani, A.S. and Gillie, M. (2004). 'Behaviour of a small composite steel frame structure in a 'long cool' and a 'short hot' fire'. *Fire Safety Journal*, **39**, 327–357.

Landau, L.D. and Lifshitz, E.M. (1987). *Fluid Mechanics*. Pergamon Press, Oxford.

Lastrina, F.A. (1970). 'Flame spread over solid fuel beds: solid and gas phase energy considerations'. PhD Thesis, Stevens Institute of Technology.

Latham, D.J., Kirby, B.R. and Thomson, G. (1987). 'The temperatures attained by unprotected structural steelwork in experimental natural fires'. *Fire Safety Journal*, **12**, 139–152.

Lattimer, B.Y. (2008). 'Heat fluxes from fires to surfaces', in *SFPE Handbook of Fire Protection Engineering*, 4th edn (eds P.J. Di Nenno *et al.*), pp. 2.303–2.336. National Fire Protection Association, Quincy, MA.

Lattimer, B.Y. and Sorathia, U. (2003). 'Thermal characteristics of fires in a non-combustible corner'. *Fire Safety Journal*, **38**, 709–745.

Laurendeau, N.M. (1982). 'Thermal ignition of methane–air mixtures by hot surfaces: a critical examination'. *Combustion and Flame*, **46**, 29–49.

Lautenberger, C., Torero, J. and Fernandez-Pello, A.C. (2006). 'Understanding materials flammability', in *Flammability Testing of Materials used in Construction, Transport and Mining* (ed. V.B. Apte), pp. 1–21. Woodhead Publishing Ltd., Cambridge.

Law, M. (1963). 'Heat radiation from fires, and building separation'. *Fire Research Technical Paper No. 5*. HMSO, London.

Law, M. (1971). 'A relationship between fire grading and building design and contents'. *Fire Research Note No. 877*.

Law, M. (1972). 'Nomograms for the fire protection of structural steelwork'. *Fire Prevention Science and Technology No. 3*. Fire Protection Association, London.

Law, M. (1986). 'A note on smoke plumes from fires in multi-level shopping malls'. *Fire Safety Journal*, **10**, 197–202.

Law, M. and O'Brien, T. (1981). *Fire Safety of Bare External Structural Steel*. Constructional Steel Research and Development Organization, London.

Lawson, D.I. (1954). 'Fire and the atomic bomb'. *Fire Research Bulletin No. 1*. HMSO, London.

Lawson, D.I. and Simms, D.L. (1952). 'The ignition of wood by radiation'. *British Journal of Applied Physics*, **3**, 288–292.

Lawson, J.R., Walton, W.D. and Twilley, W.H. (1983). 'Fire performance of furnishings in the NBS furniture calorimeter Part 1'. NBSIR83–2727, National Bureau of Standards, Gaithersburg, MD.

Le Chatelier, H. and Boudouard, O. (1898). 'Limits of flammability of gaseous mixtures'. *Bulletin de la Société Chimique (Paris)*, **19**, 485.

Lee, B.T. (1982). 'Quarter scale modeling of room fire tests of interior finish'. National Bureau of Standards, NBSIR 81–2453.

Leisch, S.O., Kauffmann, C.W. and Sichel, M. (1984). 'Smouldering combustion in horizontal dust layers'. *Proc. Combustion Institute*, **20**, 1601–1610.

Lennon, T. and Moore, D. (2003). 'The Natural Fire Safety Concept – full scale tests at Cardington'. *Fire Safety Journal*, **38**, 623–643.

Lewin, M. and Weil, E.D. (2001). 'Mechanisms and modes of action in flame retardancy of polymers', in *Fire Retardant Materials* (eds A.R. Horrocks and D. Price), pp. 31–68. Woodhead Publishing Ltd., Cambridge.

Lewis, B. (1954). *Selected Combustion Problems*. AGARD (Butterworths), p. 177.

Lewis, B. and von Elbe, G. (1987). *Combustion, Flames and Explosions of Gases*, 3rd edn. Academic Press, Orlando, FL.

Lide, D.R. (ed.) (1993/94). *Handbook of Chemistry and Physics*, 74th edn. Chemical Rubber Company, Ohio.

Lie, T.T. (1972). *Fires and Buildings*. Applied Science Publishers, London.

Lindner, H. and Seibring, H. (1967). 'Self-ignition of organic substances in lagging materials'. *Chimie-Ing. Tech*, **39**, 667.

Linnett, J.W. and Simpson, C.J.S.M. (1957). 'Limits of inflammability'. *Proceedings of the Combustion Institute*, **6**, 20–27.

Lois, E. and Swithenbank, J. (1979). 'Fire hazards in oil tank arrays in wind'. *Proceedings of the Combustion Institute*, **17**, 1087–1098.

Long, R.T., Torero, J.L., Quintiere, J.G. and Fernandez-Pello, A.C. (1999). 'Scale and transport considerations in piloted ignition of PMMA'. *Proceedings of the Sixth International Symposium on Fire Safety Science*, pp. 567–578.

Lougheed, G.D. and Yung, D. (1993). 'Exposure to adjacent structures from flames issuing from a compartment opening'. *Interflam'93: Proceedings of the Sixth International Fire Conference*, pp. 297–306. Interscience Communications, London.

Lovachev, L.A., Babkin, V.S., Bunev, V.A., V'yun, A.V., Krivulin, V.N. and Baratov, A.N. (1973). 'Flammability limits: an invited review'. *Combustion and Flame*, **20**, 259–289.

Luo, M. and Beck, V.R. (1994). 'The fire environment in a multi-room building – comparison of predicted and experimental results'. *Fire Safety Journal*, **23**, 413–438.

Lyons, J.W. (1970). *The Chemistry and Uses of Fire Retardants*. John Wiley & Sons, Inc., New York.

Ma, T.G. and Quintiere, J.G. (2003). 'Numerical simulation of axi-symmetric fire plumes; accuracy and limitations'. *Fire Safety Journal*, **38**, 467–492.

McAdams, W.H. (1954). *Heat Transmission*, 3rd edn. McGraw-Hill, New York.

McAlevy, R.F. and Magee, R.S. (1969). 'The mechanism of flame spreading over the surface of igniting condensed phase materials'. *12th Symposium (International) on Combustion*, pp. 215–227.

McCaffrey, B.J. (1979). 'Purely buoyant diffusion flames: some experimental results'. National Bureau of Standards, NBSIR 79–1910.

McCaffrey, B.J. and Rockett, J.A. (1977). 'Static pressure measurements of enclosure fires'. *Journal of Research of the National Bureau of Standards*, **82**, 107–117.

McCaffrey, B.J., Quintiere, J.G. and Harkleroad, M.F. (1981). 'Estimating room temperatures and the likelihood of flashover using fire test data correlations'. *Fire Technology*, **17**, 98–119; **18**, 122.

McCarter, R.J. (1976). 'Smouldering of flexible polyurethane foam'. *Journal of Consumer Product Flammability*, **3**, 128–140.

McCarter, R.J. (1977). 'Smouldering combustion of cotton and rayon'. *Journal of Consumer Product Flammability*, **4**, 346–357.

McCarter, R.J. (1978). 'Smouldering combustion of wood fibres: cause and prevention'. *Journal of Fire and Flammability*, **9**, 119–126.

McCartney, C.J., Lougheed, G.D. and Weckman, E.J. (2008). 'CFD investigation of balcony spill plumes in atria'. *ASHRAE Transactions*, **114**, 369–378.

McGrattan, K. (2006). *Fire Dynamics Simulator (Version 4)*. NIST Special Publication 1018. National Institute for Standards and Technology, Gaithersburg, MD.

McGrattan, K. and Miles, S. (2008). 'Modelling enclosure fires using CFD', in *SFPE Handbook of Fire Protection Engineering*, 4th edn (eds P.J. Di Nenno *et al.*), pp. 3.229–3.246. National Fire Protection Association, Quincy, MA.

McGuire, J.H. (1953). 'Heat transfer by radiation'. *Fire Research Special Report No. 2*. HMSO, London.

McIntosh, A.C., Bains, M., Crocomb, W. and Griffiths, J.F. (1994). 'Autoignition of combustible fluids in porous insulating materials'. *Combustion and Flame*, **99**, 541–550.

McIntosh, A.C. and Tolputt, T.A. (1990). 'Critical heat losses to avoid self-heating in coal'. *Combustion Science and Technology*, **69**, 133–145.

Mackinven, R., Hansel, J.G. and Glassman, I. (1970). 'Influence of laboratory parameters on flame spread over liquid surfaces'. *Combustion Science and Technology*, **1**, 293–306.

Madorsky, S.L. (1964). *Thermal Degradation of Organic Polymers*. John Wiley & Sons, Inc., New York.

Magee, R.S. and McAlevy, R.F. (1971). 'The mechanism of flame spread'. *Journal of Fire and Flammability*, **2**, 271–297.

Magnusson, S.-E. and Sundstrom, B. (1985). 'Combustible linings and room fire growth – a first analysis'. *Fire Safety Science and Engineering*, ASTM STP 882 (ed. T.Z. Harmathy), pp. 45–69. American Society for Testing and Materials, Philadelphia, PA.

Makhviladze, G.M. and Yakush, S.E. (2002). 'Large scale unconfined fires and explosions'. *Proceedings of the Combustion Institute*, **29**, 195–210.

Malalasekera, W.M.G., Versteeg, H.K. and Gilchrist, K. (1996). 'A review of research and an experimental study on the pulsation of buoyant diffusion flames and pool fires'. *Fire and Materials*, **20**, 261–271.

Malhotra, H.L. (1967). 'Movement of smoke on escape routes. Instrumentation and effect of smoke on visibility'. *Fire Research Notes No. 651, 652 and 653*.

Malhotra, H.L. (1982). *Design of Fire-resisting Structures*. Surrey University Press, Surrey.

Mannan, S. (ed.) (2005). *Lees' Loss Prevention in the Process Industries: Hazard Identification, Assessment and Control*, 3rd edn (in three volumes). Butterworth–Heinemann, Oxford.

Manton, J., von Elbe, G. and Lewis, B. (1953). 'Burning velocity measurements in a spherical vessel with central ignition'. *Proceedings of the Combustion Institute*, **4**, 358–363.

Manzello, S.L., Cleary, T.G., Shields, J.R., Maranghides, A., Mell, W.E. and Jiang, J.C. (2008). 'Experimental investigation of firebrands: generation and ignition of fuel beds'. *Fire Safety Journal*, **43**, 226–233.

Manzello, S.L., Maranghides, A. and Mell, W.E. (2007). 'Firebrand generation from burning vegetation'. *International Journal of Wildland Fire*, **16**, 458–462.

Marchant, E.W. (1976). 'Some aspects of human behaviour and escape route design'. *5th International Fire Protection Seminar*, Karlsruhe, September.

Marchant, E.W. (1984). 'Effect of wind on smoke movement and smoke control systems'. *Fire Safety Journal*, **7**, 55–63.

Margenau, H. and Murphy, G.M. (1956). *The Mathematics of Chemistry and Physics*. Van Nostrand, Princeton, NJ.

Markstein, G.H. (1975). 'Radiative energy transfer from gaseous diffusion flames'. *Proceedings of the Combustion Institute*, **15**, 1285–1294.

Markstein, G.H. (1976). 'Radiative energy transfer from turbulent diffusion flames'. *Combustion and Flame*, **27**, 51–63.

Markstein, G.H. (1979). 'Radiative properties of plastics fires'. *Proceedings of the Combustion Institute*, **17**, 1053–1062.

Markstein, G.H. (1984). 'Effects of turbulence on flame radiation from diffusion flames'. *Proceedings of the Combustion Institute*, **20**, 1055.

Markstein, G.H. and de Ris, J. (1972). 'Upward fire spread over textiles'. *Proceedings of the Combustion Institute*, **14**, 1085–1097.

Markstein, G.H. and de Ris, J.N. (1975). 'Flame spread along fuel edges'. *Journal of Fire and Flammability*, **6**, 140–154.

Martin, S. (1965). 'Diffusion-controlled ignition of cellulosic materials by intense radiant energy'. *Proceedings of the Combustion Institute*, **10**, 877–896.

Martin, S.B. and Wiersma, S.J. (1979). 'An experimental study of flashover criteria for compartment fires'. Final Report, Products Research Committee PRC No. P-77-3–1, Stanford Research Institute, International Project No. PYC 6496.

Mayer, E. (1957). 'A theory of flame propagation limits due to heat loss'. *Combustion and Flame*, **1**, 438–452.

Meidl, J. (1970). *Explosive and Toxic Hazardous Materials*. Glencoe Press, Fire Science Series.

Middleton, J.F. (1983). 'Developments in flame detectors'. *Fire Safety Journal*, **6**, 175–182.

Mikkola, E. (1992). 'Ignitability of solid materials', in *Heat Release in Fires* (eds V. Babrauskas and S.J. Grayson), pp. 225–232. Elsevier Applied Science, Barking.

Mikkola, E. and Wichman, I.S. (1989). 'On the thermal ignition of combustible materials'. *Fire and Materials*, **14**, 87–96.

Milke, J.A. (2008a). 'Analytical methods for determining fire resistance of steel members', in *SFPE Handbook of Fire Protection Engineering*, 4th edn (eds P.J. Di Nenno *et al.*), pp. 4.297–4.326. National Fire Protection Association, Quincy, MA.

Milke, J.A. (2008b). 'Smoke management by mechanical exhaust or natural venting', in *SFPE Handbook of Fire Protection Engineering*, 4th edn (eds P.J. Di Nenno *et al.*), pp. 4.387–4.412. National Fire Protection Association, Quincy, MA.

Miller, F.J. and Ross, H.D. (1992). 'Further observations of flame spread over laboratory-scale alcohol pools'. *Proceedings of the Combustion Institute*, **24**, 1703–1711.

Miller, F.J. and Ross, H.D. (1998). 'Smoke visualization of the gas-phase flow during flame spread across a liquid pool'. *Proceedings of the Combustion Institute*, **27**, 2715–2722.

Miller, R.B. (1999). 'Structure of wood', in *Wood Handbook – Wood as an Engineering Material*. Forest Products Laboratory General Technical Report FPL-GTR-113. US Department of Agriculture Forest Service, Madison, WI.

Ministry of Aviation (1962). *Report of the Working Party on Aviation Kerosene and Wide-Cut Gasoline*. HMSO, London.

Minzer, G.A. and Eyre, J.A. (1982). 'Large scale LNG and LPG pool fires'. *I Chem Eng Symposium Series No. 71*, pp. 147–163.

Miron, Y. and Lazzara, C.P. (1988). 'Hot surface ignition temperatures of dust layers'. *Fire and Materials*, **12**, 115–126.

Mitchell, N.D. (1951). *Quarterly Journal of the National Fire Protection Association*, **45**, 165.

Mitler, H.E. (1985). 'The Harvard fire model'. *Fire Safety Journal*, **9**, 7–16.

Mitler, H.E. and Emmons, H.W. (1981). Documentation for CFC V, the Fifth Harvard Computer Fire Code. *Home Fire Project Technical Report No. 45*, Harvard University, Harvard, MA.

Modak, A.T. and Croce, P.A. (1977). 'Plastic pool fires'. *Combustion and Flame*, **30**, 251–265.

Moghtaderi, B. (2006). 'Pyrolysis modeling of lignocellulosic fuels'. *Fire and Materials*, **30**, 1–34.

Moghtaderi, B., Novozhilov, V., Fletcher, D.F. and Kent, J.H. (1997). A new correlation for bench-scale piloted ignition of wood'. *Fire Safety Journal*, **29**, 41–59.

Montreal Protocol on Substances that Deplete the Ozone Layer, UNEP, Nairobi, 1987.

Moore, H. (2007). *MATLAB for Engineers*. Pearson Prentice Hall, New Jersey.

Moore, L.D. (1978). 'Full scale burning behaviour of curtains and drapes'. NBSIR78–1448, National Bureau of Standards, Gaithersburg, MD.

Moore, W.J. (1972). *Physical Chemistry*, 5th edn. Longman, London.

Moorhouse, J. (1982). 'Scaling criteria for pool fires derived from large scale experiments'. *Institute of Chemical Engineers Symposium Series No. 71*, pp. 165–179.

Morandini, F., Santoni, P.A. and Balbi, J.H. (2001). 'Fire front width effects on fire spread across a laboratory scale sloping fuel bed'. *Combustion Science and Technology*, **166**, 67–90.

Morgan, H.P., Marshall, N.R. and Goldstone, B.M. (1976). 'Smoke hazards in covered multi-level shopping malls. Some studies using a model 2-storey mall'. Building Research Establishment, BRE CP 45/76.

Morgan, H.P. and Marshall, N.R. (1979). 'Smoke control measures in a covered two-storey shopping mall having balconies and pedestrian walkways'. BRE CP11/79. Fire Research Station, Borehamwood.

Morris, W.A. and Hopkinson, J.S. (1976). 'Fire behaviour of foamed plastics ceilings used in dwellings'. Building Research Establishment, BRE CP 73/76.

Morton, B.R., Taylor, G. and Turner, J.S. (1956). 'Turbulent gravitational convection from maintained and instantaneous sources'. *Proceedings of the Royal Society (London)*, **A234**, 1–23.

Moss, J.B. (1995). 'Turbulent diffusion flames', in *Combustion Fundamentals of Fire* (ed. G. Cox), pp. 221–272. Academic Press, London.

Moulen, A.W. (1971). 'Horizontal projections in the prevention of spread of fire from storey to storey'. *Report TR52/75/397*, Commonwealth Experimental Building Station, Australia.

Moulen, A.W. and Grubits, S.J. (1979). 'Flammability testing of carpets'. Technical Record 449, Experimental Building Station, Department of Housing and Construction, Australia.

Moussa, N.A., Toong, T.Y. and Backer, S. (1973). 'An experimental investigation of flame-spreading mechanisms over textile materials'. *Combustion Science and Technology*, **8**, 165–175.

Moussa, N.A., Toong, T.Y. and Garris, C.A. (1977). 'Mechanisms of smouldering of cellulosic materials'. *Proceedings of the Combustion Institute*, **16**, 1447–1457.

Mowrer, F.W. (1990). 'Lag times associated with fire detection and suppression'. *Fire Technology*, **26**, 244–265.

Mowrer, F.W. (2008) 'Enclosure smoke filling and fire-generated environmental conditions', in *SFPE Handbook of Fire Protection Engineering*, 4th edn (eds P.J. Di Nenno *et al.*), pp. 3.247–3.270. National Fire Protection Association, Quincy, MA.

Mowrer, F.W. and Williamson, R.B. (1987). 'Estimating room temperatures from fires along walls and in corners'. *Fire Technology*, **23**, 133–145.

Mudan, K.S. (1984). 'Thermal radiation hazards from hydrocarbon pool fires'. *Progress in Energy and Combustion Science*, **10**, 59–80.

Mulholland, G.W. (2008). 'Smoke production and properties', *SFPE Handbook of Fire Protection Engineering*, 4th edn (eds P.J. Di Nenno *et al.*), pp. 2.291–2.302. National Fire Protection Association, Quincy, MA.

Mullins, B.P. and Penner, S.S. (1959). *Explosions, Detonations, Flammability and Ignition*. Pergamon Press, London.

Mundweiler, H. (1990). 'Brand im Hotel International Zurich, 16th February 1988'. *Schwietzerische Feuerwehr-Zeitung*, **116**, 144–162.

Munoz, M., Arnaldos, J., Casal, J. and Planas, E. (2004). 'Analysis of the geometric and radiative characteristics of hydrocarbon pool fires'. *Combustion and Flame*, **139**, 263–277.

Murrell, J. and Rawlins, P. (1996). 'Fire hazard of multi-layer paint surfaces'. *Proceedings of the 4th International Conference on Fire and Materials*. Interscience Communications, London.

Nakaya, I., Tanaka, T., Yoshida, K. and Steckler, K. (1986). 'Doorway flow induced by a propane fire'. *Fire Safety Journal*, **10**, 185–195.

National Bureau of Standards (1972). 'The mechanisms of pyrolysis oxidation and burning of organic materials'. *Proceedings of the 4th Materials Research Symposium*, Gaithersburg, MD.

National Fire Protection Association (1978). *NFPA UL913 Intrinsically safe process control equipment for use in hazardous locations*. NFPA, Quincy, MA.

National Fire Protection Association (1996). *Fire investigation summary: Airport terminal fire, Dusseldorf, Germany, April 11 1996*. NFPA, Quincy, MA.

National Fire Protection Association (2004). *NFPA 53: Materials, Equipment and Systems used in oxygen enriched atmospheres*. NFPA, Quincy, MA.

National Fire Protection Association (2006a). *NFPA 253: Standard method of test for critical radiant flux of floor covering systems using a radiant heat energy source*. NFPA, Quincy, MA.

National Fire Protection Association (2006b). *NFPA 286: Standard Methods of Fire Tests for evaluating contribution of wall and ceiling interior finish to room fire growth*. NFPA, Quincy, MA.

National Fire Protection Association (2008a). *NFPA Handbook*, 20th edn. NFPA, Quincy, MA.

National Fire Protection Association (2008b). *NFPA 70 National Electrical Code: Hazardous Area Classification*. NFPA, Quincy, MA.

National Fire Protection Association (2008c). *NFPA No. 30: Flammable and Combustible Liquids Code*. NFPA, Quincy, MA.

National Fire Protection Association (2009). *NFPA 92B Guide for Smoke Management Systems in Malls, Atria and Large Areas*. NFPA, Quincy, MA.

National Fire Protection Association (2010). *NFPA 72, National Fire Alarm and Signalling Code*. NFPA, Quincy, MA.

Nelson, G.L. (1998). 'CO and fire toxicity: a review and analysis of recent work'. *Fire Technology*, **34**, 38–58.

Nelson, H.E. and Mowrer, F.W. (2002). 'Emergency movement', in *SFPE Handbook of Fire Protection Engineering*, 3rd Edition (eds P.J. Di Nenno *et al.*), pp. 3-367–3-380 (National Fire Protection Association, Quincy, MA).

Neoh, K.G., Howard, J.B. and Sarofim, A.F. (1984). 'Effect of oxidation on the physical structure of soot'. *Proceedings of the Combustion Institute*, **20**, 951–957.

Newman, J.S. (1988). 'Principles for fire detection'. *Fire Technology*, **24**, 116–127.

Newton, The Rt Hon Lord (2008). *The Buncefield Incident 11 December 2005: The Final Report of the Major Incident Investigation Board*, Volumes 1 and 2. Health and Safety Executive, London. (http://www.buncefieldinvestigation.gov.uk/reports/index.htm)

Nordtest (1986). 'Surface products: Room fire tests in full scale'. Nordtest method NT Fire 025.

Nordtest (1991). 'Upholstered furniture: burning behaviour – full scale test. Edition 2'. Nordtest method NT Fire 032.

Novozhilov, V. (2001). 'Computational fluid dynamics modeling of compartment fires'. *Progress in Energy and Combustion Science*, **27**, 611–666.

Odeen, K. (1963). 'Theoretical study of fire characteristics in enclosed spaces'. *Bulletin No. 10*, Division of Building Construction, Royal Institute of Technology, Stockholm.

Office of the Deputy Prime Minister (2006). *The Economic Costs for Fire: Estimates for 2004*. ODPM, 2006.

Ohlemiller, T.J. (1985). 'Modelling of smouldering combustion'. *Progress in Energy and Combustion Science*, **11**, 277–310.

Ohlemiller, T.J. (1990). 'Smouldering combustion propagation through a permeable horizontal fuel layer' and 'Forced smoulder propagation and the transition to flaming in cellulosic insulation'. *Combustion and Flame*, **81**, 341–353; 354–365.

Ohlemiller, T.J. (1991). 'Smouldering combustion propagation on solid wood'. *Fire Safety Science*, **3**, 565–574.

Ohlemiller, T.J. (2008). 'Smouldering combustion', in *SFPE Handbook of Fire Protection Engineering*, 4th edn (eds P.J. Di Nenno *et al.*), pp. 2.229–2.240. National Fire Protection Association, Quincy, MA.

Ohlemiller, T.J. and Lucca, D.A. (1983). 'An experimental comparison of forward and reverse smoulder propagation in permeable fuel beds'. *Combustion and Flame*, **54**, 131–148.

Ohlemiller, T.J. and Rogers, F.E. (1978). 'A survey of several factors influencing smouldering combustion in flexible and rigid polymer foams'. *Journal of Fire and Flammability*, **9**, 489–509.

Ohtani, H., Hirano, T. and Akita, K. (1981). 'Experimental study of bottom surface combustion of polymethylmethacrylate'. *Proceedings of the Combustion Institute*, **18**, 591–599.

Oleskiewicz, I. (1989). 'Heat transfer from a window fire plume to a building façade'. *Proceedings of the Winter Annual Meeting of the American Society of Mechanical Engineers*, San Francisco, HTD-volume 123, pp. 163–170.

Open University (1973). 'Giant Molecules'. Science Foundation Course S100, Unit 13.

Orloff, L. (1981), 'Simplified radiation modelling of pool fires'. *Proceedings of the Combustion Institute*, **18**, 549–561.

Orloff, L. and de Ris, J. (1972). 'Cellular and turbulent ceiling fires'. *Combustion and Flame*, **18**, 389–401.

Orloff, L. and de Ris, J.N. (1982). 'Froude modelling of pool fires'. *Proceedings of the Combustion Institute*, **19**, 885–895.

Orloff, L., de Ris, J. and Markstein, G.H. (1974). 'Upward turbulent fire spread and burning of fuel surface'. *Proceedings of the Combustion Institute*, **15**, 183–192.

Orloff, L., Modak, A.T. and Alpert, R.L. (1976). 'Burning of large scale vertical surfaces'. *Proceedings of the Combustion Institute*, **16**, 1345–1354.

Orloff, L., Modak, A.T. and Markstein, G.H. (1979). 'Radiation from smoke layers'. *Proceedings of the Combustion Institute*, **17**, 1029–1038.

Östman, B.A.L. (1992). 'Smoke and soot', in *Heat Release in Fires* (eds V. Babrauskas and S.J. Grayson), pp. 233–250. Elsevier Applied Science, London.

Östman, B.A.L. and Nussbaum, R.M. (1988). 'Correlation between small-scale rate of heat release and full-scale room flashover for surface linings'. *Fire Safety Science*, **2**, 823–832.

Östman, B.A.L. and Tsantardis, L.D. (1991). 'Smoke production in the cone calorimeter and the room fire test'. *Fire Safety Journal*, **17**, 27–43.

Pagni, P.J. (1990). 'Some unanswered questions in fluid mechanics'. *Applied Mechanics Review*, **43**, 153–170.

Pagni, P.J. and Bard, S. (1979). 'Particulate volume fractions in diffusion flames'. *Proceedings of the Combustion Institute*, **17**, 1017–1025.

Palmer, K.N. (1957). 'Smouldering combustion of dusts and fibrous materials'. *Combustion and Flame*, **1**, 129–154.

Palmer, K.N. (1973). *Dust Explosions and Fires*. Chapman and Hall, London.

Panagiotou, J. and Quintiere, J.G. (2004). 'Generalising flammability of materials'. *Interflam'04: Proceedings of the 10th International Fire Conference*. Interscience Communications, London, pp. 895–905.

Pape, R. and Waterman, T.E. (1979). 'Understanding and modeling preflashover compartment fires', in *Design of Buildings for Fire Safety* (eds E.E. Smith and T.Z. Harmathy). American Society for Testing and Materials, STP 685.

Parker, W.J. (1972). 'Flame spread model for cellulosic materials'. *Journal of Fire and Flammability*, **3**, 254–269.

Parker, W.J. (1982). 'An assessment of correlations between laboratory and full-scale experiments for the FAA Aircraft Fire Safety Programme. Part 3: ASTM E84'. NBSIR 82-2564. Centre for Fire Research, NBS, Gaithersburg, MD.

Paul, K.T. (1979). Private communication.

Paul, K.T. (1983). 'Measurement of smoke in large scale fires'. *Fire Safety Journal*, **5**, 89–102.

Paulsen, O.R. and Hadvig, S. (1977). 'Heat transfer in fire test furnaces'. *Journal of Fire and Flammability*, **8**, 423–442.

Peacock, R.D., Jones, W.W., Reneke, P.A. and Forney, G.P. (2008). 'CFAST – Consolidated Model of Fire Growth and Smoke Transport (Version 6). User's Guide'. *NIST Special Publication 1041* (December 2008 revision).

Peacock, R.D., Reneke, P.A., Bukowski, R.W. and Babrauskas, V. (1999). 'Defining flashover for hazard calculations'. *Fire Safety Journal*, **32**, 331–345.

Perdue, G.R. (1956). 'Spontaneous combustion: how it is caused and how to prevent it occurring'. *Power Laundry*, **95**, 599.

Perry, R.H., Green, D.W. and Maloney, J.O. (eds) (1984). *Perry's Chemical Engineers' Handbook*. McGraw-Hill, New York.

Petrella, R.V. (1979). 'The mass burning rate and mass transfer number of selected polymers, wood and organic liquids'. *Polymer and Plastics Technology Engineering*, **13**, 83–103.

Pettersson, O., Magnuson, S.E. and Thor, J. (1976). *Fire Engineering Design of Structures*. Swedish Institute of Steel Construction, Publication 50.

Petty, S.E. (1983). 'Combustion of crude oil on water'. *Fire Safety Journal*, **5**, 123–134.

Phillips, H. (1965). 'Flame in a buoyant methane layer'. *Proceedings of the Combustion Institute*, **10**, 1277–1283.

Phylaktou, H.N. and Andrews, G.E. (1993). 'Gas explosions in linked vessels'. *Journal of Loss Prevention in the Process Industries*, **6**, 15–19.

Pipkin, O.A. and Sliepcevich, C.M. (1964). 'The effect of wind on buoyant diffusion flames'. *Industrial and Engineering Chemistry: Fundamentals*, **3**, 147–154.

Pitts, D.R. and Sissom, L.E. (1977). *Schaum's Outline Series: Theory and Problems of Heat Transfer*. McGraw-Hill, New York.

Pitts, W.M. (1991). 'Wind effects on fire'. *Progress in Energy and Combustion Science*, **17**, 83–134.

Pitts, W.M. (1994). 'Application of thermodynamic and detailed chemical kinetic modelling to understanding combustion product generation in enclosure fires'. *Fire Safety Journal*, **23**, 272–303.

Pizzo, Y., Consalvi, J.L., Querre. P., Coutin, M. and Porterie, B. (2009). 'Width effects on the early stage of upward flame spread over PMMA slabs: experimental observations'. *Fire Safety Journal*, **44**, 407–414.

Popplewell, O. (1986). *Inquiry into Crowd Safety at Sports Grounds*. The Stationery Office, London.

Potts, W.J. and Lederer, T.S. (1977). 'A method for comparing testing of smoke toxicity'. *Journal of Combustion Toxicology*, **4**, 114–162.

Powell, F. (1969). 'Ignition of gases and vapours: a review of ignition of flammable gases and vapours by friction and impact'. *Industrial and Engineering Chemistry*, **61**, 29–37.

Prahl, J. and Emmons, H.W. (1975). 'Fire induced flow through an opening'. *Combustion and Flame*, **25**, 369–385.

Pratap, R. (2006). *Getting started with MATLAB 7*. Oxford University Press, Oxford.

Prime, D.M. (1981). Private communication.

Prime, D.M. (1982). 'Ignitibility and leather upholstery'. *Cabinet Maker and Retail Furnisher*, 26 March, p. 39.

Products Research Committee (1980). Fire research on cellular plastics: the final report of the Products Research Committee.

Proulx, G. (2008). 'Evacuation time', in *SFPE Handbook of Fire Protection Engineering*, 4th edn (eds P.J. Di Nenno *et al.*), pp. 3.355–3.372. National Fire Protection Association, Quincy, MA.

Punderson, J.O. (1981). 'A closer look at cause and effect in fire fatalities – the role of toxic fumes'. *Fire and Materials*, **5**, 41–46.

Puri, R. and Santoro, R.J. (1991). 'The role of soot particle formation on the production of carbon monoxide in fires'. *Fire Safety Science*, **3**, 595–604.

Purser, D.A. (2008). 'Assessment of hazards to occupants from smoke, toxic gases and heat', in *SFPE Handbook of Fire Protection Engineering*, 4th edn (eds P.J. Di Nenno *et al.*), pp. 2.96–2.193. National Fire Protection Association, Quincy, MA.

Purser, D.A. (2010). 'Hazards from smoke and irritants', in *Fire Toxicity* (eds A. Stec and R. Hull). Woodhead Publishers Ltd., Oxford.

Purser, D.A. and Purser, J. (2001). 'HCN yields and fate of fuel nitrogen from materials under different combustion conditions in the ISO 19700 Tube Furnace and large scale fires'. *Fire Safety Science*, **9**, 1117–1128.

Putorti, A.D., McElroy, J.A. and Madrzykowski, D. (2001). 'Flammable and combustible liquid spill/burn patterns'. NIJ Report 604-00. National Institute of Justice, Washington, DC.

Putzeys, O., Bar-Ilan, A., Rein, G., Fernandez-Pello, A.C. and Urban, D.L. (2007). 'The role of secondary char oxidation in the transition from smouldering to flaming'. *Proceedings of the Combustion Institute*, **31**, 2669–2676.

Quintiere, J.G. (1975a). 'Some observations on building corridor fires'. *Proceedings of the Combustion Institute*, **15**, 163–174.

Quintiere, J.G. (1975b). 'The application and interpretation of a test method to determine the hazard of floor covering fire spread in building corridors'. *International Symposium on Fire Safety of Combustible Materials*, Edinburgh University, pp. 355–366.

Quintiere, J.G. (1976). 'Growth of fire in building compartments', in *Fire Standards and Safety* (ed. A.F. Robertson). American Society for Testing and Materials, STP 614, pp. 131–167.

Quintiere, J.G. (1979). 'The spread of fire from a compartment: a review', in *Design of Buildings for Fire Safety* (eds E.E. Smith and T.Z. Harmathy). American Society for Testing and Materials, STP 685, pp. 139–168.

Quintiere, J.G. (1981). 'A simplified theory for generalising results from a radiant panel rate of flame spread apparatus'. *Fire and Materials*, **5**, 52–60.

Quintiere, J. (1982). 'An assessment of correlations between laboratory and full-scale experiments for the FAA Aircraft Fire Safety Programme, Part 2. Rate of energy release in fires'. NBSIR82-2536. National Bureau of Standards, Gaithersburg, MD.

Quintiere, J.G. (1983). 'An assessment of correlations between laboratory and full-scale experiments for the FAA Aircraft Fire Safety Program, Part 1. Smoke'. NBSIR82-2508. National Bureau of Standards, Gaithersburg, MD.

Quintiere, J.G. (1989a). 'Scaling applications in fire research'. *Fire Safety Journal*, **15**, 3–30.

Quintiere, J.G. (1989b). 'Fundamentals of enclosure fire zone models'. *Journal of Fire Protection Engineering*, **1**, 99–119.

Quintiere, J.G. (2001). 'The effects of angular orientation on flame spread over thin materials'. *Fire Safety Journal*, **36**, 291–312.

Quintiere, J.G. (2002). 'Fire behaviour in building compartments'. *Proceedings of the Combustion Institute*, **29**, 181–193.

Quintiere, J.G. (2006). *Fundamentals of Fire Phenomena*. John Wiley & Sons, Ltd, Chichester.

Quintiere, J.G. (2008). 'Compartment fire modelling', in *SFPE Handbook of Fire Protection Engineering*, 4th edn (eds P.J. Di Nenno *et al.*), pp. 3.195–3.203. National Fire Protection Association, Quincy, MA.

Quintiere, J.G. and Grove, B.S. (1998). 'Unified analysis for fire plumes'. *Proceedings of the Combustion Institute*, **27**, 2757–2766.

Quintiere, J., Harkleroad, M. and Hasemi, Y. (1986). 'Wall flames and implications for upward flame spread'. *Combustion Science and Technology*, **48**, 191–222.

Quintiere, J.G., McCaffrey, B.J. and Kashiwagi, T. (1978). 'A scaling study of a corridor subjected to a room fire'. *Combustion Science and Technology*, **18**, 1–19.

Quintiere, J.G., Rinkinen, W.J. and Jones, W.W. (1981). 'The effect of room openings on fire plume entrainment'. *Combustion Science and Technology*, **26**, 193–201.

Rae, D., Singh, B. and Damson, R. (1964). *Safety in Mines Research Report No. 224*. HMSO, London.

Raj, P.P.K., Moussa, A.N. and Aravamudan, K. (1979). 'Experiments involving pool and vapour fires from spills of liquefied natural gas on water'. *US Coast Guard Report No. CG-D-55–79*.

Rasbash, D.J. (1951). 'The efficiency of hand lamps in smoke'. *Institute of Fire Engineers Quarterly*, **11**, 46–52.

Rasbash, D.J. (1962). 'The extinction of fires by water spray'. *Fire Research Abstracts and Reviews*, **4**, 28–53.

Rasbash, D.J. (1967). 'Smoke and toxic products produced at fires'. *Transactions and Journal of the Plastics Institute, Conference Supplement No. 2*, pp. 55–62 (January 1967).

Rasbash, D.J. (1969). 'The relief of gas and vapour explosions in domestic structures'. *The Structural Engineer*, October.

Rasbash, D.J. (1974). Personal communication.

Rasbash, D.J. (1975). 'Relevance of firepoint theory to the assessment of fire behaviour of combustible materials'. *International Symposium on Fire Safety of Combustible Materials*, Edinburgh University, pp. 169–178.

Rasbash, D.J. (1976). 'Theory in the evaluation of fire properties of combustible materials'. *Proceedings of the 5th International Fire Protection Seminar (Karlsruhe, September)*, pp. 113–130.

Rasbash, D.J. (ed.) (1984). 'Special Issue: the Stardust Club Fire, Dublin, 1991'. *Fire Safety Journal*, **7**, 205–295.

Rasbash, D.J. (1985). 'The extinction of fire with plain water: a review.' *Fire Safety Science*, **1**, 1145–1163.

Rasbash, D.J. (1986). 'Quantification of explosion parameters for combustible fuel–air mixtures'. *Fire Safety Journal*, **11**, 113–125.

Rasbash, D.J. (1991). 'Major fire disasters involving flashover'. *Fire Safety Journal*, **17**, 85–93.

Rasbash, D.J. and Drysdale, D.D. (1982). 'Fundamentals of smoke production'. *Fire Safety Journal*, **5**, 77–86.

Rasbash, D.J. and Drysdale, D.D. (1983). 'Theory of fire and fire processes'. *Fire and Materials*, **7**, 79–88.

Rasbash, D.J. and Phillips, R.P. (1978). 'Quantification of smoke produced at fires. Test methods for smoke and methods of expressing smoke evolution'. *Fire and Materials*, **2**, 102–109.

Rasbash, D.J. and Pratt, B.T. (1980). 'Estimation of the smoke produced in fires'. *Fire Safety Journal*, **2**, 23–37.

Rasbash, D.J. and Rogowski, Z.W. (1960). 'Relief of explosions in duct systems'. *1st Symposium on Chemical Process Hazards*, pp. 58–65. Institution of Chemical Engineers, London.

Rasbash, D.J. and Rogowski, Z.W. (1962). 'Venting gaseous explosions in duct systems. IV. The effect of obstructions'. *Fire Research Note No. 490*.

Rasbash, D.J., Drysdale, D.D. and Deepak, D. (1986). 'Critical heat and mass transfer at pilot ignition and extinction of a material'. *Fire Safety Journal*, **10**, 1–10.

Rasbash, D.J., Palmer, K.N., Rogowski, Z.W. and Ames, S.A. (1970). 'Gas explosions in multiple compartments', *Fire Research Note No. 847*.

Rasbash, D.J., Rogowski, Z.W. and Stark, G.W.V. (1956). 'Properties of fires of liquids'. *Fuel*, **31**, 94–107.

Rath, J., Wolfinger, M.G., Steiner, G., Krammer, G., Barontini, F. and Cozzani, V. (2003). 'Heat of wood pyrolysis'. *Fuel*, **82**, 81–91.

Rein, G. (2009). 'Smouldering combustion phenomena in science and technology'. *International Review of Chemical Engineering*, **1**, 3–18.

Rein, G., Cohen, S. and Simeoni, A. (2009). 'Carbon emissions from smouldering peat in shallow and strong fronts'. *Proceedings of the Combustion Institute*, **32**, 2489–2496.

Rein, G., Fernandez-Pello, A.C. and Urban, D.L. (2007). 'Computational model of forward and opposed smouldering in microgravity.' *Proceedings of the Combustion Institute*, **31**, 2677–2684.

Rein, G., Torero, J.L., Jahn, W., Stern-Gottfried, J., Ryder, N.L., Desanghere, S. *et al.* (2009) 'Round-robin study of *a priori* modelling predictions of the Dalmarnock Fire Test One'. *Fire Safety Journal*, **44**, 590–602.

Rich, D., Lautenberger, C., Torero, J.L., Quintiere, J.G and Fernandez-Pello, C. (2007). 'Mass flux of combustible solids at piloted ignition'. *Proceedings of the Combustion Institute*, **31**, 2653–2660.

Richter, H. and Howard, J.B. (2000). 'Formation of polycyclic aromatic hydrocarbons and their growth to soot – a review of chemical reaction pathways'. *Progress in Energy and Combustion Science*, **26**, 565–608.

Ricou, F.P. and Spalding, B.P. (1961). 'Measurements of entrainment by axi-symmetric jets'. *Journal Fluid Mechanics*, **11**, 21–32.

Roberts, A.F. (1964a). 'Calorific values of partially decomposed wood samples'. *Combustion and Flame*, **8**, 245–246.

Roberts, A.F. (1964b). 'Ultimate analysis of partially decomposed wood samples'. *Combustion and Flame*, **8**, 345–346.

Roberts, A.F. (1970). 'A review of kinetic data for the pyrolysis of wood and related substances'. *Combustion and Flame*, **14**, 261–272.

Roberts, A.F. (1971a). 'Problems associated with the theoretical analysis of the burning of wood'. *Proceedings of the Combustion Institute*, **13**, 893–903.

Roberts, A.F. (1971b). 'The heat of reaction during the pyrolysis of wood'. *Combustion and Flame*, **17**, 79–86.

Roberts, A.F. (1981/82). 'Thermal radiation hazards from releases of LPG from pressurised storage'. *Fire Safety Journal*, **4**, 197–212.

Roberts, A.F. and Quince, B.W. (1973). 'A limiting condition for the burning of flammable liquids'. *Combustion and Flame*, **20**, 245–251.

Roberts, P., Smith, D.B. and Wood, D.R. (1980). 'Flammability of paraffin hydrocarbons in confined and unconfined conditions'. *Chemical Process Hazards VII, I. Chem. E. Symposium Series* No. 58, pp. 157–169. Institution of Chemical Engineers, Rugby.

Robertson, A.F. (1975). 'Estimating smoke production during building fires'. *Fire Technology*, **11**, 80–94.

Robertson, R.B. (1976). 'Spacing in chemical plant design against loss by fire'. *Symposium on Process Industry Hazards – Accidental Release: Assessment, Containment and Control*. Institution of Chemical Engineers Symposium Series No. 47, pp. 157–173.

Rockett, J.A. (1976). 'Fire induced gas flow in an enclosure'. *Combustion Science and Technology*, **12**, 165–175.

Rockett, J.A. and Milke, J.A. (2008). 'Conduction of heat in solids', in *SFPE Handbook of Fire Protection Engineering*, 4th edn (eds P.J. Di Nenno *et al.*), pp. 1.27–1.43. National Fire Protection Association, Quincy, MA.

Rogers, F.E. and Ohlemiller, T.J. (1978). 'Minimising smoulder tendency in flexible polyurethanes'. *Journal of Consumer Product Flammability*, **5**, 59–67.

Rohsenow, W.M. and Choi, H.Y. (1961). *Heat, Mass and Momentum Transfer*. Prentice-Hall, London.

Ross, H.D. (1994). 'Ignition of and flame spread over laboratory-scale pools of pure liquid fuels'. *Progress in Energy and Combustion Science*, **20**, 17–63.

Ross, H.D. and Miller, F.J. (1999). 'Understanding flame spread over alcohol pools'. *Fire Safety Science*, **6**, 77–94.

Rothbaum, H.P. (1963). 'Spontaneous combustion of hay'. *Journal of Applied Chemistry (London)*, **13**, 291–302.

Routley, A.F. and Skipper, R.S. (1980). 'A new approach to testing materials in the NBS Smoke Chamber'. *Fire and Materials*, **4**, 98–103.

Royal, J.H. (1970). 'The influence of fuel bed thickness on flame spreading rate'. Honours Report, Stevens Institute of Technology.

Rushbrook, F. (1985). *Fire Aboard: the Problems of Prevention and Control in Ships, Port Installations and Offshore Structures*. Brown, Son and Ferguson, Glasgow.

Saito, K., Quintiere, J.G. and Williams, F.A. (1985). 'Upward turbulent flame spread'. *Fire Safety Science*, **1**, 75–86.

Saito, K., Williams, F.A., Wichman, I.S. and Quintiere, J.G. (1989). 'Upward turbulent flame spread on wood under external radiation'. *Transactions of the ASME*, **111**, 438–445.

Salig, R.J. (1981). 'The smouldering behaviour of polyurethane cushioning and its relevance to home furnishing fires'. M.Sc. Dissertation, Massachusetts Institute of Technology.

Sapko, M., Furno, A. and Kuchta, J. (1976). US Bureau of Mines, RI 8176.

Sato, K. and Sega, S. (1988). 'Smoulder spread in a horizontal layer of cellulosic powder'. *Fire Safety Science*, **2**, 87–96.

Sato, T. and Kunimoto, T. (1969). *Memoirs of the Faculty of Engineering, Kyoto University*, **31**, 47.

Sawyer, R.F. and Fristrom, R.M. (1971). 'Flame inhibition and chemistry'. AGARD Conference No. 84, AGARD CP 84–71.

Schleich, J.-B. and Cajot, L.G. (2001) 'Natural fire safety for buildings', *Interflam'01: Proceedings of the 9th International Fire Conference*. Interscience Communications, London, pp. 359–366.

Schnell, L.G. (1996). 'Flashover training in Sweden'. *Fire Engineers Journal*, **56**, 25–28.

Schult, D.A., Matkowski, B.J., Volpert, V.A. and Fernandez-Pello, A.C. (1995). 'Propagation and extinction of forced opposed flow smoulder waves'. *Combustion and Flame*, **101**, 471–490.

Seader, J.D. and Chien, W.P. (1974). 'Mass optical density as a correlating parameter for the NBS Smoke Chamber'. *Journal of Fire and Flammability*, **5**, 151–163.

Seader, J.D. and Chien, W.P. (1975). 'Physical aspects of smoke development in an NBS smoke density chamber'. *Journal of Fire and Flammability*, **6**, 294–310.

Seigel, L.G. (1969). 'The projection of flames from burning buildings'. *Fire Technology*, **5**, 43–51.

Semenov, N.N. (1928). 'Theories of combustion processes'. *Z. Phys. Chem.*, **48**, 571–582.

Setchkin, N.P. (1954). 'Self-ignition temperatures of combustible liquids'. *Journal of Research, National Bureau of Standards*, **53**, 49–66.

Shaw, A. and Weckman, E. (2010). 'Evaluation of the ignition of diesel fuels on hot surfaces.' *Fire Technology*, **46**, 407–423.

Sherratt, J. and Drysdale, D.D. (2001) 'The effect of the melt-flow process on the fire behaviour of thermoplastics', *Interflam'01: Proceedings of the 9th International Fire Conference*. Interscience Communications, London, pp. 149–159.

Shields, T.J., Silcock, G.W. and Murray, J. (1993). 'The effects of geometry and ignition mode on ignition times obtained using a cone calorimeter and ISO ignitability apparatus'. *Fire and Materials*, **17**, 25–32.

Shipp, M. (1983). 'A hydrocarbon fire standard: an assessment of existing information'. Department of Energy, Offshore Energy Technology Board, TO/R/8294.

Shokri, M. and Beyler, C.L. (1989). 'Radiation from large pool fires'. *Journal of Fire Protection Engineering*, **1**, 141–150.

Silcock, A. and Hinkley, P.L. (1974). 'Report on the spread of fire at Summerland in Douglas, Isle of Man, 2 August 1973', Building Research Establishment, BRE CP 74/74.

Simcox, S., Wilkes, N.S. and Jones, I.P. (1992). 'Computer simulation of the flows of hot gases from the fire at King's Cross Underground Station'. *Fire Safety Journal*, **18**, 49–73.

Simmons, R.F. (1995). 'Fire chemistry', in *Combustion Fundamentals of Fire* (ed. G. Cox), pp. 405–473. Academic Press, London.

Simmons, R.F. (2008). 'Premixed burning', in *SFPE Handbook of Fire Protection Engineering*, 4th edn (eds P.J. Di Nenno *et al.*), pp. 1.1561–1.166. National Fire Protection Association, Quincy, MA.

Simmons, R.F. and Wolfhard, H.G. (1955). 'The influence of methyl bromide on flames – 1. Premixed flames'. *Transactions of the Faraday Society*, **51**, 1211–1217.

Simms, D.L. (1963). 'On the pilot ignition of wood by radiation'. *Combustion and Flame*, **7**, 253–261.

Simms, D.L. and Hird, D. (1958). 'On the pilot ignition of materials by radiation'. *Fire Research Note No. 365*.

Sirignano, W.A. and Glassman, I. (1970). 'Flame spreading above liquid fuels: surface tension-driven flows'. *Combustion Science and Technology*, **1**, 307–312.

Sivathanu, Y.R. and Gore, J.P. (1992). 'Transient structure and radiation properties of strongly radiative buoyant flames'. *Journal of Heat Transfer*, **114**, 659–665.

Smith, D.A. (1992). 'Measurements of flame length and flame angle in an inclined trench'. *Fire Safety Journal*, **18**, 231–244.

Smith, D.A. and Cox, G. (1992). 'Major chemical species in buoyant turbulent diffusion flames'. *Combustion and Flame*, **91**, 226–238.

Smith, D. and Shaw, K. (1999). 'The single burning item (SBI) test: the Euroclasses and transitional arrangements', *Interflam'99: Proceedings of the 8th International Fire Conference*. Interscience Communications, London, pp. 1–9.

Smith, E.E. (1972). 'Heat release rate of building materials', in *Ignition, Heat Release and Noncombustibility of Materials* (ed. A.F. Robertson). American Society for Testing and Materials, STP 502, pp. 119–134.

Sorenson, S.C., Savage, L.D. and Strehlow, R.A. (1975). 'Flammability limits – a new technique'. *Combustion and Flame*, **24**, 347–355.

Spalding, D.B. (1955). *Some Fundamentals of Combustion*. Butterworths, London.

Spalding, D.B. (1957). 'A theory of inflammability limits and flame quenching'. *Proceedings of the Royal Society*, **A240**, 83–100.

Spearpoint, M.J. (1999). 'Predicting the ignition and burning rate of wood in the cone calorimeter using an integral model', Master's Thesis, Department of Fire Protection Engineering, University of Maryland. (See also NIST GCR 99-775.)

Stanzak, W.W. and Lie, T.T. (1973). 'Fire tests on protected steel columns with different cross sections'. National Research Council of Canada, Ottawa.

Stark, G.W.V. (1972). 'Liquid spillage fires', *4th Symposium on Chemical Process Hazards with Special Reference to Plant Design*. Institution of Chemical Engineers, NW Branch, Manchester, pp. 71–78.

Stec, A. (2010). 'Experimental methods in combustion toxicology', in *Fire Toxicity* (eds A. Stec and R. Hull). Woodhead Publishing Ltd., Oxford, pp. 217–228.

Stec, A. and Hull, R. (eds) (2010). *Fire Toxicity*. Woodhead Publishing Ltd., Oxford.

Stec, A., Hull, R., Purser, J.A. and Purser, D.A. (2009). 'Comparison of toxic product yields from bench-scale to ISO room'. *Fire Safety Journal*, **44**, 62–70.

Steckler, K.D., Baum, H.R. and Quintiere, J.G. (1984). 'Fire induced flows through room openings – flow coefficients'. *Proceedings of the Combustion Institute*, **20**, 1591–1600.

Steckler, K.D., Quintiere, J.G. and Rinkinen, W.J. (1982). 'Flow induced by a fire in a compartment'. *Proceedings of the Combustion Institute*, **19**, 913–920.

Steel Construction Institute (2009). *Buncefield Explosion Mechanism Phase 1*. HSE Research Report RR718. (Available on-line at <http://news.hse.gov.uk/2009/06/25/rr718-buncefield-explosion-mechanism-phase-1-volumes-1-and-2/>.)

Steinhaus, T., Welch, S., Carvel, R. and Torero, J.L. (2007). 'Large scale pool fires'. *Thermal Sciences Journal*, **11**, 101–117.

Sterner, E.S. and Wickstrom, U. (1990). 'TASEF – Temperature Analysis of Structures Exposed to Fire (Users Manual)'. Report 1990:05, Swedish National Testing Institute, Borås, Sweden.

Stern-Gottfried, J., Rein, G., Bisby, L.A. and Torero, J.L. (2010). 'Experimental review of the homogeneous temperature assumption in post-flashover compartment fires'. *Fire Safety Journal*, **45**, 249–261.

Stern-Gottfried, J., Rein, G. and Torero, J.L. (2009). 'Travel guide'. *Fire Risk Management*, November, pp. 12–16.

Stevens, M.P. (1999). *Polymer Chemistry: an Introduction*. Oxford University Press, New York.

Steward, F.R. (1970). 'Prediction of the height of turbulent diffusion buoyant flames'. *Combustion Science and Technology*, **2**, 203–212.

Steward, F.R. (1974a). 'Radiative heat transfer associated with fire problems', in *Heat Transfer in Fires* (ed. P.L. Blackshear), pp. 273–314. John Wiley & Sons, Inc., New York.

Steward, F.R. (1974b). 'Ignition characteristics of cellulosic materials', in *Heat Transfer in Fires* (ed. P.L. Blackshear), pp. 379–407. John Wiley & Sons, Inc., New York.

Strehlow, R.A. (1973). 'Unconfined vapour cloud explosions – an overview'. *Proceedings of the Combustion Institute*, **14**, 1189–1200.

Stull, D.R. (1971). 'Chemical thermodynamics and fire problems'. *Fire Research Abstracts and Reviews*, **13**, 161–186.

Stull, D.R. (1977). 'Fundamentals of fire and explosion'. *AIChE Monograph Series* No. 10, Volume 73. American Institute of Chemical Engineers, New York.

Sugawa, O., Momita, D. and Takahashi, W. (1997). 'Flow behaviour of ejected fire flame/plume effected by external side wind'. *Fire Safety Science*, **5**, 249–260.

Sultan, M.A. (2004). 'A comparison of heat exposure in fire resistance test furnaces controlled by plate thermometers and by shielded thermocouples', *Interflam'04: Proceedings of the Tenth International Fire Conference*, pp. 219–229. Interscience Communications, London.

Sundström, B. (1984). 'Room fire test in full-scale for surface products'. Report SP-RAPP 1984: 16, Swedish National Testing Institute, Borås, Sweden.

Sundström, B. (ed.) (1995). 'Fire Safety of Upholstered Furniture – the Final Report on the CBUF Research Programme', European Commission Measurement and Testing Report EUR 16477 EN.

Sutton, O.G. (1953). *Micrometeorology*. McGraw-Hill, New York.

Suzuki, M., Dobashi, R. and Hirano, T. (1994). 'Behaviour of fire spreading downward over thick paper'. *Proceedings of the Combustion Institute*, **25**, 1439–1446.

Suzuki, T. and Hirano, T. (1982). 'Flame propagation across a liquid fuel in an air stream'. *Proceedings of the Combustion Institute*, **19**, 877–884.

Swann, J., Hartman, J. and Beyler, C.L. (2008). 'Study of radiant smouldering ignition of plywood subjected to prolonged heating using the cone calorimeter, TGA and DSC'. *Proceedings of the 9th IAFSS Symposium*, pp. 155–166.

Takahashi, W., Tanaka, H., Sugawa, O. and Ohtake, M. (1997). 'Flame and plume behaviour in and near a corner of walls'. *Fire Safety Science*, **5**, 261–271.

Takeda, H. and Akita, K. (1981). 'Critical phenomena in compartment fires with liquid fuels'. *Proceedings of the Combustion Institute*, **18**, 519–527.

Takeno, K. and Hirano, T. (1988). 'Behaviour of combustible liquids soaked in porous beds during flame spread'. *Proceedings of the Combustion Institute*, **22**, 1223–1230.

Tan, S.H. (1967). 'Flare system design simplified'. *Hydrocarbon Processing*, **46**, 172–176.

Terai, T. and Nitta, K. (1975). *Proceedings Symposia of the Architectural Institute, Japan*. (See Zukoski *et al.*, 1981a.)

Tewarson, A. (1972). 'Some observations on experimental fires in enclosures. Part II. Ethyl alcohol and paraffin oil'. *Combustion and Flame*, **19**, 363–371.

Tewarson, A. (1980). 'Heat release in fires'. *Fire and Materials*, **4**, 185–191.

Tewarson, A. (1982). 'Experimental evaluation of flammability parameters of polymeric materials', in *Flame Retardant Polymeric Materials*, Volume 3 (eds M. Lewin, S.M. Atlas and E.M. Pearce), pp. 97–153. Plenum Press, New York.

Tewarson, A. (1984). 'Fully developed enclosure fires of wood cribs'. *Proceedings of the Combustion Institute*, **20**, 1555.

Tewarson, A. (1996). Personal communication (see Grant and Drysdale (1997)).

Tewarson, A. (2008). 'Generation of heat and gaseous, liquid and solid products in fires', in *SFPE Handbook of Fire Protection Engineering*, 4th edn (eds P.J. Di Nenno *et al.*), pp. 3.109–3.194. National Fire Protection Association, Quincy, MA.

Tewarson, A. and Ogden, S.D. (1992). 'Fire behaviour of PMMA'. *Combustion and Flame*, **89**, 237–259.

Tewarson, A. and Pion, R.F. (1976). 'Flammability of plastics. I. Burning intensity'. *Combustion and Flame*, **26**, 85–103.

Tewarson, A. and Pion, R.F. (1978). Factory Mutual Research Serial UI 1AGR1. RC.

Tewarson, A., Lee, J.L. and Pion, R.F. (1981). 'The influence of oxygen concentration on fuel parameters for fire modelling'. *Proceedings of the Combustion Institute*, **18**, 563–570.

Theobald, C.R. (1968). 'The critical distance for ignition from some items of furniture'. *Fire Research Note No. 736*.

Thomas, I.R. and Bennetts, I.D. (1999). 'Fires in enclosures with single ventilation openings – comparison of long and wide enclosures'. *Fire Safety Science*, **6**, 941–953.

Thomas, I.R., Moinuddin, K. and Bennetts, I.D. (2005). 'Fire development in a deep enclosure'. *Fire Safety Science*, **8**, 1277–1288.

Thomas, P.H. (1958). 'On the thermal conduction equation for self-heating materials with surface cooling'. *Transactions of the Faraday Society*, **54**, 60–65.

Thomas, P.H. (1960). 'Some approximations in the theory of self-heating and thermal explosion'. *Transactions of the Faraday Society*, **56**, 833–839.

Thomas, P.H. (1965). 'Fire spread in wooden cribs. Part 3. The effect of wind'. *Fire Research Note No. 600*.

Thomas, P.H. (1972a). 'Self heating and thermal ignition: a guide to its theory and applications', in *Ignition, Heat Release and Non-Combustibility* (ed. A.F. Robertson), ASTM STP 502, pp. 56–82. American Society for Testing and Materials, Philadelphia, PA.

Thomas, P.H. (1972b). 'Behaviour of fires in enclosures – some recent progress'. *Proceedings of the Combustion Institute*, **14**, 1007–1020.

Thomas, P.H. (1973). 'An approximate theory of 'hot spot' criticality'. *Combustion and Flame*, **21**, 99–109.

Thomas, P.H. (1974a). 'Effects of fuel geometry in fires'. Building Research Establishment CP 29/74.

Thomas, P.H. (1974b). 'Fires in enclosures', in *Heat Transfer in Fires* (ed. P.L. Blackshear), pp. 73–94. John Wiley & Sons, Inc., New York.

Thomas, P.H. (1975a). 'Old and new looks at compartment fires'. *Fire Technology*, **11**, 42–47.

Thomas, P.H. (1975b). 'Factors affecting ignition of combustible materials and their behaviour in fire', *International Symposium on Fire Safety of Combustible Materials*, Edinburgh University, pp. 84–99.

Thomas, P.H. (1981). 'Testing products and materials for their contribution to flashover in rooms'. *Fire and Materials*, **5**, 103–111.

Thomas, P.H. (1982). 'Modelling of compartment fires'. *6th International Fire Protection Seminar, Karlsruhe*, pp. 29–46. Vereinigung zur Forderung des Deutschen Brandschutzes, eV.

Thomas, P.H. and Bowes, P.C. (1961a). 'Some aspects of the self-heating and ignition of solid cellulosic materials'. *British Journal of Applied Physics*, **12**, 222–229.

Thomas, P.H. and Bowes, P.C. (1961b). 'Thermal ignition in a slab with one face at a constant high temperature'. *Transactions of the Faraday Society*, **57**, 2007–2016.

Thomas, P.H. and Bullen, M.L. (1979). 'On the role of $k\rho c$ of room lining materials in the growth of room fires'. *Fire and Materials*, **3**, 68–73.

Thomas, P.H. and Heselden, A.J.M. (1972). 'Fully developed fires in single compartments. A cooperative research programme of the Conseil Internationale du Bâtiment'. Conseil Internationale du Bâtiment Report No. 20, *Fire Research Note No. 923*.

Thomas, P.H. and Hinkley, P.L. (1964). 'Design of roof venting systems for single storey buildings'. *Fire Research Technical Paper No. 10*. HMSO, London.

Thomas, P.H. and Law, M. (1972). 'The projection of flames from burning buildings'. *Fire Research Note No. 921*.

Thomas, P.H. and Law, M. (1974). 'The projection of flames from buildings on fire'. *Fire Prevention Science and Technology No. 10*. Fire Protection Association, London.

Thomas, P.H. and Nilsson, L. (1973). 'Fully developed compartment fires: new correlations of burning rates'. *Fire Research Note No. 979*.

Thomas, P.H. and Webster, C.T. (1960). 'Some experiments on the burning of fabrics and the height of buoyant diffusion flames'. *Fire Research Note No. 420*.

Thomas, P.H., Baldwin, R. and Heselden, A.J.M. (1965a). 'Buoyant diffusion flames: some measurements of air entrainment, heat transfer and flame merging'. *Proceedings of the Combustion Institute*, **10**, 983–996.

Thomas, P.H., Baldwin, R. and Theobald, C.R. (1968). 'Some model scale experiments with multiple fires'. *Fire Research Note No. 700*.

Thomas, P.H., Bullen, M.L., Quintiere, J.G. and McCaffrey, B.J. (1980). 'Flashover and instabilities in fire behaviour'. *Combustion and Flame*, **38**, 159–171.

Thomas, P.H., Heselden, A.J.M. and Law, M. (1967a). 'Fully developed compartment fires: two kinds of behaviour'. *Fire Research Technical Paper No. 18*. HMSO, London.

Thomas, P.H., Hinkley, P.L., Theobald, C.R. and Simms, D.L. (1963). 'Investigations into the flow of hot gases in roof venting'. *Fire Research Technical Paper No. 7*. HMSO, London.

Thomas, P.H., Morgan, H.P. and Marshall, N. (1998). 'The spill plume in smoke control design'. *Fire Safety Journal*, **30**, 21–46.

Thomas, P.H., Simms, D.L. and Law, M. (1967b). 'The rate of burning of wood'. *Fire Research Note No. 657*.

Thomas, P.H., Simms, D.L. and Wraight, H. (1964). 'Fire spread in wooden cribs. Part 1'. *Fire Research Note No. 537*.

Thomas, P.H., Simms, D.L. and Wraight, H. (1965b). 'Fire spread in wooden cribs. Part 2. Heat transfer experiments in still air'. *Fire Research Note No. 799*.

Thomas, P.H., Webster, C.T. and Raftery, M.M. (1961). 'Some experiments on buoyant diffusion flames'. *Combustion and Flame*, **5**, 359–367.

Thomson, H.E. and Drysdale, D.D. (1987). 'Flammability of polymers I. Ignition temperatures'. *Fire and Materials*, **11**, 163–172.

Thomson, H.E. and Drysdale, D.D. (1988). 'Critical mass flowrate at the firepoint of plastics'. *Fire Safety Science*, **2**, 67–76.

Thomson, H.E., Drysdale, D.D. and Beyler, C.L. (1988). 'An experimental evaluation of critical surface temperature as a criterion for piloted ignition of solid fuels'. *Fire Safety Journal*, **13**, 185–196.

Thorne, P.F. (1976). 'Flashpoints of mixtures of flammable and non-flammable liquids'. *Fire and Materials*, **1**, 134–140.

Thyer, A.M. (2003). 'A review of data on spreading and vaporization of cryogenic liquid spills'. *Journal of Hazardous Materials*, **A99**, 31–40.

Tien, C.L., Lee, K.Y. and Stretton, A.J. (2008). 'Radiation heat transfer', in *SFPE Handbook of Fire Protection Engineering*, 4th edn (eds P.J. Di Nenno *et al*.). pp. 1.74–1.90 National Fire Protection Association, Quincy, MA.

Toal, B.R., Silcock, G.W.H. and Shields, T.J. (1989). 'An examination of piloted ignition characteristics of cellulosic materials using the ISO ignitability test'. *Fire and Materials*, **14**, 97–106.

Torero, J.L. (2008) 'Flaming ignition of solid fuels', in *SFPE Handbook of Fire Protection Engineering*, 4th edn (eds P.J. Di Nenno *et al*.), pp. 2.260–2.277. National Fire Protection Association, Quincy, MA.

Torero, J.L. and Fernandez-Pello, A.C. (1995). 'Natural convection smoulder of polyurethane foam: upward convection'. *Fire Safety Journal*, **24**, 35–52.

Torero, J.L. and Fernandez-Pello, A.C. (1996). 'Forward smoulder of polyurethane foam in a forced air flow'. *Combustion and Flame*, **106**, 89–109.

Torero, J.L., Fernandez-Pello, A.C. and Kitano, M. (1994). 'Downward smoulder of polyurethane foam'. *Fire Safety Science*, **4**, 409–420.

Tritton, D.J. (1988). *Physical Fluid Dynamics*. Oxford Science Publishers, Oxford.

Tsai, K.-C. and Drysdale, D.D. (2002). 'Flame height correlation and upward flame spread modelling'. *Fire and Materials*, **26**, 279–287.

Tse, S.D., Fernandez-Pello, A.C. and Miyasaka, K. (1996). 'Controlling mechanisms in the transition from smouldering to flaming of flexible polyurethane foam'. *Proceedings of the Combustion Institute*, **26**, 1505–1513.

Tseng, L.-K., Ismail, M.A. and Faeth, G.M. (1993). 'Laminar burning velocities and Markstein numbers of hydrocarbon/air flames'. *Combustion and Flame*, **95**, 410–426.

Tu, K.-M. and Quintiere, J.G. (1991). 'Wall flame heights with external radiation'. *Fire Technology*, **27**, 195–203.

Tyler, B.J. (2008). Personal communication.

Ural, E.A. and Garzia, H.W. (2008). 'Explosion prevention and protection', in *NFPA Handbook*, 20th edn (ed. A.E. Cote), Section 17, Chapter 8, pp. 17.141–17.165. NFPA, Quincy, MA.

US Department of Commerce (1971a). 'Standard for the flammability of children's sleepwear'. DOC FF 3–71. *Federal Register*, **36**, 14062–14073.

Vagelopoulos, C.M. and Egolfopoulos, F.N. (1998). 'Direct experimental determination of flame speeds'. *Proceedings of the Combustion Institute*, **27**, 513–519.

van de Leur, P.H.E. (1991). '*Tunnel fire simulations for the Ministry of Public Works*'. TNO Report B-91-0043.

Vandevelde, P., van Hees, P., Twilt, L., van Mierlo, R.J.M. and van de Leur, P.H.E. (1996). *Reaction to Fire of Construction Products. Area A: Test Methods*. CI/SfB (K4), European Commission Directorate General XII, Brussels.

Vervalin, C.H. (ed.) (1973). *Fire Protection Manual for Hydrocarbon Processing Plants*, 2nd edn. Gulf Publishers, Houston, TX.

Viegas, D.X. (2006). 'Parametric study of an eruptive fire behaviour model'. *International Journal of Wildland Fire*, **15**, 169–177.

Viegas, D.X. and Pita, L.P. (2004). 'Fire spread in canyons'. *International Journal of Wildland Fires*, **13**, 253–274.

Viennean, H. (1964). 'Mixing controlled flame heights from circular jets'. B.Sc. Thesis, Department of Chemical Engineering, University of New Brunswick, Fredericton, NB.

Vytenis, R.J. and Welker, J.R. (1975). 'End-grain ignition of wood'. *Journal of Fire and Flammability*, **6**, 355–361.

Wade, S.H. (1942). 'Evaporation of liquids in currents of air'. *Transactions of the Institution of Chemical Engineers*, **20**, 1–14.

Wagner, H. Gg. (1979). 'Soot formation in combustion'. *Proceedings of the Combustion Institute*, **17**, 3–19.

Wakatsuki, K., Jackson, G.S., Hamins, A. and Nyden, M.R. (2007). 'Effects of fuel absorption on radiative heat transfer in methanol pool fires'. *Proceedings of the Combustion Institute*, **31**, 2573–2580.

Walker, I.K. (1967). 'The role of water in the spontaneous heating of solids'. *Fire Research Abstracts and Reviews*, **9**, 5–22.

Wall, L.A. (1972). 'The pyrolysis of polymers', in *The Mechanisms of Pyrolysis, Oxidation and Burning of Organic Materials* (ed. L.A. Wall), NBS Special Publication 357, pp. 47–60. National Bureau of Standards, Washington, DC.

Walther, D.C., Anthenien, R.A. and Fernandez-Pello, A.C. (2000). 'Smoulder ignition of polyurethane foam: the effect of oxygen concentration'. *Fire Safety Journal*, **34**, 343–359.

Walton, W.D., Carpenter, D.J. and Wood, C.B. (2008). 'Zone computer fire models for enclosures', in *SFPE Handbook of Fire Protection Engineering*, 4th edn (eds P.J. Di Nenno *et al.*), pp. 3.222–3.228. National Fire Protection Association, Quincy, MA.

Walton, W.D. and Thomas, P.H. (2008). 'Estimating temperatures in compartment fires', in *SFPE Handbook of Fire Protection Engineering*, 4th edn (eds P.J. Di Nenno *et al.*), pp. 3.204–3.221. National Fire Protection Association, Quincy, MA.

Wang, H., Dlugogorski, B.Z. and Kennedy, E.M. (2006). 'Tests for spontaneous ignition of solid materials', in *Flammability Testing of Materials used in Construction, Transport and Mining* (ed. V.B. Apte), pp. 385–442. Woodhead Publishing Ltd., Cambridge.

Wang, Y.C. (ed.) (2002). *Steel and Composite Structures. Behaviour and Design for Fire Safety*. Spon Press, London.

Waterman, T.E. (1966). 'Determination of fire conditions supporting room flashover'. *Final Report IITRI Project M6131, DASA 1886*, Defense Atomic Support Agency, Washington, DC.

Waterman, T.E. (1968). 'Room flashover – criteria and synthesis'. *Fire Technology*, **4**, 25–31.

Weast, R.C. (ed.) (1974/75). *Handbook of Chemistry and Physics*, 54th edn. Chemical Rubber Company, Ohio.

Weatherford, W.D. and Sheppard, D.M. (1965). 'Basic studies of the mechanism of ignition of cellulosic materials'. *Proceedings of the Combustion Institute*, **10**, 897–910.

Webster, C.T. and Raftery, M.M. (1959). 'The burning of fires in rooms. Part II', *Fire Research Note* 401. JFRO, Borehamwood. (See also *Fire Research Notes 474 and 578*.)

Weckman, E.J. and Sobiesiak, A. (1988). 'The oscillatory behaviour of medium scale pool fires'. *Proceedings of the Combustion Institute*, **22**, 1299–1310.

Welker, J.R. and Sliepcevich, C.M. (1966). 'Bending of wind-blown flames from liquid pools'. *Fire Technology*, **2**, 127–135.

Wells, G. (1997). *Major Hazards and their Management*. Institution of Chemical Engineers, Rugby.

Welty, J.R., Wicks, C.E., Wilson, R.E. and Rorrer, G. (2008). *Fundamentals of Momentum, Heat and Mass Transfer*, 5th edn. John Wiley & Sons, Inc., New York.

Weng, W.G., Fan, W.C., Yang, L.Z., Song, H., Deng, Z.H., Qin, J. and Liao, G.X. (2003). 'Experimental study of backdraught in a compartment with openings of different geometries'. *Combustion and Flame*, **132**, 709–714.

Weng, W.G. and Hasemi, Y. (2008). 'A numerical model for flame spread along a combustible flat solid with charring materials with experimental validation of ceiling flame spread and upward flame spread'. *Fire and Materials*, **32**, 87–102.

White, A.G. (1925). 'Limits for the propagation of flame in inflammable gas/air mixtures. Part III. The effect of temperature on the limits'. *Journal of the Chemical Society*, **127**, 672–684.

White, D.A., Beyler, C.L., Fulper, C. and Leonard, J. (1997). 'Flame spread on aviation fuels'. *Fire Safety Journal*, **28**, 1–32.

White, D.A., Beyler, C.L., Scheffey, J.L and Williams, F.W. (1996). 'Modelling the impact of post-flashover shipboard fires on adjacent spaces'. *Interflam'96: Proceedings of the 7th International Fire Conference*. Interscience Communications, London, pp. 225–233.

Whiteley, R.H. (1994). 'Short communication: some comments concerning the measurement of smoke'. *Fire and Materials*, **18**, 57–59.

Wichman, I.S. (1992). 'Review of flame spread'. *Progress in Energy and Combustion Science*, **18**, 553.

Wickstrom, U. (1989). 'The plate thermometer – a simplified instrument for reaching harmonised fire resistance tests'. *SP-Report 1989:03*.

Wickstrom, U. (1994). 'The plate thermometer – a simplified instrument for reaching harmonised fire resistance tests'. *Fire Technology*, **30**, 195–208.

Wickstrom, U. (1997). 'The plate thermometer – practical aspects'. *SP-Report 1997:28*.

Wickstrom, U., Sundström, B. and Holmstedt, G. (1983). 'The development of a full-scale room fire test'. *Fire Safety Journal*, **5**, 191–197.

Williams, A. (1973). 'Combustion of droplets of liquid fuels: a review'. *Combustion and Flame*, **21**, 1–31.

Williams, F.A. (1974a). 'A unified view of fire suppression'. *Journal of Fire and Flammability*, **5**, 54–63.

Williams, F.A. (1974b). 'Chemical kinetics of pyrolysis', in *Heat Transfer in Fires* (ed. P.L. Blackshear), pp. 197–237. John Wiley & Sons, Inc., New York.

Williams, F.A. (1977). 'Mechanisms of fire spread'. *Proceedings of the Combustion Institute*, **16**, 1281–1294.

Williams, F.A. (1981). 'A review of flame extinction'. *Fire Safety Journal*, **3**, 163–175.

Williams, F.A. (1982). 'Urban and wildland fire phenomenology'. *Progress in Energy and Combustion Science*, **8**, 317–354.

Williams, F.A. (1988). *Combustion Theory*, 2nd edn. Addison-Wesley Publishing Co., New York.

Williams, F.A. (2000). 'Progress in knowledge of flamelet structure and extinction'. *Progress in Energy and Combustion Science*, **26**, 657–682.

Williamson, R.B., Revenaugh, A. and Mowrer, F.W. (1991). 'Ignition sources in room fire tests and some implications for flame spread evaluation'. *Fire Safety Science*, **3**, 657–666.

Wolfhard, H.G. and Simmons, R.F. (1955). 'Influence of methyl bromide on flames. Part I. Premixed flames'. *Transactions of the Faraday Society*, **51**, 1211–1217.

Wood, B.D., Blackshear, P.H. and Eckert, E.R.G. (1971). *Combustion Science and Technology*, **4**, 131.

Woodburn, P. and Drysdale, D.D. (1997). 'Fires in inclined trenches: the effects of trench and burner geometry on the critical angle'. *Fire Safety Science*, **5**, 225–236.

Woodburn, P. and Drysdale, D.D. (1998). 'Fires in inclined trenches; the dependence of critical angle on the trench and burner geometry'. *Fire Safety Science*, **31**, 143–164.

Woolley, W.D. (1971). 'Decomposition products of PVC for studies of fires'. *British Polymer Journal*, **3**, 186–193.

Woolley, W.D. (1972). 'Studies of the dehydrochlorination of PVC in nitrogen and air'. *Plastics and Polymers*, pp. 203–208.

Woolley, W.D. and Ames, S.A. (1975). 'The explosion risk of stored foam rubber'. Building Research Establishment Current Paper CP 36/75.

Woolley, W.D. and Fardell, P.J. (1982). 'Basic aspects of combustion toxicology'. *Fire Safety Journal*, **5**, 29–48.

Woolley, W.D., Ames, S.A. and Fardell, P.J. (1979). 'Chemical aspects of combustion toxicology of fires'. *Fire and Materials*, **3**, 110–120.

Woolley, W.D., Ames, S.A., Pitt, A.I. and Murrell, J.V. (1975). 'Fire behaviour of beds and bedding materials'. *Fire Research Note No. 1038*.

Woolley, W.D., Raftery, M.M., Ames, S.A. and Murrell, J.V. (1979/80). 'Smoke release from wall linings in full scale compartment fires'. *Fire Safety Journal*, **2**, 61–72.

Wotton, B.M., McAlpine, R.S. and Hobbs, M.W. (1999). 'The effect of fire front width on surface fire behavior'. *International Journal of Wildland Fire*, **9**, 247–253.

Wu, P. (1999). 'Parallel panel fire tests for flammability assessment', *Interflam'99: Proceedings of the 8th International Fire Conference*. Interscience Communications, London, pp. 605–614.

Yao, C. (1976). 'Development of large-drop sprinklers'. *FMRC Technical Report Serial No. 22476*. Factory Mutual Research Corporation.

Yao, C. (1980). 'Application of sprinkler technology'. Presented at Workshop on 'Engineering Applications of Fire Technology'. National Bureau of Standards.

Yao, C. (1997). 'Overview of sprinkler technology research'. *Fire Safety Science*, **5**, 93–110.

Yaws, C.L. (ed.) (1999). *Chemical Properties Handbook: Physical, Thermodynamic, Environmental, Transport, Safety and Health Related Properties for Organic and Inorganic Chemicals*. McGraw-Hill, New York.

Yeoh, G.H. and Yuen, K.K. (2009). *Computational Fluid Dynamics in Fire Engineering: Theory, Modelling and Practice*. Elsevier, Amsterdam.

Yih, C.S. (1952). 'Free convection due to a point source of heat', *Proceedings of the 1st US National Congress in Applied Mechanics*, pp. 941–947.

Yokoi, S. (1960). 'Study of the prevention of fire spread caused by hot upward currents'. Building Research Institute, Report No. 34, Tokyo.

You, H.Z. and Faeth, G.M. (1979). 'Ceiling heat transfer during fire plume and fire impingement'. *Fire and Materials*, **3**, 140–147.

You, H.Z. and Kung, H.C. (1985). 'Strong buoyant plumes of growing rack storage fires'. *Proceedings of the Combustion Institute*, **20**, 1547–1554.

Yuan, L.M. and Cox, G. (1996). 'An experimental study of some line fires'. *Fire Safety Journal*, **27**, 123–140.

Yule, A.J. and Moodie, K. (1992). 'A method of testing the flammability of sprays of hydraulic fluid'. *Fire Safety Journal*, **18**, 273–302.

Yumoto, T. (1971). 'An experimental study on radiation from oil tank fires'. *Combustion and Flame*, **17**, 108–110.

Zabetakis, M.G. (1965). 'Flammability characteristics of combustible gases and vapours'. US Bureau of Mines, *Bulletin 627*.

Zabetakis, M.G. and Burgess, D.S. (1961). 'Research on the hazards associated with the production and handling of liquid hydrogen', US Bureau of Mines RI5707, Pittsburgh, PA.

Zalosh, R.G. (2008). 'Explosion protection', in *SFPE Handbook of Fire Protection Engineering*, 4th edn (eds P.J. Di Nenno *et al.*), pp. 3.418–3.439. National Fire Protection Association, Quincy, MA.

Zhang, J., Shields, T.J. and Silcock, G.W.H. (1997). 'Effect of melting behaviour on upwards flame spread of thermoplastics'. *Fire and Materials*, **21**, 1–6.

Zhou, L. and Fernandez-Pello, A.C. (1990). 'Concurrent turbulent flame spread'. *Proceedings of the Combustion Institute*, **23**, 1709–1714.

Zhou, L. and Fernandez-Pello, A.C. (1993). 'Turbulent, concurrent ceiling flame spread: the effect of buoyancy'. *Combustion and Flame*, **92**, 45–59.

Zhou, Y., Yang, L., Dai, J., Wang, Y. and Deng, Z. (2010). 'Radiation attenuation characteristics of pyrolysis volatiles of solid fuels and their effect for radiant ignition model'. *Combustion and Flame*, **157**, 167–175.

Zukoski, E.E. (1975). 'Convective flows associated with room fires', Semi Annual Progress Report, National Science Foundation Grant No. GI 31892 X1, Institute of Technology, Pasadena, CA.

Zukoski, E.E. (1986). 'Fluid dynamic aspects of room fires'. *Fire Safety Science*, **1**, 1–30.

Zukoski, E.E. (1995). 'Properties of fire plumes', in *Combustion Fundamentals of Fire* (ed. G. Cox), pp. 101–219. Academic Press, London.

Zukoski, E.E., Kubota, T. and Cetegen, B. (1981a). 'Entrainment in fire plumes'. *Fire Safety Journal*, **3**, 107–121.

Zukoski, E.E., Kubota, T. and Cetegen, B. (1981b). 'Entrainment in the near field of a fire plume'. National Bureau of Standards, GCR-81–346, US Department of Commerce, Washington, DC.

Zukoski, E.E., Morehart, J.H., Kubota, T. and Toner, S.J. (1991). 'Species production and heat release rates in two-layered natural gas fires'. *Combustion and Flame*, **83**, 325–332.

Answers to Selected Problems

1.1 CO_2: 1.798 kg/m^3; C_3H_8: 1.798 kg/m^3; C_4H_{10}: 2.370 kg/m^3.

1.2 3.32 m^3.

1.3 (a) 3.26 mm Hg; (b) 28.41 mm Hg; (c) 61.38 mm Hg.

1.4 n-Hexane: 3.93 mm Hg; n-decane: 1.71 mm Hg.

1.5 iso-Octane: 2.46 mm Hg; n-dodecane: 0.19 mm Hg.

1.6 −103.8 kJ/mol.

1.7 −159.5 kJ/mol.

1.8 −1844 kJ/mol.

1.9 −2976 kJ/mol.

1.10 (a)−41.3 kJ/g pentane; (b)−2.89 kJ/g air.

1.11 12.37 g air/g propane; 12.1 g air/g pentane; 12.27 g air/g decane.

1.12 (a) 6308°C (dissociation would be dominant, preventing such high temperatures being achieved); (b) 2194°C (dissociation would be significant and keep the temperature below 2000°C); (c) 1373°C (dissociation relatively unimportant).

2.1 320 W/m^2 (to 3 d.p., 319.861 W/m^2).

2.2 Taking 319.861 W/m^2, the temperatures are 60.020°C and 59.983°C, respectively (effectively both at 60°C). Bi = 0.001 (definitely thermally thin). For the 50 mm thick barrier, the rate of heat transfer is 318.615 W/m^2, and the temperatures are 60.17°C and 59.83°C. The associated Bi = 0.009, also thermally thin. Conclusion? Because the thermal conductivity of steel is so high, substantial sections may be treated as 'thermally thin'. (See Chapter 10.)

2.3 1.71 kW (1.27 kW through the brick, 0.44 kW through the window).

2.4 (a) 0.34 kW through window; (b) 0.79 kW through brickwork; (c) 0.20 kW through brickwork.

2.5 239°C (it is likely that the brickwork would fail before a steady state is reached).

2.6 963°C (the fibre insulation board would ignite and burn long before a steady state is reached).

An Introduction to Fire Dynamics, Third Edition. Dougal Drysdale.
© 2011 John Wiley & Sons, Ltd. Published 2011 by John Wiley & Sons, Ltd.

2.7 The numerical solution gives 47.36°C, while the analytical solution is 46.81°C.

2.8 6 seconds.

2.9 202°C.

2.10 (a) 60°C; (b) 50°C; (c) 50°C (approximate values only, obtained by interpolating data from Figure 2.6(a) and 2.6(b)).

2.11 (a) 50°C; (b) 34°C; (c) 20°C (approximate values only). Compare these results with those from Problem 2.10.

2.12 The polyurethane foam and the fibre insulation board can be treated as semi-infinite solids.

2.13 (a) 54°C; (b) 115°C.

2.14 (a) 0.85 kW/m^2; (b) 1.1 kW/m^2; (c) 1.8 kW/m^2.

2.15 0.07.

2.16 13.1 kW/m^2.

2.17 10.7 kW/m^2.

2.18 The convective and radiative heat losses respectively are: at 200°C, 3.0 kW and 4.8 kW; at 400°C, 8.2 kW and 19.8 kW; at 600°C, 14.5 kW and 56.0 kW; and at 800°C, 21.7 kW and 127.8 kW.

2.19 649.5 K (taking $h = 11$ W/m^{2}K).

3.1 4.35%.

3.2 1614 K.

3.3 (a) 1.7%; (b) 3.4%.

3.4 (a) 43.5%; (b) 31%.

3.5 (a) 31.6% CO$_2$; (b) 19.9% CF$_3$Br. Much less CF$_3$Br is found to be sufficient because this species acts as a chemical suppressant.

3.6 At 200°C, $L = 1.8\%$ and $U = 10.7\%$: at 400°C, $L = 1.5\%$ and $U = 12.1\%$.

3.7 Between −26°C and 6.3°C (assuming limits unchanged from their values at 25°C).

3.8 Between 32 and 235 mm Hg (however, refer to Figure 3.8(a)).

3.9 The diagram will look very similar to Figure 3.11, but the intercepts on the vertical axis (BUTANE) should be 1.8% and 49%, and the envelope should intersect the line CA and 1.8% and 8.4%. The envelope should just touch the line CL, where L corresponds to 87% Nitrogen.

3.10 This question is set to ensure that the reader is satisfied that the empirical results in Section 3.5 are consistent with Equation (3.9).

4.1 19.9 m and 9.7 m.

4.2 (a) 654 kW (b) 1672 kW

4.3 (a) 0.612 (b) 1.566

4.4 Using Equation 40, (a) 2.06 m: (b) 3.46 m. Using Equation 41, (a) 2.02 m; (b) 3.41 m. The flame volumes are (a) 0.54 m^3, and (b) 1.39 m^3.

4.5 1.5 Hz predicted for both.

4.6 From Table 4.2, (a) 72°C; (b) 60°C. From Equation 4.35, (a) 71°C; (b) 60°C. If the virtual source is taken into account, then these become 70°C and 57°C respectively.

4.7 (a) 240°C; (b) 90°C and 64°C.

4.8 At 4 m, $u = 1.18$ m/s: at 8 m, $u = 0.797$ m/s. The response times are 359 s and 1069 s respectively. If the RTI is reduced to $25 \, m^{1/2}.s^{1/2}$, the response times are 87.5 s and 267 s respectively.

4.9 0.9 MW if fire directly below detector. Worst situation gives 2.2 MW. If fire close to one wall or in a corner expect minimum size to be 1.1 MW and 0.55 MW respectively.

4.10 (a) 372 kW; (b) 186 kW; and (c) 93 kW.

4.11 With 5 m centres, (a) 64 kW; (b) 32 kW; and (c) 16 kW. With 2.5 m centres, (a) 32 kW; (b) 16 kW; and (c) 8 kW.

4.12 (a) 73 kg/s; (b) 110 kg/s.

4.13 The values of ΔT are (a) 87 K and (b) 36 K.

4.14 The distance to the receiver is measured from the edge of the pool. Shokri and Beyler's formula gives: (a) 8.1 kW/m², and (b) 3.6 kW/m². As given, Equation 4.62 gives: (a) 15.1 kW/m², and (b) 6.7 kW/m². However, the radiation factor (ρ) will be much less than 0.3, according to Figure 4.36.

5.1 Using ΔH_c (n-hexane) $= 45$ kJ/g (see Table 1.13), rate of heat release $= 9.6$ MW.

5.2 Rate of heat release $= 1.4$ MW. Flame height ~2.1 m. For radiation calculation, need H_2O and CO_2 concentrations in the flame, to enable Hottell's method to be used.

5.3 Assuming that the oil is consumed in 120 minutes, average rate of burning $= 20.85$ kg/s (0.118 kg/m² · s); average rate of heat release $= 939$ MW.

5.4 12.5 kW/m².

5.5 The limiting regression rates (Table 5.1) give 0.0224 kg/m² · s and 0.083 kg/m² · s for methanol and heptane, respectively. (Table 5.2 gives 0.017 for methanol). The 'ideal' values of Tewarson are 0.032 kg/m² · s and 0.093 kg/m² · s respectively. The latter assume no heat losses, and are therefore slightly higher.

5.6 (a) Ethanol, 0.00253 kg/m² · s; (b) Hexane, 0.0427 kg/m² · s; (c) Benzene, 0.063 kg/m² · s.

5.7 Assuming that the liquids burn at their limiting values, the diameters of the trays are predicted to be 1.54 m and 0.459 m for alcohol and heptane respectively. According to Equation (5.11), the corresponding burning rates are too low. Arbitrarily doubling the diameters gives burning rates closer to the values required.

For ethanol, this gives 825 kW, and will require 18.5 kg of fuel (c. 23 litres). (A closer estimate could be obtained if required.)

For heptane, this gives 765 kW, and requires 10.2 kg of fuel (c. 14.6 litres). (Note: combustion will be less than 100% efficient, particularly for the heptane.)

5.8 From Equation (5.32), (a) 4 mm thick, $\dot{m}'' = 9$ g/m²s; (b) 2 mm thick, $\dot{m}'' < 0$. i.e. it cannot maintain this surface temperature. See Section 6.3.2.

6.1 13°C. Equation (3.3e) can be used to check the effect of temperature on the lower limit. The effect is not significant.

6.2 Closed cup flashpoint is less than 32°C.

6.3 56°C.

6.4 2.7% by volume.

6.5 Ignoring radiant heat losses, $t_i = 4.9$ s. Including radiant heat losses (calculated with $T_{fabric} = 200°C$ as an average) $t_i = 11.2$ s.

6.6 As in Problem 6.5 $t_i = 2.2$ s and $t_i = 4.2$ s.

6.7 These answers are approximate: (a) 270 s; (b) 100 s. They can be derived by extracting the relevant data from Table 2.2 (inverse error function) and using a spreadsheet.

6.8 The radiative loss (320°C = 593 K) is 7 kW/m². The depth of the heated layer can be approximated by $\sqrt{(\alpha t)}$, and the temperature gradient immediately below the surface estimated as $(320 - 20)/\sqrt{(\alpha t)}$. This gives the conductive losses as (a) 8.9 kW/m² and (b) 14.6 kW/m².

7.1 $V = 0.57$ mm/s (from Section 7.2.2)

7.2 Steady-state surface temperature is 116°C (ignoring re-radiation) so that $V = 0.13$ mm/s. (Equation 7.7) Re-radiation is significant and should be included in the calculation.

7.3 Steady-state surface temperature is 65.8°C (ignoring re-radiation). Thus, $V = 0.07$ mm/s.

7.4 From Figure 9.4, 2.75 m (no ceiling); 5.9 m (ceiling 950 mm above crib base); and >7 m (ceiling 660 mm above crib base).

8.1 The data plotted according to Equation (8.1) lie on a slight curve as Bi < 10 for the smaller cubes. A slight extrapolation gives $r_0 \approx 0.87$ m at 40°C for a 'slab'. The data are inadequate.

8.2 Long extrapolation required to give $r_0 \approx 3.4$ m at 40°C for a 'slab'. Spontaneous ignition of this material would be most unlikely.

9.1 Take air density at 500 K = 0.70 kg/m³ (Table 11.6), and $\Delta H_{ox} = 13$ kJ/g, to give $t = 646$ s. (The effect of subsequent cooling is ignored here.)

9.2 This involves the reverse procedure: 9.9% oxygen (by mass) remains.

9.3 (a) 1.8 MW (average kpc for plaster + brick taken for walls and ceiling); (b) 1.9 MW; (c) 1.2 MW (fibreboard thermally thick); (d) 2.8 MW.

10.1 23.2 minutes (120 s timestep), 23.5 (30 s timestep).

10.2 130°C at 30 minutes. 'Fire resistance' works out at 146 minutes.

10.3 550°C at $14\frac{1}{2}$ minutes. Maximum temperature 820°C.

10.4 76°C at 30 minutes.

10.5 45 minutes.

10.6 10.25 minutes (with a timestep of 30 s).

Author Index

Abbasi, S.A. 198, 475
Abbasi, T. 198, 475
Abdul-Rahim, F.F. 453, 488
Abecassis-Empis, C. 177, 514
Abrams, M.S. 36, 429, 475
Abu-Zaid, M. 248, 263, 477
Ahonen, A. 376, 475
Ahmed, T.A. 131, 484
Akita, K. 189, 209, 278–80, 282, 354,
        364, 366, 400, 475, 509, 518
Albini, F. 312, 325, 475
Alderson, S. 305, 475
Alexander, J.L. 238, 483
Al-Hassam, T. 273, 481
Alpert, R.L. 152, 157–9, 169, 171,
        175–6, 207–8, 292, 314, 374, 384,
        475–6, 509
Alvares, N.J. 300, 476
Alvear, D. 177, 514
American Gas Association. 164–5, 476
American Society for Testing and
        Materials 2, 23, 199–200, 235,
        238–9, 242, 244, 246–7, 249, 266,
        296, 301, 375–7, 387, 420, 442,
        447, 450–1, 453, 476–7
Ames, S.A. 116, 332, 362, 443, 513, 524
Anderson, M.K. 344–5, 477
Anderson, R.A. 442, 457, 477
Andrews, G.E. 107, 109, 116, 477, 510
Andrews, P.L. 313, 484
Anon, 332, 340, 362, 477
Anthenien, R.A. 345, 522

Anthony, E.J. 324, 477
Apte, V. 185, 477, 502
Aravamudan, K. 164, 512
Arnaldos, J. 164–5, 186, 189, 484, 497,
        507
Ashton, L.A. 422, 425, 439, 477, 484
Atkins, P. 18–20, 27, 32, 477
Atkinson, G.T. 287, 447, 477
Atreya, A. 36, 248, 260, 263, 267, 477
Avenyan, V.A. 198, 487

Babkin, V.S. 106, 110, 503
Babrauskas, V. 13, 23, 30, 159–160, 181,
        186, 193, 199, 220–1, 225, 233,
        237, 241, 256, 259, 262, 343,
        354–6, 366, 369–72, 374–6,
        382–4, 390–1, 393, 397, 406, 410,
        412, 416, 420, 432, 447–8, 454–5,
        458, 478, 510
Back, E.L. 325, 479
Back, G. 153–6, 250, 479
Backer, S. 292, 296, 507
Bains, M. 318, 329, 504
Bajpai, A.C. 70, 479
Baker, R.R. 335, 477, 479
Balbi, J.H. 312, 506
Baldwin, R. 131, 495, 520
Bamford, C.H. 263, 479, 487
Bankston, C.P. 443, 479
Baratov, A.N. 106, 110, 503
Bard, S. 78, 444, 510
Bar-Ilan, A. 343, 479, 511

An Introduction to Fire Dynamics, Third Edition. Dougal Drysdale.
© 2011 John Wiley & Sons, Ltd. Published 2011 by John Wiley & Sons, Ltd.

Barnard, J.A. 27, 92, 444, 492
Barontini, F. 218, 514
Baroudi, D. 377, 479
Bartknecht, W. 34, 106, 222, 272, 479
Bartlette, R.A. 313, 484
Batchelor, G.K. 35, 479
Baulch, D.L. 29, 479
Baum, H.R. 185, 393, 517
Beard, A.N. 177, 364, 438, 479–80, 496
Beaulieu, P.A. 256, 258, 269, 479
Beck, V.R. 177, 503
Bedford, G.K. 397, 484
Beever, P.F. 170, 317, 479
Bengtsson, L. 362, 479
Benjamin, I.A. 463, 465, 479
Bennetts, I.D. 419, 519
Bergman, T.L. 35–7, 57, 62, 123, 497
Best, R.L. 385, 462, 479
Beyler, C.L. 29, 69, 136, 150, 153–6, 163, 165–6, 168–9, 171, 250, 256, 259–60, 262–3, 270, 274, 281–3, 345, 347, 356–9, 366, 439, 454, 457–8, 479–80, 492, 516, 518, 521, 523
Bill, R.G. 170, 496
Billmeyer, F.W. 3–5, 480
Bisby, L.A. 419, 517
Bishop, S.R. 364, 480, 496
Blackshear, P.H. 145, 524
Blinov, V.I. 145–6, 182–3, 190, 480
Block, J.A. 146, 220, 480
Blomqvist, P. 457, 480
Bluhme, D.A. 256, 480
Board of Trade 323, 480
Boddington, T. 231, 480
Bohm, B. 401, 421, 480
Boonmee, N. 257, 262, 269, 270, 480
Booth, K. 83, 88, 115, 496
Botha, J.P. 98, 107–9, 480
Boucher, B. 94, 481
Boudouard, O. 90, 503
Bouhafid, A. 130, 139, 182, 481
Bowen, P.J. 194–5, 481
Bowes, P.C. 317–18, 321–9, 331, 481, 520

Bowman, C.T. 27, 29, 481
Bradley, D. 107, 109, 116, 477, 481
Bradshaw, L.S. 313, 484
Brandrup, J. 4, 481
Brannigan, F.L. 215, 481
Breden, L. 305, 475
Brenden, J.J. 211–12, 215, 218, 481–2
Brenton, J.R. 196, 273–4, 481, 492
Brindley, J. 330
British Standards Institution 86, 165, 209, 220, 228, 241, 249, 271–2, 296, 301, 303, 342, 344, 351, 375–6, 387, 420, 426–8, 430, 439, 441–3, 450, 455, 474, 481–2
Britton, L.G. 330, 482
Brosmer, M.A. 78, 182, 184, 189, 203–4, 482
Brown, A.L. 185, 502
Brown, N.J. 453, 490
Browne, F.L. 212, 215, 483
Bruce, H.D. 380, 482
Bryan, J.L. 142, 482
Bryner, N. 351, 353, 493
Brzustowski, T.A. 193, 244, 483
Buchanan, A.H. 387, 429, 438–9, 483
Budavari, S. 2, 483
Budnick, E.K. 355, 483
Building Research Establishment 352, 481, 483
Bukowski, R.W. 354–5, 369, 371–2, 510
Bullen, M.L. 148, 354, 360, 363–4, 366, 374, 381, 393, 397–404, 435, 472, 483, 520
Bunev, V.A. 106, 110, 503
Burgess, D.S. 92, 118, 184–6, 198, 483, 525
Burgoyne, J.H. 91, 191–2, 194, 235, 238, 243–5, 278, 282, 284, 484
Burke, S.P. 123–6, 484
Bustamente Valencia, T. 26, 483
Butcher, E.G. 174, 397, 422, 425, 449–50, 462, 468, 470–2, 474, 483–4
Butler, B.W. 313, 484
Butler, C.P. 219–20, 484

Cadorin, J.-F. 406, 415–16, 484
Cajot, L.G. 415, 515
Calcote, H.F. 444, 484
Calus, I.M. 70, 479
Cameron, A.J. 322–3, 481
Cameron, L.R.J. 194–5, 481
Capote, J.A. 177, 514
Carpenter, D.J. 177, 379, 459, 522
Carpentier, C. 248, 260, 267, 477
Carras, J.N. 323, 484
Carslaw, H.S. 42, 47, 251, 484
Carvel, R.O. 427, 484
Casal, J. 164–5, 186, 189, 484, 497, 499,
    507, 517
Cassanova, R.A. 443, 479
Catchpole, E.A. 312, 484
Catchpole, W.R. 312, 484
CEN, 155, 484
Cetegen, B. 131, 135–7, 140–3, 145,
    148, 150–1, 153–4, 484, 525
Chaivanov, B.B. 198, 487
Chartered Institute of Building Services
    Engineers (CIBSE) 138, 351, 474,
    484
Chase, M.W. 30, 32, 484
Chatris, J.M. 186, 189, 485
Chen, P. 346, 500
Cheng, X.F. 77, 444, 487
Chien, W.P. 448, 515–16
Chitty, R. 121, 131, 137, 139–40, 287,
    362, 485–6
Chitty, T.B. 422, 425, 484
Choi, H.Y. 35, 44, 57, 63, 515
Chuah, K.H. 131, 485
Clancey, V.J. 198, 242, 485
Cleary, T.G. 311, 313, 377, 485, 505
Cohen, L. 91, 194, 483
Cohen, L.D. 313, 484
Cohen, S. 332, 514
Coles, A. 177, 514
Colwell, J.D. 233, 246–7, 485
Consalvi, J.L. 297–8, 511
Cook, S.J. 485
Conseil Internationale du Batiment (CIB)
    423, 427, 485
Cook, S.J. 194, 485

Cooke, R.A. 217, 485
Cooper, L.Y. 356, 459, 485
Cordova, J.L. 248, 259, 266–7, 485
Corlett, R.C. 130–1, 140, 146, 182, 485
Coutin, M. 297–8, 511
Coward, H.F. 83–4, 90, 485
Cox, G. 121, 131, 137, 139–41, 147–8,
    156, 174, 177–8, 287, 466, 485–6,
    490, 501, 507, 516, 525
Cozzani, V. 218, 514
Crank, J. 263, 479
Crauford, N.L. 177, 486
Crescitelle, S. 177, 487
Croce, P.A. 189, 203, 359, 378, 380–1,
    389–90, 486, 506
Crocomb, W. 318, 329, 504
Croft, W.M. 362, 443, 486
Cullen, Lord 116, 486
Cullis, C.F. 7–10, 194, 199, 485–6
Cumber, P. 177, 486
Custer, R.L.P. 157, 171, 486

Dai, J. 269, 525
Dakka, S.M. 250, 256–7, 486
Damkohler, G. 116, 486
Damson, R. 233–4, 512
Deal, S. 366, 486
Deepak, D. 247, 266–7, 486, 513
De Haan, J.D. 217, 486
Delichatsios, M.A. 129, 157, 171, 259,
    486, 496
Dembsey, N.A. 256–8, 269, 479
Demers, D.P. 385, 462, 479
Deng, Z. 269, 362, 523, 525
Department for Communities and Local
    Government, 1, 441, 459, 487
Department of Transport 387, 487
De Paula, J. 18–20, 27, 32, 477
de Ris, J.N. 76–8, 147–8, 176, 182–4,
    189, 207, 209, 263, 285–7, 288,
    290–2, 296–9, 308–10, 400, 444,
    487, 501, 505, 509
Desanghere, S. 177, 514
Deutsches Institut für Normung (DIN)
    375, 427, 457
De Witt, D.P. 35–7, 57, 62, 123, 497
di Blasi, C. 177, 210–11, 487

di Nenno, P.J. 35, 115, 153–6, 165, 250,
    273, 476, 479, 487
Dixon-Lewis, G. 28, 101, 487
Dlugogorski, B.Z. 318, 522
Dobashi, R. 292–3, 518
Dod, R.L. 453, 490
Dong, Y. 98, 487
Dorofeev, S.B. 198, 487
Dosanjh, S.S. 333–4, 488
Dotreppe, J.-C. 415, 484
Dryer, F. 236, 238–40, 279, 491
Drysdale, D.D. 25, 29, 34, 155–6, 176,
    196, 199, 208–9, 235–7, 241,
    247–50, 256, 259–60, 263–8, 274,
    285–7, 294, 296, 310–12, 340,
    344, 349, 363–4, 384, 443–4, 447,
    453, 463, 477, 479–80, 486, 488,
    490, 492–6, 513–16, 520–1, 524
D'Souza, M.V. 146, 488
Dugger, G.L. 112, 488
Dungan, K.W. 318, 488
Dwyer, R.W. 346, 488, 500

Earl, W.L. 347, 491
Eckert, E.R.G. 145, 524
Eckhoff, R.K. 222, 489
Edwards, D.K. 74–6, 489
Efimenko, A.A. 198, 487
Egbert, R.B. 74, 497
Egerton, A.C. 335, 489
Egolfopoulos, F.N. 98, 109–10, 487, 521
Elliot, D.A. 428, 489
Emmons, H.W. 131, 176–7, 301, 308,
    312, 376–8, 390, 393, 489, 506,
    511
Endo, H. 335, 501
Enright, P.A. 24, 489
European Committee for Standardisation
    421, 489
Evans, D.D. 171, 489
Eyre, J.A. 117, 164, 493, 506

Faeth, G.M. 109, 143, 162, 521, 525
Fairley, J.A. 70, 479
Fan, W.C. 192, 362, 489, 523
Fang, J.B. 355, 382, 460, 489

Fardell, P.J. 29, 397, 457, 484, 524
Farley, J.P. 362, 492
Fay, J.A. 193–4, 489
Fennell, D. 287, 353, 489
Fernandez-Pello, A.C. 130, 139, 177, 182,
    191–2, 247–8, 259, 266–8, 277,
    287, 291–2, 294–5, 300–2,
    306–10, 333–5, 341–5, 377, 479,
    481, 485, 487, 488–9, 490–1,
    502–3, 511, 514–15, 521–2, 525
Field, P. 222, 490
Fine, D.H. 231–2, 490
Fire Protection Association 84, 475,
    477–8, 479–81, 486–90
Fishenden, M. 57–8, 490
Fleischman, C.M. 24, 360–4, 443, 453,
    489, 490
Fletcher, D.F. 263, 268, 506
Folch, J. 186–9, 484
Foley, M. 155, 250, 490
Fons, W.L. 308, 490
Forney, G.P. 177, 510
Foster, C.D. 34, 116, 490
Foster, J. 360–3, 490
Fournier, L.G. 346, 488
Frank-Kamenetskii, D.S. 41, 100, 230,
    317–18, 324, 490
Franssen, J.-M. 406, 415–16, 484
Friedman, R. 3, 31–2, 175, 197, 277–8,
    306, 352–3, 379, 459, 490–1
Fristrom, R.M. 98, 114, 491, 515
Fulper, C. 281–3, 523
Fung, F. 463, 465, 479, 491
Furno, A. 86, 113, 495, 515

Galea, E.R. 172, 496
Gann, R.G. 24, 347, 491, 501
Gardiner, A.J. 172, 498
Garo, J.P. 182, 191–2, 491
Garris, C.A. 331–2, 334–5, 339, 507
Garzia, H.W. 272, 521
Gaydon, A.G. 98–9, 107, 109, 123,
    127–9, 188, 491
GHS 237, 241–2, 491
Gilchrist, K. 140, 504
Gill, W. 185, 502

Gillie, M. 412, 428, 491, 502
Glassman, I. 195, 236, 238–40, 243–4,
    278–82, 296, 306–7, 443–5,
    490–1, 504, 516
Gojkovich, D. 362, 491
Goldstone, B.M. 473, 507
Good, A.J. 194, 485
Gore, J.P. 77, 516
Gottuk, D.T. 29, 182, 359, 362, 454,
    457–8, 491–2
Graham, T.L. 364, 492
Grant, G.B. 25, 176, 196, 274, 310–11,
    469, 492, 499
Gratkowski, M.T. 345, 347, 480
Gray, B.F. 317, 322, 492
Gray, P. 230–2, 317, 480, 490, 492
Gray, W.A. 61, 74, 76, 492
Greaney, D. 324, 477
Green, A.R. 185, 477
Green, D.W. 19, 510
Griffiths, J.G. 27, 92, 318, 329–30, 444,
    492, 504
Gritzo, L.A. 185, 502
Gross, D. 159, 161, 220–2, 389, 441, 492
Grosshandler, W.L. 76, 351, 353, 492
Grove, B.S. 136, 512
Grubits, S.J. 305, 507
Grumer, J. 184–5, 198, 483
Gugan, K. 118, 318, 335, 489, 493
Guillaume, E. 26, 483

Hadvig, S. 217, 421, 493, 510
Haggerty, W. 300, 497
Hagglund, B. 78, 146, 355, 359, 365,
    368, 371, 408, 493
Hall, A.R. 182, 191, 493
Hall, C. 3–5, 493
Hall, H. 191, 493
Halliburton, B. 322, 492
Halliday, D. 448, 493
Hamada, T. 158, 160, 496
Hamilton, D.C. 70–1, 493
Hamilton, S.E. 463, 488
Hamins, A. 140, 189, 493, 522
Hansel, J.G. 244, 278–82, 296, 491, 504
Harkleroad, M.F. 47, 156, 260, 267, 311,
    363, 366–72, 381, 477, 504, 512

Harland, W.A. 442, 477
Harmathy, T.Z. 36, 394–5, 399, 401,
    419, 426–7, 429, 439, 493, 501
Harris, R.J. 34, 85, 106, 116, 493
Harrison, A.J. 117, 493
Harrison, R. 174, 466, 493–4
Hartman, J. 262, 270, 347, 518
Hartzell, G.E. 455–6, 494
Harvey, D.I. 231, 480
Hasegawa, K. 192, 237, 494
Hasemi, Y. 131, 138, 146–7, 150,
    152–3, 155–6, 160, 162, 250, 292,
    300, 311, 466, 494, 512, 523
Hassan, M.I. 131, 494
Hawthorne, W.R. 128–9, 494, 497
Health and Safety Executive 117, 273,
    495
Heimel, S. 112, 488
Hertzberg, M. 86, 88, 106, 110, 113–14,
    495
Heselden, A.J.M. 355, 379–81, 388–9,
    394–6, 404, 419, 424, 435–6, 495,
    520
Heskestad, A.W. 453–4, 495
Heskestad, G. 29, 123, 130, 132, 135,
    137–8, 144, 149–151, 158, 160,
    170–1, 335, 340, 384, 495–6
Heywood, G.M. 172, 502
Hillstrom, W.W. 282, 496
Hinkley, P.L. 132, 137, 159, 161–2, 175,
    209, 355–6, 385, 435, 438,
    467–71, 496, 516, 520
Hirano, T. 209, 277, 283, 285–6, 291–3,
    300–1, 309, 490, 496, 509, 518
Hird, D. 271, 516
Hirschler, M.M. 6–10, 199, 292, 486, 496
Hirst, R. 83, 88, 115, 496
Hjertager, B.H. 117, 496
Hobbs, M.W. 296, 312, 524
Hoffman, N. 172, 496
Holborn, P.G. 364, 480, 496
Hollyhead, R. 227, 496
Holman, J.R. 35, 496
Holmes, F.H. 296, 496

Holmstedt, G. 194, 496, 523
Holve, D.J. 114, 497
Hopkins, D. 213–14, 497
Hopkinson, J.S. 384, 507
Hottel, H.C. 64–5, 74, 128–9, 175,
    182–3, 494, 497
Hovde, P.J. 453–4, 495
Howard, J.B. 30, 444, 508, 514
Hua, J.S. 192, 489
Huggett, C. 23, 300, 372, 497
Hull, R. 454, 458, 517

Ide, R. 217, 485
Immergut, E.H. 4, 481
Inamura, T. 190, 497
Incropera, F.P. 35, 37, 57, 62, 123, 497
Ingason, H. 163, 184, 497
Ingberg, S.H. 423, 427, 497
International Standards Organisation
    (ISO) 155, 249, 257, 259, 301, 303,
    305, 311, 351, 373–4, 376–7, 384,
    387, 420–1, 447, 450, 453, 456,
    458, 480, 482, 489, 497–8
Iqbal, N. 203–4, 497
Ishida, H. 281, 498
Ismail, M.A. 109, 521
Ito, A. 279, 287, 291, 294, 306, 498

Jackman, L.A. 172, 498
Jackson, G.S. 189, 250, 256–7, 486, 522
Jaeger, J.C. 42, 47, 251, 484
Jagger, S. 269, 499
Jahn, W. 277, 514
Jaluria, Y. 356, 499
James, J.J. 241, 499
Janssens, M.L. 20, 24, 217, 247–8, 250,
    255, 257, 260–3, 499
Jansson, R. 355, 359, 365, 368, 371, 493
Japan Institute for Safety Engineering
    164, 499
Jiang, J.C. 313, 505
Jin, T. 449–50, 456, 499
Johnson, A.L. 113, 495
Joint Fire Prevention Committee 341, 499
Joint Fire Research Organisation 381–2,
    499

Jones, G.W. 83–4, 90, 485
Jones, I.P. 177, 287, 516
Jones W.W. 151, 165, 177, 379, 510, 512
Jost, W. 125–6, 499
Joulain, P. 130, 139, 182, 481, 499
Jowsey, A. 177, 514
Joyeux, D. 177, 514

Kandola, B.S. 35, 500
Kanury, A.M. 53–7, 85, 100, 109,
    112–13, 125, 129, 194–6, 250,
    264, 270, 500
Karlsson, B. 176, 310–11, 362, 377, 378,
    479, 491, 500
Kashiwagi, T. 6–7, 12, 140, 199, 211,
    213, 250, 269, 285, 287, 291, 294,
    300–2, 366, 493, 498, 500, 512
Kashuki, K. 237, 494
Katan, L.L. 191–2, 483
Kauffman, C.W. 336, 339, 503
Kawagoe, K. 365, 387–9, 396, 406, 416,
    418, 500
Kay, T.R. 419, 501
Kaye, G.W.C. 36, 500
Kellogg, D.S. 346, 500
Keil, D.G. 444, 484
Kemp, N. 34, 488
Kennedy, E.M. 318, 522
Kennedy. I.M. 443, 500
Kent, J.H. 127, 185, 263, 268, 477, 500,
    506
Khan, M.M. 263, 500
Khudiakov, G.N. 145–6, 182–3, 190, 480
Kikuchi, R. 155, 494
Kiley, F. 259, 486
Kimmerle, G. 457, 501
Kinbara, T. 335, 501
Kirby, B.R. 411–12, 419, 422, 425, 430,
    501–2
Kitano, M. 341, 521
Klein, D.P. 355, 483
Klitgard, P.F. 366, 501
Klopovic, S. 437, 501
Klote, J.H. 463, 465, 473–4, 501
Kochurko, A.S. 198, 487
Kodur, V. 36, 429, 501

Kokkala, M. 155, 159, 162, 192, 250, 376–7, 475, 479, 501
Koseki, H. 78, 166–7, 191–2, 491, 501
Krammer, G. 218, 514
Krause, R.F. 24, 501
Krause, U. 334, 501
Kreith, F. 73, 500
Krivulin, V.N. 106, 110, 503
Kubota, T. 131, 135–7, 140, 142–3, 145, 150–1, 153–4, 175, 359, 484, 525
Kuchta, J 86, 113, 495, 515
Kudo, Y. 306, 498
Kumar, S. 172, 174, 177, 287, 466, 486, 501
Kung, H.C. 137, 145, 148–9, 502, 525
Kuntz, K. 351, 353, 493
Kuo, K.K. 100, 107, 109, 195, 196, 502
Kuwana, K. 131, 485, 494
Kuznetsov, M.S. 198, 487

Laby, T.H. 36, 500
Lam, C.S. 186, 502
Lamont, S. 412, 502
Landau, L.D. 35, 502
Lastrina, F.A. 306, 502
Latham, D.J. 312, 411–12, 422, 425, 430, 484, 502
Lattimer, B.Y. 29, 154–6, 162, 457, 491, 502
Laurendeau, N.M. 233, 247, 502
Lautenberger, C. 247, 267–8, 377, 502, 514
Lavine, A.S. 35, 37, 57, 62, 123, 497
Law, M. 36, 66, 67, 79, 174, 217, 255, 388–9, 395, 406, 424, 425–7, 430, 435–7, 502, 520
Lawson, D.I. 69, 254–5, 260–1, 502
Lawson, J.R. 372, 376, 478, 502
Lazaro, M. 177, 514
Lazzara, C.P. 326, 506
Le Chatelier, H. 90, 502
Lederer, T.S. 457, 511
Lee, B.T. 370, 371, 503
Lee, J.L. 197, 203–4, 519
Lee, K.Y. 166, 359, 444, 521
Lee, P.R. 230, 231, 317, 492

Leisch, S.O. 336, 339, 503
Lennon, T. 416, 418–19, 503
Leonard, J. 281–3, 523
Lewin, M 28, 503, 518
Lewis, B. 30, 32, 84, 88, 99–100, 106, 109, 110–13, 117, 129, 196, 228, 503, 505
Liao, G.X. 192, 362, 489, 523
Lide, D.R. 4, 187, 503
Lie, T.T. 428, 430, 503, 517
Liew, S.K. 172, 502
Liew, S.L. 177, 486
Lifshitz, E.M. 35, 502
Lindner, 330, 503
Linnett, J.W. 86, 503
Loftus, J.J. 441, 492
Lois, E. 163, 184–5, 503
Long, R.T. 248, 503
Lonnermark, A. 457, 480
Lougheed, G.D. 66, 174, 435–6, 503–4
Lovachev, L.A. 106, 110, 503
Lucca, D.A. 332–5, 338–9, 342, 345, 509
Luo, M.C. 177, 503
Lyons, J.W. 28, 207, 211, 347, 503

Ma, T.G. 149, 503
McAdams, W.H. 62, 503
McAlevy, R.F. 284–5, 292–3, 296, 300, 305–7, 503–4
McAlpine, R.S. 296, 312, 524
McCaffrey, B.J. 47, 121, 123, 130–3, 140–1, 143–5, 148–9, 171, 175, 363–4, 366–72, 374, 381, 390, 503, 512, 520
McCarter, R.J. 341, 346–7, 504
McCartney, C.J. 174, 504
McElroy, J.A. 193, 511
McGratten, K. 172, 177–8, 504
McGuire, J.H. 64–6, 146, 488, 504
McIntosh, A.C. 318, 323, 329–30, 504
Mackinven, R. 231–2, 244, 278–81, 296, 490, 504
Macmillan, A.J.R. 285–8, 296, 312, 384, 488
McRae, D.D. 346, 500
Madorsky, S.L. 5, 8–12, 199, 210–11, 504

Madrzykowski, D. 193, 351, 353, 493, 511
Magee, R.S. 284–5, 292–3, 296, 300, 305–7, 503–4
Magnusson, S.E. 52, 176, 311, 377, 405–16, 423, 427, 430, 432–4, 504, 510
Makhviladze, G.M. 131, 364, 492, 504
Malalakesera, W.M.G. 140, 504
Malan, D.H. 263, 479
Malhotra, H.L. 421, 439, 449, 477, 504
Maloney, J.O. 19, 510
Mangan, R.J. 313, 484
Manka, M.J. 347, 491
Mannan, S. 2, 79, 85, 163, 166, 191, 198, 222, 504
Manton, J. 109, 505
Manzello, S.L. 313, 325, 505
Maranghides, A. 313, 325, 505
Marchant, E.W. 351, 441, 463–4, 505
Margeneau, H. 410, 505
Markatos, N.C. 172, 496
Markstein, G.H. 76–9, 129, 203–5, 207, 285, 287–8, 290–1, 296–8, 310, 359–60, 370, 444, 505, 509
Marshall, N.R. 138, 174, 466, 473, 507, 520
Martin, S. 264, 354, 505
Masuda, D. 279, 498
Matkowski, B.J. 334, 515
Matsukov, D.I. 198, 487
Mayer, E. 101, 103, 105–6, 110, 505
Meacham, B.J. 157, 171, 486
Mehaffey, J.F. 426, 493
Meidl, J. 2, 505
Melinek, S.J. 355, 379–81, 495
Mell. W.E. 313, 325, 505
Middleton, J.F. 142, 505
Mikkola, E. 250–1, 258–9, 263, 505
Miles, L.B. 347, 491
Miles, S. 177–8, 504
Milke, J.A. 36, 174, 427, 430, 441, 466, 474, 505, 515
Miller, F.J. 278–9, 506
Miller, R.B. 210, 506, 515
Ministry of Aviation 94, 506

Minzer, G.A. 164, 506
Miron, Y. 326, 506
Mitchell, N.D. 322, 506
Mitler, H.E. 176, 177, 379, 506
Miyasaka, K. 343, 521
Modak, A.T. 76, 79, 189, 203, 207–8, 359–60, 370, 476, 492, 506, 509
Moghtaderi, B. 210, 213, 263, 268, 506
Moinuddin, K. 419, 519
Momita, D. 437, 518
Montreal Protocol 115, 272, 506
Moodie, K. 194, 525
Moore, D. 416, 418–19, 503
Moore, H. 49, 506
Moore, L.D. 376, 506
Moore, W.J. 18–20, 27, 32, 506
Moorhouse, J. 164–5, 506
Morandini, F. 312, 506
Morehart, J.H. 359, 525
Morgan, A.B. 6, 8–9, 199, 496
Morgan, H.P. 138, 172, 174, 466, 473, 498, 506, 520
Morgan, W.R. 70–1, 493
Morris, G.A. 312, 484
Morris, W.A. 384, 507
Morton, B.R. 132, 134–7, 371, 507
Moss, J.B. 78, 177, 486, 507
Moulen, A.W. 305, 439, 507
Moussa, N.A. 164, 292, 296, 331–2, 334–5, 339, 507, 512
Mowrer, F.W. 154, 157, 177, 272, 351, 357, 371, 384, 441–2, 453, 459, 490, 507–8, 524
Mudan, K.S. 78, 164, 165, 507
Mulholland, G.W. 192, 443, 446, 458, 501, 507
Muller, R. 61, 74, 76, 492
Mullins, B.P. 33, 92–3, 507
Mundwiler, H. 237, 507
Munoz, M. 164–5, 507
Murphy, G.M. 410, 505
Murray, J. 270–1, 516
Murrell, J.V. 260, 294, 507, 524

Nakabayashi, T. 300, 311, 494
Nam, D. 292, 494

Nam. S. 162
Nakaya, I. 389, 507
National Bureau of Standards (NBS) 9, 507
National Fire Protection Association (NFPA) 2, 84, 86, 155, 171, 228, 241, 273, 297, 303, 305–6, 313, 318, 320, 384–5, 423, 428, 438, 441, 459, 508
Nelson, G.L. 457, 508
Nelson, H.E. 351, 441–2, 508
Neoh, K.G. 30, 508
Newitt, D.M. 194, 483
Newman, D.L. 285, 500
Newman, J.S. 157, 508
Newton, The Rt Hon Lord 242, 508
Nilsson, L. 396, 520
Nishihata, M. 146–7, 156, 494
Nitta, K. 143, 518
Nohara, A. 300, 311, 494
Nolan, P.F. 172, 498
NORDTEST 23, 372–4, 508,
Noreikis, S.E. 285–6, 291, 309, 496
Novakov, T. 453, 490
Novozhilov, V. 177–8, 263, 268, 506, 508
Nussbaum, R.M. 376, 377, 509
Nyden, M.R. 189, 522

O'Brien, T. 79, 436, 437, 502
Odeen, K. 405, 508
Office of the Deputy Prime Minister 1, 508
Ogden, S.D. 24, 256, 258–60, 263, 501, 519
Ohlemiller, T.J. 7, 199, 211, 213, 250, 331–6, 338–9, 341–3, 345–7, 500, 508, 515
Ohtaki, M. 154, 518
Ohtani, H. 209, 509
Oleszkiewicz, I. 435, 509
Omori, A. 12, 500
Onnermark, B. 355, 359, 365, 368, 371, 493
Open University 3, 509

Orloff, L. 76, 79, 129, 147–8, 182, 207–9, 291, 296–7, 310, 359–60, 370, 486–7, 509
Östman, B.A.L. 376–7, 448, 453, 509
Oyama, H. 306, 498

Pagni, P.J. 78, 140, 141, 333, 334, 360–2, 443, 444, 488, 490, 509
Palmer, K.N. 116, 222, 333, 335–9, 343–4, 510, 513
Panagiotou, Th. 259, 267, 486, 510
Pape, R. 381–2, 510
Parker, W.J. 24, 256, 291, 294, 308–9, 377, 458, 478, 499, 510
Parnell, A.C. 174, 449–50, 462, 468, 470–2, 474, 483
Paul, K.T. 295, 510
Paulsen, O.R. 217, 421, 493, 510
Peacock, B.N. 419, 501
Peacock, R.D. 13, 23, 30, 177, 354–5, 369–72, 454, 478, 510
Peatross, M.J. 362, 457–8, 492
Penner, S.S. 33, 92–3, 507
Perdue, G.R. 325, 510
Pereverzev, A.K. 198, 487
Perry, R.H. 19, 510
Persson, H. 194, 496
Persson, L.E. 78, 146, 408, 493
Petrella, R.V. 217, 218, 510
Pettersson, O. 52, 405–16, 423, 426–7, 430, 432–4, 510
Petty, S.E. 191, 510
Phillips, H. 283, 510
Phillips, R.P. 448, 451–2, 454, 513
Phylaktou, H.N. 116, 510
Pintea, D. 415, 484
Pion, R.F. 9, 197, 199–204, 217–19, 264–5, 519
Pipkin, O.A. 163, 510
Pita, L.P. 313, 522
Pitts, D.R. 35, 37, 52, 68, 510
Pitts, W.M. 29, 131, 165, 307, 312, 457, 510–11
Pizzo, Y. 297–8, 511
Planas, E. 164–5, 186, 189, 484, 507
Popplewell, O. 352, 385, 438, 511

Porter, A. 177, 486
Porterie, B. 297, 298, 511
Potts, W.J. 457, 511
Powell, E.A. 443, 479
Powell, F. 233, 511
Prado, G. 127, 501
Prahl, J. 393, 511
Pratrap, R. 49, 511
Pratt, B.T. 454, 513
Priest, D.N. 455–6, 494
Prime, D. 341, 511
Products Research Committee 206, 505, 511
Proulx, G. 351, 441, 511
Ptchelintsev, A. 160, 162, 494
Punderson, J.O. 442, 247, 511
Puri, R. 30, 511
Purser, D.A. 69, 454, 456–8, 511, 517
Purser, J.A. 454, 457, 511, 517
Putnam, T. 313, 484
Putorti, A.D. 193, 511
Putzeys, O.M. 343, 479, 511

Qin, J. 362, 523
Querre, P. 297–8, 511
Quince, B.W. 239, 514
Quintela, J. 186, 189, 484
Quintiere, J.G. 47, 134–6, 149, 151, 156, 165, 175–7, 185, 203–4, 213–14, 247–8, 256–7, 259–60, 262, 267–71, 285–6, 288, 300–1, 303–5, 308, 311, 313–14, 356, 359, 363–72, 374, 376–7, 379, 381, 393, 396, 419, 437–8, 480, 485, 497, 503–4, 510–12, 514, 515, 517, 520–1
Quinton, P.G. 243–5, 278, 483

Rae, D. 233–4, 512
Raftery, M.M. 122, 142–4, 175, 520, 522, 524
Raj, P.P.K. 164, 512
Randsalu, E.J. 186, 502
Rasbash, D.J. 29, 76–7, 116, 117, 140, 171–2, 187–9, 205, 239–40, 247, 259, 264, 267, 350–1, 353, 358,

438, 443–4, 447–9, 451–2, 454, 512–13
Rath, J. 218, 513
Rawlins, P. 260, 294, 507
Ray, S.R. 305, 307, 490
Rein, G. 26, 177, 331–4, 343, 419–20, 437, 479, 483, 511, 514, 517
Reneke, P.A. 177, 354–5, 369–72, 478, 510
Resnick, R. 448, 493
Reszka, P. 177, 514
Revenaugh, A. 154, 272, 384, 524
Reza, A. 233, 246–7, 485
Rich, D. 247, 267, 268, 514
Richter, H. 444, 514
Ricou, F.P. 150, 514
Rinkinen, W.J. 151, 159, 165, 379, 396, 419, 501, 512, 517
Roberts, A.F. 198, 211, 215, 216, 238–9, 243–45, 278, 282, 284, 483, 514
Roberts, J.P. 364, 492
Roberts, P. 86, 514
Robertson, A.F. 389, 441, 452, 492, 512, 514, 517, 519
Robertson, R.B. 79, 163, 514
Roby, R.J. 29, 359, 454, 457–8, 492
Rockett, J.A. 36, 367, 390, 504, 514
Rogaume, T. 26, 483
Rogers, F.E. 341–3, 346, 509, 515
Rogowski, Z.W. 76–7, 116–17, 140, 187–9, 513
Rohsenow, W.M. 35, 44, 57, 63, 515
Rorrer, G. 35, 37, 43, 45, 55, 57, 62, 76, 82, 123, 173, 523
Ross, H.D. 236, 277–9, 282, 506, 515
Roth, L. 463, 465, 479
Rothbaum, H.P. 331, 515
Rothermel, R.C. 312, 484
Rotter, J.M. 428, 491
Routley, A.F. 455, 515
Royal, J.H. 292–3, 515
Rushbrook, F. 439, 515
Russo, G. 177, 487
Ryder, N.L. 177, 514

Saga, S. 335, 501
Saito, K. 131, 156, 190, 279, 300, 311,
    485, 494, 497–8, 515
Salig, R. 343, 515
Santoni, P.A. 312, 506
Santoro, R.J. 30, 291, 294, 309–10, 490,
    511
Sapko, M. 86, 515
Sarofim, A.F. 30, 64, 497, 508
Sato, K. 336–7, 515
Saunders, O.A. 57–8, 490
Savage, L.D. 83, 517
Savage, N. 83, 88, 115, 496
Sawyer, R.F. 114, 497, 515
Scheffey, J.L. 439, 523
Schifiliti, R.P. 157, 171, 486
Schleich, J.-B. 415, 515
Schmidt, M. 334, 501
Schnell, L.G. 274, 515
Schult, D.A. 334, 515
Schumann, T.E.W. 123–6, 483
Seader, J.D. 448, 515
Sega, S. 336–7, 515
Seibring, H. 330, 503
Seigel, L.G. 435, 515
Sekine, T. 388, 406, 416, 418, 500
Semenov, N.N. 228, 230, 515
Setchkin, N.P. 233, 246, 516
Shaw, A. 246, 516
Shaw, K. 455, 517
Sheppard, D.M. 264, 522
Sherratt, J. 199, 208–9, 294, 516
Shields, J.R. 313, 505
Shields, T.J. 208, 258, 270–1, 294, 516,
    521, 525
Shipp, M. 421, 427, 516
Shokri, M. 168–9, 516
Sichel, M. 336, 339, 503
Sidorov, V.P. 198, 487
Sikorski, J. 345, 347, 480
Silcock, A. 385, 438, 516
Silcock, G.W.H. 208, 258, 270–1, 294,
    521, 516, 525
Simcox, S. 177, 287, 516
Simeoni, A. 332, 514
Simmons, R.F. 27–8, 97, 113, 516, 524

Simms, D.L. 132, 137, 175, 217, 250–1,
    254–7, 260–1, 271, 308, 312,
    467–8, 470, 502, 516, 520
Simpson, C.J.S.M. 86, 503
Singh, B. 233–4, 512
Sirignano, W.A. 243, 278, 516
Sissom, L.E. 35, 37, 52, 68, 511
Sivanathu, Y.R. 77, 516
Skipper, R.S. 455, 515
Sleight, R.T. 344–5, 477
Sliepcevich, C.M. 163, 165, 185, 510, 523
Smith D.A. 131, 287, 455, 516
Smith, D.B. 86, 514
Smith, H. 170, 496
Sobiesiak, A. 142, 523
Song, H. 362, 523
Sorathia, U. 154, 155, 502
Sorenson, S.C. 83, 517
Spalding, D.B. 98, 101, 107–10, 150,
    195–6, 267, 480, 514, 517
Spearpoint, M. 174, 213–14, 263, 466,
    493–4
Spedding, G.R. 98, 487
Stanzak, W.W. 430, 517
Stark, G.W.V. 76–7, 140, 187–9, 194,
    428, 513, 517
Stavrianides, P. 145, 148–9, 502
Stec, A. 454, 457–8, 517
Steckler, K.D. 185, 389, 393, 396, 419,
    517
Steel Construction Institute 118, 517,
Steiner, G. 218, 514
Sterner, E.S. 47, 517
Stern-Gottfried, J. 177, 419–20, 514, 517
Stevens M.P. 3, 5, 517
Steward, F.R. 29, 69, 143–4, 149, 517
Strasser, A. 184–5, 198, 483
Strehlow, R.A. 83, 118, 517–18
Stretton, A.J. 166, 359, 444, 521
Stroup, D.W. 171, 459, 485, 489
Stull, D.R. 2, 33, 92, 518
Sugawa, O. 154, 437, 518
Sultan, M.A. 421, 518
Sundstrom, B. 23, 176, 311, 374–7, 381,
    504, 518, 523

Sutton, O.G. 242, 518
Suzuki, M. 292–3, 518
Suzuki, T. 283, 518
Swann, J. 262, 270, 347, 518
Swithenbank, J. 163, 184, 185, 503
Switzer, W.G. 455–6, 494

Tagivi, K.A. 190, 497
Takahashi, W. 154, 437, 518
Takashima, S. 155, 494
Takeda, H. 354, 364, 366, 400, 518
Tan, S.H. 69, 518
Tanaka, H. 154, 518
Tanaka, T. 389–90, 489
Tatum, P. 153–6, 250, 479
Taylor, G. 132, 134–7, 371, 507
Taylor, G.M. 115, 273, 487
Tazawa, K, 300–1, 496
Terai, T. 143, 518
Tewarson, A. 9, 13, 23–5, 181, 197,
    199–207, 217–19, 247–8, 256,
    258–9, 260, 264–5, 267–8, 401,
    444, 446, 518–19
Theobald, C.R. 131–2, 137,159, 161–2,
    175, 355–6, 382, 435, 467–8, 470,
    496, 519–20
Thomas, A. 194, 483
Thomas, I.R. 419, 519
Thomas, G.O. 273, 481
Thomas, P.H. 122, 131–2, 137–8,
    142–4, 148, 163, 165, 174–5, 217,
    220, 277, 290, 296, 308, 312–13,
    321–2, 324–5, 327, 349, 353–4,
    357, 363–4, 366, 371–2, 374, 381,
    388, 393–6, 399–404, 406, 419,
    424, 435–7, 466–8, 470, 483, 501,
    519–520, 522
Thompson, I. 457, 477
Thompson, V. 177, 486
Thomson, G. 411–12, 422, 425, 430, 502
Thomson, H.E. 247–9, 256, 259–60,
    263–5, 267–8, 488, 520
Thor, J. 52, 405–16, 423, 427, 430,
    432–4, 510
Thorne, P.F. 19, 330, 521
Thyer, A.M. 194, 198, 521

Tien, C.L. 36, 63–4, 69, 74, 78, 166,
    182, 184, 189, 203–4, 359, 444,
    482, 521
Toal, B.R. 258, 521
Tokunaga, T. 131, 150, 152–3, 466, 494
Tolputt, T.A. 323, 504
Tomlinson, L.N. 419, 501
Toner, S.J. 359, 525
Toong, T.Y. 292, 296, 331–2, 334–5,
    339, 507
Torero, J.L. 26, 177, 247, 248, 250–1,
    256–7, 266–9, 333–5, 341–5, 377,
    419–20, 477, 483, 485–6, 502–3,
    514, 517, 521
Townsend, S.E. 325–26, 328–9, 481
Tritton, D.J. 35, 521
Tsai, K.-C. 156, 521
Tsantardis, L.D. 448, 509
Tse, S.D. 343, 521
Tseng, L.-K. 109, 521
Tu, K.-M. 311, 521
Turan, O.F. 437, 501
Turner, J.S. 132, 134–7, 371, 507
Twardus, E.M. 193, 244, 483
Twilley, W.H. 372, 376, 458, 478, 502
Twilt, L. 303, 521
Tyler, B.J. 241, 521

Umino, P.Y. 292, 496
Ural, E.A. 272, 521
Urban, D.L. 333–4, 343, 479, 511, 514
US Department of Commerce 296, 481,
    521, 525
Usmani, A. 412, 428, 491, 502

Vagelopoulos, C.M. 98, 109–10, 487, 521
van de Leur, P.H.E. 303, 427, 521
Vandevelde, P. 303, 521
van Hees, P. 303, 521
van Mierlo, R.J.M. 303, 521
Vantelon, J.-P. 130, 139, 182, 191–2,
    481, 491
Versteeg, H.K. 140, 504
Vervalin, C.H. 191, 521
Viegas, D.X. 313, 522
Vienneau H. 146, 522

Volpert, V.A. 334, 515
von Elbe, G. 30, 32, 84, 88, 99–100, 106, 109–11, 117, 129, 196, 228, 300, 497, 503, 505
Vytenis, R.J. 263, 522
V'yun, A.V. 106, 110, 503

Wade, S.H. 242, 522
Wagner, H.Gg. 127, 444, 501, 522
Wainman, D.E. 419, 501
Wakamatsu, T. 160, 162, 494
Wakatsuki, K. 189, 522
Walker, I.K. 331, 522
Walker, J. 448, 493
Wall, L.A. 8, 522
Walther, D.C. 248, 266–7, 345, 485, 522
Walton, W.D. 142, 177, 349, 372, 374, 376, 379, 459, 478, 502, 522
Wang, F.-J. 131, 494
Wang, H. 318, 522
Wang, Y. 269, 525
Wang, Y.C. 428–9, 522
Ward, E.J. 292, 313, 476
Waterman, T.E. 285–6, 291, 309, 354, 365, 381–2, 485, 496, 510, 522
Watson, A.A. 442, 477
Waymack, B.E. 346, 500
Weast, R.C. 16, 18, 112, 200, 238, 488, 522
Weatherford, W.D. 264, 522
Webster, C.T. 122, 142–4, 175, 290, 296, 435, 520, 522
Weckman, E.J. 142, 174, 185, 246, 502, 504, 516, 523
Weckman, H. 376, 475
Weddell, D.S. 129, 494
Weil, E.D. 28, 503
Weinberg, F.J. 335, 489
Welker, J.R. 165, 185, 263, 522–3
Wells, G. 194, 523
Welty, J.R. 35, 37, 43, 45, 55, 57, 62, 76, 82, 123, 173, 523
Weng, W.G. 292, 362, 523
Werner, K. 211, 213, 250, 500
Westenberg, A.A. 98, 491
White, A.G. 110, 523

White, D.A. 182, 281–3, 439, 492, 523
Whiteley, R.H. 453, 523
Wichman, I.S. 250–1, 258–9, 277, 300, 506, 515, 523
Wicks, C.E. 35, 37, 43, 45, 55, 57, 62, 76, 82, 123, 173, 523
Wickstrom, U. 49, 421, 517, 523
Wiersma, S.J. 354, 505
Wilkes, N. 177, 287, 516
Williams, A. 28, 195, 487, 532
Williams, F.A. 9, 47, 55, 57, 100, 106, 128, 132, 156, 274, 277, 287, 291–2, 294–5, 300, 308, 310–12, 333, 377, 490, 515, 523–4
Williams, F.W. 362, 439, 492, 523
Williams-Leir, G. 235, 483
Williamson, R.B. 154, 272, 356, 360–2, 366, 371, 384, 390–1, 393, 397, 406, 410, 412, 416, 420, 432, 443, 453, 478, 490, 501, 507, 524
Wilson, R.E. 35, 37, 43, 45, 55, 57, 62, 76, 82, 123, 173, 523
Wolfhard, H.G. 97–9, 107, 109, 113, 123, 127–9, 188, 491, 516, 524
Wolfinger, M.G. 218, 514
Wood, B.D. 145, 524
Wood, C.B. 177, 379, 459, 522
Wood, D.R. 86, 514
Woodburn, P. 287–8, 312, 524
Woolley, W.D. 29, 332, 362, 443, 457, 524
Wotton, B.M. 296, 312, 524
Wraight, H.G. 159, 161–2, 209, 308, 312, 355–6, 435, 438, 496, 520
Wu, Y. 287, 477

Yakush, S.E. 131, 504
Yang, L. 269, 362, 525
Yang, L.Z. 362, 523
Yang, T.C. 140, 493
Yao, C. 171–2, 524
Yaws, C.L. 84, 234
Yeoh, G.H. 78, 524
Yih, C.S. 132, 135, 150, 525
Ying, S.J. 131, 489
Yokobayashi, Y. 155, 160, 162, 494

Yokoi, S. 136, 138, 435, 439, 525
Yoshida, K. 389, 508
Yoshida, M. 155, 162, 292, 300, 311, 494
You, H.Z. 137, 143, 162, 525
Young, B.C. 323, 484
Yuan, L.M. 147, 156, 525
Yuen, K.K. 78, 524
Yule, A.J. 194, 525
Yumoto, T. 184, 525
Yumoto, Y. 189, 475
Yung, D. 66, 435–6, 503

Zabetakis, M.G. 83–9, 92–3, 95–7, 112, 118, 186, 194, 483, 525
Zahria, R. 406, 490
Zalosh, R.G. 34, 118, 198, 525
Zhang, J. 208, 294, 330, 525
Zhou, L. 292, 306, 525
Zhou, Y. 269, 525
Zicherman, J.B. 292, 496
Zinn, B.T. 443, 479
Zukoski, E.E. 121–2, 131, 134–8, 140, 142–6, 148, 150–1, 153–4, 158, 175, 356, 357, 359, 457, 484, 525

Subject Index

Abel Closed Cup apparatus 241
Absorptivity 62, 70, 74
Activation energy 9–10, 229–30, 274
Adiabatic 20, 30
Adiabatic flame temperature 30–3
Afterglow 347
Alligatoring, *see* Crocodiling
Arrhenius 9, 230
Autoignition of liquids 233, 245–7
Autoignition temperatures 87, 112, 229,
  233, 234

Backdraught 360–3
Bernoulli's Equation 390
Biot number 42–4, 49, 54, 220, 230–1,
  252–3, 321, 329, 432
Black body 37, 60–2
BLEVE 118, 198
Blow-out/-off 274
B-number 195–7, 204, 267
Boilover 182, 191–2
Bouger's Law, see Lambert-Beer Law
Boundary Layer 52–7, 207, 285
Boussinesq approximation 134
Bradford Football Stadium Fire 352, 385,
  438
Building separation 66–7
Buoyancy 17, 35–6, 52, 54, 57, 105,
  121–2, 130, 132, 134, 142–3,
  157–8, 175, 285, 334, 343, 389,
  391, 459–62
Buoyant plumes 132–8, *see also* Line
  plumes

Burners,
  Bunsen 107–8, 121
  counterflow diffusion flame 127–8
  porous bed 131, 133
  porous disc 108–9
Burning
  droplets 194–7
  dusts 221–2, 335–8, 340, 343
  items of furniture 366, 372–5, 381–4
  liquids 182–98
  mists and sprays 194–5
  solids 181, 199–222
  wood 209–21
Burning intensity 203
Burning rate, *see also* Regression rate
  182, 184, 199–209
  effect of orientation 207–9
  effect of wind 185–9
  in compartments 353, 388, 396–404
  wood 209–21
  wood cribs 220–21
Burning velocity 100–1, 106–14
  effect of mixture composition
    110–11
  effect of pressure 112–13
  effect of suppressants 113–15
  effect of temperature 111–12
  effect of turbulence 116–18
  measurement of 106–9

Calorimetry
  Bomb 20, 206, 266
  Oxygen consumption 23–4, 372–5

An Introduction to Fire Dynamics, Third Edition. Dougal Drysdale.
© 2011 John Wiley & Sons, Ltd. Published 2011 by John Wiley & Sons, Ltd.

Carpets
  Flame spread characteristics 301–3,
    305–6, 438
CBUF 374–6, 381
Ceiling jet 157–63
Cellulose 5, 209–13
Chain reactions 27
  branching 28
Char 4, 6, 9, 200, 206, 210–20
  burning of 215, 217, 221, 334, 347
  depth of 219, 222
Chemical suppression
  flaming 96–7, 113–15, 272
  smouldering 347
CIB 379–82, 394, 404–6, 424, 427
Cleveland Open Cup apparatus 238,
    244–5
Combustibility ratio 205–7
Combustible liquids 237, 241, 244,
    246
Compartment fires
  course of 349–50
  pre-flashover 349–58, 465
  post-flashover 387–439
  temperatures 404–18
Compartmentation 420, 438–9
Conduction 36
  steady state 38–40
  transient 40–52, 250–68
Cone calorimeter 23–4, 26, 249, 256–7,
    260, 266, 271–2, 374, 376–7
Configuration factor 38, 64–72, 168, 184,
    188
Convection 37, 52–8, 243
Corner test 272, 373, 376–7, 384
Cribs, *see* Wood cribs
Critical radiant heat flux 255–60
Criticality 212, 227–37, 269–71, 274,
    318 *et seq.*
Crocodiling 215
Cross-radiation 69–72, 209, 220, 297,
    298, 345, 384–5
Cryogenic liquids 194, 197–8

Damkohler number 269, 274
Deflagration 83

Detectors
  heat 157, 169–71, 251
  infra-red 79, 141–2
  smoke 138, 147, 251
  ultra-violet 59
Detonation 117–18
Diffusion 123–6
Droplets
  burning of 194–7
Dusts 221–2, 325–8, 336–40

Electromagnetic spectrum 59–61
Emissive power 37, 60, 63
Emissivity 37, 60, 76–9
  of flames 76–9, 184
  of hot gases 72–6
Enthalpy 20
Entrainment 121, 134–6, 142, 144,
    149–54, 158–9, 165, 174, 466,
    472–3
Equivalence ratio 29–30
  Global equivalence ratio (GER) 454,
    457
Error function 45–6
Exchange area 69
Explosion Limits, *see* Flammability limits
Explosions 33–4, 94, 116–18, 237
  dust 221–2
  mist 194
  prevention 272–3
  smoke 332, 362, 443
  suppression 272–3
  venting 34
External flaming 397, 400–4, 435–7
Extinction 128, 272–4, 306–7

Fabrics,
  flame spread 290–1
  ignition 251
Fick's law of diffusion 123
Fire growth 71–2, 79, 151, 157, 165
Fire load 423–6
Fire plume 130–2, 139–65, 355–7
Firepoint,
  of liquids 236, 238–45
  of solids 247–52, 277, 290, 295,
    299–300, 302

Firepoint equation 240, 264, 266–8,
    274
Fire resistance 387, 420–1, 424, 427
    Calculation 427–35
    equivalent time exposure 424–7
    predicted 406
    tests 420–1
Fire retardants 207, 211–12, 218,
    264–5
Fire severity 387, 402, 404, 416,
    420–7
Fires from spontaneous combustion
    317–18, 321–4, 325–31
Fire spread,
    between compartments 435, 437–9
    between fuel 'packets' 382
Fire storm 131
Fire whirl 130–1
Flame arresters 104, 227, 272
Flame deflection
    by compartment boundaries 151–63,
        353, 356–7
    by wind 163–5
Flame emissivity, *see* emissivity 76–9,
    184–9, 203–5, 408
Flame extension 159–63, 209, 437
Flame flicker 123, 140–2
Flame height 125, 128–9, 140, 142–7,
    183, 285
Flame projection, *see* External flaming,
Flame speed 106, 282–3
Flame spread 277–314
    concurrent and counter-current spread
        279, 284–5, 287–8
Flame spread over liquids 277–84
    effect of air movement (wind) 283
    effect of temperature 278–80, 281–3
    effect of width 281
    when soaked into sand/soil 281
Flame spread over solids 284–307
    effect of edges and corners 297–8
    effect of fuel temperature 299
    effect of k, ρ and c 294–6
    effect of orientation and direction
        284–92, 314

effect of oxygen concentration 297,
    299, 305, 307
effect of pressure 304
effect of radiant heat 299–304,
    313–14
effect of thermal capacity of the
    atmosphere 299
effect of thickness 292–4
effect of width 296–8
effect of wind (air flow) 304–7, 312
modelling 307–11
through open fuel beds 312–13
Flame suppression, *see* Chemical
    suppression
Flame temperatures,
    diffusion 77, 129, 148–9
    premixed 30–3
Flames–diffusion 11, 121–32
    buoyant 121–2, 130, 139–47
    cellular 209
    extinction 273–4
    laminar 123–8
    momentum jet 121–2, 128–9
    smoke formation 443
    turbulent jet 121–2, 128–9
Flames–premixed 12, 83, 97–118
    extinction 272–3
    heat losses from 101–6
    propagation 99–101, 282–3
    structure 97–101
Flame volume 147–8
Flammable liquids 237
Flammability diagrams 94–7
Flammability limits 32, 83–97, 101,
    235–7
    characterization of 88–91
    measurement of 83–8
    of dusts 222
    of mists 87, 91, 194
    of mixtures 90–1, 94–7
    pressure dependence 92–3
    temperature dependence 87, 91–4
Flash fire 242
Flashover 161, 274, 313, 349–54
    conditions necessary for 349–78
    critical rate of heat production 365–73

Flashover (*continued*)
  definition 351–2
  factors affecting 382–5
  time to 351–2, 378–82
Flashpoint 237–45
  Closed Cup 237–9
  effect of altitude/atmospheric pressure
    238
  of alcohols 240
  of solids 247
  Open Cup 235, 238–9
  upper flashpoint 237–8
Forest fire spread 296, 304, 307–8,
  312
Fourier Number 42–3, 52, 54
Fourier's Law 36, 56
Frank Kamenetskii's δ 231–2, 318,
  321–7
Free radicals 3, 27–30, 226
Froude modeling 143, 148, 175–6,
Froude number 54, 121–2, 130, 144,
  175–6
Fuel control 389, 394–7, 412, 415
Fuels 2–6
Fully developed fire 387–439
Fundamental burning velocity, *see*
  Burning velocity

Gas phase combustion 26–30
Glowing combustion 262, 270–1, 347
Grashof number 17, 54–5, 57–8, 132,
  143
Grey body 60–1, 70
Growth period 351–4, 382–5
  see also Compartment fire,
    pre-flashover fire

Halons 115
Heat of combustion 19–26, 181, 205–6
  of "air" 21, 25, 30, 397
  of "oxygen" 21, 23–4
  of wood 215–16
Heat of formation 20–22
Heat of gasification, *see* Heat required to
  produce the volatiles,
Heat required to produce the volatiles
  12–14, 199–203

effect on compartment fire behavior
    397–401
  of wood 217–18
Heat transfer 35–79, *see also* Conduction,
    Convection and Radiation
  infinite slab 41–45
  role in flame spread 285, 290–6,
    299–302, 307–13
  role in ignition of solids 250–9
  semi-infinite solid 45–8
Heat transfer from flames
  surface ignition 271–2
  to burning surface 183–4, 199–209
  to ceilings 162–3
  to walls 154–6
Hess' Law 21–2
Highly flammable liquids 94, 237–8,
  241
Home Fire Project 378
'Hot spots' 233, 324–5, 345
Hot zone formation 191–2

Ideal gas constant 14–15
Ideal gas law 14
Ignitability limits 87
Ignition, piloted 225–68
  Gases 83–7, 226, 238
  Liquids 235–45
  Solids 247, 268, 271–2
Ignition, spontaneous
  of gases 225–34
  of liquids 244–7
  of solids (surface) 269–71
  within solids 317–31
Ignition of solids,
  ease of 248–9, 251, 257, 260,
    264–7
  effect of moisture 263
  effect of thickness 251–2
  surface 271
  time to 251–2, 256–9
Ignition sources 225
  electrical sparks 85–7, 226–7
  hot surfaces 229, 233–4, 245–7
  mechanical sparks 227, 229, 233
  pyrophoric sparks 233

Induction period 228–9, 318, 321–4
Inerting 90–1, 96–7, 110, 113–15, 272–3
Inhibition, *see also* Chemical suppression 28, 97, 272–4
Intensity of normal radiation 62–4
Intrinsic safety 86

King's Cross Underground Station fire 287, 353, 384
Kirchhoff's Law 60, 74, 77

Lagging fires 318, 328–30
Lambert-Beer Law 72–4, 448
Lambert's cosine law 62–4
Latent heat of evaporation 13, 184–6, 195–6, 240, 264, 268
Le Chatelier's Law 90, 237
Lewis number 54, 195
Lignin 5, 210–12, 217
Limiting adiabatic flame temperature 30–3, 90
Limiting (minimum) oxygen concentration 95–6

Mass transfer 195–7, 204, 267
MGM Grand Hotel fire 385, 462
Microbiological heating 330–1
Minimum ignition energy 85–7, 226, 228
Minimum ignition temperatures, *see also* Autoignition temperatures
Mole 14
Molecular weight 3, 14
Montreal Protocol 115

Neutral plane 389–90, 392–3, 438, 460–3, 473
Normalised heat load 426
Numerical methods 48–52, 427–35
Nusselt number 54–8

Opening factor 396, 405, 411, 413, 416–17, 433
Optical density 447–53, 455, 465, 467–8
Oxygen depletion 350, 360, 456
Oxygen-enriched atmospheres 297, 299, 305, 307

Oxygen consumption calorimetry 23–4, 372–4

Pensky-Martens apparatus 235
Planck's distribution law 61–2
Polymers 3–11
    burning of 199–209
Pool fires 182–92, 207–9
Post-flashover fire 387–439
Prandtl number 54–8
Pre-flashover fire 299–300, 349–85
Pressure modeling 175–6
Pressurised liquids 197–8
Pyrolysis 2, 199, *see* Thermal decomposition

Q^* 122–3, 130, 136, 138, 143, 145–6, 175
Quenching diameter, *see* Quenching distance
Quenching distance 102–4, 228, 273

Radiation 37, 59–79
    from hot gases 72–6
    from hot smoke 79, 352, 358–60
    from luminous flames 76–9, 166–9, 184–9
    from non-luminous flames 74–6, 166
    in flame spread 299–305
    in post-flashover fires 397–403
    in pre-flashover fires 354–8, 360, 382–4
Radiation blocking 169, 184, 189, 354, 400
Raoult's law 18, 235–7
Rate of burning of wood, *see* Burning rate
Rate coefficient 9
Rate of heat release 13, 23–4, 26, 181, 205, 372–5
    in compartments 368, 407
Reaction kinetics 26–30
Regression rate (liquids) 182–5, 189–92
Response time index (RTI) 45, 170, 351
Reynold's number 54–6, 123, 128, 176
Richardson number 158

Self-heating 228–32, 318–27
  haystacks 330
Semi-infinite solid 45–8, 253
Smoke 441–58
  control systems 469–74
  explosions 332, 362, 443
  load 452
  measurement 447–50, 453
  methods of test 442, 450–4
  movement 459–65
  pressurization 473–4
  production 443, 447
  rate of production 459, 465–9
  reservoirs 471–2
  stratification 132, 134
  toxicity 441–2, 455–8
  venting 463, 465, 470–3
  visibility 449–50
Smoke point 77, 444–6
Smouldering 7, 221, 317–18, 325, 328,
    329, 331–47
  chemical requirements 346–7
  in cellular polymers 341–2
  in dusts 336–40
  in fibreboard etc. 339–40
  initiation 344–5
  in leather 340
  in rubber latex 331–2, 340, 344
  smoke production 332, 443
  transition to flaming 342–4
Soot particles 29–30, 76–8, 128–9, 188,
    443–6
Soot volume fraction 77–8, 444
Specific extinction area 453
Spill fires 193–4
Spontaneous heating, *see* Self heating
Spontaneous ignition of solids
  bulk 317–28
  surface 269–71
Spread of flame, *see* Flame spread
Sprinklers 148, 170–2, 351, 470,
    472
Stack effect 18, 460–3, 474
Station Nightclub Fire 353
Stefan-Boltzmann constant 37, 60
Stoichiometric 20, 26, 393–5

Surface spread of flame tests 302
Surface-tension-driven flows 243–4, 278

Temperature-time curves 406–18, 420 *et
    seq.*
  hydrocarbon curve 421, 427
  parametric 416–18
  standard curve 420–3, 427–31
Thermal conduction length 47, 258
Thermal conductivity 36–52, 215
Thermal decomposition 2, 6–12
  of wood 210–13
Thermal diffusivity 37, 41, 54, 56, 266
Thermal explosion theory 226–32,
    317–25
Thermal inertia 47–8
  effect on fire growth 209, 266, 368,
    370, 381–2
  effect on fire resistance tests 421
  effect on flame spread 295, 300
  effect on ignition 248, 253, 257–8,
    260–2, 268
Thermal properties 4, 37
Thermal radiation, *see* Radiation
Thermal stability, *see* Thermal
    decomposition
Thermal thickness 45–7, 218–19, 258,
    368
Thermite reaction 233
Thermoplastics 4, 6, 199–200, 205–8,
    294, 384, 421–2
Thermosetting resins 5–6, 200, 342
Thick fuels 45
  flame spread 294
  ignition 253
Thin fuels 42–4
  flame spread 292–4
  ignition 251–2
Total flooding 272–3
Toxic gases 359, 441–2, 455–8
  asphyxiants 456, 458
  carbon monoxide 359, 455–8
  fractional effective dose 456
  hydrogen cyanide 455–7
Turbulence
  effect on explosions 116–17

effect on flame spread 306
effect on jet flames 128

Vapour pressure 18–19, 93, 94,
    235–9
Ventilation control 387–96,
    398–404, 412
Ventilation factor 388–9
    derivation of 389–93
Ventilation parameter,
    see Ventilation factor
View factor, *see* Configuration factor
Virtual origin 134, 136–7, 150
Virtual source, *see* Virtual origin

Volatiles 2, 6–12, 181, 199, 211–13,
    269
    critical flow rate 240, 263–8
    reactivity 264, 268

Wall linings 300–4, 310–11, 375–8,
    380–2, 397
Water (extinction) 272–4
Wick ignition 243–5, 278–9
Wildland fires, *see* Forest fire spread
Wood 209–23
    cribs 220–1, 312, 379–80, 356–7,
        379, 382
    ignition of 253–7

Printed and bound by CPI Group (UK) Ltd, Croydon, CR0 4YY

27/10/2024

14580152-0001